Non-Thermal Technologies for the Food Industry

Depending on the mechanisms involved in non-thermal technologies (such as ozonization, irradiation, ultrasound processing, plasma processing, and advanced oxidative processes), interaction with food molecules differs, which might lead to desirable reactions. *Non-Thermal Technologies for the Food Industry: Advances and Regulations* explores the possibility of using non-thermal technologies for various purposes such as shelf-life extension, reduced energy consumption, adhesion, and safety improvement. Further, it reviews the present status of these technologies, international regulations, and sustainability aspects in food processing including global case studies.

Features:

1. Provides a comprehensive overview of all the non-thermal processing technologies that have potential for use within food manufacturing
2. Covers novel disinfectant technologies and packaging methods for non-thermal processing
3. Includes electro-spraying and electrospinning; low-temperature drying techniques, cold plasma techniques, hydrodynamic cavitation, oscillating magnetic field processing, and so forth
4. Focus on topics such as the valorization of agri-food wastes and by-products and sustainability
5. Reviews ClO_2 in combined/hybrid technologies for food processing

This book is aimed at researchers and graduate students in food and food process engineering.

Non-Thermal Technologies for the Food Industry

Advances and Regulations

Edited by
C. Anandharamakrishnan, V. R. Sinija and R. Mahendran

CRC Press
Taylor & Francis Group
Boca Raton London New York

CRC Press is an imprint of the
Taylor & Francis Group, an **Informa** business

Designed cover image: C. Anandharamakrishnan, V. R. Sinija and R. Mahendran

First edition published 2024
by CRC Press
2385 NW Executive Center Drive, Suite 320, Boca Raton FL 33431

and by CRC Press
4 Park Square, Milton Park, Abingdon, Oxon, OX14 4RN

CRC Press is an imprint of Taylor & Francis Group, LLC

ISBN: 9781032399737 (hbk)
ISBN: 9781032416984 (pbk)
ISBN: 9781003359302 (ebk)

DOI: 10.1201/9781003359302

Typeset in Times LT Std
by KnowledgeWorks Global Ltd.

Contents

Preface

Years ago, non-thermal processing techniques were believed to be laboratory techniques that have no scope for commercialization in food industry. Nevertheless, scientists all over the world put their constant efforts to explore the possibility of utilizing these technologies in large scale to overcome the reliance on thermal techniques with advanced technological and manufacturing supports. At present, non-thermal technologies are no more an option instead a demand due to the concerns raised by modern fast-growing world for sustainable, safe, and high-quality food products. High-pressure processing, pulsed electric field, and ultrasound processing are some of the nonthermal techniques that became commercially successful for selected food products and food applications. However, not all the nonthermal technologies are commercialized to industrial level; even for the commercialized techniques, researchers are figuring out their multifaceted applications. Therefore, we aimed to deliver a book that has all the basics and latest inventions of each nonthermal technique along with their applications, environmental impact, and regulatory guidelines.

The current version of the book not only discusses about the conventional nonthermal techniques such as ultraviolet processing, high voltage pulsed electric field, ozone technology, irradiation, ultrasound processing, high-pressure processing, and membrane processing; but also covers some of the emerging areas such as processed water technology, electro-spraying/spinning, subcritical fluid technology, and low-temperature drying. In addition, efforts were also made to deliver the comprehensive details about some of the emerging applications of cold plasma technology, biological control agents, and novel disinfectants. Hence, we believe that the 23 chapters of this book will enrich the knowledge of the readers with up-to-date technological advances and sow them the seed of curiosity to explore more on the budding applications of nonthermal processing.

We are grateful to all the 37 authors who collaborated with us and gave their time, effort, and ideas to successfully compile the book. Their knowledge, experience, and ideas to create these highly useful chapters will make the book a great resource in the field of nonthermal processing for many years to come. Furthermore, we are grateful to all reviewers who dedicated their valuable time and experience to improve the contents of the book.

We tried our level best to avoid the overlapping of chapter contents to enrich the reading experience, however, overlapping of certain fundamental information is inevitable in a book that covers different perspectives of nonthermal processing. This will, however, aid the readers to understand the essential information of all the nonthermal treatments from the authors perspectives.

Though most of the nonthermal techniques are residue free green technologies, their environmental impact and scalability status are important to adapt them as a sustainable technology in industries. Furthermore, researchers should also need to understand the possible future guidelines for practicing these technologies at industrial level. In this regard, our book will truly inspire the researchers to continue bringing their new ideas to the field of nonthermal processing and utilize the technologies in an effective manner that provide opportunity for the process scalability and environmental safety.

C. Anandharamakrishnan
V. R. Sinija
R. Mahendran

Editor Biographies

Dr. C. Anandharamakrishnan

Dr. C. Anandharamakrishnan is presently the Director of the CSIR-National Institute for Interdisciplinary Science and Technology (CSIR-NIIST), Thiruvananthapuram, Kerala, India, from 2022. Earlier, he served as Director of National Institute of Food Technology, Entrepreneurship and Management, Thanjavur (NIFTEM-T) (formerly known as Indian Institute of Food Processing Technology (IIFPT), Thanjavur), Tamil Nadu, during the period April 2016–2022. He has done his doctoral research in Chemical Engineering with a Specialization in Food Engineering at Loughborough University in the United Kingdom. During his PhD, he was awarded the prestigious Commonwealth Scholarship Programme of the UK Government. His research endeavors are well documented in the form of 180 impact factor publications, 2 international patents, and 7 Indian patents. He is also the author and editor of 10 books and 68 book chapters. He serves as the Associate Editor of *International Journal of Food Engineering*, Academic Editor of *PLOS ONE* journal, International Editorial Advisory Board member for *Drying Technology Journal*, Editorial Board member of *Journal of Food Process Engineering*, and various international journals, apart from being a peer reviewer for more than 20 international journals.

Affiliations and Expertise

Director, CSIR-National Institute for Interdisciplinary Science and Technology (CSIR-NIIST), Thiruvananthapuram, Kerala, India (anandh@niist.res.in)

Dr. V. R. Sinija

Dr. Sinija currently working as Professor and Chairman for the working group on Nonthermal Processing at National Institute of Food Technology Entrepreneurship and Management (NIFTEM-T), Thanjavur, has acquired her PhD (Agricultural & Food Engineering) and M. Tech. (Post-Harvest Engineering) degrees from the prestigious Indian Institute of Technology, Kharagpur, after her B. Tech. (Agricultural Engineering) degree from Kerala Agricultural University. She was awarded with National Doctoral Fellowship by AICTE (All India Council for Technical Education), New Delhi, for carrying out her research work. She has 18 years of teaching and research experience in the field of food processing. Dr. Sinija is a member of Academic Council, Board of Studies, and Board of Examinations of various institutions across the country. Her research works have been published as papers in 79 peer-reviewed journals, 20 popular articles, 5 book chapters, and more than 85 international and national conference proceedings and 3 patents. She is also a reviewer of various national and international journals. She has received the prestigious Prof. Jeevan Singh Sidhu Award for the year 2018 by the Association of Food Scientists and Technologists (India) for Teaching Excellency. She has also received the Bharat Vikas Award from Institution of Self Reliance, Bhubaneswar, for outstanding performance in food research and incubation. Also, for the year 2022, she has been conferred the Dr. V Subramanian Best Scientist Award from NIFTEM, Thanjavur.

Dr. R. Mahendran

Dr. R. Mahendran is presently working as a Professor at National Institute of Food Technology, Entrepreneurship and Management (NIFTEM), Thanjavur. He has acquired graduation in B. Tech (Chemical Engineering) and M. Tech (Industrial Biotechnology) from SASTRA University, and PhD in Food Process Engineering from Tamil Nadu Agricultural University, Coimbatore, India. Prior to his association with NIFTEM, Thanjavur, he worked as a Scientist in TIFAC, Department of Science and Technology, Government of India at New Delhi from 2006 to 2010.

Dr. Mahendran also heading the department, Centre of Excellence in Nonthermal Processing, one of the unique centers in the world. Mahendran's research focus includes cold plasma applications on agri-food, food structure shape transformation, and novel nonthermal processing on food systems. He is the recipient of Dr. V Subramanian Best Scientist Award from NIFTEM, Thanjavur, for the year 2021. Mahendran has authored over 100 publications in reputed journals, apart from popular articles and book chapters. He is also a member of various professional bodies of food & chemical engineering. He serves as guest editor and editorial board member for reputed food processing journals.

List of Contributors

Anbarasan Rajan
National Institute of Food Technology,
  Entrepreneurship and Management
Thanjavur, India

Annet Mary Anto
National Institute of Food Technology,
  Entrepreneurship and Management
Thanjavur, India

B. Kamalapreetha
National Institute of Food Technology,
  Entrepreneurship and Management
Thanjavur, India

Kevin Britto
National Institute of Food Technology,
  Entrepreneurship and Management
Thanjavur, India

C. Anandharamakrishnan
National Institute of Food Technology,
  Entrepreneurship and Management
Thanjavur, India

Chaitanya Chivate
National Institute of Food Technology,
  Entrepreneurship and Management
Thanjavur, India

D. V. Chidanand
National Institute of Food Technology,
  Entrepreneurship and Management
Thanjavur, India

Arunkumar Elumalai
National Institute of Food Technology,
  Entrepreneurship and Management
Thanjavur, India

J. A. Moses
National Institute of Food Technology,
  Entrepreneurship and Management
Thanjavur, India

K. Najma
National Institute of Food Technology,
  Entrepreneurship and Management
Thanjavur, India

Yuvraj Bhosale Khasherao
National Institute of Food Technology,
  Entrepreneurship and Management
Thanjavur, India

L. Kavitha
National Institute of Food Technology,
  Entrepreneurship and Management
Thanjavur, India

L. Mahalakshmi
National Institute of Food Technology,
  Entrepreneurship and Management
Thanjavur, India

M. Dharini
National Institute of Food Technology,
  Entrepreneurship and Management
Thanjavur, India

Malini Buvaneswaran
National Institute of Food Technology,
  Entrepreneurship and Management
Thanjavur, India

Addanki Mounika
National Institute of Food Technology,
  Entrepreneurship and Management
Thanjavur, India

Pramila Murugesan
National Institute of Food Technology,
  Entrepreneurship and Management
Thanjavur, India

N. Bhanu Prakash Reddy
National Institute of Food Technology,
  Entrepreneurship and Management
Thanjavur, India

Shubham Nimbkar
National Institute of Food Technology,
  Entrepreneurship and Management
Thanjavur, India

P. Santhoshkumar
National Institute of Food Technology,
  Entrepreneurship and Management
Thanjavur, India

P. T. Atchaya
National Institute of Food Technology,
    Entrepreneurship and Management
Thanjavur, India

Thivya Perumal
National Institute of Food Technology,
    Entrepreneurship and Management
Thanjavur, India

K. Krishna Priya
National Institute of Food Technology,
    Entrepreneurship and Management
Thanjavur, India

R. Mahendran
National Institute of Food Technology,
    Entrepreneurship and Management
Thanjavur, India

R. B. Ramyaa
National Institute of Food Technology,
    Entrepreneurship and Management
Thanjavur, India

Ashish Rawson
National Institute of Food Technology,
    Entrepreneurship and Management
Thanjavur, India

S. Akalya
National Institute of Food Technology,
    Entrepreneurship and Management
Thanjavur, India

S. Anandakumar
National Institute of Food Technology,
    Entrepreneurship and Management
Thanjavur, India

S. Chittibabu
National Institute of Food Technology,
    Entrepreneurship and Management
Thanjavur, India

S. Shanmugasundaram
National Institute of Food Technology,
    Entrepreneurship and Management
Thanjavur, India

S. Susmitha
National Institute of Food Technology,
    Entrepreneurship and Management
Thanjavur, India

S. K. Reshmi
National Institute of Food Technology,
    Entrepreneurship and Management
Thanjavur, India

Prakyath Shetty
National Institute of Food Technology,
    Entrepreneurship and Management
Thanjavur, India

Subaitha Z. Afrose
National Institute of Food Technology,
    Entrepreneurship and Management
Thanjavur, India

U. K. Aruna Nair
National Institute of Food Technology,
    Entrepreneurship and Management
Thanjavur, India

Neeta S. Ukkunda
National Institute of Food Technology,
    Entrepreneurship and Management
Thanjavur, India

V. Athira
National Institute of Food Technology,
    Entrepreneurship and Management
Thanjavur, India

V. Monica
National Institute of Food Technology,
    Entrepreneurship and Management
Thanjavur, India

V. R. Sinija
National Institute of Food Technology,
    Entrepreneurship and Management
Thanjavur, India

1 Introduction to Nonthermal Processing

R. Mahendran, V. R. Sinija, and C. Anandharamakrishnan

As the world population continues to grow, the demand for food products increases, which also raises concerns regarding food safety and quality. Food processing plays a crucial role in meeting these demands while ensuring food safety and quality (Rajan et al., 2023). Nonthermal food processing technologies have emerged as an alternative to traditional thermal processing methods, such as pasteurization, sterilization, and thermal cooking. Nonthermal processing methods use techniques that do not involve heating or cooling to achieve microbial reduction, such as high-pressure processing (HPP), pulsed electric field (PEF), ultrasound, cold plasma, and irradiation. These technologies aim to reduce the energy input required for food processing, maintain the nutritional and sensory properties of foods, and reduce the formation of harmful byproducts that can occur during thermal processing (Chhanwal et al., 2019).

HPP is a nonthermal processing method that applies high pressure to food products to eliminate microorganisms/hazardous enzymes' activity. This method is widely used in the food industry for products such as ready-to-eat meals, fruit juices, and meats (Anandharamakrishnan, 2013). PEF processing uses electrical pulses to disrupt microbial cells, reducing microbial load in food products and also inactivating the enzymes causing undesirable changes to the food products (Velusamy et al., 2023). This technology is used in the production of fruit juices, dairy products, and alcoholic beverages. Ultrasound technology is another nonthermal processing method that uses high-frequency sound waves to disrupt microbial cells. This technology has been used in the food industry to reduce microbial load in liquid foods such as milk, fruit juices, and beer apart from other applications such as ultrasound-assisted extraction of various components from raw material for enhancing the drying rate. Irradiation is a nonthermal processing method that uses ionizing radiation to kill microorganisms in food products. This technology is used in the production of spices, fruits, and vegetables.

Nonthermal processing methods offer several advantages over traditional thermal processing methods. They are faster, require less energy, and have minimal impact on the nutritional and sensory properties of food products. Additionally, these technologies can extend the shelf life of food products and reduce the risk of foodborne illnesses. However, nonthermal processing methods also have some drawbacks. For instance, some of these technologies can be expensive to implement, and there are still concerns regarding the safety of some of these methods. Further research is required to fully understand the safety and efficacy of nonthermal processing methods.

1.1 SAFETY ASPECTS OF NONTHERMAL PROCESSING

Nonthermal food processing technologies have emerged as an alternative to traditional thermal processing methods to achieve microbial reduction and food preservation. While nonthermal technologies offer several advantages, such as reduced energy input, minimal impact on nutritional and sensory properties, and longer shelf life of food products, there are also safety considerations that need to be taken into account (Drishya et al., 2022). One of the primary safety considerations of nonthermal food processing is the potential for incomplete microbial inactivation (Manoharan et al., 2021). Unlike thermal processing methods, which have well-established time and temperature combinations to achieve a certain level of microbial reduction, nonthermal processing methods rely on

DOI: 10.1201/9781003359302-1

other factors, such as pressure, electric fields, or radiation exposure (Anandharamakrishnan, 2013; Manoharan & Radhakrishnan, 2022; Velusamy et al., 2023). Therefore, it is essential to validate the effectiveness of nonthermal processing for each food product and processing condition to ensure the elimination of harmful microorganisms. Another safety consideration is the potential for unintended chemical changes in food products (Dharini et al., 2023). Nonthermal processing methods can induce changes in food components, such as enzymes, vitamins, and pigments, that may affect the nutritional and sensory properties of the food. Additionally, some nonthermal processing methods, such as irradiation, can cause the formation of new chemical compounds that are not present in unprocessed food. Therefore, it is important to evaluate the safety of these changes and assess their impact on human health.

Furthermore, nonthermal processing methods may also lead to changes in the physical properties of food products, such as texture (Anbarasan, Jaspin et al., 2022), color (Rajan et al., 2023), and flavor (Dharini et al., 2022). These changes may impact consumer acceptability, leading to decreased sales or product rejection. Thus, it is important to consider the sensory quality of food products and the impact of nonthermal processing on consumer preferences (Anbarasan, Gomez Carmona, & Mahendran, 2022). Additionally, the safety of nonthermal processing methods can be affected by the quality of the food product itself. Factors such as pH, water activity, and microbial load can influence the efficacy of nonthermal processing methods. Therefore, it is crucial to consider the food matrix and its characteristics when developing nonthermal processing protocols. In conclusion, nonthermal processing technologies offer several benefits for food preservation and safety. However, safety considerations, such as microbial inactivation, unintended chemical changes, sensory quality, and food matrix, need to be taken into account when developing and implementing nonthermal processing methods. Careful validation and assessment of the safety and efficacy of nonthermal processing are necessary to ensure the production of safe and high-quality food products.

1.2   GLOBAL REGULATION ON NONTHERMAL PROCESSING

Nonthermal food processing technologies are used to reduce the risk of foodborne illnesses and improve the shelf life of food products. However, the use of these technologies is regulated by different regulatory bodies around the world to ensure their safety and efficacy. In this section, we will discuss the regulatory frameworks required for nonthermal food processing. In the United States, nonthermal food processing technologies are regulated by the Food and Drug Administration (FDA) and the United States Department of Agriculture (USDA). Both agencies have established regulations and guidelines for the use of nonthermal processing methods. For instance, the FDA has issued guidance on HPP, PEF, and irradiation, which outlines the safety and labeling requirements for these technologies. The USDA has also issued regulations for the use of irradiation in meat and poultry products. In the European Union (EU), the use of nonthermal processing technologies is regulated by the European Food Safety Authority (EFSA) and the European Commission. The EFSA assesses the safety of novel food processing technologies, including nonthermal technologies, and provides scientific opinions on their safety and efficacy. The European Commission, on the other hand, regulates the use of these technologies and sets guidelines for their use in the food industry. In Canada, nonthermal food processing technologies are regulated by the Canadian Food Inspection Agency (CFIA) and Health Canada. The CFIA ensures that the use of nonthermal technologies complies with Canadian food safety regulations, while Health Canada evaluates the safety of these technologies and sets guidelines for their use in the food industry. In Australia and New Zealand, nonthermal food processing technologies are regulated by Food Standards Australia New Zealand (FSANZ). FSANZ evaluates the safety and efficacy of these technologies and sets guidelines for their use in the food industry. In conclusion, the use of nonthermal food processing technologies is regulated by different regulatory bodies around the world. These bodies set guidelines and regulations for the use of these technologies to ensure their safety and efficacy. Food processors and manufacturers are required to comply with these regulations and guidelines to ensure the production of safe and high-quality food products.

The biggest challenge facing the world is how to feed an ever-increasing population without affecting planet Earth in an adverse way to preserve the resources for future generations. The regulatory criteria are pressuring the food industry to modify their processing procedures (Hartley et al., 2020; King et al., 2017). In the 20th century, food requirements created significant environmental issues that harmed people's well-being and potentially restricted development (Arshad et al., 2022). People should understand that the current generation must conserve environmental and natural resources as much as possible for the good of future generations by utilizing scarce natural resources to sustain society (Bengtsson et al., 2018; Sánchez-Zarco et al., 2020). Therefore, the conservation of water, energy, and natural resources should be an integral aspect of global ethics.

Nonthermal processing has emerged as a promising, safe, and effective technique for extracting bioactive compounds from food residues. Nonthermal processing helps in implementing a circular economy and meeting the United Nations approved Sustainable Development Goals (SDGs).

1.3 SUMMARY

Nonthermal processing technologies have emerged as a viable alternative to traditional thermal processing methods. These technologies offer several benefits, including faster processing times, reduced energy requirements, and minimal impact on the nutritional and sensory properties of food products. While there are still concerns regarding the safety and efficacy of these methods, ongoing research is helping to address these issues and further advance the use of nonthermal processing methods in the food industry.

REFERENCES

Anandharamakrishnan, C. (2013). Computational Fluid Dynamics Modeling for High Pressure Processing. In *Computational Fluid Dynamics Applications in Food Processing* (pp. 59–62). Springer New York. https://doi.org/10.1007/978-1-4614-7990-1_6

Anbarasan, R., Gomez Carmona, D., & Mahendran, R. (2022). Human taste-perception: Brain computer interface (BCI) and its application as an engineering tool for taste-driven sensory studies. *Food Engineering Reviews, 14*(3), 408–434. https://doi.org/10.1007/s12393-022-09308-0

Anbarasan, R., Jaspin, S., Bhavadharini, B., Pare, A., Pandiselvam, R., & Mahendran, R. (2022). Chlorpyrifos pesticide reduction in soybean using cold plasma and ozone treatments. *LWT, 159*, 113193. https://doi.org/10.1016/j.lwt.2022.113193

Arshad, R. N., Abdul-Malek, Z., Roobab, U., Ranjha, M. M. A. N., Režek Jambrak, A., Qureshi, M. I., Khan, N., Manuel Lorenzo, J., & Aadil, R. M. (2022). Nonthermal food processing: A step towards a circular economy to meet the sustainable development goals. *Food Chemistry: X, 16*, 100516. https://doi.org/10.1016/j.fochx.2022.100516

Bengtsson, M., Alfredsson, E., Cohen, M., Lorek, S., & Schroeder, P. (2018). Transforming systems of consumption and production for achieving the sustainable development goals: Moving beyond efficiency. *Sustainability Science, 13*(6), 1533–1547. https://doi.org/10.1007/s11625-018-0582-1

Chhanwal, N., Bhushette, P. R., & Anandharamakrishnan, C. (2019). Current perspectives on non-conventional heating ovens for baking process—A review. *Food and Bioprocess Technology, 12*(1), 1–15. https://doi.org/10.1007/s11947-018-2198-y

Dharini, M., Jaspin, S., Jagan Mohan, R., & Mahendran, R. (2022). Characterization of volatile aroma compounds in cold plasma-treated milk. *Journal of Food Processing and Preservation, 46*(11), 1–13. https://doi.org/10.1111/jfpp.17059

Dharini, M., Jaspin, S., & Mahendran, R. (2023). Cold plasma reactive species: Generation, properties, and interaction with food biomolecules. *Food Chemistry, 405*(PA), 134746. https://doi.org/10.1016/j.foodchem.2022.134746

Drishya, C., Yoha, K. S., Perumal, A. B., Moses, J. A., Anandharamakrishnan, C., & Balasubramaniam, V. M. (2022). Impact of nonthermal food processing techniques on mycotoxins and their producing fungi. *International Journal of Food Science & Technology, 57*(4), 2140–2148. https://doi.org/10.1111/ijfs.15444

Hartley, K., van Santen, R., & Kirchherr, J. (2020). Policies for transitioning towards a circular economy: Expectations from the European Union (EU). *Resources, Conservation and Recycling, 155*, 104634. https://doi.org/10.1016/j.resconrec.2019.104634

King, T., Cole, M., Farber, J. M., Eisenbrand, G., Zabaras, D., Fox, E. M., & Hill, J. P. (2017). Food safety for food security: Relationship between global megatrends and developments in food safety. *Trends in Food Science and Technology, 68*, 160–175. https://doi.org/10.1016/j.tifs.2017.08.014

Manoharan, D., & Radhakrishnan, M. (2022). Computational cold plasma dynamics and its potential application in food processing. *Reviews in Chemical Engineering, 38*(8), 1089–1105. https://doi.org/10.1515/revce-2021-0005

Manoharan, D., Stephen, J., & Radhakrishnan, M. (2021). Study on low-pressure plasma system for continuous decontamination of milk and its quality evaluation. *Journal of Food Processing and Preservation, 45*(2). https://doi.org/10.1111/jfpp.15138

Rajan, A., Boopathy, B., Radhakrishnan, M., Rao, L., Schlüter, O. K., & Tiwari, B. K. (2023). Plasma processing: A sustainable technology in agri-food processing. *Sustainable Food Technology, 1*(1), 9–49. https://doi.org/10.1039/D2FB00014H

Sánchez-Zarco, X. G., Mora-Jacobo, E. G., González-Bravo, R., Mahlknecht, J., & Ponce-Ortega, J. M. (2020). Water, energy, and food security assessment in regions with semiarid climates. *Clean Technologies and Environmental Policy, 22*(10), 2145–2161. https://doi.org/10.1007/s10098-020-01964-2

Velusamy, M., Rajan, A., & Radhakrishnan, M. (2023). Valorisation of food processing wastes using PEF and its economic advances—Recent update. *International Journal of Food Science & Technology, 58*(4), 2021–2041. https://doi.org/10.1111/ijfs.16297

2 Ultraviolet Processing of Food Products

L. Kavitha, S. Susmitha, K. Najma, and D. V. Chidanand

2.1 INTRODUCTION

Promoting global sustainability and good health begins with the supply of safe and wholesome food. Almost 600 million individuals have been sickened by foodborne illnesses due to the presence of pathogenic microorganisms (bacteria, fungi, and viruses) in food, and each year, 420,000 people die because of these infections (Singh et al., 2021). Due to concerns about quality, safety, and toxicity, modern consumers are encouraging food manufacturers to shift to processing methods that are more affordable, nutritionally stable, and environmentally friendly (Afraz et al., 2020). The development of a wide range of innovative technologies by the food industry has the potential to enhance or replace existing processing methods used to produce high-quality food products (Singh et al., 2021). Decontamination methods involving thermal inactivation degrade the organoleptic qualities of food and are undesirable for temperature-sensitive commodities. Additionally, chemical processes give rise to issues about toxicity and unintended residuals. As a result, alternative techniques for disinfecting food have been explored and assessed, which intend to avoid using heat and chemicals (Keklik et al., 2012).

Recently, a non-thermal decontamination method called UV light treatment was employed to extend the shelf life and increase food safety. UV radiation has wavelengths between 200 and 280 nm, is generally regarded as non-ionizing radiation, and possesses germicidal effects (Chacha et al., 2021). Since everyone is aware of how heat affects the nutritional value and sensory aspects of food, there is a paradigm change taking place in the food business toward minimal and non-thermal technologies. This has led researchers in the field of food processing to create a variety of technologies as alternatives to the traditional thermal process, such as ohmic heating, radiofrequency (RF), microwave (MW), infrared (IR) heating, high-pressure processing (HPP), ultrasound (US), pulsed electric field (PEF), pulsed light (PL), cold plasma (CP), oscillating magnetic field (OMF), etc. (Anbarasan et al., 2022; Rajan & Radhakrishnan, 2023; Rawson et al., 2011; Velusamy et al., 2023). These techniques are used at ambient and sub-lethal temperatures eliminating the adverse effects caused by thermal technologies. The above-mentioned innovative technologies are all their own distinct fields of study (Mandal et al., 2020). The absorption spectrum of the reactants decides the type of light source to be utilized because light radiation is absorbed to carry out a photochemical reaction. It is necessary to understand the unique absorption spectra at various wavelengths to examine how radiation affects a molecule. This spectrum is obtained by applying several wavelengths of radiation to a solution containing a known concentration of the substance and then measuring the absorbance that results. The greatest absorbance peaks in this absorption spectrum typically indicate the best region for working with wavelengths (Falguera et al., 2011).

Additionally, ultraviolet (UV) light irradiation can be used as a disinfectant to lower the bacteria population in food. The so-called UV-C region has the most efficient wavelengths, particularly at 254 nm, while at 320 nm, its efficiency is practically non-existent. When bacteria, viruses, fungi, and other microbes are exposed to UV radiation, it alters their DNA in a way that prevents them from reproducing. Microorganisms respond differently to radiation depending on their species, strain, culture, and growth stage. Additionally, a significant factor is the type and makeup of the irradiated food (Falguera, Pagán, & Ibarz, 2011).

DOI: 10.1201/9781003359302-2

The United States Food and Drug Administration (USFDA) and the United States Department of Agriculture (USDA) have determined that UV-C radiation at 253.7 nm is acceptable for its usage in food processing. They have also allowed its use as a substitute treatment to reduce pathogens and other microbes. Code 2ICFR179.41, published by the USFDA, authorized the use of UV-C light in the preparation, processing, and handling of food (Gunter-Ward et al., 2018). This chapter provides information about the principle of UV, the type of lamps used, the inactivation mechanism of microbes and enzymes, and its application in the context of various food products.

2.2   FUNDAMENTAL PRINCIPLE OF UV TECHNOLOGY

The spectrum of different electromagnetic waves traveling across space includes light as just one component. From radio waves with wavelengths of one meter or more to X-rays with wavelengths of less than a billionth of a meter, the electromagnetic spectrum provides a wide range (Gunter-Ward et al., 2018). Disinfection is the process of reducing surface microorganisms to an acceptable level by public interest. UV radiation falls between the 200 and 400 nm regions of the electromagnetic spectrum. It is primarily classified into wavelength-based regions: UV-C (200–280 nm), UV-B (280–320 nm), and UV-A (320–400 nm). Cells can be damaged by UV-C light at a wavelength of about 254 nm, and DNA absorption data show that this wavelength is the most germicidal. The vacuum UV (100–200 nm) can only be transmitted in a vacuum because it is absorbed by practically all materials. UV radiation was frequently produced by low and medium-pressure mercury vapor lamps (Yemmireddy et al., 2022).

In a continuous UV system (Figure 2.1), the food item can be moved or flown through a quartz tube (for liquids) or a conveyor while being exposed to a UV lamp for the required time, dosage, and intensity. The shadow effect, which arises when some portions of the food surfaces do not receive enough UV light because of their curvatures, can be eliminated by integrating more lamps around the food product (Keklik et al., 2012).

2.2.1   UV LIGHT GENERATORS

The capability of UV equipment to efficiently treat food relies on the proper balancing of the UV supply radiating characteristics with the unique needs of UV applications. Low pressure (LP), low pressure, high output (LPHO), and medium pressure UV lamps are the three main types of UV lamps that are extensively utilized (MP). A UV transmitting envelope consists of a vitreous silica

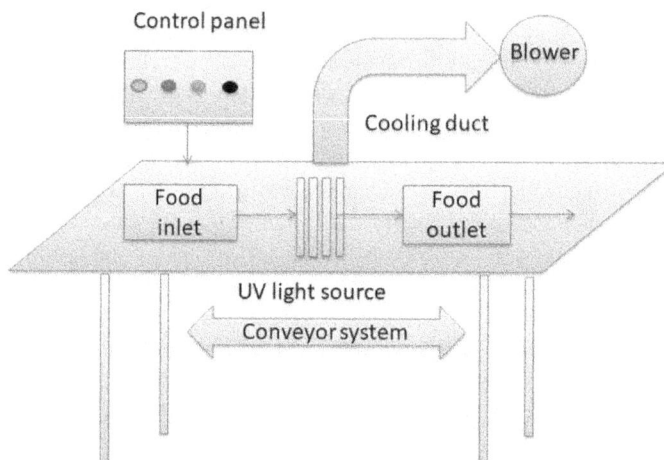

FIGURE 2.1   A continuous UV system.

glass tube, which makes up vapor-discharge lamps sealed at both ends. Each end of the envelope has an electrode were sealed off from outer surroundings. The envelope consists of mercury and inert gas inside (Koutchma, 2009).

Alternating current (AC) is a typical source of electricity for the lamp. A ballast must also be present for stable operation. The UV-transmitting sleeve separates the treated liquid from the UV lamps inside a commercial UV reactor, which can have an open or closed channel. The UV transmittance is used to monitor the dose delivery by the reactor, UV-intensity sensors, and flow meters (Gunter-Ward et al., 2018).

2.2.1.1 Mercury Lamps

Most often, elemental mercury (Hg) and inert gas are present in mercury-emission lamps (e.g., argon). The fact that mercury is most flammable metal element and can be activated in the gas phase at temperatures appropriate to the lamp structures explains why it is so common. The so-called avalanche effect, a chain reaction that underlies the electrical discharge, is also made possible by the fact that it has ionization energy that is low enough (Piskarev et al., 2013). The most prevalent filler gas is argon, whose ionization energy is 15.8 eV, while the energy of the lowest activated metastable state is 11.6 eV. If there is a collision, the energy of this metastable state may be lost, but if there is a collision with a mercury atom, mercury may ionize, which will cause light to be released. Therefore, the fundamental function of the filling gases is to stimulate the beginning of the activation-ionization of the mercury as well as to assist the beginning of the discharge (Gunter-Ward et al., 2018).

2.2.1.2 Excimer Lamps

Excimer lamps, which apply an electric potential to a mixture of rare gases across a dielectric barrier, are another source of monochromatic radiation (Naunovic et al., 2008). An excimer lamp has a characteristic wavelength that is "tuneable" since it depends on the gas mixture utilized in the system (Keklik et al., 2012).

Excimer lamps typically consist of two coaxial tubes constructed of a dielectric material, commonly quartz that is coaxial and contains a rare gas or a combination of rare gases/halogens at pressures near to atmospheric pressure. The development of an exciting molecular complex, which under normal circumstances lacks a stable ground state, is caused by the application of an electrical potential across the two dielectric barriers (Bergman, 2021). An excimer is a type of molecular complex that breaks down in a matter of nanoseconds, dissipating its excitation (binding) energy as photons with a certain wavelength. The gas mixture inside the dielectric barrier determines the wavelength of radiation (Naunovic et al., 2008).

In comparison to other types of UV sources, excimer lamps may operate at significantly lower surface temperatures, which may be favorable in terms of fouling behavior. An excimer lamp has a lifespan of six months and an electrical efficiency that ranges from 10% to 35% (Keklik et al., 2012).

2.2.1.3 Broadband Pulsed Lamps

This method creates a pulse of intense light emission that lasts for around 100 nanoseconds by discharging stored AC through a high-speed switch. The emission's wavelength makeup is comparable to that of solar light. The pulsed UV devices have higher treatment rates and can emit high-intensity UV that can penetrate opaque liquids better than mercury lamps. Although it is not yet well established in the market, this technology is more promising (Koutchma, 2019).

2.2.1.4 Microwave Powdered Electrodeless Mercury Lamps

The innovative MW-powered electrodeless mercury UV lamp technology seems to dispense with the necessity for electrodes in UV lamps. Instead of using electrodes, a magnetron produces MW radiation that passes through a waveguide and into the quartz lamp that contains the gas filling. Comparison with the other mercury lamps, the directed MW energy excites the argon atoms, which in turn excites the mercury atoms, causing them to emit radiation as they transition from excited states to lower energy ones (Horikoshi et al., 2009). The same temperatures and pressures as of the standard LPM lamp are used by electrodeless lamps. There are two types of commercially available UV water-disinfection systems that use MW lamps to disinfect drinking water and small flows of wastewater (Horikoshi et al., 2007).

The following are some features of employing MW-powered lights over regular lamps with electrodes: they warm up rapidly, the UV lamp deterioration process is stopped, and their lamp life is around three times longer than that of electrode lamps (Gutierrez et al., 2006). Since there aren't any electrical connections in the water and no electrodes are being utilized, there won't be any electrical connections to break down, which also avoids the corrosion problem. Additionally, bulbs can be turned on and off at any time without deterioration. The food business has not yet utilized this technology. However, expertise in the food sector has to take into consideration of its aforementioned benefits (Koutchma, 2019).

2.3 INACTIVATION MECHANISM

2.3.1   On Microorganisms

UV is a non-thermal inactivation technique acknowledged for its germicidal efficiency and helps in preventing spoilage of microorganisms and foodborne pathogens. The inactivation of microorganisms by UV light involves numerous pathways that may vary depending on the wavelengths employed (Gayán, García-Gonzalo, et al., 2014). The cell inactivation can be accomplished either implicitly or explicitly by absorbing light that strikes the DNA of the microbial cell. The mechanism responsible for microbial inactivation by UV light is its direct impact through lesions that obstruct the replication of DNA (Brem et al., 2017).

Since DNA absorbs UV-C radiation at its highest level between 200 and 280 nm, this area of the UV spectrum is regarded as the most deadly and germicidal for many microorganisms, particularly viruses and bacteria (Koutchma, 2009). The UV light generated in the 250–260 nm range has a stronger effect because the nitrogenous bases of thymine and cytosine in DNA absorb most of the UV light at these wavelengths. As a result of this process, photo products including such pyrimidine 6-4 pyrimidone and cyclobutane pyrimidine dimers, are formed which then interfere with the gene expression of DNA (Guerrero-Beltr·n & Barbosa-C·novas, 2004).

Pyrimidine dimer is the most significant photoproduct because it's generated between pyrimidine molecules that are adjacent to each other on the DNA strand, thus resulting in disruption of transcription and replication of DNA, ultimately triggering cell death (Salcedo et al., 2007). Additionally, ultrastructural modifications can be brought on by indirect photochemical effects including the production of free radicals (Cutler & Zimmerman, 2011).

Various parameters that influence the efficiency of the UVC radiation for germicidal action on food surfaces are the processing conditions, the radiation dose utilized, wavelength used, product qualities, time duration, characteristics of the product surface, types of microorganisms, and the profile of the targeted microbe. Hence, it is crucial to understand these parameters and how they affect microbial survival or death to select the appropriate radiation and achieve the desired inactivation (Gayán, Condón, et al., 2014; Guerrero-Beltr·n & Barbosa-C·novas, 2004).

UV radiation is more effective in inactivating Gram-negative bacteria than Gram-positive bacteria, followed by yeast and viruses (López et al., 2005). The production of pyrimidine dimers is responsible for the antimicrobial action of UV radiation on bacteria. Bacterial intrinsic characteristics, like cell wall thickness, might be the reason for variances in UV resistance among bacteria. Since gram-positive bacteria have a dense peptidoglycan structure on their cell walls are often more resistant to UV-C, which makes it harder for UV photons to enter the cells (Beauchamp & Lacroix, 2012).

Yeast has large cells and a smaller percentage of pyrimidine in its genetic material, which is resistant to UV damage. It might be photons will be absorbed by other substances before damaging DNA (Gayán, Condón, et al., 2014). A virus's ability to withstand denature is increased by the complexity of its capsid structure. Viruses are inactivated by disruption to their genomes and/or by actions on their viral proteins (Garcia & Barardi, 2019). Compared to vegetative cells, bacterial spores have higher UV resistance due to the dehydrated condition of the core which inhibits the dimerization of pyrimidine and the thick covering of proteins on spores (Gayán, Condón, et al., 2014).

Consequently, many microbes may evolve a means of repairing UV-induced DNA damage after exposure to UV radiation. Hence, whenever UV-damaged cells are subjected to greater than 330 nm wavelength, photo-reactivation, and dark healing must be taken into account (Guerrero-Beltr·n & Barbosa-C·novas, 2004).

2.3.2 ON ENZYMES

Peroxidase (POD) and polyphenol oxidase (PPO) are the deteriorative endogenous enzymes that significantly diminish the perishability of various fruits and vegetable commodities. PPO enzyme contains copper and catalyzes the oxidation of several phenolic compounds. The o-dihydroxy phenols oxidized to form o-quinones result in the generation of undesired brown pigments, which induce enzymatic browning (Koutchma, 2009). As a result, it is critical to inactivate the synthesis of melanin, which causes color degeneration. Similarly to PPO, POD causes enzymatic browning when interacting with phenolic substances. It catalyzes the oxidation of different substances when hydrogen peroxidase is present (Taze et al., 2020).

There are two ways that UV radiation might influence enzymes. The direct photooxidation that results from the absorbing of UV rays by a linked chromophore group or proteins can produce radicals or excited states through photoionization. Furthermore, when the transfer of energy of molecular oxygen by chromophores groups or proteins, the indirect oxidation of protein occurs by the singlet oxygen generated (Davies, 2005). These mechanisms can cause oxidative stress, which can alter the protein's three-dimensional structure and cause it to lose its catalytic function (Lante et al., 2013). For instance, UV light was more effective for polyphenol oxidase inactivation in reducing the unintentional oxidation of newly cut vegetables (Manzocco et al., 2009). Falguera et al. (2011) reported that after 15 minutes and 100 minutes of exposure to radiation with an incident energy of $3.88*10^{-7}$ E/min, enzymes including PPO and POD were completely inactivated in apple juice.

Furthermore, the combination technique with moderate heat treatment can improve the effectiveness of UV light on enzyme inactivation. Aguilar et al. (2018) assessed the effect of UV-Vis (250–740 nm) on the inactivation of an enzyme in three varieties of peach juices at temperatures of 25°C and 45°C. The PPO and POD activities were found to be effectively inactivated by UV-Vis irradiation. While POD activity was decreased by 60% at 45°C, PPO was entirely inactivated.

2.4 APPLICATION OF UV LIGHT

2.4.1 MILK PRODUCTS

Milk is a highly perishable food commodity due to the presence of nutrients and moisture which helps in the proliferation of microorganisms. Consumers are demanding safe food without compromising the organoleptic and nutritional properties. Milk undergoing thermal treatments such as pasteurization, sterilization, and ultra-high-temperature processing can help in destroying pathogens and microorganisms to a certain extent. However, most of this might affect the nutritional properties and can even lead to unfavorable sensory changes. Taking this into consideration, non-thermal processing is gaining popularity due to its inactivation power without affecting the organoleptic properties (Shabbir et al., 2021).

The use of UV as a non-thermal technology is a recognized physical method for decontamination which makes use of the germicidal property to inactivate pathogens and spillage microorganisms. Additionally, it is a non-ionizing radiation which does not emit radioactivity and residual products unlike gamma radiation (Delorme et al., 2020). UV sources with low pressure are most favorable for the application in milk because this won't affect the organoleptic and nutritional properties. According to the European Food Safety Authority (EFSA), the conversion of 7-dehydrocholesterol to vitamin D3 might occur when milk is exposed to UV radiation in low doses thereby increasing the vitamin D3 content in milk. Additionally, no detectable variations in the nutrient content in UV-treated milk have been observed (Dominique et al., 2017). Table 2.1 shows the studies conducted on various milk and milk products.

Pulsed UV light is a modified and improved form of UV-C light that is commonly used here which uses equipment with lamps that periodically emit high power UV light for 1 s to 0.1 s in the 200–1100 nm wavelength range (Gómez-López et al., 2007). Pulsed UV light causes thymine dimers to produce in the bacteria's DNA, ultimately leading to the death of bacterial cells. In addition, the pulsed UV-light system can also work in tandem with the

TABLE 2.1
The Effect of Ultraviolet Light on Various Dairy Products

Milk and Milk Products	Type of UV Used	Treatment Condition	Result	References
Raw milk	Pulsed light	20, 30, or 40 mL/min flow rate 3 number of pulses Distance of 8 cm from the light source	Inactivation of *Staphylococcus aureus* 0.55- to 7.26-log10 CFU/mL reduction	(Krishnamurthy et al., 2007)
Goat milk	Ultraviolet irradiation	Wavelength of 254 nm Time: 18 seconds	Alteration in chemical and sensory properties Increase in TBA	(Matak et al., 2007)
Whole milk Semi-skim milk	Laboratory scale UV device	1000 mJ/mL of UV dose 16 pulses	Inactivation of *Mycobacterium avium subsp. Paratuberculosis* in milk (0.5- to 1.0-log10 reduction)	(Altic et al., 2007)
Processed milk	Pulsed UV light	0.7, 2.1, 4.2, 8.4 and 11.9 J/cm^2 pulses	Less than 4.2 shows no significant change in protein oxidation 8.4 and 11.9 J/cm^2 show increase in carbonyl amount	(Fernández et al., 2014)
Fresh kashar cheese	Pulsed UV light	Time: 45 s Distance of exposure: 13 cm Fluence of 44 J/cm^2	1.62 (CFU/cm^2) log10 reductions in *S. aureus* 3.02 (CFU/cm^2) log10 reductions in *E. coli O157:H7*	(Keklik et al., 2019)
Fiordilatte cheese	UV-C light	Fluences of 0.1, 0.6, 1.2 and 6.0 kJ/m^2 for 750 s	Prolonged shelf life to nearly 80%	(Lacivita et al., 2016)
Skim cow milk Whole cow milk Sheep milk	Short wave UV-C light	UV-C dose of 2.37 ± 0.126 J/mL Thermal treatment of 110 °C for 30 s	Inactivation of *B. subtilis* 6 log CFU/mL in skim cow milk; 2.9 log CFU/mL in whole cow milk; 1.1 log CFU/mL in sheep milk	(Ansari et al., 2019)

pasteurization equipment to ensure the safety of the product. Pathogens like *S. aureus* that are resistant to heat treatment may be rendered inactive by delivering lethal dose of pulsed UV light (Krishnamurthy et al., 2007). However, some changes in chemical and sensory properties were detected in goat milk. This is due to the formation of thiobarbituric acid (TBA) reactive substances upon oxidation and reduction reactions on exposure of 1.3 m J/cm^2 UV dose at 254 nm (Matak et al., 2007).

2.4.2 SPICES

Over the last few years, several studies have been conducted on the effect and application of UV light on various spices. UV was found to be an effective technology in the microbial reduction of various spices from the experiments assessed by various authors. Nyhan et al. (2021) investigated the application of UV-LED devices on the surface inactivation of microorganisms such as *Listeria monocytogenes, Escherichia coli, Bacillus subtilis*, and *Salmonella Typhimurium*. In this study, UV-C LED emitted at a wavelength of 270 nm for 40 s showed a significant reduction in microbial count (say, 0.75–3 log10 colony forming units per mL) in powdered seasonings. Onion powder, garlic powder, cheese, and chili powder were utilized as powdered seasoning. In addition, Negi et al. (2021) mentioned the effect of UV light on adulterant authentication in spices such as saffron and turmeric. UV light on absorption results in the rise of energy content in atoms and molecules which aids in the identification of banned dyes such as Sudan dyes in saffron. It can detect adulterants up to 1 to 5 ppm and is referred to as a low-cost high sensitive adulterant authentication technique.

Nicorescu et al. (2013) investigated the inactivation of *Bacillus subtilis* vegetative cells on samples of ground caraway, ground red pepper, and ground black pepper. The selected spices were artificially inoculated with *Bacillus subtilis*. He observed that on treatment with PL of ten pulses (10 J/cm^2), a significant reduction of 1 log CFU/ml was assessed in red pepper and 0.8 log reduction in caraway and ground black pepper. However, this leads to microbial cell wall disruption. In contrast, PL was applied to a liquid medium (0.9% saline solution) which is inoculated with 8.7 log of B. subtilis vegetative cells. The initial microbial population was completely reduced by a single pulse at 0.6 J/cm^2. SEM and DNA extraction tests revealed that contrast to the spice samples, the bacterial cells in suspension were not damaged, but they showed altered cell shape, enlarged vacuoles, and distorted cell membranes.

Dogu-Baykut and Gunes (2019) discovered the influence of the fluidized bed ultraviolet (UV-C) system in the microbial decontamination of thyme (*Thymus vulgaris L.*). These authors found that thyme treated with UV-C at 254 nm wavelength and 205.6 J/cm^2 doses rendered at an intensity of 26.7 m W/cm^2 showed a reduction of 1.8 log CFU/g in total aerobic mesophilic bacteria, 1.3 log CFU/g of total yeast/mold count, and 0.3 log CFU/g of *Bacillus cereus* without affecting the overall physical, chemical and sensorial properties. Cheon et al. (2015) performed experiments on powdered red pepper to inactivate foodborne pathogens by using UV-C irradiation in tandem with mild heat treatment. The UV-C application in powdered red pepper at 20.4 kJ/m^2 for 10 min with mild heat treatment at 65°C reduces the level of *E. coli* O157:H7 and *S. Typhimurium* by 2.88 and 3.06 log CFU/g, respectively without affecting the color value. However, moisture and capsaicinoid contents reduced significantly when heated at 65°C due to drying of the sample at a higher temperature.

Dogu-Baykut and Gunes (2022) reported the microbial and color qualities of black pepper seeds found a significant reduction in microbial load. The samples exposed to 28.8 J/cm^2 UV dose and the ozone treatments of 15 ppm for 1 hour can degrade the total aerobic mesophilic bacteria count (TAMB) and *E. coli* by 0.8 log CFU/g. However, the combined treatment can reduce the *E. coli* count by 1.7 log without much significant difference in TAMB and color of the sample. From these experiments, we can conclude that UV light alone or used as a hurdle with other technologies can significantly reduce the microbial load in spices.

2.4.3 Fruits and Vegetables

Fruits and vegetables play a vital role in a wholesome and healthy lifestyle. Therefore, providing safe and hygienic produce is a crucial factor. But most of these are affected by various causative agents such as *Salmonella, Shigella spp., Campylobacter, Escherichia coli O157:H7, Listeria monocytogenes, Yersinia enterocolitica, Staphylococcus aureus, Clostridium spp., Bacillus cereus* leading to foodborne illness (Gayas et al., 2019). Table 2.2 summarizes the effect of UV light as a non-thermal treatment for the inactivation of these microorganisms on fresh produce.

TABLE 2.2

The Studies Conducted on the Effect of Ultraviolet Light on Fruits and Vegetables

Fruits and Vegetables	Types of UV	Treatment Conditions	Result	References
Fresh cut apples	Pulsed light	UV dose of 8 and 16 J cm^{-2} 15 days/5°C	Inactivation of natural microorganisms No significant difference in the phenolic and vitamin C content	(Llano et al., 2016)
Blueberries	Water assisted pulsed light	9 J/cm^2 3 days/5°C	Inactivation of Salmonella by 4.4 log CFU No significant impact on the antioxidant, anthocyanin, color, and other health compounds	(Cao et al., 2017)
Mangoes	High intensity pulsed light	Fluences between 3.6 and 10.8 J/cm^2	Increase in the vitamin C and carotenoid content by 10 to 40% Increase in the vitamins B1, B3 and B5 by 10 to 25% Degradation of vitamin B6	(Braga et al., 2019)
Avocados	Intense light pulses (ILP)	Pulses of 15 or 30 at 0.4 J/cm^2 per pulse and kept undisturbed for 15 days at 5°C	15 pulses significantly reduced 2.61 log CFU/g of *L. innocua* and 2.90 log CFU/g of *E. coli* 30 pulses reduced 2.97 log CFU/g of *L. innocua* and 3.3 log CFU/g of *E. coli.* Texture and firmness are affected on treating with 30 pulses	(Ramos-Villarroel et al., 2011)
Garlic	UV-C light	Low UV dose of 2.0 kJ/m^2 for 15 days at 0°C	1 log reduction of microbial load; increase in phenols; increase of flavonoids; maintenance of color and texture	(M. H. Park & Kim, 2015)
Leafy vegetables (spinach, leek, cabbage)	UV-C light	2.0 kJ/m^2 for 15 days at 0°C	Multiple UV-C cycles delayed the breakdown of chlorophyll while maintaining larger quantities of soluble protein and ascorbic acid than a single cycle	(Liao et al., 2016)
Button mushroom	UV-C light	1.0 kJ/m^2 21 days 4°C	Increase in antioxidant activity Increase in phenols Increased chance of browning	(Wu et al., 2016)

Llano et al. (2016) assessed the effects on the quality and antioxidant attributes of fresh-cut apples through the combined treatment of PL and a quality-stabilizing pre-treatment. Exposure to UV dose at 8 and 16 J cm^{-2} over 15 days at a temperature of 5°C was effective in the inactivation of natural microflora present in the apple without causing browning and antioxidant degradation. Moreover, phenolic compounds and vitamin C content were maintained in this treatment. Hence, PL is considered an emerging tool in extending the shelf life of fresh fruits like apples. Cao et al. (2017) evaluated the influence of water-assisted and dry PL on the inactivation of Salmonella in blueberries and found that water-assisted PL showed an effective result in the degradation of Salmonella by 4.4 log CFU compared to that of the dry PL, which reduced microbial load by 0.9 log CFU/mL. On observing the physiochemical properties, blueberries on water-assisted PL treatment minimally affect the antioxidant, anthocyanin, color, and other biologically active compounds.

Braga et al. (2019) investigated the use of high-intensity PL as an alternative tool for drying mangoes. The mangoes subjected to pre-treatment at fluences between 3.6 and 10.8 J/cm^2 resulted in an increase in vitamin C and carotenoid content by 10 to 40% and vitamins B1, B3, and B5 by 10 to 25%. While the light-sensitive vitamin B6 is degraded by 40–50% irrespective of the doses exposed, and most of the vitamins become degraded by exposing it to more than 10.8 J/cm^2. Therefore, PL at an optimized dose can be used as an effective pre-treatment technique in the drying of fresh produce by conferring better nutritional benefits.

Ramos-Villarroel et al. (2011) investigated the effect of intense light pulses (ILP) in the inactivation of *Listeria innocua* and *Escherichia coli* in avocados. The effect of avocados incoluted with *L. innocua* or *E. coli* placed in cylinders covered with plastic trays, sealed using oxygen-permeable polypropylene films exposed to pulses of 15 or 30 at 0.4 J/cm^2 per pulse, and kept undisturbed for 15 days at 5°C. It is observed that this application of ILP with 15 pulses significantly reduced 2.61 log CFU/g of *L. innocua* and 2.90 log CFU/g of *E. coli* while on treatment with 30 pulses reduced 2.97 log CFU/g of *L. innocua* and 3.3 log CFU/g of *E. coli*. However, the use of 30 pulses can affect the physiochemical properties, such as texture, and can cause the browning of avocados. In addition to microbial load, other physiochemical properties were also studied. In another study, Park and Kim (2015) assessed the application of low-dose UV-C at a fluence of 2.0 kJ m^{-2} for 15 days at 0°C in garlic and observed that the reduction of microbial load by 1 log, there cause an increase in the flavonoid and phenolic content. Moreover, it all maintains the texture and color of UV-C treated garlic.

Liao et al. (2016) evaluated the impact of UV-C treatment multiple times for maintaining post-harvest qualities in three leafy vegetables (spinach, leek, and cabbage). The results showed that after exposing leafy vegetables to UV-C treatment at a fluence of 2.46 kJ m^{-2} stored at 4°C for 5 days, soluble protein and ascorbic acid content was increased. Moreover, this treatment aid in the delay of chlorophyll degradation from leafy vegetables for nearly 5 days at 4°C giving freshness to the produce. Likewise, the study conducted by Wu et al. (2016) evaluated the impact of UV-C (1.0 kJ m^{-2}) on the quality of button mushrooms and their effect on the biosynthesis of melanin. The observation of UV-C treatment indicated that both antioxidant and phenolic activity were improved during the storage of button mushrooms for 21 days at 4°C. The scavenging activity of DPPH (2,2-diphenyl-1-picryl-hydrazyl-hydrate) was found to be highest in the stipe, and the phenols were more in the gill. However, a significant effect on the conversion of phenolic compounds takes place which results in the browning of mushrooms.

2.4.4 POST-HARVEST TREATMENT ON CEREALS AND PULSES

Cereals and pulses are the richest sources of starch and are consumed as a stable food throughout the world. To maintain the health benefits and nutritional content, they are being undergone non-thermal treatments. These treatments will degrade the harmful hazards and render them safe for consumption. UV light is an effective non-thermal treatment that uses its germicidal property to degrade eggs and larvae from causing infestation (Srivastava & Mishra, 2021). Various studies have been conducted to find the potency of UV light on cereals and pulses. For example, Kumar et al.

(2020) used the application of UV-C at the wavelength of 254 nm to produce significant changes in the physiochemical properties of wheat. The disintegration of wheat protein and gluten thereby losing the elasticity and compactness was observed. In addition to this, there occurs an increase in total volatile basic nitrogen (TVBN) content, a change in color, and a reduction in pH content.

Wang et al. (2016) assessed the degradation and detoxification of aflatoxin B1 (AFB1) and aflatoxin B2 (AFB2) by the application of PL in rice bran and rough rice. Samples were inoculated with Aspergillus flavus followed by exposing UV pulses of 0.52 J/cm^2 at 80 s and 15 s for detoxifying and degrading aflatoxin B1 and B2. Rough rice on treatment for 80 s causes a reduction of AFB1 by 75.0% and AFB2 by 39.2% while the treatment for 15 s reduced AFB1 and AFB2 in rice bran by 90.3% and 86.7%, respectively. This study reported the application of PL as an effective tool in the detoxification of aflatoxin in rice bran and rough rice.

In the study conducted by Ferreira et al. (2018), maize irradiated with a UV-C dose of 10 J/cm^2 shows a reduction in the live adult weevils but the significant reduction of carotenoids negatively resulted in the antioxidant capacity. However, failed to identify the best UV dose for the control of adult weevils in stored grain. Hidema et al. (2005) reported the effect of UV-B applied at an intensity of 1.0 W m^{-2} and wavelength of 300 nm in rice. They observed an increase in the storage protein content and a 2% rise in the N2 content affecting positively the grain size. In addition, the study conducted in lentils showed significant changes in the protein and carbohydrate content on treatment with high-intensity UV light at 3.99 W/cm^2 intensity. This can also favor the yield of the plant by using appropriate dosage and intensity.

2.4.5 PACKAGING

UV light has been successfully incorporated into the packaging to prevent microorganism contamination. Various studies were conducted to find the potency of UV light in packaging. For instance, Xie and Hung (2018) evaluated the bactericidal property of polymer films through the application of photocatalysis (UV-A/photocatalyst) using 5% TiO$_2$ at a UV intensity of 1.30 ± 0.15 m W/cm^2 for 2 hours. On observation among the three biodegradable films used (cellulose acetate, polycaprolactone, and polylactic acid), TiO$_2$ incorporated cellulose acetate shows a significant reduction of 1.6 log CFU/mL with the highest bactericidal activity compared to other films. The study suggests the incorporation of TUV (titanium assisted UV photocatalysis) in cellulose acetate film has the highest potency of antimicrobial properties which can be used successfully in the field of packaging.

Researchers have reported the application of PL in the sterilization of packaging materials. It shows various applications on packaging materials such as barrier protection, antimicrobial properties, migration potential, mechanical properties, and microbial decontamination (Marangoni Junior et al., 2020). Keklik et al. (2009) evaluated the inactivation of *Listeria monocytogenes* and compared the log reduction in vacuum-packed and unpacked chicken frankfurters. In the unpacked sample, 0.3 log reduction was found on treating with UV from a distance of 13 cm at 5 s and 1.9 log reduction was observed when the distance was reduced to 5 cm by increasing time to 60 s. However, on treating with the packaged sample, 0.1 log reduction was found at a distance of 13 cm at 5 s, but the results were the same 1.9 log reduction at 5 cm for 60 s. In addition to this, De Moraes et al. (2020) studied the potency of combined treatment of PL with antimicrobial starch films on Cheddar cheese surface for the inactivation of *Listeria innocua*. A 4.5 log reduction of *Listeria innocua* was observed on the surface of cheddar cheese after 3 days of storage at 4°C on the combined treatment with PL applied with 6 pulses at a fluence of 6.14 J/cm^2. Kramer et al. (2019) also demonstrated the application of pl at fluences of 0.35 J/cm^2 to 3.6 J/cm^2 in the inactivation of *Listeria innocua* in meat products packed with (poly acetate/polyethylene) PA/PE vacuum bags. In their study, 1 log reduction was observed in packed ham and 3 to 4 log reduction in packed chicken which is maintained for 12 days stored at 5°C.

Furthermore, the efficacy of UV-C irradiation was studied by Ha et al. (2016) in sliced cheese packaged with various films of 0.07 mm, 0.10 mm, and 0.13 mm thickness to inactivate the foodborne pathogens (*Escherichia coli O157:H7, Salmonella Typhimurium,* and *Listeria monocytogenes*). They observed a drastic log reduction of more than 3 log CFU/g in sliced cheese packed with polypropylene (PP) and polyethylene (PE) films of thickness 0.7 mm on treatment with UV-C at 3.04 m W/cm^2 for 1 minute. Another study was investigated by Chawengkijwanich and Hayata (2008) in which a 3 log CFU/ml reduction of *Escherichia coli* was observed with the development of TiO$_2$ powder-coated packaging. The two black light bulbs illuminating the UV source at a wavelength of 300–400 nm with 20 W power at 1 m W/cm^2 intensity are used for the inactivation in the study. Overall, the study demonstrates the titanium dioxide-assisted UV photocatalysis packaged film can be used as an effective applicant in the decontamination of fresh produce.

2.5   EFFECT OF UV LIGHT ON QUALITY OF FOODS

UV light is a novel food processing technique that includes various benefits over traditional processes, including the efficient inactivation of a broad range of harmful and spoilage bacteria, as well as little loss of sensory and nutritional qualities in food products. It also doesn't produce waste or have any hazardous impacts (Gayán, Condón, et al., 2014). Food properties such as constituent, absorptivity, color, the level of suspended particles, and the microbial load are significant factors impacting how well UV radiation disinfects food (Koutchma, 2009).

Liu et al. (2018) investigated the potential of UV light treatment to increase the level of phenolics and flavonoids and to induce pathway of phenylpropanoid in tomato fruit. The UV-C radiation of 0.4 J cm^{-2} resulted in enhanced TPC of 12.82% when compared with control and also induced the key enzyme pathway of phenylpropanoid. The total amount of phenolics in UV-C treated fruit was greatest that is 246.42 mg kg^{-1}. A study on the quality characteristics of UV-C treated pineapple at a dosage of 4.5 kJ/m^2 throughout a duration of 0–90 s, accompanied by a storage period at 10°C, revealed that the treated pineapple was more capable of maintaining its firmness qualities than that of the untreated sample (Pan & Zu, 2012).

Since penetrability is often constrained, the efficacy of UV light highly depends on the material intended for the process. The capacity of UV radiation to permeate solid food products depends on several parameters such as surface, physical, and chemical characteristics including viscosity, roughness, color, thickness, optical qualities, dirtiness, or density (Koca et al., 2018). Edward et al. (2014) conducted an experiment on oyster mushrooms to study the effects of UV-A and UV-C throughout the range of temperatures of 25–100°C. In comparison with UV-A, the UV-C subjected mushroom exhibited a greater rise in loss factor and loss modulus. The results showed that UV-C light had a stronger effect on the mechanical characteristics of oyster mushrooms than UV-A radiation.

Due to the potential activation of several damaging photochemical processes after the production of free radicals, UV radiation may negatively influence the composition of food, perhaps reducing the amount of beneficial dietary components (Yongyue et al., 2011). Adverse reactions of UV resulting in modification of the composition, oxidization of lipids, diminishes vitamin content, and produces off-odor and color change (Csapó et al., 2019).

2.6   SYNERGISTIC EFFECT OF UV LIGHT COMBINED WITH OTHER TECHNOLOGIES

Combining UV radiation with several other technologies or treatment methods can have additive or synergistic benefits since it is an easy-to-use, low-cost technology. Araby et al. (2022) studied the effectiveness of UV-C and chitosan nanoparticles alone and their combined action to inactivate the Listeria Monocytogenes and Escherichia coli O157:H7 in pomegranate juice. The combination of these techniques entirely inactivated these pathogens for a UV dosage period of 30 minutes at 0.356 J/cm^{-2} in vitro. Additionally, it is a prospective barrier for improved nutrient, nutritional value, and health-related constituent preservation. During the storage period, the microbial reduction was highest in sequential application than alone. In comparison to the control, asparagus treated with a combination of UV-C at 1 kJ/m² and ozonated water for 3 s at 3 ppm had enhanced cutting energy across 4 days of storage (Tiecher et al., 2013). Table 2.3 provides information about the application of combined action of UV and other technologies/methods in various food products.

Ha and Kang (2014) observed the effect of UV radiation and near-infrared (NIR) heating in Infant food powder for inactivation of *Cronobacter sakazakii* and analyzed the quality parameters such as sensory and color change parameters. *C. sakazakii* was reduced by 2.79 log-units CFU after receiving a combined NIR-UV treatment for 7 min. The primary element causing the synergistic fatal impact of the combination of NIR-UV radiation was the fundamental inactivation processes that disrupted the cell membrane of bacteria. Infant formula powder that was concurrently NIR-UV treated did not differ substantially ($P > 0.05$) from the control in terms of color values or sensory qualities.

TABLE 2.3
Synergistic Effect of UV with Other Technologies in Different Food Products

Combined Effect of UV with Other Methods	Food Products	Process Condition	Results	References
UV + thermal treatment	Vegetables	UV dosage of 58.2 mJ/cm² at 60°C	• Inactivation of oxidative enzymes (PPO and POD) achieved • Synergistic effect • Low energy required	(Sampedro et al., 2014)
Super atmospheric oxygen packaging + UV-C + heat treatment	Pomegranate arils (fresh cut)	Dipping in hot water for 30 s at 50°C; UV-C dosage of 4.54 kJ m⁻²; Initial oxygen level 90 kPa	• Reduction in PPO and POD activity • Maintained antioxidant compounds • Preserved superoxide dismutase and catalase	(Maghoumi et al., 2013)
UV treatment + high hydrostatic pressure (HPP)	Vacuum packed tilapia	Dosage of UV 0.103±0.002 J/cm²; HPP – 220 MPa at 25°C for 10 min	• Favorable impact on lipid oxidation and cadaverine levels • Retarded bacterial growth • Delayed the rise in pH, biogenic amines, and ammonia	(Monteiro et al., 2018)
UV treatment + modified atmospheric packaging (MAP)	Fresh cut carrots	Dosage of UV-C; 2 kJ m⁻²; MAP – 80% O₂, 10% N₂, and 10% CO₂	• Delayed process of senescence • Lessened surface whitening • Retention of physiological quality • Reduced growth of microbes	(Li et al., 2021)
UV treatment + mild heat treatment	Lemon and melon blend	UV-C dosage of 2.461 J/mL and heat treatment at 72°C for 71 s	• *E. coli* K12 was reduced by >6 log10 CFU/mL • Improved performance in quality preservation • Storage life was extended from 2 to 30 days	(Kaya et al., 2015)

2.7 FUTURE TRENDS

Further research into this subject is thus required to establish UV technology as an affordable and practical substitute for thermal methods and surface decontamination of various food products. To apply a consistent treatment to the food products, it is necessary to develop validation methodologies, understand the inactivation kinetics for spoilage and bacteria frequently found in foods, and optimize the process parameters. Further studies need to be conducted on the UV photoproducts found in food products.

2.8 SUMMARY

Microorganisms and enzymes in different food products have been demonstrated to be effectively inactivated by UV radiation, a non-thermal and eco-friendly technique that appears to have no negative effects on the product's qualities. It effectively retains sensory and nutritional qualities of food more than traditional processing methods. The UV light procedure has restricted use since it has low penetration power, requires transparent products to be processed, requires proper mixing, and demands an even food surface.

In addition to improving microbiological hygiene and shelf life, UV radiation can retain food quality attributed and improve the concentration of specific nutrients. The treatment parameters such as light intensity, wavelength, treatment chamber design, and duration of exposure, have a substantial impact on the quality characteristics of the treated products. As a result, careful consideration must be given while choosing the processing parameters for the product that will be irradiated.

REFERENCES

Afraz, M. T., Khan, M. R., Roobab, U., Noranizan, M. A., Tiwari, B. K., Rashid, M. T., Inam-ur-Raheem, M., Hashemi, S. M. B., & Aadil, R. M. (2020). Impact of novel processing techniques on the functional properties of egg products and derivatives: A review. *Journal of Food Process Engineering*, *43*(12), e13568.

Aguilar, K., Garvín, A., & Ibarz, A. (2018). Effect of UV–Vis processing on enzymatic activity and the physicochemical properties of peach juices from different varieties. *Innovative Food Science & Emerging Technologies*, *48*, 83–89.

Altic, L. C., Rowe, M. T., & Grant, I. R. (2007). UV light inactivation of Mycobacterium avium subsp. paratuberculosis in milk as assessed by FASTPlaqueTB phage assay and culture. *Applied and Environmental Microbiology*, *73*(11), 3728–3733. https://doi.org/10.1128/AEM.00057-07

Anbarasan, R., Boopathy, B., Stephen, J., & Radhakrishnan, M. (2022). Cold plasma disinfestation of Callosobruchus maculatus infested soybeans: Its subsequent impact on soymilk extraction yield and quality. *Journal of Food Process Engineering*. https://doi.org/10.1111/jfpe.14246

Ansari, J. A., Ismail, M., & Farid, M. (2019). Investigate the efficacy of UV pretreatment on thermal inactivation of Bacillus subtilis spores in different types of milk. *Innovative Food Science and Emerging Technologies*, *52*(January), 387–393. https://doi.org/10.1016/j.ifset.2019.02.002

Araby, E., Abd El-Khalek, H. H., & Amer, M. S. (2022). Synergistic effects of UV-C light in combination with chitosan nanoparticles against foodborne pathogens in pomegranate juice with enhancement of its health-related components. *Journal of Food Processing and Preservation*, *46*(7), e16695.

Beauchamp, S., & Lacroix, M. (2012). Resistance of the genome of Escherichia coli and Listeria monocytogenes to irradiation evaluated by the induction of cyclobutane pyrimidine dimers and 6-4 photoproducts using gamma and UV-C radiations. *Radiation Physics and Chemistry*, *81*(8), 1193–1197.

Bergman, R. S. (2021). Germicidal UV sources and systems. *Photochemistry and Photobiology*, *97*(3), 466–470.

Braga, T. R., Silva, E. O., Rodrigues, S., & Fernandes, F. A. N. (2019). Drying of mangoes (Mangifera indica L.) applying pulsed UV light as pretreatment. *Food and Bioproducts Processing*, *114*, 95–102.

Brem, R., Guven, M., & Karran, P. (2017). Oxidatively-generated damage to DNA and proteins mediated by photosensitized UVA. *Free Radical Biology and Medicine*, *107*, 101–109.

Cao, X., Huang, R., & Chen, H. (2017). Evaluation of pulsed light treatments on inactivation of Salmonella on blueberries and its impact on shelf-life and quality attributes. *International Journal of Food Microbiology*, *260*, 17–26.

Chacha, J. S., Zhang, L., Ofoedu, C. E., Suleiman, R. A., Dotto, J. M., Roobab, U., Agunbiade, A. O., Duguma, H. T., Mkojera, B. T., & Hossaini, S. M. (2021). Revisiting non-thermal food processing and preservation methods—Action mechanisms, pros and cons: A technological update (2016–2021). *Foods*, *10*(6), 1430.

Chawengkijwanich, C., & Hayata, Y. (2008). Development of TiO$_2$ powder-coated food packaging film and its ability to inactivate Escherichia coli in vitro and in actual tests. *International Journal of Food Microbiology*, *123*(3), 288–292.

Cheon, H.-L., Shin, J.-Y., Park, K.-H., Chung, M.-S., & Kang, D.-H. (2015). Inactivation of foodborne pathogens in powdered red pepper (Capsicum annuum L.) using combined UV-C irradiation and mild heat treatment. *Food Control*, *50*, 441–445.

Csapó, J., Prokisch, J., Albert, C., & Sipos, P. (2019). Effect of UV light on food quality and safety. *Acta Univ Sapientiae Alimentaria*, *12*, 21–41.

Cutler, T. D., & Zimmerman, J. J. (2011). Ultraviolet irradiation and the mechanisms underlying its inactivation of infectious agents. *Animal Health Research Reviews*, *12*(1), 15–23.

Davies, M. J. (2005). The oxidative environment and protein damage. *Biochimica et Biophysica Acta (BBA)-Proteins and Proteomics*, *1703*(2), 93–109.

De Moraes, J. O., Hilton, S. T., & Moraru, C. I. (2020). The effect of pulsed light and starch films with antimicrobials on Listeria innocua and the quality of sliced cheddar cheese during refrigerated storage. *Food Control*, *112*, 107134.

Delorme, M. M., Guimarães, J. T., Coutinho, N. M., Balthazar, C. F., Rocha, R. S., Silva, R., Margalho, L. P., Pimentel, T. C., Silva, M. C., Freitas, M. Q., Granato, D., Sant'Ana, A. S., Duart, M. C. K. H., & Cruz, A. G. (2020). Ultraviolet radiation: An interesting technology to preserve quality and safety of milk and dairy foods. *Trends in Food Science and Technology*, *102*(June), 146–154. https://doi.org/10.1016/j.tifs.2020.06.001

Dogu-Baykut, E., & Gunes, G. (2019). Ultraviolet (UV-C) radiation as a practical alternative to decontaminate thyme (Thymus vulgaris L.). *Journal of Food Processing and Preservation*, *43*(6), e13842.

Dogu-Baykut, E., & Gunes, G. (2022). Effect of ultraviolet (UV-C) light and gaseous ozone on microbial and color qualities of whole black pepper seeds (*Piper nigrum* L.). *Carpathian Journal of Food Science & Technology*, *14*(2), 122–131.

Dominique, T., Bresson, J. L., Barbara, B., Tara, D., Tait, S. F., Marina, H., Ernst, K. I. H., Inge, M., McArdle, H. J., & Androniki, N. (2017). Safety of hydroxytyrosol as a novel food pursuant to Regulation (EC) No 258/97. *EFSA Journal*, *15*(3), e04728. https://doi.org/10.2903/j.efsa.2017.4728

Edward, T. L., Kirui, M. S. K., Omolo, J. O., Ngumbu, R. G., & Odhiambo, P. M. (2014). Effect of ultraviolet-A (UV-A) and ultraviolet-C (UV-C) light on mechanical properties of oyster mushrooms during growth. *Journal of Biophysics*, *2014*, 687028. https://downloads.hindawi.com/archive/2014/687028.pdf

Falguera, V., Pagán, J., Garza, S., Garvín, A., & Ibarz, A. (2011). Ultraviolet processing of liquid food: A review. Part 1: Fundamental engineering aspects. *Food Research International*, *44*(6), 1571–1579.

Falguera, V., Pagán, J., & Ibarz, A. (2011). Effect of UV irradiation on enzymatic activities and physico-chemical properties of apple juices from different varieties. *LWT-Food Science and Technology*, *44*(1), 115–119.

Fernández, M., Ganan, M., Guerra, C., & Hierro, E. (2014). Protein oxidation in processed cheese slices treated with pulsed light technology. *Food Chemistry*, *159*, 388–390. https://doi.org/10.1016/j.foodchem.2014.02.165

Ferreira, C. D., Ziegler, V., Schwanz Goebel, J. T., Lang, G. H., Elias, M. C., & de Oliveira, M. (2018). Quality of grain and oil of maize subjected to UV-C radiation (254 nm) for the control of weevil (Sitophilus zeamais Motschulsky). *Journal of Food Processing and Preservation*, *42*(2), e13453.

Garcia, L. A. T., & Barardi, C. R. M. (2019). Performance of a storage tank coupled with UV light on enteric virus inactivation in drinking water. *Water Supply*, *19*(4), 1103–1109.

Gayán, E., Condón, S., & Álvarez, I. (2014). Biological aspects in food preservation by ultraviolet light: A review. *Food and Bioprocess Technology*, *7*(1), 1–20.

Gayán, E., García-Gonzalo, D., Álvarez, I., & Condón, S. (2014). Resistance of Staphylococcus aureus to UV-C light and combined UV-heat treatments at mild temperatures. *International Journal of Food Microbiology*, *172*, 30–39.

Gayas, B., Munaza, B., & Sidhu, G. K. (2019). Ultraviolet Light Treatment of Fresh Fruits and Vegetables. In *Processing of Fruits and Vegetables* (pp. 83–100). Apple Academic Press.

Gómez-López, V. M., Ragaert, P., Debevere, J., & Devlieghere, F. (2007). Pulsed light for food decontamination: A review. *Trends in Food Science and Technology*, *18*(9), 464–473. https://doi.org/10.1016/j.tifs.2007.03.010

Guerrero-Beltr·n, J. A., & Barbosa-C·novas, G. V. (2004). Advantages and limitations on processing foods by UV light. *Food Science and Technology International*, *10*(3), 137–147.

Gunter-Ward, D. M., Patras, A., S. Bhullar, M., Kilonzo-Nthenge, A., Pokharel, B., & Sasges, M. (2018). Efficacy of ultraviolet (UV-C) light in reducing foodborne pathogens and model viruses in skim milk. *Journal of Food Processing and Preservation*, *42*(2), e13485.

Gutierrez, R. L., Bourgeous, K. N., Salveson, A., Meir, J., & Slater, A. (2006). Microwave UV: A new wave of tertiary disinfection. *Proceedings of the Water Environment Federation*, *2006*(10), 2853–2864.

Ha, J.-W., Back, K.-H., Kim, Y.-H., & Kang, D.-H. (2016). Efficacy of UV-C irradiation for inactivation of food-borne pathogens on sliced cheese packaged with different types and thicknesses of plastic films. *Food Microbiology*, *57*, 172–177.

Ha, J.-W., & Kang, D.-H. (2014). Synergistic bactericidal effect of simultaneous near-infrared radiant heating and UV radiation against Cronobacter sakazakii in powdered infant formula. *Applied and Environmental Microbiology*, *80*(6), 1858–1863.

Hidema, J., Zhang, W., Yamamoto, M., Sato, T., & Kumagai, T. (2005). Changes in grain size and grain storage protein of rice (Oryza sativa L.) in response to elevated UV-B radiation under outdoor conditions. *Journal of Radiation Research*, *46*(2), 143–149.

Horikoshi, S., Abe, M., & Serpone, N. (2009). Novel designs of microwave discharge electrodeless lamps (MDEL) in photochemical applications. Use in advanced oxidation processes. *Photochemical & Photobiological Sciences*, *8*(8), 1087–1104.

Horikoshi, S., Kajitani, M., Sato, S., & Serpone, N. (2007). A novel environmental risk-free microwave discharge electrodeless lamp (MDEL) in advanced oxidation processes: Degradation of the 2, 4-D herbicide. *Journal of Photochemistry and Photobiology A: Chemistry*, *189*(2–3), 355–363.

Kaya, Z., Yıldız, S., & Ünlütürk, S. (2015). Effect of UV-C irradiation and heat treatment on the shelf life stability of a lemon–melon juice blend: Multivariate statistical approach. *Innovative Food Science & Emerging Technologies*, *29*, 230–239.

Keklik, N M, Demirci, A., & Puri, V. M. (2009). Inactivation of Listeria monocytogenes on unpackaged and vacuum-packaged chicken frankfurters using pulsed UV-light. *Journal of Food Science*, *74*(8), M431–M439.

Keklik, Nene M., Elik, A., Salgin, U., Demirci, A., & Koçer, G. (2019). Inactivation of Staphylococcus aureus and Escherichia coli O157:H7 on fresh kashar cheese with pulsed ultraviolet light. *Food Science and Technology International*, *25*(8), 680–691. https://doi.org/10.1177/1082013219860925

Keklik, N.M., Krishnamurthy, K., & Demirci, A. (2012). Microbial Decontamination of Food by Ultraviolet (UV) and Pulsed UV Light. In *Microbial Decontamination in the Food Industry* (pp. 344–369). Elsevier. https://doi.org/10.1533/9780857095756.2.344

Koca, N., Urgu, M., & Saatli, T. E. (2018). Ultraviolet light applications in dairy processing. *Technological Approaches for Novel Applications in Dairy Processing*, *1*, 1–20.

Koutchma, T. (2009). Advances in ultraviolet light technology for non-thermal processing of liquid foods. *Food and Bioprocess Technology*, *2*(2), 138–155.

Koutchma, T. (2019). Ultraviolet Light in Food Technology: Principles and Applications. In *Ultraviolet Light in Food Technology: Principles and Applications* (Vol. 2). CRC press. https://doi.org/10.1201/9780429244414

Kramer, B., Wunderlich, J., & Muranyi, P. (2019). Inactivation of Listeria innocua on packaged meat products by pulsed light. *Food Packaging and Shelf Life*, *21*, 100353. https://doi.org/10.1016/j.fpsl.2019.100353

Krishnamurthy, K., Demirci, A., & Irudayaraj, J. M. (2007). Inactivation of Staphylococcus aureus in milk using flow-through pulsed UV-light treatment system. *Journal of Food Science*, *72*(7). https://doi.org/10.1111/j.1750-3841.2007.00438.x

Kumar, A., Rani, P., Purohit, S. R., & Rao, P. S. (2020). Effect of ultraviolet irradiation on wheat (Triticum aestivum) flour: Study on protein modification and changes in quality attributes. *Journal of Cereal Science*, *96*, 103094.

Lacivita, V., Conte, A., Manzocco, L., Plazzotta, S., Zambrini, V. A., Del Nobile, M. A., & Nicoli, M. C. (2016). Surface UV-C light treatments to prolong the shelf-life of Fiordilatte cheese. *Innovative Food Science and Emerging Technologies*, *36*, 150–155. https://doi.org/10.1016/j.ifset.2016.06.010

Lante, A., Tinello, F., & Lomolino, G. (2013). Effect of UV light on microbial proteases: From enzyme inactivation to antioxidant mitigation. *Innovative Food Science and Emerging Technologies*, *17*, 130–134. https://doi.org/10.1016/j.ifset.2012.11.002

Li, L., Li, C., Sun, J., Xin, M., Yi, P., He, X., Sheng, J., Zhou, Z., Ling, D., & Zheng, F. (2021). Synergistic effects of ultraviolet light irradiation and high-oxygen modified atmosphere packaging on physiological quality, microbial growth and lignification metabolism of fresh-cut carrots. *Postharvest Biology and Technology, 173*, 111365.

Liao, C., Liu, X., Gao, A., Zhao, A., Hu, J., & Li, B. (2016). Maintaining postharvest qualities of three leaf vegetables to enhance their shelf lives by multiple ultraviolet-C treatment. *LWT, 73*, 1–5.

Liu, C., Zheng, H., Sheng, K., Liu, W., & Zheng, L. (2018). Effects of postharvest UV-C irradiation on phenolic acids, flavonoids, and key phenylpropanoid pathway genes in tomato fruit. *Scientia Horticulturae, 241*, 107–114.

Llano, K. R. A., Marsellés-Fontanet, A. R., Martín-Belloso, O., & Soliva-Fortuny, R. (2016). Impact of pulsed light treatments on antioxidant characteristics and quality attributes of fresh-cut apples. *Innovative Food Science & Emerging Technologies, 33*, 206–215.

López, M. A., Palou, E., Barbosa, C. G., Tapia, M. S., & Cano, M. P. (2005). Ultraviolet Light and Food Preservation. In *Novel Food Processing Technologies.* (pp. 405–421). CRC Press

Maghoumi, M., Gómez, P. A., Mostofi, Y., Zamani, Z., Artés-Hernández, F., & Artés, F. (2013). Combined effect of heat treatment, UV-C and superatmospheric oxygen packing on phenolics and browning related enzymes of fresh-cut pomegranate arils. *LWT-Food Science and Technology, 54*(2), 389–396.

Mandal, R., Mohammadi, X., Wiktor, A., Singh, A., & Singh, A. P. (2020). Applications of pulsed light decontamination technology in food processing: An overview. *Applied Sciences (Switzerland), 10*(10), 3606. https://doi.org/10.3390/app10103606

Manzocco, L., Quarta, B., & Dri, A. (2009). Polyphenoloxidase inactivation by light exposure in model systems and apple derivatives. *Innovative Food Science & Emerging Technologies, 10*(4), 506–511.

Marangoni Junior, L., Cristianini, M., & Anjos, C. A. R. (2020). Packaging aspects for processing and quality of foods treated by pulsed light. *Journal of Food Processing and Preservation, 44*(11), e14902.

Matak, K. E., Sumner, S. S., Duncan, S. E., Hovingh, E., Worobo, R. W., Hackney, C. R., & Pierson, M. D. (2007). Effects of ultraviolet irradiation on chemical and sensory properties of goat milk. *Journal of Dairy Science, 90*(7), 3178–3186. https://doi.org/10.3168/jds.2006-642

Monteiro, M. L. G., Mársico, E. T., Mano, S. B., da Silveira Alvares, T., Rosenthal, A., Lemos, M., Ferrari, E., Lázaro, C. A., & Conte-Junior, C. A. (2018). Combined effect of high hydrostatic pressure and ultraviolet radiation on quality parameters of refrigerated vacuum-packed tilapia (Oreochromis niloticus) fillets. *Scientific Reports, 8*(1), 1–11.

Naunovic, Z., Lim, S., & Blatchley III, E. R. (2008). Investigation of microbial inactivation efficiency of a UV disinfection system employing an excimer lamp. *Water Research, 42*(19), 4838–4846.

Negi, A., Pare, A., & Meenatchi, R. (2021). Emerging techniques for adulterant authentication in spices and spice products. *Food Control, 127*, 108113.

Nicorescu, I., Nguyen, B., Moreau-Ferret, M., Agoulon, A., Chevalier, S., & Orange, N. (2013). Pulsed light inactivation of Bacillus subtilis vegetative cells in suspensions and spices. *Food Control, 31*(1), 151–157.

Nyhan, L., Przyjalgowski, M., Lewis, L., Begley, M., & Callanan, M. (2021). Investigating the use of ultraviolet light emitting diodes (UV-LEDS) for the inactivation of bacteria in powdered food ingredients. *Foods, 10*(4), 797.

Pan, Y.-G., & Zu, H. (2012). Effect of UV-C radiation on the quality of fresh-cut pineapples. *Procedia Engineering, 37*, 113–119.

Park, M. H., & Kim, J. G. (2015). Low-dose UV-C irradiation reduces the microbial population and preserves antioxidant levels in peeled garlic (Allium sativum L.) during storage. *Postharvest Biology and Technology, 100*, 109–112. https://doi.org/10.1016/j.postharvbio.2014.09.013

Piskarev, I. M., Ivanova, I. P., & Trofimova, S. V. (2013). Comparison of chemical effects of UV radiation from spark discharge in air and a low-pressure mercury lamp. *High Energy Chemistry, 47*(5), 247–250.

Rajan, A., & Radhakrishnan, M. (2023). Green Technologies for Sustainable Food Production and Preservation: An Overview of Ohmic Heating, Infrared Heating and UV Light Technology. In *Reference Module in Food Science.* Elsevier. https://doi.org/10.1016/b978-0-12-823960-5.00066-4

Ramos-Villarroel, A. Y., Martín-Belloso, O., & Soliva-Fortuny, R. (2011). Bacterial inactivation and quality changes in fresh-cut avocado treated with intense light pulses. *European Food Research and Technology, 233*(3), 395–402.

Rawson, A., Patras, A., Tiwari, B. K., Noci, F., Koutchma, T., & Brunton, N. (2011). Effect of thermal and non thermal processing technologies on the bioactive content of exotic fruits and their products: Review of recent advances. *Food Research International, 44*(7), 1875–1887. https://doi.org/10.1016/j.foodres.2011.02.053

Salcedo, I., Andrade, J. A., Quiroga, J. M., & Nebot, E. (2007). Photoreactivation and dark repair in UV-treated microorganisms: effect of temperature. *Applied and Environmental Microbiology*, *73*(5), 1594–1600.

Sampedro, F., Phillips, J., & Fan, X. (2014). Use of response surface methodology to study the combined effects of UV-C and thermal processing on vegetable oxidative enzymes. *LWT-Food Science and Technology*, *55*(1), 189–196.

Shabbir, M. A., Ahmed, H., Maan, A. A., Rehman, A., Afraz, M. T., Iqbal, M. W., Khan, I. M., Amir, R. M., Ashraf, W., Khan, M. R., & Aadil, R. M. (2021). Effect of non-thermal processing techniques on pathogenic and spoilage microorganisms of milk and milk products. *Food Science and Technology (Brazil)*, *41*(2), 279–294. https://doi.org/10.1590/fst.05820

Singh, H., Bhardwaj, S. K., Khatri, M., Kim, K.-H., & Bhardwaj, N. (2021). UVC radiation for food safety: An emerging technology for the microbial disinfection of food products. *Chemical Engineering Journal*, *417*, 128084.

Srivastava, S., & Mishra, H. N. (2021). Ecofriendly nonchemical/nonthermal methods for disinfestation and control of pest/fungal infestation during storage of major important cereal grains: A review. *Food Frontiers*, *2*(1), 93–105. https://doi.org/10.1002/fft2.69

Taze, B. H., Akgun, M. P., Yildiz, S., Kaya, Z., & Unluturk, S. (2020). UV processing and storage of liquid and solid foods: Quality, microbial, enzymatic, nutritional, organoleptic, composition and properties effects. *Innovative Food Processing Technologies: A Comprehensive Review*, 277–305. https://doi.org/10.1016/b978-0-08-100596-5.22938-7

Tiecher, A., de Paula, L. A., Chaves, F. C., & Rombaldi, C. V. (2013). UV-C effect on ethylene, polyamines and the regulation of tomato fruit ripening. *Postharvest Biology and Technology*, *86*, 230–239.

Velusamy, M., Rajan, A., & Radhakrishnan, M. (2023). Valorisation of food processing wastes using PEF and its economic advances–Recent update. *International Journal of Food Science & Technology*, *58*(4), 2021–2041. https://doi.org/10.1111/ijfs.16297

Wang, B., Mahoney, N. E., Pan, Z., Khir, R., Wu, B., Ma, H., & Zhao, L. (2016). Effectiveness of pulsed light treatment for degradation and detoxification of aflatoxin B1 and B2 in rough rice and rice bran. *Food Control*, *59*, 461–467. https://doi.org/10.1016/j.foodcont.2015.06.030

Wu, X., Guan, W., Yan, R., Lei, J., Xu, L., & Wang, Z. (2016). Effects of UV-C on antioxidant activity, total phenolics and main phenolic compounds of the melanin biosynthesis pathway in different tissues of button mushroom. *Postharvest Biology and Technology*, *118*, 51–58.

Xie, J., & Hung, Y.-C. (2018). UV-A activated TiO2 embedded biodegradable polymer film for antimicrobial food packaging application. *LWT*, *96*, 307–314.

Yemmireddy, V., Adhikari, A., & Moreira, J. (2022). Effect of ultraviolet light treatment on microbiological safety and quality of fresh produce: An overview. *Frontiers in Nutrition*, *9*, 871243. https://www.frontiersin.org/articles/10.3389/fnut.2022.871243/full

Yongyue, L. U., Xin, G., & Ling, Z. (2011). Effect of temperature on the development of the mealybug, Phenacoccus solenopsis Tinsley (Hemiptera: Pseudococcidae). *Scientific Research and Essays*, *6*(31), 6459–6464.

3 High Voltage Pulsed Electric Fields in Food Processing and Preservation

N. Bhanu Prakash Reddy, Annet Mary Anto,
P. T. Atchaya, V. R. Sinija, and S. Anandakumar

3.1 INTRODUCTION

The world population is expected to reach 8.5 billion by 2030 and an addition of close to 2 billion in the next 70 years, i.e., by 2100. Increasing life expectancy, reduced child mortality, and improved medical infrastructure ensure population growth positively (World Population Prospects, 2022). Rapid urbanization, income level differences, changing diet structures, climate change, and food losses enhance the existing threat to global food security (UNFAO, 2009). It is equally important to look at meeting the consumers' food demand and looking after the nutrition point of view. A recent report called 'The State of Food Security and Nutrition in the World 2022' by FAO and other global organizations highlighted the importance of food nutrition with some concerning figures; for instance, around 8.9 to 10.5% of the world's population was short of food and nutrition, 29.3% were food insecure in 2021, it was warned about the actions to be taken in that report (FAO et al., 2022).

Keeping global food insecurity and malnutrition in mind, one needs to look deeply into the critical role of food processing in meeting food security in terms of quantity and nutrition globally. The majority of the world's population consumes cereals such as rice, wheat, and maize as their staple food. But, for meeting overall nutrition, along with cereals (having longer shelf-life), fruits, vegetables, milk and milk products, meat and aqua products, and other brewery-based products are consumed to meet the recommended daily intake (RDA) suggested by the different regulatory authorities all over the world. When looking after fruits, vegetables, meat, and milk products, their little shelf-life is very; thus, they had to be eaten in and around the places where they were produced. But with the advancement of food technology, traditionally, heat treatment has been used to kill harmful microorganisms and other undesirable enzymes in food to extend their shelf-life and make them available to people at longer distances from the place of production. However, traditional heat-based methods may destroy many essential heat-sensitive nutrients and sensory properties of the food. Also, they have a high energy requirement, which raises the price of the finished product as a whole (Bi et al., 2013). Alternatives to heat and chemical processing are thus required, leading to products with equivalent or superior flavour, nutritional content, and shelf life. Moreover, with the inception of COVID-19, consumer preference has been shifting towards raw or minimally processed foods.

These requirements can be satisfied only when the energy in different forms other than heat is used that counter the disadvantages of thermal processing. Some modern technologies that use other forms of energy are microwave processing, radiofrequency processing, ohmic heating, etc. However, due to their unique mechanism, these technologies generate heat during food product processing. Thus, scientists had to go to other processing forms that neither use heat as an energy source nor produce heat during the processing. Some alternative methods, called non-thermal technologies, have been adapted widely to process food during the 20th century; some of these non-thermal technologies were commercialized towards the last quarter of the 1900s, for instance, high-pressure processing (HPP).

 DOI: 10.1201/9781003359302-3

One such technology which was commercialized for a few food products in the 1990s is pulsed electric field (PEF) processing. Initially, electric fields were used in industries to inactivate the microorganisms in the 1900s in milk, where electric fields were used to generate heat due to the resistance offered by the food product, which in turn was used to sterilize the milk.

This chapter focuses on the basics of PEF processing, PEF equipment design aspects, effects and applications of high-intensity PEF on different food products and processing industries to process and preserve the food, and packaging requirements for the PEF-treated food products.

3.1.1 Methods of Applying Electricity to the Food Product

Electricity can be applied to the food product in various forms, which can be either direct or indirect, by converting the electric field to some other form, for instance, microwaves or radiowaves.

a. *Ohmic Heating:* In this process, an electric current is passed through the food product, which then acts as a resistance to the current, generating heat. Ohmic heating is a fast and efficient method for cooking and pasteurizing food, as it heats the food evenly throughout the product (Jaeger et al., 2016).

b. *Radio Frequency (RF) Heating:* This method uses electromagnetic energy to generate heat within the food product. The RF waves penetrate the food and generate heat by causing the molecules to oscillate, which leads to even heating of the food (Guo et al., 2019).

c. *Microwave Heating:* Microwaves use electromagnetic energy to generate heat within the food product by causing water molecules to vibrate and generate heat. This method is commonly used for cooking, defrosting, and pasteurizing food products (Guo et al., 2017).

d. *High Voltage Arc Discharge:* This is a process where an electric arc is generated between two electrodes, creating a high-temperature plasma that is directed towards the food product. The plasma generates heat, which is used for cooking and sterilizing the food product. This method is commonly used for the sterilization of packaged food products (Dalvi-Isfahan et al., 2016). When high voltages are discharged through liquids, several physical effects, for example, intense shock waves and chemical compounds are produced, resulting in bacterial inactivation. High-voltage arc discharges also inhibit the activity of enzymes, which are linked to oxidation reactions driven by free radicals and atomic oxygen. However, the main disadvantages of this electrical approach include contamination of the treated food by electrolytic chemical compounds and destruction of food particles by shock waves. As a result, this approach is unsuitable for use in the food sector (Barbosa-Canovas et al., 1999).

e. *Low Voltage Electrolysis:* In this method, low voltage electricity (5–10 V/cm) is applied in the form of alternating current (AC) pulses to a solution containing salt or acid through oppositely charged electrodes fixed one at each end, creating a flow of ions. The ion flow causes changes in the pH and oxidation-reduction potential of the solution, which can be used to modify the texture, flavour, and colour of food products. This method also helps in bacterial count reduction (Barbosa-Canovas et al., 1999). For instance, Tao et al. (2020) reported bacterial reduction and cell membrane disruption in anaerobic digestion-activated sludge.

f. *Induction Heating:* Induction heating uses magnetic fields to generate heat within the food product. A magnetic field is applied to the food, which induces an electrical current, generating heat. This method is commonly used for cooking and processing food products (El-Mashad & Pan, 2017).

Other than the abovementioned modes, other technologies also use electric fields while processing. For instance, *electrostatic spraying* uses an electric field to generate charged droplets of liquid, which are sprayed onto the food product. The charged droplets adhere to the food surface, creating a thin and even coating. This method is commonly used for the application of oils, flavourings, and preservatives onto food products. *Electrostatic defrosting* uses an electric field to generate heat within frozen food products, accelerating defrosting. This method is commonly used in the frozen food industry to improve the efficiency of the defrosting process.

To minimize the heating effects and nutritional and sensory losses that occur during ohmic, microwave, and RF heating, an alternative method had been developed called high voltage 'pulsed electric fields (PEF)'. PEF is a method that applies short pulses of high-voltage electricity to food products, which disrupts the cell membranes of microorganisms and bacteria, leading to their inactivation. Energy loss from heating the food is minimized due to treatment at an ambient or sub-ambient temperature for only milli to microseconds. When electrical energy is supplied in the form of short pulses, it breaks the bacterial cell membrane mechanically without significantly heating the food products. This method is commonly used for pasteurizing and preserving food products (Arshad et al., 2021; Barbosa-Cánovas et al., 1999a).

PEF technology offers the possibility of lowering energy usage while also providing consumers with microbially safe, minimally processed, healthy, and fresh-like foods. Cold sterilization of liquid foods such as juices, cream soups, milk, and egg products is one of the potential applications of PEF technology.

3.1.2 BASICS OF HIGH-INTENSITY PEF

High-intensity PEF is a non-thermal food preservation technology that uses short pulses of high voltage electrical energy to process food. The basic concepts related to high-intensity PEF technology are:

- *Application of High Voltage Pulses:* High-intensity PEF technology works by applying short-duration, high-voltage electrical pulses (typically ranging from 10 to 100 kV/cm) to the food placed between two electrodes, i.e., a high voltage electrode and a ground electrode.
- *Generation of Electric Fields:* The application of high voltage pulses generates intense electric fields within the food product, which cause the formation of small pores in the cell membrane of the microorganisms present in the food.
- *Microbial Inactivation:* These pores disrupt the cellular integrity of the microorganisms, leading to their inactivation or death. The degree of microbial inactivation depends on various factors, such as the electric field strength, pulse duration, and the properties of the food.
- *Minimal Thermal Effects:* High-intensity PEF technology does not rely on heat to kill microorganisms, unlike traditional thermal processing methods, which can affect the taste, texture, and nutritional value of the food. The minimal thermal effects of PEF result in better retention of food quality attributes, such as flavour, texture, and nutritional value.
- *Efficiency and Safety:* PEF technology offers several advantages over traditional processing methods, including reduced processing times, increased efficiency, and improved food safety. However, the application of PEF technology requires careful control and monitoring of the processing parameters to ensure that the food product is treated effectively while maintaining its quality and safety.

Food materials with greater water contents, such as liquid foods and meat, are generally considered good electrical conductors as they contain greater amounts of ions that act as carriers of electrical charge. It is important to have a sequential set-up of several electronic components to generate a high-intensity electric field in pulse form within a food. A large amount of current must be passed in very short periods (milliseconds to microseconds) and a greater pulse OFF time (up to several milliseconds); the pulse generation is achieved through slow charging and quick discharging

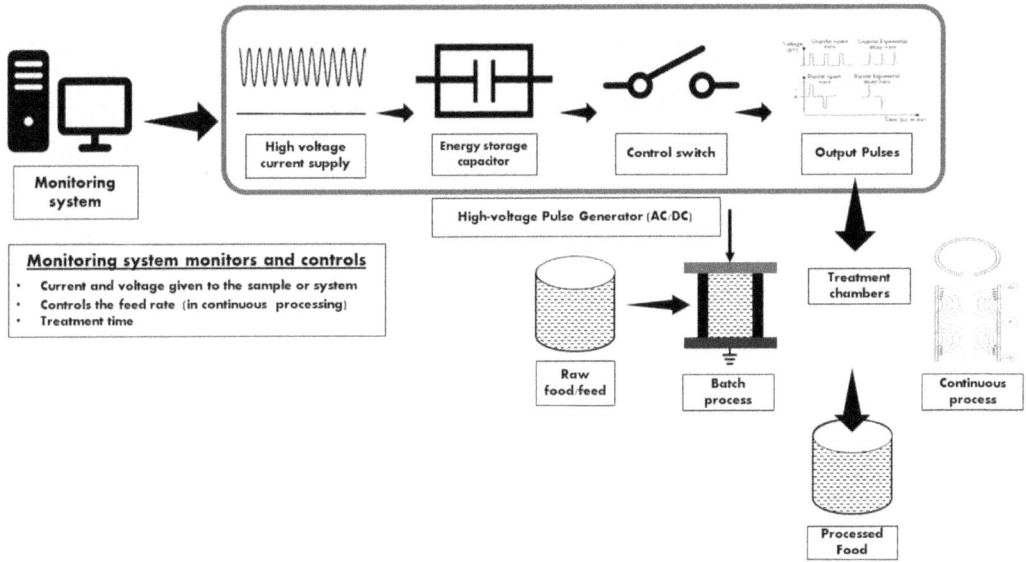

FIGURE 3.1 Basic elements of the PEF treatment system.

of the capacitor. The basic elements of the PEF processing system are shown in Figure 3.1, where the high-voltage current is supplied to the capacitor to store and discharge the energy, which is governed by control switches to produce pulses, the generated pulsed energy is given to the food product with the help of a treatment chamber having a high-voltage electrode and a ground electrode either in a continuous feed form or batch feed. The processed food is collected by cooling (if required) in a collecting tank and finally packed in an aseptic environment.

The key principle behind food processing and preservation with the help of the PEF technique is its unique mechanism of destroying the undesired activity of microorganisms and enzymes and making some structural changes in the food products that, at times, are beneficial. The effective working mechanism on biological molecules in PEF processing is often called electro-permeabilization or electroporation (Kotnik et al., 2019; Ranjha et al., 2021; Toepfl, Siemer, Saldaña-Navarro et al., 2014; Vanga et al., 2021). As mentioned in the earlier statements, this mechanism happens when an external electric field is applied for a very short duration and in pulsed form. Despite the all-round and detailed knowledge about electroporation mechanism, it is universally accepted that 'electroporation' consists majorly four different stages (Saulis, 2010); they are (a) *Charging of the cell plasma membrane:* Cell plasma membrane gets charged when an external electric field is applied increases the transmembrane potential. This first stage lasts from nano to milliseconds depending on the PEF process parameters and the cell size. (b) *Pore initiation stage*: Creation of small metastable hydrophilic pores when an electric pulse is applied (c) *Pore evolution stage:* The stage at which the pore population evolves during an electric field application (within nano- to milliseconds). The number and/or size of pores alters throughout this period. (d) *Post-treatment stage*: This stage lasts milli-seconds to hours, or even more. Once the electric field is removed, several intracellular compounds' leakage, extracellular substances' entry, pore resealing, and several other processes occur, which recover the membrane from electroporation, called reverse electroporation (Chemat et al., 2017).

On the other hand, enzymes are inactivated during the PEF treatment with different mechanisms that are poorly understood. According to available scientific evidence, the electrochemical and thermal impacts of PEF, either separately or in combination, cause modifications to the structure and conformation of enzymes, which may result in inactivation. In general, enzymes are globular proteins with catalytic activity dependent on the natural conformation of their active site and the conformation of surrounding proteins. The structural stability and catalytic capabilities of enzymes are maintained by a delicate balance in the complex network of non-covalent interactions, namely *ion pairing, electrostatic forces, van der Waals forces, hydrogen bonding, hydrophobic effect,* and covalent interactions (*disulphide bonds*) (Martín-Belloso et al., 2014; Wang et al., 2022). Moreover, PEF-induced electrolysis and free-radical generation may cause localized pH alterations near electrodes in various food systems and amino acid residue oxidation that are crucial for enzyme activity and stability. The increased instantaneous temperature during pulsing and isolated hot areas in the PEF treatment chamber may also contribute to PEF-induced enzyme denaturation (Buckow et al., 2013; Niu et al., 2020).

3.1.3 ENERGY REQUIREMENTS IN PEF

The energy requirements for PEF processing of foods depend on several factors, including the type of food, the desired level of microbial inactivation, and the processing parameters, such as the electric field strength and pulse duration.

The energy required for PEF processing is typically expressed in specific energy, which is the energy delivered per unit mass of the food. The specific energy required for PEF processing varies depending on the food product and the microorganisms present. Generally, higher specific energies are required for more resistant microorganisms or foods with thicker or more complex cell structures. The equipment used for PEF processing also affects the energy requirements. PEF systems are designed to deliver high-voltage, short-duration pulses of electrical energy to the food product. The energy transfer efficiency from the PEF system to the food product depends on the system's design and the food's properties. Efficiency can be improved by optimizing the system design, such as the electrode geometry and pulse shape, and by using appropriate processing conditions, such as pulse frequency and duration. Improving energy transfer efficiency can reduce the energy requirements for PEF processing and help make the process more cost-effective.

The huge power level generation requires a specific and robust design when we talk about the electric fields exceeding 10 kV/cm, which is greater than 0.1 GW power levels. Lower amounts of energy are stored in capacitors, producing higher energy levels for short durations. These energy levels accelerate the charged particles, inducing thermal, mechanical, chemical, electromagnetic, or breakdown effects. These effects help modify the biological molecules from their natural form with native properties (Löffler, 2022).

Overall, the energy requirements for PEF processing of foods can vary widely depending on the specific application. Optimizing the process parameters and equipment design can help minimize the energy required while achieving the desired level of microbial inactivation. The application of PEF technology in food processing and preservation varies with the requirement, which can be accomplished by manipulating the process parameters, for instance, specific energy and electric field intensity (EFI) (Huang & Wang, 2009). The treatment patterns for achieving the optimum results in some of the food processing are as follows: Liquid food preservation (2–30 kV/cm, 10–200 kJ/kg, and 1–20 μs pulse width), enzyme inactivation (10–70 kV/cm, 100–250 kJ/kg, and 1–20 μs pulse width), extraction of bioactive compounds (0.5–5 kV/cm, 1–15 kJ/kg, and 10–100 μs pulse width), dehydration (0.2–3 kV/cm, 1–20 kJ/kg, 20–200 μs pulse width), and pre-treatment purposes (0.5–2 kV/cm, 0.24–0.96 kJ/kg, and 20–200 μs pulse width) (Arshad, Abdul-Malek, et al., 2020; Chemat et al., 2017).

3.2 DESIGN CONCEPTS OF PEF EQUIPMENT

The design concepts of PEF equipment are important in determining the effectiveness and efficiency of the equipment. Some key points to consider while designing high-intensity PEF equipment are discussed below:

3.2.1 High-Voltage Pulse Generators

The pulse generator is a key component of HIPEF equipment that generates high-voltage, high-intensity electrical pulses. It should be designed to deliver pulses with a specific amplitude, duration, and frequency to achieve the desired processing effect. While designing the pulse generator, one must remember that the machine should be consistent enough to perform efficiently in a continuous operation, economical at the industrial level, and safe for the operator. Various designs of pulse generators are used to generate pulses, some of them are (a) transmission line-based generator, (b) transformer-based generator, (c) gas-filled switch-based topology, (d) Marx generators, (e) semiconductor-based generators, and (f) solid-state H-bridge topology. These pulse generators have been discussed in detail by Arshad, Abdul-Malek, et al. (2020).

Overall, PEF processing is a highly specialized field requiring sophisticated equipment to generate the required high-voltage pulses. The choice of a high-voltage pulse generator depends on the application's specific requirements, such as pulse duration, repetition rate, and pulse shape. Each type of generator has advantages and disadvantages, and choosing the appropriate generator requires careful consideration of these factors.

The modulator's capabilities are commonly described in peak power (the greatest pulse voltage and current transmitted to the treatment chamber) and pulse frequency. Furthermore, the system must run within the average power supplied from the DC power supply, which means that:

*Peak power (MW) * Pulse frequency (Hz) * Pulse width (μs or ns) < average power*

3.2.2 Control Switches

Control switches are an essential component and the most expensive components of PEF system design, as they enable the generation and control of high-voltage pulses. The control switches to turn the high voltage ON and OFF and shape the pulse waveform. Capacitive storage deals with closed switches, while inductive storage needs an opening switch. There are several types of control switches used in PEF system design (for an easy understanding point of view (Arshad, Abdul-Malek et al., 2020) classified into two major segments such as *material (gas or liquid) filled switches* and *solid-state switches*).

3.2.3 Processing Chambers

The treatment chamber is where the sample to be processed is placed. It should be designed to accommodate the sample and allow for uniform exposure to the electrical pulses. The design should also minimize any potential for arcing or sparking. Depending on the mode of treatment in the PEF process (either 'static' or 'continuous'), various designs of chambers are used. In the case of static treatment, all the material (with greater conductivity, i.e., solid or liquid) is subjected to the electric field for a predetermined period, whereas in continuous treatment, the fluid food is passed through the treatment system with the help of a peristaltic pump or a positive displacement pump.

The treatment chamber in static and continuous PEF treatment plays a significant role, affecting the whole treatment and its outcomes. For example, the treatment homogeneity, peak EFI, and, therefore, process efficiency (Masood et al., 2018). Some of the important points while designing a treatment chamber are the configuration of electrodes and the shape of the insulator. Some of the key points to be kept in mind while designing and treating a food product with the designed chamber are (i) to avoid the direct contact between liquid foods and electrodes to get rid of electrochemical reactions, sometimes even at the cost of the energy wastage to keep the product safe ad longer life of the electrodes by avoiding them from corrosion; (ii) round edged electrodes are to be given priority over the sharp-edged ones as the later ones probe sparking at higher electric fields; and (iii) the dielectric breakdown strength of gases is significantly lesser than fluids. There are good chances of partial discharge production when treated beyond the gas bubbles' dielectric barrier. So this should be taken care of (Arshad, Abdul-Malek, et al., 2020; Pataro et al., 2015).

Moreover, while designing a PEF treatment system, it is equally important to consider the configuration of the electrodes in the treatment chambers. Generally, three basic configurations have been used to treat the foods, such as (1) parallel electrodes, (2) co-axial electrodes, and (3) co-linear electrodes (Arshad, Abdul-Malek, et al., 2020; Buckow et al., 2011; Huang & Wang, 2009; Masood et al., 2018). Parallel electrode configuration creates the most uniform electric field between the electrodes, provided the distance between the electrodes is much smaller than the area. In general, two electrodes are recommended, while sometimes four electrodes separated by an insulating material are also used; however, no considerable additional advantages are seen with four electrode configurations. Parallel plate configuration offers less additional resistance in the chamber (Arshad, Abdul-Malek, et al., 2020; Masood et al., 2018). The EFI (kV/cm) resistance (R), and capacitances (C) of a parallel plate chamber are

$$EFI = \frac{V}{d}$$

$$R = \sigma \frac{d}{A}$$

$$C = \frac{\varepsilon_o \varepsilon_r A}{d}$$

Where 'σ' is the conductivity of the food (Siemens per meter (S/m)), 'ε_o' is the relative permittivity or dielectric constant of the free space ($8.84*10^{-8}$ micro Faraday per centimetre or μF/cm), 'ε_r' is the relative permittivity of the food material, 'V' is the voltage applied (kV), 'd' is the gap between the two electrodes (cm), and 'A' is the area of the electrode (cm^2) (Arshad, Abdul-Malek, et al., 2020; Barbosa-Cánovas et al., 1999b).

The co-axial treatment chamber comprises two concentric electrodes separated by a small gap. The product to be treated is passed through the gap between the electrodes and exposed to the high-voltage pulses. The design of the co-axial chamber allows for more efficient use of the electric field, resulting in better treatment uniformity and higher processing rates. It also reduces the risk of electrode erosion and contamination of the product. The impedance of the processing chamber is controlled by the conductivity of the liquid food and also changes the output waveform of the pulse generator. This change in resistance is an inevitable obstacle to the advancement of PEF processing systems. The rise in temperature of the treated sample mainly occurs for two main reasons, Joule heating and arcing. This configuration can be used at high electric field strength with less temperature rise of the treated material. This low-temperature rise is due to the absence of the electric arc that normally accompanies heat (Arshad, Malek, et al., 2020; Huang & Wang, 2009).

The EFI (kV/cm) resistance (R), and capacitances (C) of a co-axial chamber are

$$EFI = \frac{V}{r * \ln \frac{R_1}{R_2}}$$

$$R = \frac{\ln \frac{R_2}{R_1}}{2\pi\sigma l}$$

$$C = \frac{2\pi\varepsilon_o \varepsilon_r l}{\ln \frac{R_2}{R_1}}$$

Where 'r' is the radius at which the electric field is measured, 'R_1' and 'R_2' are the radii of inner and outer electrodes, respectively, and 'l' is the length of the electrode (Arshad, Abdul-Malek, et al., 2020; Barbosa-Cánovas et al., 1999b).

In a collinear chamber, the product to be treated is passed through a narrow channel between two parallel electrodes that are aligned along the same axis. The electric field is applied perpendicular to the direction of product flow. Collinear chambers are simpler in design and less expensive than co-axial chambers. However, they are less efficient in terms of field distribution and require a higher voltage to achieve the same level of inactivation as a co-axial chamber. The non-uniform distribution of electric field strength and temperature is controlled by the physical design of the insulator placed between the electrodes. This electrode configuration provides high throughput and lower electrode cross-section, resulting in high impedance development, favourable for continuous food processing (Arshad, Abdul-Malek, et al., 2020; Buckow et al., 2011; Masood et al., 2018).

PEF treatment using a co-axial treatment chamber has been shown to be effective in preserving the quality and nutritional value of various food products while extending their shelf life. Designs of various static and continuous chambers with their line diagrams were discussed in detail by Huang and Wang (2009) in their review.

3.2.4 ELECTRODES

Electrodes act as a *pulse delivery system* which delivers high-intensity electrical pulses to the food product. They should be designed to minimize energy loss and ensure that the electrical pulses are delivered uniformly and consistently. When we go for the processing of any food material, it is of utmost importance to consider the materials coming into contact with the food while processing, as the interaction of food with foreign material has some effect on the final quality and safety aspects of the processed food material. Here, during PEF treatment, the materials used for electrodes and the whole treatment chamber should be inert in nature, easy to clean, food grade, and sterilizable. Some of the best materials for electrode manufacturing are metals with higher electrical conductivity and lower resistances and that offer lower energy wastage in the heating form, such as gold, platinum, etc., which are very precious/costly to prefer (Toepfl et al., 2007). These materials have longer lifetimes, unlike carbon/graphite-based electrodes, even though the latter ones do not contaminate the food product (Tokuşoğlu & Swanson, 2014). However, stainless steel is considered suitable due to its economic viability, high corrosion resistance, lower processing costs over its lifetime, and it doesn't release any toxic materials into the food material. Most of the commercially available PEF processing systems utilize electrodes made of stainless steel, aluminium, gold-plated, and silver-plated (Arshad, Abdul-Malek, et al., 2020; Pataro et al., 2017).

One of the major factors we must consider while designing an electrode material is electrochemical reactions that may occur during high-intensity electric field application. These electrochemical reactions induce corrosion of electrodes, and over a longer period, the electrodes and the treatment chamber undergo a natural deterioration due to their repetitive use at higher intensities. With the longer treatment times, when food material is directly in contact with the food material, some metal particles are released into the food product from electrodes. Arcing is also a major problem when dealing with high voltages (Mihindukulasuriya & Jayaram, 2020). Therefore, to counter this unwanted issue, extend the electrodes' lifetime, and keep processed food safe with good quality, we need to use some insulating materials coated on the surface of the materials. However, the limitation with insulation is energy wastage, but when compared with repeated replacement of the electrode due to its corrosion and food safety point of view. Polyvinyl chloride (PVC), polyetherimide, polysulfone, pyrex-glass, and polypropylene have been used as insulators on electrodes in treatment chambers. The appropriate electrode type and configuration selection depend on the specific application and material being treated. Factors such as the electrical conductivity of the material, the desired electric field strength, and the processing capacity of the system all play a role in electrode selection (Arshad, Abdul-Malek, et al., 2020; Huang & Wang, 2009).

3.2.5 COOLING SYSTEMS

When food is processed with PEF treatment, a certain amount of heat is generated, and the increase in the temperature of the food (ΔT) is

$$\Delta T = \frac{Q}{\rho_f C_p} = \frac{E^2 n \tau \sigma}{\rho_f C_p}$$

Where 'Q' is the energy input and ρ_f and C_p are the density and specific heat of the food, respectively. 'E' is the electric field, 'n' is the number of pulses, 'τ' is the pulse duration, and 'σ' is the electrical conductivity of the food (Barbosa-Cánovas et al., 1999a).

Cooling systems are an important component of PEF processing. They help dissipate the heat (though very little) generated during processing at higher electric field intensities. They also prevent the product from overheating, which helps preserve its natural nutritional and organoleptic properties. Several cooling systems can be used in PEF processing by using cooler liquids in heat exchangers, such as (a) *Water cooling*, as water has a high heat capacity and can be circulated to dissipate the heat. It is often used in conjunction with stainless steel electrodes, which can be submerged in a water bath during processing. (b) *Air cooling* is particularly for smaller-scale processing equipment. Air is circulated over the electrodes and the product to remove heat generated by the electric field. (c) *Liquid nitrogen-cooling* in which liquid nitrogen can be used as a coolant in PEF processing, particularly for high-intensity processing applications. Liquid nitrogen is circulated to cool the electrodes and prevent the product from overheating. (d) *Chilled glycol-water mixtures* are used where precise temperature control is required.

The cooling system must be carefully designed and optimized to ensure the product remains at a safe temperature throughout the PEF processing. Sometimes a considerable amount of heat is generated, which can impact the processing performance and equipment longevity.

3.2.6 MEASUREMENT DEVICES

Various measurement devices are used to monitor and control the PEF process. Parameters like voltage and current were given to the treatment chamber, the temperature of the food material, and a flow meter to find the flow rate and pressure sensors (in continuous mode). Here are some of the commonly used measurement devices in PEF processing:

  a. *Oscilloscope:* This device is used to measure the magnitude of voltage and current levels given to the treatment chamber; it also helps confirm critical process parameters like pulse width, pulse OFF time, frequency, waveform shapes, etc.
  b. *Temperature Sensors:* These sensors are used to measure the temperature of the material before, during, and after the PEF process. Temperature is a critical parameter that affects the efficiency of the PEF process and the quality of the treated material.
  c. *Flow Meters:* These devices are used to measure the flow rate of the material through the PEF processing chamber. Flow rate is an essential parameter that determines the residence time of the material in the PEF processing chamber.
  d. *Pressure Sensors:* These sensors are used to measure the pressure inside the PEF processing chamber. Pressure is an essential parameter determining the food material's behaviour during the PEF process.

3.2.7 SAFETY FEATURES AND SCALABILITY

HIPEF equipment generates high voltage and high-intensity electrical pulses, which can pose a safety risk to operators. The equipment should be designed with safety features, such as grounding and insulation, to protect operators from electrical shock.

The design of HIPEF equipment should be scalable to accommodate different sample sizes and processing volumes. This flexible design allows for flexibility in processing different types of samples and accommodating different processing needs.

Overall, the design concepts of HIPEF equipment should focus on delivering uniform and consistent electrical pulses to the treatment chamber while ensuring safety, scalability, and efficient processing performance.

3.3 EFFECT OF PEF IN PROCESSING AND PRESERVATION

PEF processing is adopted in various sectors in food processing and applied to serve different purposes, as mentioned in Table 3.1 and the treatment affects food products in many ways. The same has been discussed vividly in the following sub-sections.

3.3.1 Physical Structure

PEF treatment for solid foods is typically an additional technique for enhancing specific processes, including pressing, decanter, centrifugation, osmotic dehydration, drying, and diffusion (Asavasanti et al., 2011). The structure and shape of the tissue, cell size, the organization of the intercellular space, and the presence of starch or other elements are all factors that impact a plant's tissue's textural properties. The efficiency of the PEF can be deduced by determining the textural property of

TABLE 3.1
Application of PEF Processing in Food Industry

Sl No	Name of the Application of PEF	Some of the Past Studies Related to the PEF Application	Reference
1	Physical structure modification	Tenderization of beef *semitendinosus* muscle	(Jeong et al., 2023)
		Tenderization of beef topside	(Bolumar et al., 2022)
		Modification of rheological, functional, and structural properties of wheat gluten	(Zhang et al., 2021)
		PEF assisted modification of texture, structure, digestibility, and flavour of rice	(Bai et al., 2021)
		Effect of PEF on assembly structure of α-amylase and pectin electrostatic complexes	(Jin et al., 2020)
2	Enzyme inactivation	High-intensity PEF on oxidative enzyme inactivation of broccoli juice	(Sánchez-Vega et al., 2020)
		Inactivation of lipoxygenase in soymilk	(Li et al., 2008)
		Peroxidase inactivation in carrot juice	(Quintão-Teixeira et al., 2013)
		Ascorbic acid oxidase (AAO) inactivation in carrots	(Leong & Oey, 2014)
		Inactivation of PPO of pear by using PEF	(Zhao et al., 2010)
3	Microbial inactivation	Inactivation of *Saccharomyces cerevisiae* in various buffer media	(Montanari et al., 2019)
		Escherichia coli in water	(Guionet et al., 2014)
4	Extraction	Bioactive compounds' extraction from beetroot	(Nowacka et al., 2019)
		Extraction of anthocyanins from purple-fleshed potato	(Puértolas et al., 2013)
		Extraction of bioactive compounds from custard apple leaves	(Ahmad Shiekh et al., 2021)
5	As a pre-treatment to drying, frying, and other processing operations	Pre-treatment to drying and enhancing extraction of polyphenols in tea leaves	(Liu et al., 2019)
		Pre-treatment to freeze-drying of apple	(Nowak & Jakubczyk, 2022)
		PEF-assisted fermentation of *Hanseniaspora sp.* yeast	(Al Daccache et al., 2020)
		Pre-treatment to osmotic dehydration of mango	(Zongo et al., 2022)
6	Pasteurization	Apple juice pasteurization	(Heinz et al., 2003)

the treated food products. Consequently, PEF-induced texture changes in solid tissues are crucial components for maximizing the effectiveness of PEF treatment, whether used alone or in combination with other food treatment techniques. High-intensity PEF alters the plant tissue's visco-elastic qualities and eliminates the cellular turgor component of a texture from cells (Paoplook & Eshtiaghi, 2013). The average size of the pores in the untreated samples was often less than that of the induced pores. The PEF-induced pores' sizes were less than the thickness of the cell wall. According to the examination of textural characteristics, the intensity of PEF treatment tended to reduce turgor pressure and failure stress. This shows that electro-plasmolysis affects sample cell wall integrity in addition to plasmalemma membranes.

When the fruits and vegetables are concerned, PEF treatment showed some promising signs by enabling the enhancement of mass transfer, helping in the tenderization of meat, fish and other seafood, and destroying the firm structure of many foods, which helps in greater extraction of components like oils, bioactive compounds, colours, etc. (Bocker & Silva, 2022b; El Kantar et al., 2018; Gómez et al., 2019). The most important criterion for determining the freshness of potato chips is their crispness, which is also reflected in their hardness (Salvador et al., 2009). The tissue of the potato, particularly the carrot, remained rather hard after the PEF treatment despite some loss in the rigidity of tissue structure brought on by the PEF. The elasticity modulus and the fracture stress for carrot and apple tissues had a significant inverse relationship with the duration of PEF treatment. There were minor changes in the potato tissue's elasticity modulus and fracture stress. By applying high intensity PEF strength of 2.5 kV/cm, it was noted that there was a significant increase in the crispness of the French fries with an increase in PEF strength. No statistically significant difference existed between its varying moisture contents (Lee et al., 2022). Due to the difference in partial vapour pressure between the product and the frying oil, moisture evaporates as the food is fried. The cell membrane permeabilization in potato cubes treated with PEF improved the diffusion of liquid from the core to the surface, thickened the surface vapour layer, raising the water vapour pressure, and thus improved the crispness (Janositz et al., 2011).

3.3.2 MICROBIAL INACTIVATION

PEF processing has been proven to be effective in inactivating microorganisms in foods (Figure 3.2). The electric field generated by the pulses can induce pores on the cell membranes of microorganisms, leading to their inactivation. The mechanism behind the inactivation of bacteria, yeast, and mould by PEF treatment of food is related to the ability of the electric pulse to induce pores in the microbial cell membrane. Pores allow intracellular components to escape, disrupting the cell's normal functioning and ultimately leading to its inactivation. It is believed to be related to the poration effect. Electroporation is the process by which an electrical impulse creates temporary pores in the cell membrane, allowing the intracellular contents to escape.

The effectiveness of PEF treatment against microorganisms depends on several factors, including the intensity and duration of the electrical impulse, food properties, and initial microbial load. The field strength and number of pulses determine the extent of electroporation and damage to the cell membrane. Food matrices can affect the penetration of electrical impulses into products and subsequent microbial inactivation. The initial microbial load may also influence the efficacy of the PEF treatment. This is because the higher the number of microorganisms, the higher the electric field strength and the longer the processing time. In addition to cell membrane damage, other mechanisms may contribute to microbial inactivation through this PEF treatment. These include generating reactive oxygen species, changing food pH and temperature, and disrupting cellular structures such as ribosomes and DNA (Buckow et al., 2013). Overall, the mechanism behind the inactivation of bacteria, yeast, and mould by PEF treatment in food is related to the ability of electrical impulses to induce damage to cell membranes and disrupt normal cell function.

PEF has been shown to be effective in reducing the microbial load of a variety of food products, including fruits, vegetables, juices, and dairy products. Here are some of the potential effects of PEF on microorganisms. For instance, *Bacteria*, PEF has been shown to be effective in inactivating a wide range of bacteria, including Gram-negative and Gram-positive bacteria. PEF has been shown

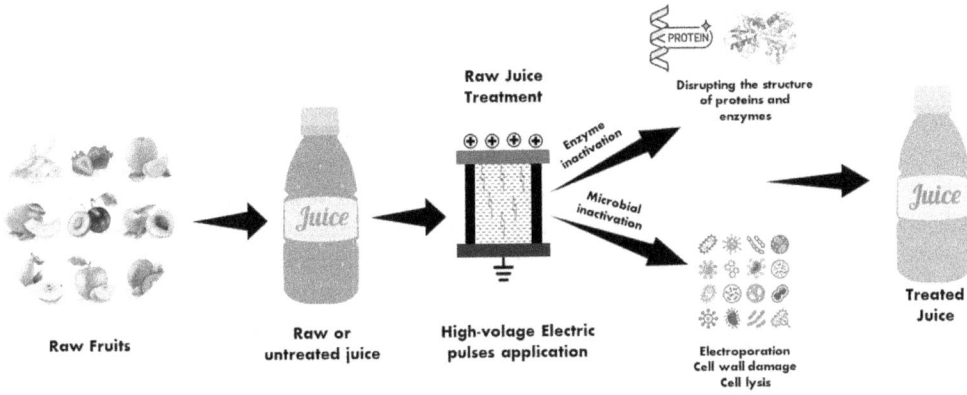

FIGURE 3.2 Schematic diagram of major effects of PEF treatment on fruit juices.

to be particularly effective in reducing the levels of spoilage bacteria in foods, such as *Lactobacillus* and *Pseudomonas* (Shamsi et al., 2008). *Yeasts and moulds*, PEF can also be effective in inactivating yeasts and moulds, which are common contaminants in food products. PEF has been shown to be particularly effective in reducing the levels of yeasts and moulds in fruit juices, dairy products, and other liquid foods (Al Daccache et al., 2020; Huang et al., 2013; Simonis et al., 2019). Some of the past research works related to microbial inactivation are discussed in Table 3.2.

TABLE 3.2
Effects of Pulsed Electric Field (PEF) Processing on Microorganisms

Sl. No.	Food Product	Microorganism	PEF Conditions	Inactivation Results	Reference
1	Blueberry juice	*Escherichia coli*	• Electric field intensity (EFI): 20, 25, 30, and 35 kV/cm • Treatment time: 27, 54, and 82 μs, unipolar pulses	5.12-log reduction in *E. coli* was observed at 35 kV/cm for 82 μs	(Chen et al., 2014)
2	Carrot juice	Bacteria, *Coliform,* and total plate count	• EFI: 25 kV/cm • Total treatment time – 280 μs • Pulse width – 2.5 μs • Flow rate 60 mL/min	99% of total plate count was reduced after treatment, aerobic bacteria and coliform counts were decreased by 80% and 95%, respectively	(Xiang et al., 2014)
3	Growth medium with yeast strain	*Saccharomyces cerevisiae*	• EFI: 25 and 50 kV/cm • Treatment time: 1–5 seconds • Initial inoculum level 4 or 6 log cfu/mL, pH of the medium: 4 or 6 • Preheating temperature: 50°C	1.26 and 1.68 log cfu/mL reduction was observed at 25 and 50 kV/cm, respectively, after 5 seconds	(Montanari et al., 2019)
4	Acid whey	*Saccharomyces sp. Lactobacillus sp.*	• EFI: 39 kV/cm, 92 kV/cm, and 95 kV/cm • Pulse duration 60 ns, 90 ns, 1000 ns • Pulse number up to 100 pulses	Complete inactivation was seen after PEF treatment and a significant reduction in bacterial cells, i.e., by 1.47 log reduction was observed	(Simonis et al., 2019)
5	Low-alcohol red wine	Aerobic bacteria, yeast, and lactic acid bacteria	• EFI: 20–50 kV/cm • 1 μs square wave • 500–1500 Hz at frequency	87.6%, 77.5%, and 93.1% of reduction in aerobic bacteria, yeast, and lactic acid bacteria were observed, respectively, at 40 kV/cm	(Puligundla et al., 2018)

3.3.3 ENZYME INACTIVATION

PEF is a non-thermal preservation technique that employs high voltage to inactivate enzymes and make foods that are safe for consumption from microorganisms while maintaining a fresh flavour and taste with minimal nutritional loss. PEF effects on enzymes can activate/enhance, inactivate, and limit their activity. The inherent property of the enzymes and treatment conditions of the PEF plays a major role in the stability of the enzymes. Protein unfolding or denaturation is generally the cause of enzyme deactivation. High-intensity PEF can disrupt non-covalent and covalent connections already present in enzymes, leading to new kinds of interactions within peptide chains (Wang et al., 2014).

The inactivation of enzymes by the PEF mechanism is still complicity in understanding. However, it is widely known that enzymes are proteins whose catalytic activity depends on the natural configuration of their active site and the conformation of the surrounding proteins. In the chemical structure of proteins, the amino acid group in enzyme proteins produces highly asymmetric spatial distributions of charge that result in strongly polarised and charged areas (Laberge, 1998). The structural stability and catalytic properties of enzymes are affirmed by a complicated network of non-covalent interactions (disulphide bonds), covalent interactions (electrostatic forces, ion pairing, Van der Waals forces, hydrogen bonding, and hydrophobic interaction effect), and combinations of the both (Zaidi et al., 2014). Changes in the three-dimensional molecular structure of enzymes may result in changes in enzyme activity if interactions produce a shift. Intense electric fields are thought to produce protein denaturation and unfolding, the dissolution of covalent connections, and oxidation-reduction processes, such as those involving sulfide groups and disulfide bonds. Moreover, it is understood that chemical interactions involving charges, dipoles, or induced dipoles in electric fields affect the conformational state of proteins. Due to the difficulties of combining the substrate with the active site, charged groups and structures are particularly vulnerable to various electric field perturbations. These alterations alter their structure, which results in a loss of activity (Barsotti & Cheftel, 1999).

A copper-containing enzyme known as polyphenol oxidase (PPO) is present in many higher plants. It is responsible for browning, off-flavour, and vitamin loss, thereby deteriorating the quality of fruits, vegetables, and processed food products. Similarly, peroxidase (POD) is an oxidoreductase enzyme which can catalyze the oxidation of various natural substances found in plants, especially those with aromatic groups. Making highly energetic free radicals that can extract hydrogen radicals from such materials is the mechanism of action. POD is more versatile than any other enzyme as it catalyzes unwanted reactions. This enzyme changes processed foods' flavour, colour, texture, nutrients, and taste. Many other enzymes are also responsible for quality deterioration, such as pectin methyl esterase (PME), polygalacturonase (PG), lipoxygenase (LOX), and others. Various research findings were reported that PEF treatment could inactivate these deteriorative enzymes (Zhao et al., 2012).

For instance, when PEF treatment and heating the apple juices to 60°C were coupled, the effect of PEF processing on a laboratory scale resulted in the total deactivation of PPO (Chilling et al., 2008) (Table 3.3). In a different investigation, PPO activities in apple extract at 24.6 kV/cm and pear extract at 22.3 kV/cm for 6 ms total treatment duration decreased to 3.15% and 38.0% from the initial value. Applying PEF at 25 kV to berry fruits such as blueberry and red raspberry showed that PPO activity decreased by about 13.5% after the treatment (Roobab et al., 2022). To better preserve fruit juices, PEF can inactivate the enzymes (Roobab et al., 2022). Preheating to 50°C and a PEF treatment period of 100 s at 40 kV/cm resulted in the greatest decrease in POD activity (68%) of freshly made apple juice (Riener et al., 2008). The PEF at the 42 kV/cm voltage for 1036 μs obtained maximum reduction (88%) in the enzyme lipoxygenases in the soybean (Riener et al., 2009). The enzyme PME was 97% inactivated in grapefruit juice when treated at the high voltage of 20–40 kV/cm for 25–100 μs (Riener et al., 2009). The enzyme alkaline phosphatases in bovine milk were inactivated by about 30–67% when treated for the electric field of 35 kV/cm for 19.6 μs (Shamsi et al., 2008). Some of the past research works related to enzymatic inactivation are discussed in Table 3.3.

TABLE 3.3

Effect of Pulsed Electric Field Treatment on Enzymes

Sl. No.	Enzyme	Food Product	PEF Conditions	Inactivation %	Reference
1	PPO	Apple juice	40 kV/cm and 100–200 pulses	100	(Ertugay et al., 2013)
2	PPO	Carrot juice	765.5 kJ/kg of energy was given to the food product at 80°C	~68	(Mannozzi et al., 2019)
3	PPO	Broccoli juice	15–35 kV/cm, 500–2000 pulses with an energy input ranging from 88.7 to 1931.7 mJ/m^3	63.89	(Sánchez-Vega et al., 2020)
4	POD	Carrot juice	35 kV/cm for 1500 μs using 6-μs bipolar pulses at 200 Hz	93.30	(Quitão-Teixeira et al., 2009)
5	POD	Carrot juice	765.5 kJ/kg of energy was given to the food product at 80°C	~89	(Mannozzi et al., 2019)
6	LOX	Broccoli juice	15–35 kV/cm, 500–2000 pulses with an energy input ranging from 88.7 to 1931.7 mJ/m^3	31.29	(Sánchez-Vega et al., 2020)
7	PME	Tomato juice	4.0–12.5 kV/cm, total treatment times of 0–12 ms, and the pulse width – 15 μs at a frequency of 300 Hz	100	(Andreou et al., 2016)
8	PG	Tomato juice	4.0–12.5 kV/cm, total treatment times of 0–12 ms, and the pulse width – 15 μs at a frequency of 300 Hz	100	(Andreou et al., 2016)
9	Ascorbic acid oxidase (AAO)	Carrot puree	0.2–1.2 kV/cm, pulse width – 20 μs, square wave bipolar waveform with a frequency of 5–300 Hz and 130–9000 pulses	62	(Leong & Oey, 2014)

3.3.4 PROTEIN AND VITAMIN

Proteins are very important components of foods that provide nutritional, functional, and structural properties primarily determined by proteins' molecular structure (Perez & Pilosof, 2004). Protein responses to various forms of external stress, such as heat, chemicals, and pressure, have been studied. However, the effect of PEF on protein conformation is still an area that has not been fully elucidated. Structural changes in proteins from microstructural (primary structure to quaternary structure) to subsequent macroscopic changes (protein aggregation) are induced by PEF treatment which was vastly discussed by Zhao et al. (2012). Protein unfolding is usually the first step for protein modification and enzyme inactivation. These conformational changes can be covalent bond changes leading to chemically modified proteins (primary structure) or non-covalent changes leading to conformational changes (secondary to quaternary) in certain protein structures (Manas & Vercet, 2006). The unfolding of protein molecules upon PEF treatment is the initial step for subsequent protein interaction and aggregation. Protein aggregation is thought to occur most easily through partially unfolded protein intermediates (Zhao et al., 2012).

However, protein-based foods such as liquid eggs and milk typically contain various proteins, and interactions between different proteins may result in different protein aggregation performances. It is not yet known whether the mechanism of PEF for proteins differs from that of heat treatment. One possible explanation for the non-thermal effects of the electric field on proteins is that the PEF causes a rapid increase in temperature that returns to baseline temperature faster than normal thermometry can detect. Increased temperature can change protein conformation and alter protein activity (Budi et al., 2005). Data on differences between PEF signalling pathways and heat treatment of proteins are lacking. Enzyme sensitivity to PEF was independent of its thermostability, suggesting that the enzyme-inactivation mechanism of PEF may be different from heat. However, different protein aggregation behaviour occurred under heat treatment. It is mainly characterized by forming protein aggregates of high molecular weight.

Moreover, weak non-covalent bonds play a much more important role in forming protein aggregates under heat treatment than under PEF treatment (Giteru et al., 2018). Minimizing the effects of PEF treatment on protein is essential for PEF treatment of protein-based foods. Therefore, the effects of PEF treatment on proteins can be minimized by modifying the processing chamber configuration and design to increase the homogeneity of the electric field distribution or by providing active cooling. Egg yolk's microstructure, water-soluble protein content, lipoprotein matrix, and mechanical properties appear to be less affected by PEF than by heat pasteurization (66°C for 4.5 minutes) (Marco-Molés et al., 2011), which is mildly altered by PEF treatment.

Moreover, the effects of PEF on vitamins can vary depending on the type of vitamin, the food matrix, and the processing conditions used. Here are some of the potential effects of PEF on different vitamins. For instance, vitamin C (ascorbic acid) is a relatively unstable nutrient easily degraded by exposure to heat, oxygen, and other processing methods. Some studies have found that PEF can reduce the vitamin C content of certain fruit and vegetable juices, such as strawberry juice (Odriozola-Serrano et al., 2009), watermelon juice (Oms-Oliu et al., 2009), broccoli juice (Sánchez-Vega et al., 2015). The extent of the vitamin C loss may depend on the processing conditions, such as the pulse intensity, frequency, and treatment time.

Similarly, B vitamins, such as thiamin, riboflavin, and niacin, are water-soluble vitamins important for energy metabolism and other cellular functions. Some studies have shown that PEF can reduce the thiamin content of milk and other dairy products (Shabbir et al., 2021). However, the effects of PEF on other B vitamins may be less pronounced and may depend on the food matrix and processing conditions. On the other hand, Vitamin E is a fat-soluble vitamin that acts as an antioxidant and helps protect cell membranes from oxidative damage. Some studies have suggested that PEF can increase the vitamin E content of certain plant-based oils besides increase in the yield, such as olive oil, maize gem oil, and soybean oil (Guderjan et al., 2005). This may be because PEF can help release vitamin E from the oil droplets and make it more bioavailable. Some of the previous works on PEF processing of various fruits and vegetables and the effects on physico-chemical and nutritional properties are mentioned in Table 3.4.

TABLE 3.4
Effect of PEF Processing on Different Physicochemical Properties of Food Products

Sl No	Food Product	PEF Conditions	Effects of the Treatments	Reference
1	Apple juice	30 and 40 kV/cm, 50–200 pulses with a heat pretreatment up to 40°C	• Total phenol content (TPC) was lost by 26% and 57% at 30 and 40 kV/cm treatment with 200 pulses, respectively	(Ertugay et al., 2013)
2	Carrot juice	35 kV/cm for 1500 µs using 6-µs bipolar pulses at 200 Hz	• 95.1% of vitamin C was retained after PEF treatment • β-carotene was increased significantly from 30 mg/L to 37 mg/L • TPC was increased by 7 mg/L additionally with PEF treatment • Antioxidant activity was reduced slightly which was insignificant	(Quitão-Teixeira et al., 2009)
3	Spinach juice	9 kV/cm, 1 kHz, 80-µs pulses with total treatment time of 335 µs	• TPC, total flavonoid content (TFC), total flavonols, anthocyanin content, and carotenoids were enhanced by 8.49%, 5.81%, 21.8%, 8.23%, and 5.97%, respectively	(Faisal Manzoor et al., 2021)
4	Grapefruit juice	20 kV/cm pulses 600 µs at 1 kHz with a flow rate of 80 mL/min	• No significant change in pH, total soluble solids (TSS), acidity, and electrical conductivity was observed • A considerable decrease in the viscosity and an increase in cloud value was seen after the PEF treatment	(Aadil et al., 2015)

(Continued)

TABLE 3.4 (*Continued*)
Effect of PEF Processing on Different Physicochemical Properties of Food Products

Sl No	Food Product	PEF Conditions	Effects of the Treatments	Reference
5	Grapefruit juice	20 kV/cm pulses 600 μs at 1 kHz with a flow rate of 80 mL/min	• An increase in TFC, total flavonols, and TPC by 9.53%, 34.25%, and 10.92%, respectively was observed • An increase of 27.75% and 18.11%, respectively in total antioxidant capacity and DPPH radical scavenging activity was observed	(Aadil et al., 2018)
6	Carrot juice	765.5 kJ/kg of energy was given to the food product at 80°C	• Lower colour change and an increase in antioxidant activity were observed in PEF-treated sample	(Mannozzi et al., 2019)
7	Grape juice	0, 7, and 14 kV/cm intensity, 1 kHz, 40 mL/min, and 500 μs treatment time	• No significant change in pH, TSS, and colour when compared to the treated sample • TPC, TFC, and % DPPH inhibition were increased by 8.99%, 83.32%, and 17.25% with the PEF treatment of 14 kV/cm	(Rahaman et al., 2020)
8	Tomato juice	4.0–12.5 kV/cm, total treatment times of 0–12 ms, and the pulse width – 15 μs at a frequency of 300 Hz	• PEF (8 kV/cm, 6 ms) pretreated tomato juice yielded very close concentrated product compared to that of untreated ones during water evaporation (at 60°C and 0.5 bar vacuum) to obtain concentrated tomato product (5000 cp viscosity)	(Andreou et al., 2016)
9	Blueberry juice	Electric field intensity: 20, 25, 30, and 35 kV/cm. Treatment time: 27, 54, and 82 μs, unipolar pulses	• Vitamin C and anthocyanin content of blueberry juice was reduced by 14.78% and 3.64%, respectively • No significant change was observed in reducing sugar, total acid, TPC, and TSS after PEF treatment • Nutrient retention was higher than heat treatment	(Chen et al., 2014)
10	Apple juice	30 kV/cm, 4, 6, and 8 cycles (each cycle consisting of 50 pulses), and storage time of 24–72 hours	• Vitamin C is retained almost completely after the treatment and after 74 hours of storage also • TPC was increased till six cycles of treatment and the same is retained more in PEF-treated juice which was not seen in raw juice to retain TPC over 3 days of storage • Antioxidant values of the treated juice were mostly retained over storage period compared to the control	(Dziadek et al., 2019)

3.3.5 EXTRACTION

Extraction is one of the essential unit operations widely used in many applications in the food industry. The main processing steps aim to extract foods such as sugars, oils, and juices or specific components such as polyphenols, carotenoids, and pigments. The process is called squeezing when mechanical force is used to extract and separate liquids such as juice. Solid-liquid extraction or leaching is the extraction of useful soluble components from raw materials by dissolving them in a predominantly liquid solvent so that the components can later be separated from the liquid and recovered. Components obtained by expression or solid/liquid extractions are generally located within the cells, so they must cross the cell envelopes to be released. Traditionally, some pre-treatments such as milling, heating, or using enzymes are applied to disrupt the cell envelopes and facilitate the subsequent extraction. However, applying the abovementioned treatments has significant drawbacks, such as excessive denaturation of the cell envelope, causing the release of cell debris and requiring incompatibility with subsequent purification techniques, high energy costs, or extraction of labile substances. Therefore, a novel method like PEF was tried as a pre-treatment, which disrupts the cell wall by using high-intensity electric energy in pulsed forms and thus reduces the negative implications of the other methods (Martínez et al., 2020). Carotenoids are pigments found in fruits and vegetables that act as antioxidants and have other health benefits. Some studies have shown that PEF can increase the bioavailability of carotenoids in certain foods, such as tomato juice and carrot juice (Roohinejad et al., 2014). This may be because PEF can help break down the cell walls and release the carotenoids from the plant cells.

3.3.6 Drying and Frying

High-moisture foods are spoiled very quickly because of the availability of water to the spoilage micro-organisms, which helps in the growth and replication of the microorganisms. The same leads to huge losses in the food industry, such as the fruit and vegetable processing industry. So, it is very much important to kill the deteriorative microbes. One way to do the same is by reducing moisture availability to microbes, which can be done by drying the food. Traditionally, moisture removal is accomplished by giving heat energy to the food product. However, some disadvantages of drying are higher energy consumption (operational costs), nutritional degradation, environmental damage, etc. keeping these in mind, continuous efforts are made to enhance the drying rate to counter the disadvantages of prolonged drying, which helps in better quality dried or dehydrated products. Various thermal and non-thermal methods, such as microwave, ohmic heating, ultrasound, HPP, etc., have been tried as pre-treatment before drying. The application of PEF technology was promising for food processing due to its potential for continuous application with very little heating of the medium, short-time treatment, instant distribution throughout a conductive food system and low energy requirements (Ade-Omowaye et al., 2001).

PEF treatment, a high-intensity, short-duration electrical stimulation, is currently applied to plant cells to increase membrane permeability. Cell membranes are generally very thin and do not require strong electric fields to generate transmembrane potentials. This increased membrane permeability appears to restore the equilibrium between the electrochemical and potential differences between the cytoplasmic and extracellular environments. Membrane electro-permeabilization occurs when the normal cell transmembrane potential (60–110 mV) or 'normal resting potential' is exceeded. Electrical disruption is the first step in permeabilization, followed by intracellular penetration processes such as releasing intracellular fluid, diffusion of solutes, and sealing membranes when treatment is mild. Although different cellular materials exhibit different morphological structures and electrophysical properties, the similarity of voltage-dependent membrane rupture indicates that the same mechanism is involved in the electro-permeabilization of cell membranes, i.e., charging separated membranes to critical values (Ade-Omowaye et al., 2001). PEF has unique properties that selectively damage cells.

The membrane is the only part affected, with the lipid portion being the site of electrical interaction. The degree of permeabilization depends on the cell membrane properties (size, conductivity, shape, direction) and the electrical pulse parameters (field strength, pulse amplitude, pulse shape, pulse duration, number of pulses). This mechanism is known as 'reversible electrical breakdown' because the conductivity of the cell increases. These advantages make PEF an attractive candidate for use in operations that require cell disruption as a preliminary step to improve mass transfer in further processing, such as drying and dehydration (Rahaman et al., 2019). For instance, the drying time reduction was observed when onion slices were pre-treated with PEF-treatment (1.07 kV/cm and 4 kJ/kg) and subsequently dried at 45°C by 30% compared to thermal drying with no pre-treatment (Ostermeier et al., 2018).

A similar mechanism was involved during frying to enhance the effectiveness of frying by enabling the inner portion of the food material to get enough access to hot oil to fry; this is because of the irreversible pores formed in the cell membranes.

3.3.7 Sensory Properties

While PEF treatment can extend the shelf life of foods by reducing microbial load and preserving nutritional content, it can also have effects on the sensory properties of the food product. PEF treatment of fruits and vegetables can result in a reduction in the bitter compounds, resulting in a milder flavour. The electrical pulses can cause the breakdown of pigments, which can affect the color of fruits and vegetables. PEF treatment of fruit juices can also lead to a reduction in the color intensity due to the breakdown of pigments. While PEF treatment can preserve the nutritional content of foods by reducing the microbial load, it can also lead to the breakdown of certain nutrients, such as vitamins and antioxidants (Buckow et al., 2013). Few studies on the effect of PEF treatment on sensory properties of apple juice (Bi et al., 2013), pomegranate juice (Evrendilek, 2017), non-dairy milk (Bocker & Silva, 2022a), and wine (Arcena et al., 2021) revealed that there was no significant change in the sensory acceptance of respective foods than the fresh ones, also, when compared with thermal treatment, PEF-treated products were rated higher in terms of natural sensory feel.

However, the extent of nutrient loss depends on the processing conditions, and optimizing the PEF treatment conditions can help to minimize the loss of nutritional content. Overall, PEF treatment can have effects on the sensory properties of food products, including texture, flavour, colour, and nutritional content. Optimizing the PEF treatment conditions can help to minimize the effects on sensory properties while still achieving the desired level of microbial inactivation.

3.4 THE POTENTIAL OF PEF APPLICATION IN FOOD INDUSTRIES

3.4.1 FRUIT AND VEGETABLE PROCESSING

Fruits and vegetables are crucial for the human diet because they provide essential vitamins and minerals. Especially fruits are converted into various forms, including juice, pulp, mash, and canned fruits; vegetables with higher water are sometimes converted into purees and pulps for raw material in various food products. However, their quality constantly deteriorates if unprocessed. In general, the quality of the fruit juice is assessed by its physico-chemical, nutritional, sensorial, enzymatic, and microbiological aspects. In fruits and vegetables, quality characteristics like colour, flavour, texture, sensory and nutritional quality depend on the endogenous activity of enzymes like PPO, POD, PME, PG, LOX, and others; deteriorative activity of pathogenic microorganisms that affect the final quality of food product prepared. Therefore, to contain the activity of microbes and enzymes, the products should be processed to inactivate the unnecessary activity post-processing. Traditionally, this is done by combining heat treatment, pH change, and chemical inhibitor utilization. Traditional heat treatment methods are usually used in preserving fruit juices, which is an economical way to ensure enzymatic inactivation and microbiological protection. But, these techniques can deteriorate the food product's quality or induce undesirable effects due to the additives, which is not the required motive. Therefore, PEF uses alternative energy (electrical form) to inactivate enzymes and microorganisms. Some of the applications of PEF technology in fruit and vegetable processing are compiled in Table 3.4.

3.4.2 MEAT

Most conventional meat preservation techniques, such as drying, frozen storage, and salting, have a detrimental impact on the product's microstructure compared to fresh meats. PEF treatment's ability to permeate cell membranes allows it to adjust meat quality attributes such as texture, colour, and water-holding capacity and improve mass transfer procedures such as curing and brining (McDonnell et al., 2014). According to Lyu et al. (2016), PEF treatment altered the temporal flavour profiles of meaty and oxidized flavour characteristics. All PEF-treated samples had caramelized juicy, livery, and meaty flavour characteristics. PEF treatment of beef has been found to improve mass transfer during drying and brining. Pore generation in animal tissue, such as flesh, may be accomplished with electric field strengths ranging from 1 to 10 kV/cm with energy input ranging from 0.5 to 10 kJ/kg (Toepfl, Siemer, & Heinz, 2014). Moreover, better micro diffusion of brine and water binding agents resulted in improved water binding during cooking.

The overall integrity of muscle cells mostly determines meat tenderness. Unlike other tenderization procedures, PEF processing has no negative side effects, such as significant structural and oxidative alterations or off-flavour development. PEF therapy of muscles before age may aid in the tenderization process by causing early activation of calpains via calcium ion release from cell organelles due to increased membrane permeability. Electrical tenderization approaches do not always give homogeneous muscle tenderization, but PEF is a stand-alone treatment that may be performed on various muscles pre- or post-rigour (Suwandy et al., 2015). Like other culinary products such as cheeses and wines, beef meats can improve their texture following an ageing period. Because of the enhanced rate of proteolysis, ageing in conjunction with PEF treatment appears to be an acceptable choice for improving meat softness (Gómez et al., 2019).

PEF-induced effects are more complicated in multicellular systems composed of several cellular types with diverse features in various tissues. Meat is a complex mixture of connective tissue, fatty tissue, vascular and neurological structures, and multinucleated muscle cells. PEF processing affects muscle cell membranes, influencing the interaction of fatty acids and cell membrane phospholipids, resulting in a prooxidant effect in meat. These interactions can produce unwanted chemicals that decompose into secondary products, causing off-flavours and aromas in meat and lowering its sensory and nutritional quality (Faridnia et al., 2015).

3.4.3 FISH

The PEF technology may also valorize fish processing businesses' by-products. It is an effective approach for increasing the water-holding attributes of fish products and drying fish. The PEF technology may also valorize fish processing businesses' by-products. In a research study, Zhou et al. (2012) developed a more effective technique for extracting calcium from fishbone using high-intensity PEF. When the findings are compared to the ultrasonic approach, it is correct to conclude that PEF enabled better extraction efficiency in a shorter time. Also, Zhou et al. (2017) investigated the extraction efficiency of PEF technology in obtaining protein from mussels. PEF application under approximated optimal circumstances (EFS of 20 kV/cm, pulse number of 8, and enzymolysis period of 2 h) enhanced protein extraction yield to 77.08%.

Furthermore, one of the disadvantages of this developing technology is its inefficiency against the decrease of naturally existing enzymes in fish. Similar to the ohmic heating process, the electrical conductivity of the product is an important characteristic that limits the usage of PEF to substances with moderate conductivity.

3.4.4 DAIRY

PEF, when combined with sub-pasteurization temperatures, can achieve microbial inactivation levels in milk comparable to traditional thermal pasteurization. The inactivation of vegetative bacteria by PEF treatment is typically governed by the applied electrical field intensity, treatment temperature, and total treatment duration (Walter et al., 2016). PEF processing can be used to enhance the refrigerated shelf life of raw milk before it is further processed in the dairy industry. According to (Craven et al., 2008), PEF was particularly successful in the inactivation of *Pseudomonas spp.*, extending the shelf-life of milk by up to 8 days at 4°C. The combination of PEF, moderate heat, and antimicrobials resulted in much larger microbial reductions than the sum of each treatment's reductions.

Mammalian milk comprises various indigenous and exogenous enzyme systems with varying functions, process stability, and influence on dairy products. Some enzymes, such as alkaline phosphatase, lactoperoxidase, or lipase, are employed as markers of thermal processing, while others are studied for their positive or negative impacts on dairy product quality. PEF-induced electrolysis and free radical production may also result in oxidation-reduction processes involving amino acid residues required for specific enzymes' function and stability. It has been proposed that the synergistic effects of both electrochemical and thermal effects associated with PEF processing result in reduced enzyme function. PEF treatments do not significantly damage thermosensitive food constituents such as proteins and vitamins. Also, no significant changes were observed in the retention of water-soluble thiamine and riboflavin or fat-soluble retinol, cholecalciferol (vitamin D), and a-tocopherol (vitamin E) in skim milk, SMUF, or fresh bovine whole milk subjected to

PEF treatment. Compared to thermally pasteurized milk, milk treated with PEF and at a moderate temperature (50°C) enhanced rentability. Cheddar cheese made from PEF-treated milk was tougher and springier than cheese made from conventionally pasteurized milk. However, the adhesiveness and cohesiveness of Cheddar cheese made from PEF-treated milk were the same (Yu et al., 2012).

3.4.5 WINE

The application of PEF treatment in the wine industry can help improve the extraction of colour and aroma compounds from grape skins and reduce the microbial load of must and wine. Some of the effects of PEF treatment in the wine industry are *enhanced extraction of colour and flavour compounds, microbial load reduction, enhanced durability, reduced energy consumption and cost effective, and accelerated ageing.* PEF treatment can improve the extraction of color and flavour compounds from grape skins. The electrical impulse breaks down the cell walls of the grape skin, releasing compounds that contribute to color and aroma (Arcena et al., 2021). PEF processing can also reduce the extraction time required to reach the desired extraction level, helping reduce production costs (Comuzzo et al., 2018). PEF treatment can reduce the microbial load of must and wine. Electrical impulses can cause cell membrane damage, disrupt the normal functioning of microorganisms, and lead to microbial inactivation. PEF treatment can also reduce the need for traditional preservatives such as sulfur dioxide, which can be a concern for those with sensitivities. PEF treatment can extend the shelf life of wine by reducing the microbial load and preserving wine quality (Puligundla et al., 2018). PEF-treated wines have been shown to have improved sensory characteristics and are more stable over time compared to conventionally treated wines (Arcena et al., 2021). Moreover, PEF processing can reduce the energy consumption required for winemaking by reducing the need for traditional extraction techniques such as maceration and fermentation. PEF treatment can also reduce the energy required for filtration and purification by improving the extraction of color and flavour compounds. So, it is expected to be potentially cost-effective for large-scale industrial production (Arshad et al., 2021).

Overall, PEF treatment has several potential benefits to the wine industry, including improved extraction of colour and aroma compounds, reduced microbial load, improved shelf life, reduced energy consumption, and cost efficiency. However, further research is needed to optimize processing conditions for different grape cultivars and wine styles to achieve desired levels of extraction and microbial inactivation without compromising final product quality.

3.5 PACKAGING STANDARDS FOR PEF-TREATED PRODUCTS

PEF treatment is a food processing technique that uses high-voltage electric pulses to preserve and enhance the quality of food products. When it comes to packaging PEF-treated products, the following standards and guidelines should be followed:

- *Barrier Properties*: The packaging material should provide an adequate barrier against moisture, oxygen, light, and other external factors that can degrade the quality of the PEF-treated product. The barrier properties of the packaging material depend on the type of product being packaged.
- *Sterility*: The packaging material should be sterilized to ensure that the PEF-treated product remains safe and free from contamination during storage and transportation. The packaging material should also be compatible with the sterilization process used.
- *Mechanical Properties*: The packaging material should have appropriate mechanical properties to withstand handling, transportation, and storage stresses. This includes resistance to puncture, tearing, and compression.
- *Shelf Life*: The packaging material should help extend the PEF-treated product's shelf life by protecting it from external factors that can cause deterioration.
- *Regulatory Compliance*: The packaging material and process should comply with relevant food safety and packaging regulations in the region where the product would be marketed.

Overall, the choice of packaging material will depend on the specific requirements of the PEF-treated product, the intended shelf life, and the intended market. Here are a few examples of packaging materials that can be used for PEF-treated products:

- *Vacuum Packaging*: This type of packaging removes the air from the package, creating a vacuum. This type of packaging is often used for PEF-treated fruits and vegetables as it helps to extend their shelf life by slowing down the growth of microorganisms that can cause spoilage. For example, polyethylene and vacuum bags with aluminium foil as an inner layer are used as packaging material.
- *Modified Atmosphere Packaging (MAP)*: This type of packaging alters the atmosphere inside the package to extend the product's shelf life. MAP is often used for PEF-treated meat products, such as beef, pork, and chicken, as it helps to maintain their colour, flavour, and texture. Barrier films (multi-layered), cans, pouches, and other vacuum packaging materials are used for MAP packaging of various food commodities treated with PEF, like liquid foods, aqua products, purees, and other foods whose quality is significantly influenced by external environmental factors such as oxygen, carbon dioxide, moisture, etc.
- *High-Barrier Film*: This type of packaging material provides an excellent barrier against oxygen, moisture, and other external factors that can cause the degradation of the product. The high-barrier film is often used for PEF-treated beverages, sauces, and fruits and vegetables.
- *Aluminium Foil*: This is a type of packaging material that provides an excellent barrier against light, oxygen, and moisture. Aluminium foil is often used for PEF-treated dairy products, such as cheese, butter, and some meat products.
- *Glass Packaging*: This type of transparent packaging makes it ideal for products that need to be visible to the consumer. Glass packaging is often used for PEF-treated juices and other beverages, as well as some sauces and condiments.

3.6   FUTURE TRENDS

PEF processing is a non-thermal food preservation technology that has gained significant interest in the recent past. Here are some future trends in the PEF processing of food products:

- *Combination Technologies:* To enhance its effectiveness, PEF processing can be combined with other food preservation technologies such as HPP, thermal processing, and ultraviolet (UV) irradiation. Combination technologies may offer better preservation effects, improve quality attributes, and reduce energy requirements.
- *Development of New PEF Equipment:* PEF equipment manufacturers are developing new, more efficient equipment with lower energy consumption and can process a wider range of food products. Additionally, PEF equipment that can be easily integrated into existing food processing lines would be developed to facilitate the adoption of PEF processing.
- *More Studies on PEF Processing:* More studies are expected to be conducted on PEF processing to understand better its effects on different types of food products and the long-term effects on product quality and safety. Further studies would help identify optimal processing conditions, develop predictive models for process optimization and increase consumer acceptance.
- *Increased Adoption of PEF Processing:* As the demand for safe, nutritious, and minimally processed food increases, PEF processing is expected to become more widely adopted by the food industry. This increased research and development efforts, improved equipment design, and increased production efficiency.
- *Expansion of PEF Applications:* PEF processing may be applied to a broader range of food products, including fruits and vegetables, meat, poultry, and seafood. Moreover, PEF technology can be used to enhance the extraction of bioactive compounds from plant-based materials and the development of novel food formulations.

From a scientific perspective, there are some key areas to focus on in future. They are

- *Optimization of Processing Parameters:* Scientists are focusing on optimizing PEF processing parameters such as voltage, pulse duration, and frequency to improve the effectiveness of the treatment while minimizing potential negative impacts on the quality and safety of food products. This optimization could lead to improved process efficiency and reduced energy consumption.
- *Modelling of PEF Treatment Effects:* Researchers are developing predictive models better to understand the effects of PEF processing on food products. These models could help identify optimal processing conditions, determine the effects of different PEF parameters on product quality and safety, and reduce the need for extensive experimental studies.
- *Application to Non-Food Industries:* PEF processing has potential applications in non-food industries such as pharmaceuticals, cosmetics, and biotechnology. Scientists are exploring how PEF processing can be applied to these industries to improve product quality and efficiency and reduce environmental impact.
- *Integration with Other Processing Methods:* PEF processing can be combined with other methods, such as thermal, high-pressure, or enzymatic processing, to create more efficient and effective preservation methods. Scientists are exploring these combinations to optimize processing outcomes and improve the quality and safety of food products.
- *Understanding the Mechanism of Action:* Despite extensively studying PEF processing, its underlying mechanism of action is not completely understood. Scientists are exploring the effects of PEF processing on the molecular and cellular level to understand its impact on food products better. Understanding the mechanism of action could lead to further process optimization and improvement of product quality and safety.

REFERENCES

Aadil, R. M., Zeng, X. A., Ali, A., Zeng, F., Farooq, M. A., Han, Z., Khalid, S., & Jabbar, S. (2015). Influence of different pulsed electric field strengths on the quality of the grapefruit juice. *International Journal of Food Science and Technology, 50*(10), 2290–2296. https://doi.org/10.1111/ijfs.12891

Aadil, R. M., Zeng, X. A., Han, Z., Sahar, A., Khalil, A. A., Rahman, U. U., Khan, M., & Mehmood, T. (2018). Combined effects of pulsed electric field and ultrasound on bioactive compounds and microbial quality of grapefruit juice. *Journal of Food Processing and Preservation, 42*(2), e13507. https://doi.org/10.1111/jfpp.13507

Ade-Omowaye, B. I. O. I. O., Angersbach, A., Taiwo, K. A. A., & Knorr, D. (2001). Use of pulsed electric field pre-treatment to improve dehydration characteristics of plant based foods. *Trends in Food Science & Technology, 12*(8), 285–295. https://doi.org/10.1016/S0924-2244(01)00095-4

Ahmad Shiekh, K., Odunayo Olatunde, O., Zhang, B., Huda, N., & Benjakul, S. (2021). Pulsed electric field assisted process for extraction of bioactive compounds from custard apple (Annona squamosa) leaves. *Food Chemistry, 359.* https://doi.org/10.1016/J.FOODCHEM.2021.129976

Al Daccache, M., Koubaa, M., Maroun, R. G., Salameh, D., Louka, N., & Vorobiev, E. (2020). Pulsed electric field-assisted fermentation of Hanseniaspora sp. yeast isolated from Lebanese apples. *Food Research International (Ottawa, Ont.), 129.* https://doi.org/10.1016/J.FOODRES.2019.108840

Andreou, V., Dimopoulos, G., Katsaros, G., & Taoukis, P. (2016). Comparison of the application of high pressure and pulsed electric fields technologies on the selective inactivation of endogenous enzymes in tomato products. *Innovative Food Science and Emerging Technologies, 38*, 349–355. https://doi.org/10.1016/j.ifset.2016.07.026

Arcena, M. R., Leong, S. Y., Then, S., Hochberg, M., Sack, M., Mueller, G., Sigler, J., Kebede, B., Silcock, P., & Oey, I. (2021). The effect of pulsed electric fields pre-treatment on the volatile and phenolic profiles of Merlot grape musts at different winemaking stages and the sensory characteristics of the finished wines. *Innovative Food Science & Emerging Technologies, 70*, 102698. https://doi.org/10.1016/j.ifset.2021.102698

Arshad, R. N., Abdul-Malek, Z., Munir, A., Buntat, Z., Ahmad, M. H., Jusoh, Y. M. M., Bekhit, A. E.-D., Roobab, U., Manzoor, M. F., & Aadil, R. M. (2020). Electrical systems for pulsed electric field applications in the food industry: An engineering perspective. *Trends in Food Science & Technology, 104*(July), 1–13. https://doi.org/10.1016/j.tifs.2020.07.008

Arshad, R. N., Abdul-Malek, Z., Roobab, U., Munir, M. A., Naderipour, A., Qureshi, M. I., El-Din Bekhit, A., Liu, Z.-W., & Aadil, R. M. (2021). Pulsed electric field: A potential alternative towards a sustainable food processing. *Trends in Food Science & Technology, 111*(February), 43–54. https://doi.org/10.1016/j.tifs.2021.02.041

Arshad, R. N., Malek, Z. A., Ahmad, M. H., Buntat, Z., Kumara, C. L. G. P., & Abdulameer, A. Z. (2020). Coaxial treatment chamber for liquid food treatment through pulsed electric field. *Indonesian Journal of Electrical Engineering and Computer Science, 19*(3), 1169. https://doi.org/10.11591/ijeecs.v19.i3.pp1169-1176

Asavasanti, S., Ristenpart, W., Stroeve, P., & Barrett, D. M. (2011). Permeabilization of plant tissues by monopolar pulsed electric fields: Effect of frequency. *Journal of Food Science, 76*(1). https://doi.org/10.1111/j.1750-3841.2010.01940.x

Bai, T.-G., Zhang, L., Qian, J.-Y., Jiang, W., Wu, M., Rao, S.-Q., Li, Q., Zhang, C., & Wu, C. (2021). Pulsed electric field pretreatment modifying digestion, texture, structure and flavor of rice. *LWT, 138*(November 2020), 110650. https://doi.org/10.1016/j.lwt.2020.110650

Barbosa-Cánovas, G. V., Góngora-Nieto, M. M., Pothakamury, U. R., & Swanson, B. G. (1999). Fundamentals of High-Intensity Pulsed Electric Fields (PEF). In *Preservation of Foods with Pulsed Electric Fields* (pp. 1–19). Academic Press.

Barbosa-Cánovas, G. V., Góngora-Nieto, M. M., Pothakamury, U. R., & Swanson, B. G. (1999a). Design of PEF Processing Equipment. *Preservation of Foods with Pulsed Electric Fields*, 20–46. https://doi.org/10.1016/b978-012078149-2/50003-9

Barbosa-Cánovas, G. V., Góngora-Nieto, M. M., Pothakamury, U. R., & Swanson, B. G. (1999b). Fundamentals of High-Intensity Pulsed Electric Fields (PEF). In *Preservation of Foods with Pulsed Electric Fields* (pp. 1–19). https://doi.org/10.1016/b978-012078149-2/50002-7

Barsotti, L., & Cheftel, J. C. (1999). Food processing by pulsed electric fields. II. Biological aspects. *Food Reviews International, 15*(2), 181–213. https://doi.org/10.1080/87559129909541186

Bi, X., Liu, F., Rao, L., Li, J., Liu, B., Liao, X., & Wu, J. (2013). Effects of electric field strength and pulse rise time on physicochemical and sensory properties of apple juice by pulsed electric field. *Innovative Food Science and Emerging Technologies, 17*, 85–92. https://doi.org/10.1016/j.ifset.2012.10.008

Bocker, R., & Silva, E. K. (2022a). Innovative technologies for manufacturing plant-based non-dairy alternative milk and their impact on nutritional, sensory and safety aspects. *Future Foods, 5*, 100098. https://doi.org/10.1016/j.fufo.2021.100098

Bocker, R., & Silva, E. K. (2022b). Pulsed electric field assisted extraction of natural food pigments and colorings from plant matrices. *Food Chemistry: X, 15*(October 2021), 100398. https://doi.org/10.1016/j.fochx.2022.100398

Bolumar, T., Rohlik, B.-A. A., Stark, J., Sikes, A., Watkins, P., & Buckow, R. (2022). Investigation of pulsed electric field conditions at low field strength for the tenderisation of beef topside. *Foods, 11*(18). https://doi.org/10.3390/foods11182803

Buckow, R., Baumann, P., Schroeder, S., & Knoerzer, K. (2011). Effect of dimensions and geometry of co-field and co-linear pulsed electric field treatment chambers on electric field strength and energy utilisation. *Journal of Food Engineering, 105*(3), 545–556. https://doi.org/10.1016/j.jfoodeng.2011.03.019

Buckow, R., Ng, S., & Toepfl, S. (2013). Pulsed electric field processing of orange juice: A review on microbial, enzymatic, nutritional, and sensory quality and stability. *Comprehensive Reviews in Food Science and Food Safety, 12*(5), 455–467. https://doi.org/10.1111/1541-4337.12026

Budi, A., Legge, F. S., Treutlein, H., & Yarovsky, I. (2005). Electric field effects on insulin chain-B conformation. *Journal of Physical Chemistry B, 109*(47), 22641–22648. https://doi.org/10.1021/jp052742q

Chemat, F., Rombaut, N., Meullemiestre, A., Turk, M., Perino, S., Fabiano-Tixier, A. S., & Abert-Vian, M. (2017). Review of Green Food Processing techniques. Preservation, transformation, and extraction. *Innovative Food Science and Emerging Technologies, 41*(February), 357–377. https://doi.org/10.1016/j.ifset.2017.04.016

Chen, J., Tao, X. Y., Sun, A. D., Wang, Y., Liao, X. J., Li, L. N., & Zhang, S. (2014). Influence of pulsed electric field and thermal treatments on the quality of blueberry juice. *International Journal of Food Properties, 17*(7), 1419–1427. https://doi.org/10.1080/10942912.2012.713429

Chilling, S. U. S., Chmid, S. A. S., Enry, H. J. A., Chieber, A. N. S., & Arle, R. E. C. (2008). Comparative study of pulsed electric field and thermal processing of apple juice with particular deactivation. *Journal of Agricultural and Food Chemistry, 56*(12), 4545–4554.

Comuzzo, P., Marconi, M., Zanella, G., & Querzè, M. (2018). Pulsed electric field processing of white grapes (cv. Garganega): Effects on wine composition and volatile compounds. *Food Chemistry, 264*, 16–23. https://doi.org/10.1016/J.FOODCHEM.2018.04.116

Craven, H. M. M., Swiergon, P., Ng, S., Midgely, J., Versteeg, C., Coventry, M. J. J., & Wan, J. (2008). Evaluation of pulsed electric field and minimal heat treatments for inactivation of pseudomonads and enhancement of milk shelf-life. *Innovative Food Science and Emerging Technologies, 9*(2), 211–216. https://doi.org/10.1016/j.ifset.2007.11.002

Dalvi-Isfahan, M., Hamdami, N., Le-Bail, A., & Xanthakis, E. (2016). The principles of high voltage electric field and its application in food processing: A review. *Food Research International, 89*, 48–62. https://doi.org/10.1016/j.foodres.2016.09.002

Dziadek, K., Kopeć, A., Dróżdż, T., Kiełbasa, P., Ostafin, M., Bulski, K., Oziembłowski, M. (2019). Effect of pulsed electric field treatment on shelf life and nutritional value of apple juice. *Journal of Food Science and Technology, 56*(3), 1184–1191. https://doi.org/10.1007/s13197-019-03581-4

El Kantar, S., Boussetta, N., Lebovka, N., Foucart, F., Rajha, H. N., Maroun, R. G., Louka, N., & Vorobiev, E. (2018). Pulsed electric field treatment of citrus fruits: Improvement of juice and polyphenols extraction. *Innovative Food Science and Emerging Technologies, 46*, 153–161. https://doi.org/10.1016/j.ifset.2017.09.024

El-Mashad, H. M., & Pan, Z. (2017). Application of induction heating in food processing and cooking. *Food Engineering Reviews, 9*(2), 82–90. https://doi.org/10.1007/s12393-016-9156-0

Ertugay, M. F., Başlar, M., & Ortakci, F. (2013). Effect of pulsed electric field treatment on polyphenol oxidase, total phenolic compounds, and microbial growth of apple juice. *Turkish Journal of Agriculture and Forestry, 37*(6), 772–780. https://doi.org/10.3906/tar-1211-17

Evrendilek, G. A. (2017). Impacts of pulsed electric field and heat treatment on quality and sensory properties and microbial inactivation of pomegranate juice. *Food Science and Technology International, 23*(8), 668–680. https://doi.org/10.1177/1082013217715369

Faisal Manzoor, M., Ahmed, Z., Ahmad, N., Karrar, E., Rehman, A., Muhammad Aadil, R., Al-Farga, A., Waheed Iqbal, M., Rahaman, A., & Zeng, X. A. (2021). Probing the combined impact of pulsed electric field and ultra-sonication on the quality of spinach juice. *Journal of Food Processing and Preservation, 45*(5), 1–11. https://doi.org/10.1111/jfpp.15475

FAO, IFAD, UNICEF, WFP, & WHO. (2022). The State of Food Security and Nutrition in the World 2022. In *The State of Food Security and Nutrition in the World 2022.* https://doi.org/10.4060/cc0639en

Faridnia, F., Ma, Q. L., Bremer, P. J., Burritt, D. J., Hamid, N., & Oey, I. (2015). Effect of freezing as pre-treatment prior to pulsed electric field processing on quality traits of beef muscles. *Innovative Food Science and Emerging Technologies, 29*, 31–40. https://doi.org/10.1016/j.ifset.2014.09.007

Giteru, S. G., Oey, I., & Ali, M. A. (2018). Feasibility of using pulsed electric fields to modify biomacromolecules: A review. *Trends in Food Science and Technology, 72*(November 2017), 91–113. https://doi.org/10.1016/j.tifs.2017.12.009

Gómez, B., Munekata, P. E. S., Gavahian, M., Barba, F. J., Martí-Quijal, F. J., Bolumar, T., Campagnol, P. C. B., Tomasevic, I., & Lorenzo, J. M. (2019). Application of pulsed electric fields in meat and fish processing industries: An overview. *Food Research International, 123*(March), 95–105. https://doi.org/10.1016/j.foodres.2019.04.047

Guderjan, M., Töpfl, S., Angersbach, A., & Knorr, D. (2005). Impact of pulsed electric field treatment on the recovery and quality of plant oils. *Journal of Food Engineering, 67*(3), 281–287. https://doi.org/10.1016/j.jfoodeng.2004.04.029

Guionet, A., Joubert-Durigneux, V., Packan, D., Cheype, C., Garnier, J. P., David, F., Zaepffel, C., Leroux, R. M., Teissié, J., & Blanckaert, V. (2014). Effect of nanosecond pulsed electric field on Escherichia coli in water: Inactivation and impact on protein changes. *Journal of Applied Microbiology, 117*(3), 721–728. https://doi.org/10.1111/JAM.12558

Guo, C., Mujumdar, A. S., & Zhang, M. (2019). New development in radio frequency heating for fresh food processing: A review. *Food Engineering Reviews, 11*(1), 29–43. https://doi.org/10.1007/s12393-018-9184-z

Guo, Q., Sun, D.-W., Cheng, J.-H., & Han, Z. (2017). Microwave processing techniques and their recent applications in the food industry. *Trends in Food Science and Technology, 67*(2), 236–247. https://doi.org/10.1016/j.tifs.2017.07.007

Heinz, V., Toepfl, S., & Knorr, D. (2003). Impact of temperature on lethality and energy efficiency of apple juice pasteurization by pulsed electric fields treatment. *Innovative Food Science and Emerging Technologies, 4*(2), 167–175. https://doi.org/10.1016/S1466-8564(03)00017-1

Huang, K., & Wang, J. (2009). Designs of pulsed electric fields treatment chambers for liquid foods pasteurization process: A review. *Journal of Food Engineering, 95*(2), 227–239. https://doi.org/10.1016/j.jfoodeng.2009.06.013

Huang, K., Yu, L., Liu, D., Gai, L., & Wang, J. (2013). Modeling of yeast inactivation of PEF-treated Chinese rice wine: Effects of electric field intensity, treatment time and initial temperature. *Food Research International, 54*(1), 456–467. https://doi.org/10.1016/j.foodres.2013.07.046

Jaeger, H., Roth, A., Toepfl, S., Holzhauser, T., Engel, K.-H., Knorr, D., Vogel, R. F., Bandick, N., Kulling, S., Heinz, V., & Steinberg, P. (2016). Opinion on the use of ohmic heating for the treatment of foods. *Trends in Food Science & Technology, 55*, 84–97. https://doi.org/10.1016/j.tifs.2016.07.007

Janositz, A., Noack, A. K., & Knorr, D. (2011). Pulsed electric fields and their impact on the diffusion characteristics of potato slices. *LWT, 44*(9), 1939–1945. https://doi.org/10.1016/j.lwt.2011.04.006

Jeong, S.-H., Jung, Y.-M., Kim, S., Kim, J.-H., Yeo, H., & Lee, D.-U. (2023). Tenderization of beef semi-tendinosus muscle by pulsed electric field treatment with a direct contact chamber and its impact on proteolysis and physicochemical properties. *Foods, 12*(3), 430. https://doi.org/10.3390/foods12030430

Jin, W., Wang, Z., Peng, D., Shen, W., Zhu, Z., Cheng, S., Li, B., & Huang, Q. (2020). Effect of pulsed electric field on assembly structure of α-amylase and pectin electrostatic complexes. *Food Hydrocolloids, 101*, 105547. https://doi.org/10.1016/j.foodhyd.2019.105547

Kotnik, T., Rems, L., Tarek, M., & Miklavcic, D. (2019). Membrane electroporation and electropermeabilization: Mechanisms and models. *Annual Review of Biophysics, 48*, 63–91. https://doi.org/10.1146/ANNUREV-BIOPHYS-052118-115451

Laberge, M. (1998). Intrinsic protein electric fields: Basic non-covalent interactions and relationship to protein-induced Stark effects. *Biochimica et Biophysica Acta – Protein Structure and Molecular Enzymology, 1386*(2), 305–330. https://doi.org/10.1016/S0167-4838(98)00100-9

Lee, S. H., Shahbaz, H. M., Jeong, S. H., Hong, S. Y., & Lee, D. U. (2022). Effect of pulsed electric field treatment on cell-membrane permeabilization of potato tissue and the quality of French fries. *Italian Journal of Food Science, 34*(3), 13–24. https://doi.org/10.15586/ijfs.v34i3.2180

Leong, S. Y., & Oey, I. (2014). Effect of Pulsed Electric Field Treatment on Enzyme Kinetics and Thermostability of Endogenous Ascorbic Acid Oxidase in Carrots (Daucus carota cv. Nantes). In *Food Chemistry* (Vol. 146). Elsevier Ltd. https://doi.org/10.1016/j.foodchem.2013.09.096

Li, Y. Q., Chen, Q., Liu, X. H., & Chen, Z. X. (2008). Inactivation of soybean lipoxygenase in soymilk by pulsed electric fields. *Food Chemistry, 109*(2), 408–414. https://doi.org/10.1016/j.foodchem.2008.01.001

Liu, Z., Esveld, E., Vincken, J.-P., & Bruins, M. E. (2019). Pulsed electric field as an alternative pre-treatment for drying to enhance polyphenol extraction from fresh tea leaves. *Food and Bioprocess Technology, 12*(1), 183–192. https://doi.org/10.1007/s11947-018-2199-x

Löffler, M. J. (2022). Generation and Application of High Intensity Pulsed Electric Fields. In *Pulsed Electric Fields Technology for the Food Industry: Fundamentals and Applications* (pp. 55–106). https://doi.org/10.1007/978-3-030-70586-2_2

Lyu, F., Shen, K., Ding, Y., & Ma, X. (2016). Effect of pretreatment with carbon monoxide and ozone on the quality of vacuum packaged beef meats. *Meat Science, 117*, 137–146. https://doi.org/10.1016/J.MEATSCI.2016.02.036

Manas, P., & Vercet, A. (2006). Effect of Pulsed Electric Fields on Enzymes and Food Constituents. In J. Raso & V. Heinz (Eds.), *Pulsed Electric Fields Technology for the Food Industry, Fundamentals and Applications* (pp. 131–152). Springer New York LLC.

Mannozzi, C., Rompoonpol, K., Fauster, T., Tylewicz, U., Romani, S., Rosa, M. D., & Jaeger, H. (2019). Influence of pulsed electric field and ohmic heating pretreatments on enzyme and antioxidant activity of fruit and vegetable juices. *Foods, 8*(7). https://doi.org/10.3390/foods8070247

Marco-Molés, R., Rojas-Graü, M. A., Hernando, I., Pérez-Munuera, I., Soliva-Fortuny, R., & Martín-Belloso, O. (2011). Physical and structural changes in liquid whole egg treated with high-intensity pulsed electric fields. *Journal of Food Science, 76*(2), C257–C264. https://doi.org/10.1111/j.1750-3841.2010.02016.x

Martín-Belloso, O., Marsellés-Fontanet, Á. R., & Elez-Martínez, P. (2014). Enzymatic Inactivation by Pulsed Electric Fields. In *Emerging Technologies for Food Processing* (pp. 155–168). Elsevier. https://doi.org/10.1016/B978-0-12-411479-1.00009-7

Martínez, J. M., Delso, C., Maza, M., Alvarez, I., & Raso, J. (2020). Utilising pulsed electric field processing to enhance extraction processes. *Innovative Food Processing Technologies: A Comprehensive Review*, 281–287. https://doi.org/10.1016/b978-0-12-815781-7.22435-9

Masood, H., Diao, Y., Cullen, P. J., Lee, N. A., & Trujillo, F. J. (2018). A comparative study on the performance of three treatment chamber designs for radio frequency electric field processing. *Computers and Chemical Engineering, 108*, 206–216. https://doi.org/10.1016/j.compchemeng.2017.09.009

McDonnell, C. K., Allen, P., Chardonnereau, F. S., Arimi, J. M., & Lyng, J. G. (2014). The use of pulsed electric fields for accelerating the salting of pork. *LWT, 59*(2P1), 1054–1060. https://doi.org/10.1016/j.lwt.2014.05.053

Mihindukulasuriya, S. D. F., & Jayaram, S. H. (2020). Release of electrode materials and changes in organoleptic profiles during the processing of liquid foods using pulse electric field treatment. *IEEE Transactions on Industry Applications, 56*(1), 711–717. https://doi.org/10.1109/TIA.2019.2955659

Montanari, C., Tylewicz, U., Tabanelli, G., Berardinelli, A., Rocculi, P., Ragni, L., & Gardini, F. (2019). Heat-assisted pulsed electric field treatment for the inactivation of saccharomyces cerevisiae: Effects of the presence of citral. *Frontiers in Microbiology, 10*(2), 1737. https://doi.org/10.3389/fmicb.2019.01737

Niu, D., Zeng, X. A., Ren, E. F., Xu, F. Y., Li, J., Wang, M. S., & Wang, R. (2020). Review of the application of pulsed electric fields (PEF) technology for food processing in China. *Food Research International, 137*(September), 109715. https://doi.org/10.1016/j.foodres.2020.109715

Nowacka, M., Tappi, S., Wiktor, A., Rybak, K., Miszczykowska, A., Czyzewski, J., Drozdzal, K., Witrowa-Rajchert, D., & Tylewicz, U. (2019). The impact of pulsed electric field on the extraction of bioactive compounds from beetroot. *Foods, 8*(7), 244. https://doi.org/10.3390/foods8070244

Nowak, D., & Jakubczyk, E. (2022). Effect of pulsed electric field pre-treatment and the freezing methods on the kinetics of the freeze-drying process of apple and its selected physical properties. *Foods, 11*(16), 2407. https://doi.org/10.3390/foods11162407

Odriozola-Serrano, I., Soliva-Fortuny, R., & Martín-Belloso, O. (2009). Impact of high-intensity pulsed electric fields variables on vitamin C, anthocyanins and antioxidant capacity of strawberry juice. *LWT – Food Science and Technology, 42*(1), 93–100. https://doi.org/10.1016/j.lwt.2008.05.008

Oms-Oliu, G., Odriozola-Serrano, I., Soliva-Fortuny, R., & Martín-Belloso, O. (2009). Effects of high-intensity pulsed electric field processing conditions on lycopene, vitamin C and antioxidant capacity of watermelon juice. *Food Chemistry, 115*(4), 1312–1319. https://doi.org/10.1016/j.foodchem.2009.01.049

Ostermeier, R., Giersemehl, P., Siemer, C., Töpfl, S., & Jäger, H. (2018). Influence of pulsed electric field (PEF) pre-treatment on the convective drying kinetics of onions. *Journal of Food Engineering, 237*, 110–117. https://doi.org/10.1016/j.jfoodeng.2018.05.010

Paoplook, K., & Eshtiaghi, M. N. (2013). Impact of high electric field pulses on cell disintegration and oil extraction from palm fruit mesocarp. *International Journal of Agriculture and Environmental Research, 2*(3), 363–369.

Pataro, G., Barca, G. M. J., Donsì, G., & Ferrari, G. (2015). On the modeling of electrochemical phenomena at the electrode-solution interface in a PEF treatment chamber: Methodological approach to describe the phenomenon of metal release. *Journal of Food Engineering, 165*, 34–44. https://doi.org/10.1016/j.jfoodeng.2015.05.009

Pataro, G., Donsì, G., & Ferrari, G. (2017). Modeling of Electrochemical Reactions during Pulsed Electric Field Treatment. In *Handbook of Electroporation* (Vol. 2, pp. 1059–1088). https://doi.org/10.1007/978-3-319-32886-7_5

Perez, O. E., & Pilosof, A. M. R. (2004). Pulsed electric fields effects on the molecular structure and gelation of β-lactoglobulin concentrate and egg white. *Food Research International, 37*(1), 102–110. https://doi.org/10.1016/j.foodres.2003.09.008

Puértolas, E., Cregenzán, O., Luengo, E., Álvarez, I., & Raso, J. (2013). Pulsed-electric-field-assisted extraction of anthocyanins from purple-fleshed potato. *Food Chemistry, 136*(3–4), 1330–1336. https://doi.org/10.1016/J.FOODCHEM.2012.09.080

Puligundla, P., Pyun, Y. R., & Mok, C. (2018). Pulsed electric field (PEF) technology for microbial inactivation in low-alcohol red wine. *Food Science and Biotechnology, 27*(6), 1691–1696. https://doi.org/10.1007/S10068-018-0422-1

Quitão-Teixeira, L. J., Odriozola-Serrano, I., Soliva-Fortuny, R., Mota-Ramos, A., & Martín-Belloso, O. (2009). Comparative study on antioxidant properties of carrot juice stabilised by high-intensity pulsed electric fields or heat treatments. *Journal of the Science of Food and Agriculture, 89*(15), 2636–2642. https://doi.org/10.1002/jsfa.3767

Quintão-Teixeira, L. J., Soliva-Fortuny, R., Mota Ramos, A., & Martín-Belloso, O. (2013). Kinetics of peroxidase inactivation in carrot juice treated with pulsed electric fields. *Journal of Food Science, 78*(2), 222–228. https://doi.org/10.1111/1750-3841.12019

Rahaman, A., Siddeeg, A., Manzoor, M. F., Zeng, X. A., Ali, S., Baloch, Z., Li, J., & Wen, Q. H. (2019). Impact of pulsed electric field treatment on drying kinetics, mass transfer, colour parameters and micro-structure of plum. *Journal of Food Science and Technology*, *56*(5), 2670–2678. https://doi.org/10.1007/s13197-019-03755-0

Rahaman, A., Zeng, X. A., Farooq, M. A., Kumari, A., Murtaza, M. A., Ahmad, N., Manzoor, M. F., Hassan, S., Ahmad, Z., Bo-Ru, C., Jinjing, Z., & Siddeeg, A. (2020). Effect of pulsed electric fields process-ing on physiochemical properties and bioactive compounds of apricot juice. *Journal of Food Process Engineering*, *43*(8), 1–8. https://doi.org/10.1111/jfpe.13449

Ranjha, M. M. A. N. N., Kanwal, R., Shafique, B., Arshad, R. N., Irfan, S., Kieliszek, M., Kowalczewski, P. Ł., Irfan, M., Khalid, M. Z., Roobab, U., & Aadil, R. M. (2021). A critical review on pulsed electric field: A novel technology for the extraction of phytoconstituents. *Molecules*, *26*(16), 4893. https://doi.org/10.3390/molecules26164893

Riener, J., Noci, F., Cronin, D. A., Morgan, D. J., & Lyng, J. G. (2008). Combined effect of temperature and pulsed electric fields on apple juice peroxidase and polyphenoloxidase inactivation. *Food Chemistry*, *109*(2), 402–407. https://doi.org/10.1016/j.foodchem.2007.12.059

Riener, J., Noci, F., Cronin, D. A., Morgan, D. J., & Lyng, J. G. (2009). Combined effect of tempera-ture and pulsed electric fields on pectin methyl esterase inactivation in red grapefruit juice (Citrus paradisi). *European Food Research and Technology*, *228*(3), 373–379. https://doi.org/10.1007/s00217-008-0943-6

Roobab, U., Abida, A., Chacha, J. S., Athar, A., Madni, G. M., Ranjha, M. M. A. N., Rusu, A. V., Zeng, X.-A. A., Aadil, R. M., & Trif, M. (2022). Applications of innovative non-thermal pulsed electric field tech-nology in developing safer and healthier fruit juices. *Molecules*, *27*(13), 4031. https://doi.org/10.3390/molecules27134031

Roohinejad, S., Everett, D. W., & Oey, I. (2014). Effect of pulsed electric field processing on carotenoid extractability of carrot purée. *International Journal of Food Science & Technology*, *49*(9), 2120–2127. https://doi.org/10.1111/IJFS.12510

Salvador, A., Varela, P., Sanz, T., & Fiszman, S. M. (2009). Understanding potato chips crispy texture by simultaneous fracture and acoustic measurements, and sensory analysis. *LWT*, *42*(3), 763–767. https://doi.org/10.1016/j.lwt.2008.09.016

Sánchez-Vega, R., Elez-Martínez, P., & Martín-Belloso, O. (2015). Influence of high-intensity pulsed electric field processing parameters on antioxidant compounds of broccoli juice. *Innovative Food Science and Emerging Technologies*, *29*, 70–77. https://doi.org/10.1016/j.ifset.2014.12.002

Sánchez-Vega, R., Rodríguez-Roque, M. J., Elez-Martínez, P., & Martín-Belloso, O. (2020). Impact of critical high-intensity pulsed electric field processing parameters on oxidative enzymes and color of broccoli juice. *Journal of Food Processing and Preservation*, *44*(3), 1–10. https://doi.org/10.1111/jfpp.14362

Saulis, G. (2010). Electroporation of cell membranes: The fundamental effects of pulsed electric fields in food processing. *Food Engineering Reviews*, *2*(2), 52–73. https://doi.org/10.1007/S12393-010-9023-3

Shabbir, M. A., Ahmed, H., Maan, A. A., Rehman, A., Afraz, M. T., Iqbal, M. W., Khan, I. M., Amir, R. M., Ashraf, W., Khan, M. R., & Aadil, R. M. (2021). Effect of non-thermal processing techniques on patho-genic and spoilage microorganisms of milk and milk products. *Food Science and Technology (Brazil)*, *41*(2), 279–294. https://doi.org/10.1590/fst.05820

Shamsi, K., Versteeg, C., Sherkat, F., & Wan, J. (2008). Alkaline phosphatase and microbial inactivation by pulsed electric field in bovine milk. *Innovative Food Science & Emerging Technologies*, *9*(2), 217–223. https://doi.org/10.1016/j.ifset.2007.06.012

Simonis, P., Kersulis, S., Stankevich, V., Sinkevic, K., Striguniene, K., Ragoza, G., & Stirke, A. (2019). Pulsed electric field effects on inactivation of microorganisms in acid whey. *International Journal of Food Microbiology*, *291*, 128–134. https://pubmed.ncbi.nlm.nih.gov/30496942/

Suwandy, V., Carne, A., van de Ven, R., Bekhit, A. E. D. A., & Hopkins, D. L. (2015). Effect of pulsed electric field on the proteolysis of cold boned beef M: longissimus lumborum and M: Semimembranosus. *Meat Science*, *100*, 222–226. https://doi.org/10.1016/j.meatsci.2014.10.011

Tao, Z., Wang, D., Yao, F., Huang, X., Wu, Y. Y., Wu, Y. Y., Chen, Z., Wei, J., Li, X., & Yang, Q. (2020). Influence of low voltage electric field stimulation on hydrogen generation from anaerobic digestion of waste activated sludge. *Science of The Total Environment*, *704*, 135849. https://doi.org/10.1016/j.scitotenv.2019.135849

Toepfl, S., Heinz, V., & Knorr, D. (2007). High intensity pulsed electric fields applied for food preserva-tion. *Chemical Engineering and Processing: Process Intensification*, *46*(6), 537–546. https://doi.org/10.1016/j.cep.2006.07.011

Toepfl, S., Siemer, C., & Heinz, V. (2014). Effect of High-Intensity Electric Field Pulses on Solid Foods. In *Emerging Technologies for Food Processing* (pp. 147–154). Elsevier. https://doi.org/10.1016/B978-0-12-411479-1.00008-5

Toepfl, S., Siemer, C., Saldaña-Navarro, G., & Heinz, V. (2014). Overview of Pulsed Electric Fields Processing for Food. In *Emerging Technologies for Food Processing* (pp. 93–114). Elsevier. https://doi.org/10.1016/B978-0-12-411479-1.00006-1

Tokuşoğlu, Ö., & Swanson, B. G. (2014). *Improving Food Quality with Novel Food Processing Technologies* (1–460). https://doi.org/10.1201/b17780

UNFAO. (2009). How to Feed the World in 2050. *Insights from an Expert Meeting at FAO.* http://www.fao.org/wsfs/forum2050/wsfs-forum/en/

Vanga, S. K., Wang, J., Jayaram, S., & Raghavan, V. (2021). Effects of pulsed electric fields and ultrasound processing on proteins and enzymes: A review. *Processes, 9*(4), 1–16. https://doi.org/10.3390/pr9040722

Walter, L., Knight, G., Ng, S. Y., & Buckow, R. (2016). Kinetic models for pulsed electric field and thermal inactivation of Escherichia coli and Pseudomonas fluorescens in whole milk. *International Dairy Journal, 57*, 7–14. https://doi.org/10.1016/j.idairyj.2016.01.027

Wang, H., Wang, N., Chen, X., Wu, Z., Zhong, W., Yu, D., & Zhang, H. (2022). Effects of moderate electric field on the structural properties and aggregation characteristics of soybean protein isolate. *Food Hydrocolloids, 133.* https://doi.org/10.1016/j.foodhyd.2022.107911

Wang, J., Wang, K., Wang, Y., Lin, S., Zhao, P., & Jones, G. (2014). A novel application of pulsed electric field (PEF) processing for improving glutathione (GSH) antioxidant activity. *Food Chemistry, 161*, 361–366. https://doi.org/10.1016/j.foodchem.2014.04.027

World Population Prospects. (2022). World Population Prospects 2022: Summary of Results. In *World Population Prospects.* https://www.un.org/development/desa/pd/sites/www.un.org.development.desa.pd/files/wpp2022_summary_of_results.pdf

Xiang, B., Sundararajan, S., Mis Solval, K., Espinoza-Rodezno, L., Aryana, K., & Sathivel, S. (2014). Effects of pulsed electric fields on physicochemical properties and microbial inactivation of carrot juice. *Journal of Food Processing and Preservation, 38*(4), 1556–1564. https://doi.org/10.1111/jfpp.12115

Yu, L. J., Ngadi, M., & Raghavan, V. (2012). Proteolysis of cheese slurry made from pulsed electric field-treated milk. *Food and Bioprocess Technology, 5*(1), 47–54. https://doi.org/10.1007/s11947-010-0341-5

Zaidi, K. U., Ali, A. S., Ali, S. A., & Naaz, I. (2014). Microbial tyrosinases: Promising enzymes for pharmaceutical, food bioprocessing, and environmental industry. *Biochemistry Research International, 2014.* https://doi.org/10.1155/2014/854687

Zhang, C., Yang, Y.-H. H., Zhao, X.-D. D., Zhang, L., Li, Q., Wu, C., Ding, X., & Qian, J.-Y. Y. (2021). Assessment of impact of pulsed electric field on functional, rheological and structural properties of vital wheat gluten. *LWT – Food Science and Technology, 147*(April), 111536. https://doi.org/10.1016/j.lwt.2021.111536

Zhao, W., Yang, R., Hua, X., Zhang, W., Tang, Y., & Chen, T. (2010). Inactivation of polyphenoloxidase of pear by pulsed electric fields. *International Journal of Food Engineering, 6*(3). https://doi.org/10.2202/1556-3758.1853

Zhao, W., Yang, R., & Zhang, H. Q. (2012). Recent advances in the action of pulsed electric fields on enzymes and food component proteins. *Trends in Food Science and Technology, 27*(2), 83–96. https://doi.org/10.1016/j.tifs.2012.05.007

Zhou, Y., He, Q., & Zhou, D. (2017). Optimization extraction of protein from mussel by high-intensity pulsed electric fields. *Journal of Food Processing and Preservation, 41*(3), e12962. https://doi.org/10.1111/jfpp.12962

Zhou, Y., Sun, X., Zhong, K., Evans, D. G., Lin, Y., & Duan, X. (2012). Control of surface defects and agglomeration mechanism of layered double hydroxide nanoparticles. *Industrial and Engineering Chemistry Research, 51*(11), 4215–4221. https://doi.org/10.1021/ie202302n

Zongo, P. A., Khalloufi, S., Mikhaylin, S., & Ratti, C. (2022). Pulsed electric field and freeze-thawing pretreatments for sugar uptake modulation during osmotic dehydration of mango. *Foods (Basel, Switzerland), 11*(17). https://doi.org/10.3390/FOODS11172551

4 Cold Plasma Applications in Food Structure Transformation

Anbarasan Rajan and R. Mahendran

4.1 INTRODUCTION

In recent years, additive manufacturing in food production has showcased the novel way of producing food products with different structural designs. 3D printing and sessile drop drying are the two major additive manufacturing techniques that are used for producing these innovative structurally designed food products. Both these methods produce flat 2D structures that are capable of transforming into 3D structures over a period of time (fourth dimension) by interacting with external stimuli. Due to the time dependent 3D structural change, this kind of additive manufacturing of food is called as 4D printing of food or 4D food (Ratish Ramanan and Mahendran 2021). The major advantage of 4D food over conventionally prepared designed foods (extruded and mold based) is that it reduces the package volume and, thus, the transportation cost of the product. Further, the 4D food manufacturing also provides opportunity to customize the structure, texture, aesthetic value, and sensory perception of the food products; and to reduce food material wastage (Oral et al. 2021; Chen et al. 2022). To design a 4D food, the flat 2D food sheet obtained from 3D printing/sessile drop dying needs to have diversely arranged particles positioned at all directions. Further, to control the shape transformation, the flat sheets need to be coated with edible constraint materials (i.e. ethyl cellulose) which aid structural changes of flat sheet by controlling its microstructural interaction with the stimuli (Ratish Ramanan and Mahendran 2021). The stimuli used for shape-shifting can be water, oil, drying, or pH. When any of these stimuli interact with constraint material coated 2D food structure, it converts the 2D shape into a 3D structure owing to the anisotropic swelling/coloring nature (e.g., the 2D material having constraint coated area absorb less water than the uncoated areas) (Gupta and Radhakrishnan 2019). Basically, the shape-changing behavior of the flat food sheets can be governed by three laws; the first one explains the "relative expansion" between active (constrain free area) and static (constraint coated) parts of the flat sheets due to their anisotropic nature. This relative extension induces stress gradient in the 2D food material and converts it into a 3D structure by folding or bending it. When it comes to the second law, it explains the relative expansion caused in 2D structure due to mass diffusion, molecular transformation, thermal expansion, and organic growth. While the third law explains the time constant dependent shape change of 4D material (Momeni and Ni 2020). Thus, it is clear that designing the flat 2D structure is the key to shape transformation of 4D foods. Hence, it is essential to vary the anisotropic nature of flat sheets' surfaces by creating defined high active and static regions by increasing (producing more surface area) and reducing (blocking stimuli response by coating) the stimuli responses of those areas. Nevertheless, it is also important to have intact-adhesive constraint material coating on 2D structure surface to avoid detachment of constraint material and to produce defined shape change. Furthermore, the shape change becomes faster when the uncoated surfaces of the 2D structure react more with the given stimuli. Therefore, 4D printing processes require pre-treatments that increase the surface reactivity of flat 2D food structures with stimuli and improve the adhesive nature of coated constraint material. In this regards, Stephen et al. (2021) found cold plasma (CP) treatment to be affective in keeping the constraint material adhesive bond with the 2D structure surface and also to increase the water and oil absorption (Stephen, Manoharan, and Radhakrishnan 2021; Anbarasan et al. 2022). Hence, this chapter is aimed to discuss the possibilities and opportunities of CP on

DOI: 10.1201/9781003359302-4

aiding the shape transformation process along with some of the basic concepts (stimuli and influencing factors) involved in it. Therefore, the next section focuses on the different stimuli involved in shape transformation and the significance of 2D structure surface nature on shape change.

4.2 SHAPE TRANSFORMATION BASICS

4D foods are constructed in additive manufacturing methods to obtain defined food layer compositions for achieving the predetermined shapes. Hence, food materials with elasic and flow behavior are preferred for 4D printing or sessile drop drying to produce shape changing 2D structures. Therefore, food gel (hydrogel) is used widely for preparing 4D foods as this high moisture 3-D polymeric network fulfils the required rheological and structural demands. These polysaccharides and protein complexes are either printed and dried or directly dried through sessile drop method to obtain 2D structures (Figure 4.1). Once the 2D structure is obtained, the dried gel will be coated

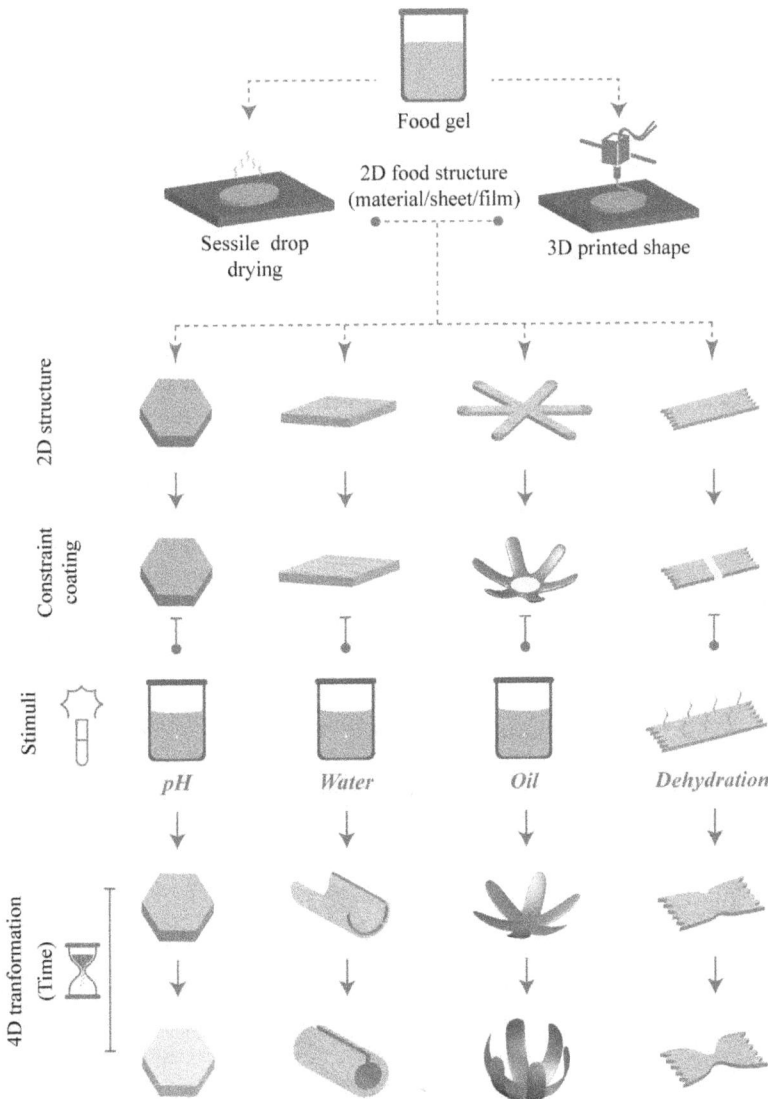

FIGURE 4.1 4D shape transformation of 2D structure under different stimuli environments.

with constraint material to control its relative extension of the structure (Ratish Ramanan and Mahendran 2021). Meanwhile, it is not necessary to coat the 3D printed 2D food structure (unlike sessile drop dried samples) with constraint material. Instead, the shape change can be controlled by providing groove-based mechanical pressing. However, in both the cases, once the dried structures are ready, they can be introduced into the new environment (stimuli) to transform their shapes. For 4D shape transformation, stimuli such as water, oil, pH, and dehydration are currently used (Figure 4.1); however, it is not restricted to the above stated list. The following subsections discuss the mechanism involved in shape transformation of 2D structure under different stimuli environment and also state the importance of surface character of 2D food material on shape transformation.

4.2.1 pH

Food product containing coloring agents (especially natural edible plant pigments) changes their chemical form when changing their pH. These chemical form modification results in food materials color change. For example, anthocyanin present in material changes its color from red to green when its pH shifts from acidic to alkali. Similarly, curcumin changes its color from yellow to red under alkali condition (Musso, Salgado, and Mauri 2017; He, Zhang, and Guo 2020; Gaikwad et al. 2022). The occurrence of color change in a food product starts only on the contact surface when it introduced in a different pH medium. Evidently, He et al. (2020) observed color changes in the mashed potatoes under various pH conditions. Thus, it can be hypothesized that food products with more contact (surface) area can achieve fast and intense color change as compared to the same sized product with lesser contact area. Hence, increasing the surface area of 4D foods with suitable surface treatment can produce better and faster color changing materials.

4.2.2 Water

Currently, water triggered shape transformation is gaining attention in various research fields including drug delivery, architecture, and biomimetic. When a 4D food absorbs water, it changes the material form by expanding its volume or by changing its structure. To obtain great programmability of the final shape, it is necessary to control the food material's relative expansion using constraint materials (Wang et al. 2017). As a shape triggering stimuli, water can roll, fold, twist, curl, or bend the food material and it occurs due to relative expansion and shrinkage of the 2D structure. In case of water triggered 4D transformation, ethylcellulose is the most preferred constraint material (insoluble fiber) that can be used for food material. Firstly, It controls water diffusion into the food by acting as a water barrier, and secondly, it forms layer on the food surface to provide mechanical support for obtaining 3D structure (Teng, Zhang, and Mujumdar 2021). For better results, the uncoated 2D food surface should absorb more water; while the coated constraint material should remain intact and resist the water absorption. Both of these scenarios can be achieved by increasing the surface area of the stimuli and constraint material contact surfaces of 2D food structures (Ratish Ramanan and Mahendran 2021). Further, the shape change behavior can be controlled by medium temperature and food material compositions (Boopathy et al. 2021; Ratish Ramanan and Mahendran 2021).

4.2.3 Oil

Oil triggered shape change can be called as "oleomorphic" which implies oil specific shape. There were very few studies that analysed the oil triggered shape changes of food materials. However, using oil for 4D shape change is similar to that of water stimuli induced shape change only. Hence, the relative extension caused by the differences in the oil absorption of 2D food surfaces is the key for converting 2D shape into a 3D shape. As compared to other types, oil triggered shape change might be holding higher commercial values in snack industry (i.e. chips) in future. However, the 2D food has to be coated with constraint to program the shape under oil stimuli (Stephen, Manoharan, and Radhakrishnan 2021). Nevertheless, coating with the constraint material that held the food material shape in water is not a suitable option for oil stimuli. Thus, cellulose acetate is used for oil triggered shape transformation studies (Stephen, Manoharan, and Radhakrishnan 2021; Cheeyattil et al. 2022). As like water stimuli, oil induced shape change is also influenced by concentration of starch in food material (composition), oil temperature, and surface texture of the dried 2D food substance.

4.2.4 DEHYDRATION

When a 2D flat food sheet is dehydrated, it causes water loss which induces moisture gradient in the 2D food surface. Due to this moisture gradient, the relative extension (stress and strain) on food samples causes them to deform (i.e. bend, curl, and curve) (C. Chen et al. 2021). To dehydrate the shape changing food materials, drying methods such as microwave, IR, or hot air oven drying can be used (Liu et al. 2021). This mass transfer process can be accelerated by changing the surface topography of the food produced (Bao et al. 2021). Once the surface is modified, the increased surface area would allow more water vapor to transfer (Zhang et al. 2019; Bao et al. 2021) from the active areas of shape changing material which increases the moisture gradient and leads to better 3D shape transformation.

4.3 CP ABILITY TO INDUCED SURFACE MODIFICATION

CP is a well-known technique for surface modification of polymer films and other packaging material for improving their surface functionalities. These surface alterations occur as the result of two mechanisms (Figure 4.2). Firstly, the high energy UV photons, electrons, free radical, ions, and other excited and non-excited atoms or molecules induce covalent bond breakages and subsequent interactions with active groups present in air. Secondly, the bombardment of high energy particles on the object's surface changes it's morphology (Dong et al. 2018). In films and packaging materials, these modifications increase the binding nature of adhesive materials. Meanwhile, they also

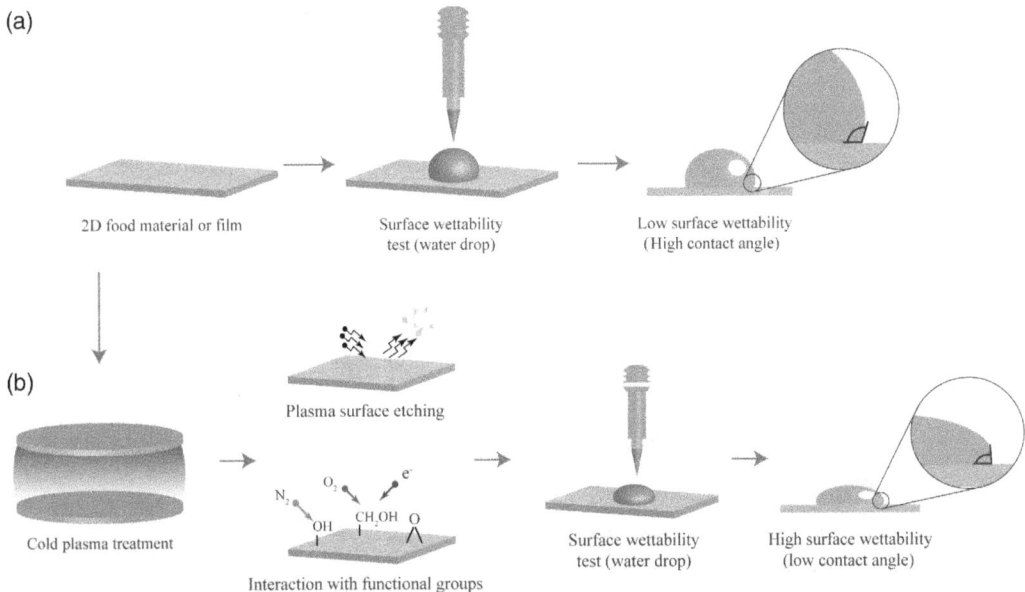

FIGURE 4.2 Difference between the surface wettability of (a) untreated and (b) CP treated 2D food material.

improve the surface wettability of solid material and lead to higher water absorption (Bahrami et al. 2022; Anbarasan et al. 2022; Rajan et al. 2023).

4.3.1 Effect of CP on Relative Expansion of 2D Food Material

The importance of plasma treatment on relative expansion of a given 2D material under a specific stimulus is represented in Figure. 4.3. Firstly, when the flat shape-changing materials (Figure 4.3: a-i) are coated with a constraint material, they loosely bind with their untreated smooth

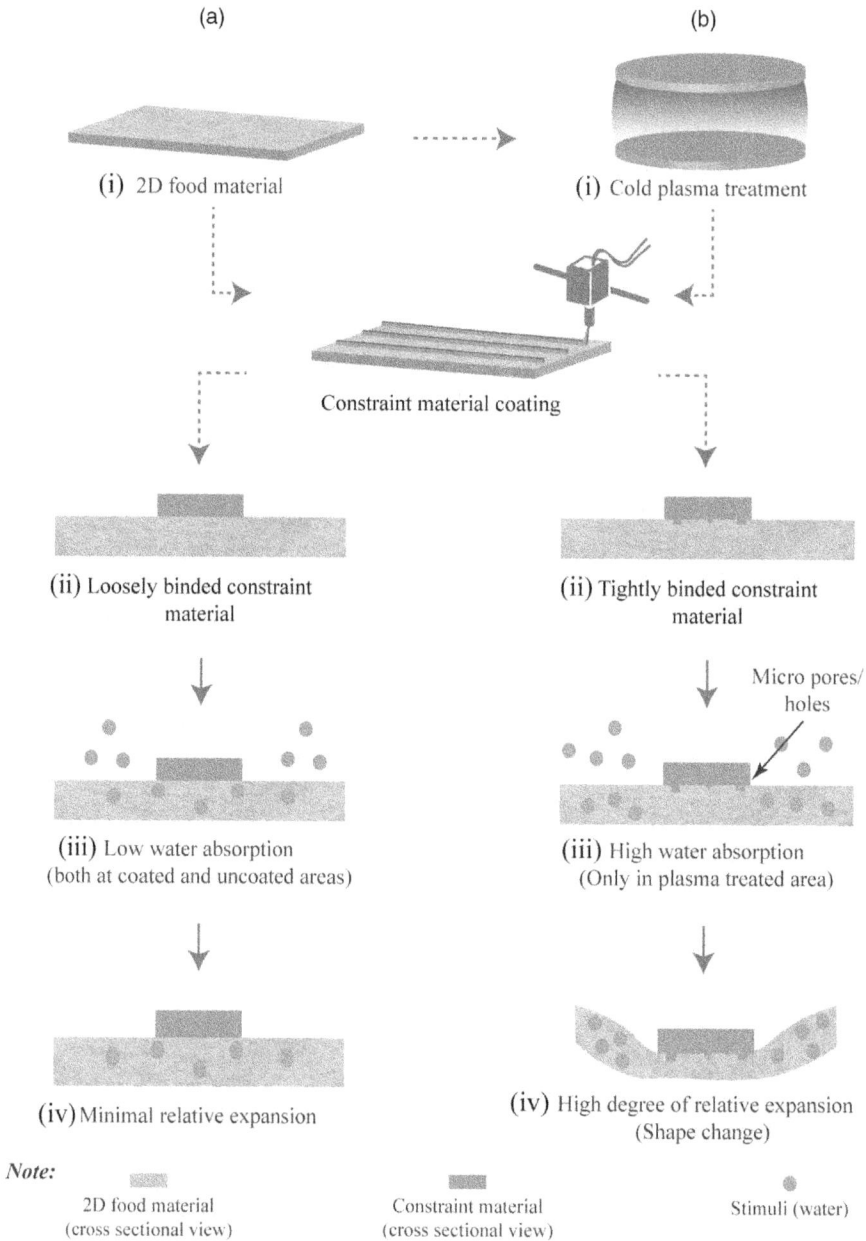

(a)   (b)

(i) 2D food material   (i) Cold plasma treatment

Constraint material coating

(ii) Loosely binded constraint material   (ii) Tightly binded constraint material

Micro pores/ holes

(iii) Low water absorption   (iii) High water absorption
(both at coated and uncoated areas)   (Only in plasma treated area)

(iv) Minimal relative expansion   (iv) High degree of relative expansion
                                              (Shape change)

Note:
2D food material   Constraint material   Stimuli (water)
(cross sectional view)   (cross sectional view)

FIGURE 4.3 Relative expansion (under water stimulus) difference between (a) untreated and (b) CP treated 2D food films.

surfaces (Figure 4.3: a-ii). In addition, the constraint will not penetrate into the 2D structures due to their lesser absorbing nature. Hence, when these food materials are introduced into a stimulus, the constraint material allows it (i.e. water) to enter into the food matrix present (Figure 4.3: a-iii). Meanwhile, the uncoated area also absorbs the stimulus in a lesser rate. Therefore, untreated 2D flat food material will not achieve high degree of relative expansion and results in minimal changes in its shape (Figure 4.3: a-iv). In order to counter the relative extension issue, the 2D food film surface can be treated with CP (Figure 4.3: b-i). In this way, the surface roughness created by CP allows the constraint material to flow inside the micro holes (Figure 4.3: b-ii) of food and forms strong binding with them (Gupta et al. 2020). Due to the intact binding, the stimuli comes in contact with the constraint coated areas will not enter into the food matrix. In contrast, the uncoated area allows more stimuli to enter into the food matrix (Figure 4.3: b-iii) through the CP created surface micro holes. Due to the anisotropic nature of plasma treated 2D structures, the relative extension increases between the coated and uncoated areas of the food structure which leads to 4D shape transformation (Figure 4.3: b-iv).

4.4 STUDIES ON CP ASSISTED FOOD STRUCTURE TRANSFORMATION

Carbohydrate rich grains/flours are the essential raw material for the preparation of hydrogels due to their amylose and amylopectin content. Wheat flour is one such agriculture product that has the ability to produce hydrogels (3D network) required for 4D food preparation. For this purpose, Gupta et al. (2020) prepared a flour suspension (10% wt/vol) by mixing wheat flour with distilled water and gelatinized it by heating to 90°C. The prepared gelatinized suspension (hydrogel) was then dried as sessile drops (1 ml each) to obtain flat xerogels (Figure 4.4: a-i). Though these flat structures are capable of changing their structure under a stimulus, their relative extension has to be controlled by coating constraint material on them. Thus, the author selected ethyl cellulose as a constraint material to coat the xerogels. However, to increase the stability of coating on the xerogel surface, the author subjected the xerogel to CP treatment prior to constraint coating. After treating the xerogels in CP (power-7.52 W, time-5 min), ethyl cellulose was coated on them (Figure 4.4: a-ii). The resulted, coated structure showed better 4D transformation (Figure 4.4: a-iii) than the untreated samples due to the higher water swelling gradient achieved in the CP treated xerogel. Similar 4D transformation was achieved in an earlier study where Stephen et al. (2021) converted a 2D flour xerogel into a 3D structure by using oil as a trigger stimulus. Initially, the xerogels (Figure 4.4: b-i) were prepared by drying (sessile drop method) the gelatinized corn suspension (5% wt/vol) at room temperature. Then, the prepared xerogels (rectangular) top surfaces were treated using CP at 1 kV voltage level for a period of 5 min. During CP treatment, the surface pores formed on the xerogels were later utilized to create a strong binding with cellulose acetate (constraint material). Thus, 10% constraint material was prepared and coated on the top surfaces of xerogels (Figure 4.4: b-ii). after that, the xerogels were dried and immersed in 220°C hot coconut oil. As the result of anisotropic oil absorption behavior, the rectangular xerogel curled into a spiral 3D structure (Figure 4.4: b-iii) after 2 sec of frying. From the experiment, the author observed two significant advantages of using CP treated xerogels over untreated xerogels. Firstly, the treatment increased the relative extension of xerogels due to increased oil absorption. Secondly, it assisted the constraint to bind tightly to the xerogels and prevented it from splitting from their surfaces. Therefore, the author continued experimenting on CP treatment for achieving improved structural change behavior of xerogel. Hence, Jaspin et al. (2021) decided to experiment with two different stimuli to analyze the tapioca xerogels 4D transformation. The required tapioca xerogels (Figure 4.4: c-i) for the study were obtained by drying (sessile drop) the gelatinized (65°C) tapioca suspension (5% wt/vol) at room temperature. After, drying, the flat xerogels were treated in low pressure CP at 1 kV power level for a period of 5 min to achieve required surface modification to hold the constraint materials on their surfaces. These xerogels were then coated with ethyl cellulose (Figure 4.4: c-ii) in a wheel pattern and immersed in hot water after drying. Due to high water absorption in the xerogel's bottom layer

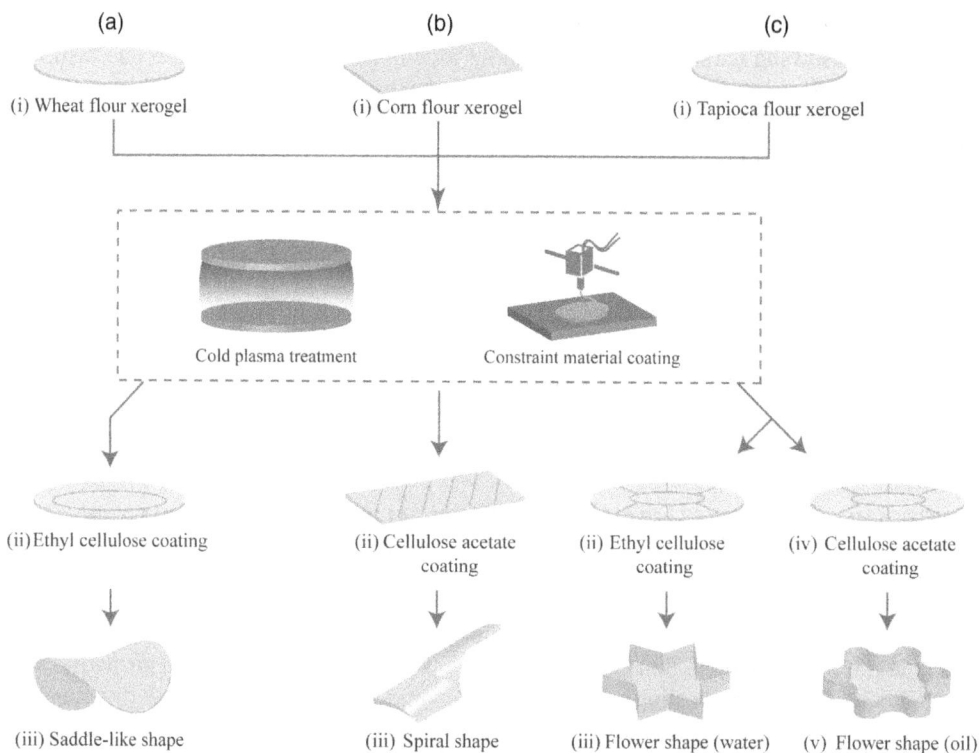

FIGURE 4.4 CP assisted 4D transformation of (a) wheat xerogel (Gupta et al. 2020); (b) corn xerogel (Stephen, Manoharan, and Radhakrishnan 2021) and (c) tapioca xerogels (both in water and oil) (Jaspin et al. 2021).

(pores), it expanded more than the top layer. Further, the insoluble hydrophobic coating material restricted the relative extension of the structure and led to a programmed 3D structural change (Figure 4.4: c-iii) after 30 sec. Further, along with water triggered 4D transformation, the author also evaluated the oil triggered 3D structural changes of tapioca xerogels. For that, the CP treated xerogels were coated with cellulose acetate or oil repellent (Figure 4.4: c-iv) and introducing into hot oil (220°C). Due to its anisotropic nature, the xerogel changes its 2D structure into 3D structure (Figure 4.4: c-v) within 2 sec. Therefore, CP treatment is a proven surface modification technique for 4D foods. Hence, future researches can focus on exploring the opportunities of CP treatment to enhance the 4D food products performances. One of the important aspects of plasma treatment of xerogels is that its ability to retain quality of bio-actives present in xerogels. Though Cheeyattil et al. (2022) observed a reduction in the curcumin content of fortified xerogels, it was observed only after water induced shape transformation process. Hence, there is no evidence of plasma caused quality deterioration in shape transforming xerogels. Further research in this aspect would provide more clarity on the retention of bio-actives in plasma treated xerogels. The next section discusses about the importance of food ingredients' property modification on their stimuli responses and the significance of CP treatment on assisting these property changes.

4.5 EFFECT OF INGREDIENTS FUNCTIONALITY ON STRUCTURE TRANSFORMATION AND SIGNIFICANCE OF CP TREATMENT

4.5.1 WATER ABSORPTION AND SWELLING

For 4D food materials, water absorption and swelling behavior are the essential properties that aid them to change their structure from 2D to 3D (Jaspin et al. 2021; Boopathy et al. 2022). To increase the water absorption of food ingredients, it is important to disintegrate the starch granules present in them (Manchun et al. 2012). Since CP has the potential to disintegrate the native starch, it is possible to enhance the water absorption in the selected surface areas of xerogel. Thus, any 4D food with lower degree of relative expansion (even after coating them with constraint material) can be treated in CP prior to constraint material coating to increase its anisotropic water absorption and swelling. In the case of xerogel, the increased water or oil absorption is due to the hydrophilic nature of CP treated surfaces (Thirumdas et al. 2017).

4.5.2 SOLUBILITY OF STARCH

Xerogel flours with high water solubility can produce lumps-free xerogel. Hence, the flour used for xerogel preparation is either optimized for its concentration (in the hydrogel) or modified for its solubility increase (Stephen et al. 2021; Thirumdas et al. 2017). Among these two methods, concentration optimization can be performed by repeated experimental trials; whereas, solubility increase requires certain treatments for the flours. In this regard, CP treatment can be performed as a suitable starch modification technique to increase the flour solubility. Plasma's potential to induce starch depolymerization is considered as the major cause for the solubility increase. The small starch fragments produced after treatment are highly soluble than its native starches and the starch oxidation caused by plasma species further increases the flour solubility. However, based on treatment intensity (i.e. voltage), the solubility enhancement also varies (Thirumdas et al. 2017).

4.5.3 AMYLOSE CONTENT

Though CP treatment has some beneficial effects to the shape transformation process, certain changes caused by this technique on the flour property might reduce the xerogel performance. For example, amylose presents in flour helps in absorbing more water (Li et al. 2020); in contrast, when the same flour treated with CP, the starch depolymerization process reduces the available amylose content (Thirumdas et al. 2017). Thus, it cannot be concluded that a plasma treated xerogel (or it's flour) will always produce a better 4D food.

4.6 CHALLENGES IN CP APPLICATION IN FOOD STRUCTURE TRANSFORMATION AND FUTURE PROSPECTS

When it comes to 4D foods, they are flat and thin in nature which gives them high surface area for a given xerogel/2D structure mass. Meanwhile, the currently available CP treatment systems are smaller in size and will not accommodate bulk quantities of xerogels within the closed environment. Further, using low pressure CP system for xerogel treatment, slowdowns the overall process due to the come up time (vacuum building) and release time (vacuum release) involved in it. Since most of the available studies on xerogels are performed only using low pressure system, the merits and demerits associated with atmospheric pressure CP treatments are still unknown. Further, CP treatment is not an approved method for food products (FDA) till date; thus, commercialization of this technique depends on FDA regulations. Nevertheless, CP also provides opportunity to modify the ingredient characteristics of 4D foods.

4.7 SUMMARY

Currently, consumers are looking for food products that provide adventurous experience in terms of both appearances and sensory perspective. Hence, 4D foods are designed to attract consumers and also to reduce the production and supply chain cost. In this regard, integrating a novel processing technique like CP in the production line of xerogels makes the process greener and sustainable than other conventional techniques. When the regulations are formed for CP processing, it is possible to introduce plasma treated 4D food products in consumer market. Until then, it is important to experiment with plasma operating conditions to determine the correlation between plasma characteristic and product quality.

REFERENCES

Anbarasan, R., S. Jaspin, B. Bhavadharini, Akash Pare, R. Pandiselvam, and R. Mahendran. 2022. "Chlorpyrifos Pesticide Reduction in Soybean Using Cold Plasma and Ozone Treatments." *LWT* 159 (April): 113193. https://doi.org/10.1016/j.lwt.2022.113193.

Bahrami, Roya, Rezvan Zibaei, Zahra Hashami, Sara Hasanvand, Farhad Garavand, Milad Rouhi, Seid Mahdi Jafari, and Reza Mohammadi. 2022. "Modification and Improvement of Biodegradable Packaging Films by Cold Plasma; a Critical Review." *Critical Reviews in Food Science and Nutrition* 62 (7): 1936–50. https://doi.org/10.1080/10408398.2020.1848790.

Bao, Tao, Xin Hao, Mohammad Rezaul Islam Shishir, Naymul Karim, and Wei Chen. 2021. "Cold Plasma: An Emerging Pretreatment Technology for the Drying of Jujube Slices." *Food Chemistry* 337 (February): 127783. https://doi.org/10.1016/j.foodchem.2020.127783.

Boopathy, Bhavadharini, Anbarasan Rajan, Jaspin Stephen, and Mahendran Radhakrishnan. 2022. "Development and Characterisation of Structurally Reforming Engineered Flat-rice Xerogel for Hot Water Cooking." *International Journal of Food Science & Technology*, October. https://doi.org/10.1111/ijfs.16128.

Boopathy, Bhavadharini, Jaspin Stephen, Anbarasan Rajan, and Mahendran Radhakrishnan. 2021. "Evaluation of Temperature and Concentration on the Development of Rice Hydrogel and 2D Xerogel." *Journal of Food Processing and Preservation*: e15853. https://doi.org/10.1111/jfpp.15853.

Cheeyattil, Subith, Anbarasan Rajan, and Mahendran Radhakrishnan. 2022. "Curcumin-Infused Xerogel-Based Nutraceutical Development and Its 4D Shape-Shifting Behavior." *Journal of Food Science*. https://doi.org/10.1111/1750-3841.16438.

Cheeyattil, Subith, Anbarasan Rajan, Jaspin Stephen, and Mahendran Radhakrishnan. 2022. "Study on the Optimization of Barley Flour Xerogel and Its Programed Oleomorphic 3D Shape-Shifting." *Journal of Food Process Engineering*, October. https://doi.org/10.1111/jfpe.14197.

Chen, Chen, Min Zhang, Chaofan Guo, and Huizhi Chen. 2021. "4D Printing of Lotus Root Powder Gel: Color Change Induced by Microwave." *Innovative Food Science & Emerging Technologies* 68: 102605. https://doi.org/10.1016/j.ifset.2021.102605.

Chen, Xiaohuan, Min Zhang, Xiuxiu Teng, and Arun S. Mujumdar. 2022. "Recent Progress in Modeling 3D/4D Printing of Foods." *Food Engineering Reviews* 14 (1): 120–33. https://doi.org/10.1007/s12393-021-09297-6.

Dong, Shuang, Peng Guo, Yue Chen, Gui yun Chen, Hui Ji, Ye Ran, Shu hong Li, and Ye Chen. 2018. "Surface Modification via Atmospheric Cold Plasma (ACP): Improved Functional Properties and Characterization of Zein Film." *Industrial Crops and Products* 115 (May): 124–33. https://doi.org/10.1016/j.indcrop.2018.01.080.

Gaikwad, P. S., Chayanika Sarma, Anbarasan Rajan, S. Anandakumar, and B. K. Yadav. 2022. "Development of Bacterial Cellulose Film Coated with Mixed Colorimetric Indicator for Tracking the Freshness/Spoilage of Ready-to-Cook Idli Batter." *Coloration Technology* 138 (1): 28–37. https://doi.org/10.1111/cote.12564.

Gupta, Vidhi, T. K. Ranjitha Gracy, Jaspin Stephen, and Mahendran Radhakrishnan. 2020. "Cold Plasma-Assisted Shape-Shifting of a Flat Two-Dimensional Wheat Xerogel and Its Morphological Behavior." *Journal of Food Process Engineering* 43 (9). https://doi.org/10.1111/jfpe.13456.

Gupta, Vidhi, and Mahendran Radhakrishnan. 2019. "Development of Programmed Shape-Shifting Wheat Xerogel Discs and Characterization of Its Hydromorphic Behavior." *Colloids and Surfaces A: Physicochemical and Engineering Aspects* 583 (December): 123928. https://doi.org/10.1016/j.colsurfa.2019.123928.

He, Chang, Min Zhang, and Chaofan Guo. 2020. "4D Printing of Mashed Potato/Purple Sweet Potato Puree with Spontaneous Color Change." *Innovative Food Science and Emerging Technologies* 59 (January): 102250. https://doi.org/10.1016/j.ifset.2019.102250.

Jaspin, S., R. Anbarasan, M. Dharini, and R. Mahendran. 2021. "Structural Analysis of Tapioca Xerogel and Its Water and Oil Triggered Shape Change." *Food Structure* 30 (October): 100226. https://doi.org/10.1016/j.foostr.2021.100226.

Li, Caili, Sushil Dhital, Robert G. Gilbert, and Michael J. Gidley. 2020. "High-Amylose Wheat Starch: Structural Basis for Water Absorption and Pasting Properties." *Carbohydrate Polymers* 245 (October): 116557. https://doi.org/10.1016/j.carbpol.2020.116557.

Liu, Zhenbin, Chang He, Chaofan Guo, Fengying Chen, Bhesh Bhandari, and Min Zhang. 2021. "Dehydration-Triggered Shape Transformation of 4D Printed Edible Gel Structure Affected by Material Property and Heating Mechanism." *Food Hydrocolloids* 115 (June): 106608. https://doi.org/10.1016/j.foodhyd.2021.106608.

Manchun, S., J. Nunthanid, S. Limmatvapirat, and P. Sriamornsak. 2012. "Effect of Ultrasonic Treatment on Physical Properties of Tapioca Starch." *Advanced Materials Research* 506: 294–97. https://doi.org/10.4028/www.scientific.net/AMR.506.294.

Momeni, Farhang, and Jun Ni. 2020. "Laws of 4D Printing." *Engineering* 6 (9): 1035–55. https://doi.org/10.1016/j.eng.2020.01.015.

Musso, Yanina S., Pablo R. Salgado, and Adriana N. Mauri. 2017. "Smart Edible Films Based on Gelatin and Curcumin." *Food Hydrocolloids* 66 (May): 8–15. https://doi.org/10.1016/j.foodhyd.2016.11.007.

Oral, M. O., A. Derossi, R. Caporizzi, and C. Severini. 2021. "Analyzing the Most Promising Innovations in Food Printing. Programmable Food Texture and 4D Foods." *Future Foods* 4 (December): 100093. https://doi.org/10.1016/j.fufo.2021.100093.

Rajan, Anbarasan, Bhavadharini Boopathy, Mahendran Radhakrishnan, Lakshminarayana Rao, Oliver K. Schlüter, and Brijesh K. Tiwari. 2023. "Plasma Processing: A Sustainable Technology in Agri-Food Processing." *Sustainable Food Technology* 1 (1): 9–49. https://doi.org/10.1039/D2FB00014H.

Ratish Ramanan, K., and R. Mahendran. 2021. "Morphogenesis and Characterization of Wheat Xerogel Structure and Insights into Its 4D Transformation." *Food Structure* 28 (December 2020): 100170. https://doi.org/10.1016/j.foostr.2020.100170.

Stephen, Jaspin, Dharini Manoharan, Bhavadharini Boopathy, Anbarasan Rajan, and Mahendran Radhakrishnan. 2021. "Investigation of Hydrogel Temperature and Concentration on Tapioca Xerogel Formation." *Journal of Food Process Engineering* 44 (11): e13833. https://doi.org/10.1111/jfpe.13833.

Stephen, Jaspin, Dharini Manoharan, and Mahendran Radhakrishnan. 2021. "Corn Morphlour Hydrogel to Xerogel Formation and Its Oleomorphic Shape-Shifting." *Journal of Food Engineering* 292 (September 2020): 110360. https://doi.org/10.1016/j.jfoodeng.2020.110360.

Teng, Xiuxiu, Min Zhang, and Arun S Mujumdar. 2021. "4D Printing: Recent Advances and Proposals in the Food Sector." *Trends in Food Science & Technology* 110 (April): 349–63. https://doi.org/10.1016/j.tifs.2021.01.076.

Thirumdas, Rohit, A. Trimukhe, R. R. Deshmukh, and U. S. Annapure. 2017. "Functional and Rheological Properties of Cold Plasma Treated Rice Starch." *Carbohydrate Polymers* 157 (February): 1723–31. https://doi.org/10.1016/j.carbpol.2016.11.050.

Wang, Wen, Lining Yao, Teng Zhang, Chin Yi Cheng, Daniel Levine, and Hiroshi Ishii. 2017. "Transformative Appetite: Shape-Changing Food Transforms from 2D to 3D by Water Interaction through Cooking." In *Conference on Human Factors in Computing Systems – Proceedings*, 2017-May:6123–32. New York, NY, USA: ACM. https://doi.org/10.1145/3025453.3026019.

Zhang, Xiao Lin, Chong Shan Zhong, Arun S. Mujumdar, Xu Hai Yang, Li Zhen Deng, Jun Wang, and Hong Wei Xiao. 2019. "Cold Plasma Pretreatment Enhances Drying Kinetics and Quality Attributes of Chili Pepper (*Capsicum Annuum L.*)." *Journal of Food Engineering* 241 (January): 51–57. https://doi.org/10.1016/j.jfoodeng.2018.08.002.

5 Processed Water

Types, Generation, and Its Applications on Food Preservation

M. Dharini, V. Monica, R. B. Ramyaa, and R. Mahendran

5.1 INTRODUCTION

Globally, a major challenge faced in the area of perishable food processing industry is the maintenance of safety and food wastage. Perishables like fruits, vegetables, and meat contain mostly 70% of water, which makes them more susceptible to microbial contaminants. Wastage of these perishables takes place in each step of the supply chain (from farm to fork) (Kumar et al., 2020). One key area at which the major contamination starts is the initial period of processing and it further gets complicated while transportation, handling, and storage (Bilek & Turantaş, 2013). The presence of dirt and soil in case of fruits and vegetables and extraneous matters like skin, feather, and blood in meat and seafood products forms a forerunner for the spoilage of perishables. Washing is the first and foremost step in almost all perishable processing in food industry (Acoglu & Omeroglu, 2021). The primary aim of washing is to remove external contaminants on the surface of the food material. The secondary aim is to reduce the microbial load on the external surface of the food material such that it can aid to maintaining the quality of the food till it reaches the next stage of processing. Apart from dirt removal and microbial reduction, washing can also aid in reducing the pesticide and harmful chemicals on the food surface in the case of fruits and vegetables.

5.2 NEED FOR PROCESSED WATER

Plain water washing presents a significant advantage of contacting all the external nooks and corners of food material irrespective of the shape and size; water was also reported to be the most crucial and common first step that can be done on large-scale (Esua et al., 2021). Water forms a significant part in food science and technology in terms of its domination in maintaining taste, flowability, and preservation (Shirahata et al., 2012). Water washing can remove the dirt and extraneous matter from the surface of the food material and can avoid any further microbial contamination, nevertheless, water washing alone cannot eliminate the existing microbial load and chemicals on the food surface (Monteiro et al., 2021). These perishables could have already been contaminated and are highly liable to spoilage during transport, storage, and processing (Haji et al., 2020). Apart from which, any contamination (pollution) or presence of biological substance in water itself could affect the end quality of the product. Thus, water must be processed to remove extraneous matter before used for processing (Shirahata et al., 2012).

Further, owing to the importance of raw fruits and vegetables consumption which can prevent chronic illnesses, there is an increasing trend to consume fruits and vegetables as salads (fresh or minimally processed); however, it also poses a major threat that challenges the food scientists on the safety of the product (Ali et al., 2018; Deng et al., 2020). Rather than providing health beneficial nutrients, the consumption of raw food materials also increases its vulnerability of causing food borne illness in individuals (Horvitz & Cantalejo, 2014). Additionally, for the production of salads, the fruits and vegetables must be minimally processed; this could also increase the deterioration through cross contamination (Botondi et al., 2021). Henceforth, to facilitate the purification of water

 DOI: 10.1201/9781003359302-5

and to maintain the safety of fresh foods and minimally processed foods, new technologies have been used in the industries. Instead of using chemical methods of preservation, the plain water can be activated to increase the quality of the food. This activation process not only ensures the purity of water, rather it also aids in maintaining the safety of commodities during application. Such that in the past decade, water has been activated using various chemical and non-thermal methods to wash the fresh and minimally processed produces (Alvaro et al., 2009; Esua et al., 2021). The major advantage of using processed water is that it only acts on the surface of the food material (physical treatment); thus, it will only have minimal effect of the nutritional quality of food (Bilek & Turantaş, 2013). The objective of this book chapter is to scientifically broaden the knowledge on recent technologies that have minimal effect on the nutrition which can extend the shelf life and quality of the food material (Botondi et al., 2021).

5.3 PROCESSED WATER

Activation or functionalization of water presents various functions. Water is an inorganic compound which consists of two hydrogen and oxygen atom bounded covalently to each other. Activation of water can be done through splitting up the water molecule by supplying energy to it; a potential carrier for producing non-conventional chemical reaction. Once activated, the physicochemical and oxidation-reduction (OR) potential of water increases which aids in increasing the biological activity (Aider et al., 2012). The water can be activated through different ways of processing; following are some of the most important methods used for producing processed water.

5.3.1 CHLORINATED WATER

Chlorinated water treatment is the simple and most widely used commercial washing method in industries for decontamination purposes; for more than a decade, chlorinated water has been used for the disinfestation of commodities (Betts & Everis, 2005). At present, 75% of the fresh and minimally processed food industry uses chlorine as a disinfectant for microbial (pathogenic and spoilage) reduction in foods. The easy availability and the least cost for generation of chlorinated water along with its promising effect on reducing contagious diseases caused by some enteric virus and vegetative bacteria makes it a majorly used processed water (Seo et al., 2019). Sodium hypochlorite is one widely available form of chlorine which when dissolves in water, forms hypochlorite ion and hypochlorous acid (Marín et al., 2020). The formed compounds can alter the pH of water; that is, increase in hypochlorite ion increases pH above 7, whereas the hypochlorous acid decreases the pH below 7. However, the ability of hypochlorous acid has been reported to be 80% more effective than the hypochlorite ion. The mechanism behind the inactivation of microorganisms using hypochlorous acid is that it restricts the oxidation of glucose in enzymes and inhibits the metabolism of carbohydrates.

The efficacy of chlorinated water depends on intrinsic parameters such as concentration, pH, exposure time, and temperature and extrinsic parameters such as the nature of the product, presence of organics in food, and the initial microbial load on the product (Luu, 2019). In fresh cut fruits and vegetables industry, the maximum concentration of chlorine used was reported to be 50 to 200 ppm for a maximum exposure time of 5 min; the exposure time longer than 5 min didn't have any increase in the efficiency of processing. Similarly, another study proved that reduced chlorine concentration of 400 and 800 ppm was effective in reducing the inoculated strains of Shiga toxin–producing *Escherichia coli* and two strains of *Salmonella* in wheat flour. They also reported that chorine water at this concentration didn't had any effect on flour; the quantities above 800 ppm reduced the quality of flour (Lin et al., 2022). Another study on iceberg lettuce and strawberries also proved that low concentration of chlorine (10 mg L^{-1}) is sufficient for attaining microbial reduction and the higher concentrations didn't have any significant increase in microbial reduction (Chen & Hung, 2018). A study on chlorinated water treatment on mung bean sprouts was found to reduce

E. coli, Salmonella spp., and *Listeria monocytogenes* by 3 to 7 log reduction (Iacumin & Comi, 2019). 10 to 20 ppm chlorine was recommended in a study to avoid cross contamination of *E. coli O157:H7* in the processing of fresh-cut romaine lettuce; nevertheless, 30 ppm chlorine was found necessary for large scale processing. Further the study also disclosed that ORP of more than 650 mV increased the cross contamination; thus, the study recommended not crossing 650 mV ORP for reducing the contamination (Fu et al., 2018). Though chlorinated water reduces the microbial load immediately after processing, the microbial load can further increase during storage (Gu et al., 2018).

Apart from dosage and exposure time, the presence of organic matter reduces the total available chlorine in the water; thus, constantly supplementation of chlorine in the water is necessary to maintain its efficiency (Fu et al., 2018). The reduction in chlorine also alters the pH of water. Further, the formation of carcinogenic by-products in chlorinated water is posing a major problem in food processing which questions its beneficial effect on maintaining safety (Ali et al., 2018; Wang et al., 2019). When chlorinated water reacts with organic compounds in water and food, it leads to the generation of harmful by-products such as trihalomethanes and haloacetic acids (Praeger et al., 2016); their regular consumption has been reported to cause adverse effects on human health (Mazhar et al., 2020). In the case of using chlorinated water for irrigation, it has been reported that the remaining chlorine in water can also be toxic for the growth of plants; it ultimately proved that the processed water used for disinfestation when discharged into the environment can affect the natural plant growth (Lonigro et al., 2017). As the efficiency of microbial decontamination of any washing technology depends mainly on the susceptibility of microorganisms, residues formed during processing, quality (sensorial, textural, and color properties), and composition of food material (Bhilwadikar et al., 2019), the need for producing alternative technologies (that is eco-friendly and non-chemical) increased.

5.3.2 ELECTROLYZED WATER (EW)

Electrolyzation is a process of passing electricity into the electrolytic cell separated as anode and cathode by an ion exchange diaphragm (Lee et al., 2006). When electricity is passed through the electrode, oxidized water is generated near the cathode side and reduced water near the anode side. In general, bactericidal activity can be reduced by adding electrolytes such as NaCl before beginning the electrolysis process (Esua et al., 2021). Thus, when electrolyzed with the addition of salt, negatively charged Cl^- from salt and OH^- from water get attracted toward the anode, whereas the positively charged particles move toward the cathode. At the anode, Cl^- and OH^- combine to form hypochlorous acid, chlorine gas, and hydrochloric acid; thereby producing reduced (acidic) EW. Contrarily, at the cathode, positive ions such as Na^+ and H^+ ions gain electrons to form sodium hydroxide and hydrogen gas, which in turn leads to the generation of oxidized (alkaline or basic) EW. pH of both the type of EW differs based on the acidic and basic nature. Apart from these two EWs, there are neutral and slightly acidic EW which are produced similar to acidic EW; however, in this method of generation, the separating membrane is not provided between the electrodes (Iram et al., 2021; Xuan & Ling, 2019). It is the simple method of processing EW; in this type, the water generated has higher pH, low concentration of chlorine, and less OR potential (Wang et al., 2019). In general, the most used are the acidic EW and slightly acidic EW. Acidic EW has low pH, more OR potential, and increased chlorine concentration, whereas in slightly acidic EW the concentration of chlorine is less. Alkaline EW has a soapy (detergent like) nature and the change in pH and OR potential contributes to the mechanism of action (Xuan & Ling, 2019). Figure 5.1 represents the pH, OR potential, and chlorine concentrations of all the types of EW.

The mechanism of microbial reduction using EW is accredited to the compounds and reactive species produced during electrolysis, which in turn causes oxidative damage to the microbial cell wall. The combination of change in pH and OR potential synergistically acts on the microbial load and affects the homeostasis. Results show that the OR potential affects the intracellular enzymes present in the microbial cell, which can alter the electron flow in the cells (Xuan & Ling, 2019).

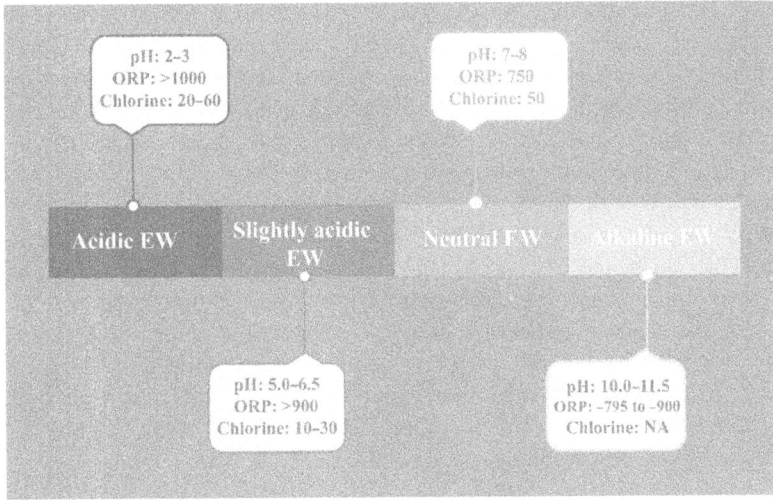

FIGURE 5.1 pH, oxidative reduction potential (ORP), and chlorine concentration in various EWs.

EW has already approved in countries like the US, Korea, and Japan as an effective food additive to maintain the safety of the food material (Xinyu et al., 2020). EW has been observed to have a more significant effect on microbial cells compared to the original chlorinated water (Possas et al., 2020). Similar to chlorinated water, the parameters associated with the efficiency of EW were assessed based on parameters such as pH, OR potential, exposure time, surface texture, and the organic compounds present in water and product. Along with the presence of organic matter in water, EW has also been reported to cause erosion of the container used for processing, which forms as a limiting factor for cleaning applications (Yan et al., 2021). It has also been reported that the EW must be generated at the time of application; the storage of water might reduce the viability of the water. However, one particular study proved that storage of EW in closed containers at a reduced temperature of 4°C might aid in preserving the viability (Wang et al., 2019). The study also unraveled that the residual chlorine in EW was less when applied on meat products in comparison with the chlorinated water.

Application of EW in food processing includes microbial reduction, pesticide reduction, and inhibition of enzymatic browning (Rahman et al., 2016). The microbial load in mango cubes dipped in neutral EW was found to significantly reduce the microbial load similar to commercial treatment; the chlorine off-flavor, which is dominant in the case of chlorinated water, was also less in the usage of EW (Almeida et al., 2021). EW has also been found effective in reducing the microbial load on food contact surfaces. A study on application of neutral and slightly acidic EW on stainless steel coupons proved that both the EWs significantly reduced the microbial load; however, slightly acidic EW was unraveled to be better compared to the neutral EW (Possas et al., 2020). The reduced pH and high OR potential could have acted synergistically on the microbial load. In another study, slightly acidic water was used for washing the pomfrets and the slightly acidic EW ice was used for storage of the fish. The study proved that the combining EW and ice for safe storage was efficient in reducing the microbial load and maintaining quality compared to tap water storage (Huang et al., 2021). Slightly acidic EW didn't have any effect on water absorption and germination of alfalfa seeds; however, it reduced the *Enterobacteriaceae* level in the seeds (Zhang et al., 2020).

Strong acidic EW treatment on sweet potato slices proved to be effective in significantly reducing the browning index and also the polyphenol oxidase enzyme in it (Liu et al., 2021). EW can also be used for pesticide reduction on fruits and vegetables. The acidic and alkaline EW were studied for their effect on pesticide and fungicide reduction; acidic EW reduced pyrethroid and organophosphates, whereas alkaline water was found more effective for fungicide reduction

(Liu et al., 2021). The mechanism behind the pesticide reduction by EW is owing to their chemical and physical properties. The breakage of double bonds in pesticides due to the pH and OR potential of EW was reported to be one of the reasons for pesticide reduction (Wang & Han, 2019). EW can also be combined with other methods to improve their capability. The combination of EW with chemicals such as calcium oxide has been found to increase the microbial reduction compared to single treatment and the point to be noted is that this combination method is commercially used in fresh fruit industry (Chen et al., 2019).

Though the EW also contains hypochlorous acid, which is an active compound in chlorinated water, the concentration of the chlorine compounds present in EW is less. Once these chlorine compounds react with organic matter, they will get converted into ordinary water; thus, unlike chlorinated water, the negative effect of EW is less compared to chlorinated water (Luan et al., 2021). However, the EW also comes with a few demerits; corrosion of processing equipment and the leakage of chlorine gas that could cause irritation of eyes at the site of production are the major disadvantages of EW (Rahman et al., 2016; Xuan & Ling, 2019). The compilation of the effect of different EW on the food preservation has been drafted in Table 5.1.

TABLE 5.1
Effect of Different EW on the Preservation of Food Commodities

Commodity	Process Parameters	Research Finding	Reference
Acidic EW			
Sunflower and roselle seeds	Low chlorine concentration – 79 mg/L, pH – 9.37 and 650 mV High chlorine concentration – 1692 mg/L, pH – 12.11 and 460 mV Exposure time – 5, 10, and 15 min	2.47 and 3.20 log CFU/g of *Salmonella* in sunflower seeds at both concentrations at 15 min exposure 1.9 and 2.19 log CFU/g of *Salmonella* in roselle seeds at both concentrations at 15 min exposure There was no change in the germination percentages	(Inpitak & Udompijitkul, 2022)
Food contact surface	Three types of acidic EW prepared at three NaCl concentrations 0.1, 1, and 3 g/L which produced chlorine concentration of 10.53, 44.74, and 134 mg/L, respectively	Almost all biofilms were eliminated at chlorine concentration of 134 mg/L on glass, stainless steel, and polystyrene	(Yan & Xie, 2021)
Fresh fuyan longans	pH – 2.5 Chlorine concentration – 80 mg/L Exposure time – 10 min	Firmness of the fruit was restored during storage. Reduced damage to the cellulose and hemicellulose was observed in EW treated fruits	(Sun et al., 2022)
Alkaline EW			
Fresh nectarine	Chlorine concentration – 200 mg/L pH – 6.7	The surface decay of the fruit for less compared to chlorinated water. The treatment maintained firmness, total soluble solids, and color of the fruit; however, the titratable acidity decreased slightly	(Belay et al., 2021)
Wheat flour	pH – 8.5, 9.5, 10.5, 11.5, and 12.5	Maximum reduction of deoxynivalenol and Fusarium species in wheat flour was found at 9.5 pH. The treatment didn't affect the quality and functional properties of the flour	(Lyu et al., 2018)
Granny smith apple	Chlorine concentration – 200 and 300 mg/L Treatment time – 10 and 15 min	1.9 log reduction in total aerobic mesophiles and the treatment also maintained the physicochemical properties of the apple	(Nyamende et al., 2022)

(Continued)

TABLE 5.1 (*Continued*)
Effect of Different EW on the Preservation of Food Commodities

Commodity	Process Parameters	Research Finding	Reference
Slightly Acidic EW			
Fresh cut cauliflower	pH – 5.5 to 5.6 chlorine content – 20 to 22 mg/L OR potential – 900 to 950 mV	2.21 and 1.17 log reduction in total bacterial load and yeast and mold growth, respectively NO detrimental reduction in quality Only the total flavonoid content reduced significantly after treatment	(Akther et al., 2022)
Shrimp and crab	pH – 5.0 to 6.5 2% HOCl Exposure time – 1, 3, 5, 10, and 15 min	Microbial reduction increased with increase in exposure time. Maximum of 2.59 and 2.32 log reduction in *Vibrio parahaemolyticus* in shrimp and crab, respectively	(Roy et al., 2021)
Cabbage rot	Concentration – 5, 15, and 25 ppm Temperature – 10, 20, and 30°C Time – 60, 120, and 180 s	5.94 log reduction in *Pectobacterium carotovorum subsp. Carotovorum* was observed at 25 ppm chlorine, 10°C and 180 s exposure time	(Song et al., 2021)
Solid waste storage unit	pH – 5.0–6.5 Available chlorine concentration – 50 to 100 mg/L	Slightly acidic EW reduced the bacterial and fungi load by 358 and 378 colony forming units, respectively. However, the volatile compounds were also reduced by 21.4 to 88.3% in EW treated solid waste storage facility	(Guo et al., 2021)
Alkaline EW			
Food contact surfaces	pH – 12.2 to 12.5 OR potential – –40/–90 mV	A total of 2.64 log reduction in *Salmonella spp.* 3.29 log reduction of *E. coli* 2.41 log reduction of *Listeria spp.* and 1.16 log reduction of *Staphylococcus aureus*	(Tomasello et al., 2021)
Granny smith apple	Chlorine concentration – 200 and 300 mg/L Dipping duration – 10 and 15 min	2 and less than 1 log reduction of aerobic mesophilic bacteria and yeasts and molds, respectively	(Tomasello et al., 2021)
Neutral EW			
Stainless steel	Tap water-based water pH – 7 and 8 OR potential 800 and 900 mV Chlorine concentration – 2.5 and 3.5 mg/L Treatment time – 20, 40, and 60 min	0.47, 0.36, and 0.40 log reduction in *E. coli*, *Salmonella typhimurium*, and *Listeria monocytogenes*, respectively, at 60 min treatment time. The microbial reduction increased with increase in exposure time	(Jee & Ha, 2021)
Fresh spinach leaves	50 and 85 ppm of chlorine in the activated water pH – 7.0 to 7.4 OR potential – 850 to 870 mV	Microbial reduction was more compared to tap water washing and control	(Ogunniyi et al., 2021)
Poultry meat	Chlorine concentration – 100 and 200 μg/mL Exposure time – 10 min	The microbial reduction was more in the case of 200 μg/mL concentration. The microbial load was also at bay during storage days (4 days)	(Moghassem Hamidi et al., 2021)

5.3.3 Ozonized Water (OW)

Ozone (O$_3$) is a triatomic molecule of atomic oxygen, also known as the allotropic form of oxygen (O$_2$), which is a powerful oxidizing agent with high reactivity and ability to spontaneously disintegrate into oxygen molecules (Premjit et al., 2022). Ozone is highly detectable at 0.01–0.05 ppm level. The USFDA approved ozone as "Generally Recognized as Safe (GRAS)" in 2001 as a disinfectant or sanitizer in the food and food processing sector (Dos Santos et al., 2021). OW is a promising technology for antifungal, antiviral, antibacterial, and antiparasitic treatments (Pandiselvam et al., 2019). It is generally formed by splitting up of oxygen molecule into atoms and combining with oxygen (O$_2$) molecules [(O) + (O$_2$) → (O$_3$)]. Ozone is commercially produced by electric discharge (Corona discharge), photochemical (UV light), thermal, electrolytic, chemical, and chemonuclear processes (Yang et al., 2013). OW is generated using ozone generators based on the Corona Discharge Method, where ozone produced is discharged into water. High-purity O$_2$ cylinders are fixed to produce ozone and no other additional chemicals are used (Premjit et al., 2022).

The application of OW includes germination/plant growth, natural treatment of water, surface disinfection, microbial reduction, quality maintenance, and pesticide reduction in fruits and vegetable processing, meat and poultry, and beverage processing (Patil et al., 2010) (Figure 5.2). OW has the potential to destroy bacteria, viruses, fungi, parasites, and other microorganisms. This is due to the OR potential of ozone (2.08 eV), which causes cell lysis and damage to nucleic acids. When water combines with ozone, it also initiates the production of reactive species and the ozone disintegrates when the production of these species stops. Factors such as pH, exposure time, ozone concentration, half-life, temperature, and flow rate are responsible for the efficient action of OW. To increase the efficiency of Ozone treatment; pH of solution is lowered, foods are exposed at low ozone concentrations, and disinfection treatment is done at low temperatures (as high temperatures accelerate degradation of ozone) (Premjit et al., 2022). Ozone decomposition occurs along with the formation of free radical species such as superoxide, hydroperoxyl, and hydroxyl as by-products (da Silva et al., 2015). Protein oxidation, breaking of the peptide bond, protein cross-linking, and other amino acid chain reactions are influenced by ozone concentration. The polyunsaturated fatty acids are oxidized to acid peroxides by ozone. These changes lead to the leakage of cellular compounds, and finally, lysis occurs (BiaŁoszewski et al., 2010).

Apart from disinfection activity, ozone is used as a preservative in many food industries. It reduces the microorganisms and extends the shelf life of foods and food products. Table 5.2 shows the application of OW and its subsequent changes in food commodities (Premjit et al., 2022). Wu et al. (2007) studied the effect of ozonated water on the removal of four pesticides in water and on the surface of the vegetable – *Brassica rapa* (Pak Choi). It was observed that 1.4 mg/L and 2 mg/L

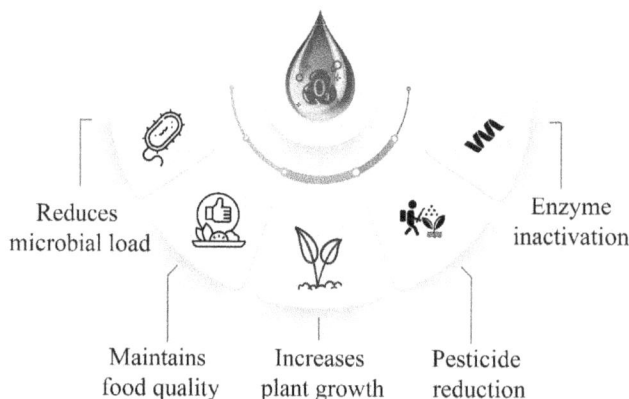

FIGURE 5.2   Applications of ozonized water in food preservation.

TABLE 5.2

Various Effects of Ozonized Water and Plasma Activated Water on the Microbial Reduction on Food Commodities

Food Commodity	Ozone Treatment	Treatment	Observations	References
Fresh cut cabbage	1.4 mg/L for 1, 5, and 10 min	OW was applied on fresh cut cabbage and stored at 4°C.	Inhibition of coliforms, aerobic bacteria, and yeasts at 10 min treatment time.	(Liu et al., 2021)
Winter-drawn Strawberry fruits	0.1 mg/L for 2 min	Physical, biochemical, fruit decay, and shelf life study for 14 days by dipping the strawberries for 2 min.	Ambient temperature storage – 21% weight loss, firmness is increased by 16%, and 15% change in fruit color. Lower temperature storage – 21% weight loss, firmness increased by 19%, and 46% lesser change in fruit color.	(Nayak et al., 2020)
Chicken drumsticks	8 mg/L	Chicken drumsticks were soaked and sprayed by OW or OW–LA. The microbial load was determined.	Six sequential soaking and seven spraying cycles with OW lowered Salmonella contamination on the skin and inner layers and light changes in skin color.	(Megahed et al., 2020)
Pacific white shrimp (*Litopenaeus vannamai*)	OW (1 mg/L for 10 min) with MAP at 100% CO_2	OW with MAP was done to study the shelf life of Pacific white shrimp (*Litopenaeus vannamai*) in cold storage at 4°C.	The shrimp were kept for 24 days with good sensory attributes, better physiochemical properties, and a lower melanosis index.	(Gonçalves & Lira Santos, 2019)

PAW

Food Product	PAW Source	Treatment Condition	Physiochemical Characteristics	Reference
Chinese bayberries	Low-electrode dielectric barrier discharge	Immersing in 1600 mL of PAW for 0.5 minutes	Reduced fruit decay by 50% Firmness remained unchanged (2.81 N) Color index increased from 8.2 to 12.2 TSS reduced to 11.45% 1.1 log CFU/g	(Ma et al., 2016)
Mung bean sprouts	Gliding arc discharge in air	Immersing in PAW 30 for 30 minutes	Log reduction of Total aerobic bacteria: 2.32 Total yeast and fungi: 2.84	(Xiang et al., 2019)
Fresh cut kiwi fruits	Atmospheric pressure cold air microplasma array	Spraying 1 mL of PAW	After 8 days Firmness ≈190 g Color value (b* value) ≈11.75 After 8 days 1.8 log reduction of *S. aureus*	(Zhao et al., 2019)
Tomatoes	Air plasma jet	Immersing in PAW for 15 minutes	Color value: 2.77 ± 1.34 Chlorothalonil and thiram: 85.3% and 79.47%	(Ali et al., 2021)

* is provided to indicate the b value representing CIELAB color scale

of ozone for 30 mins reduced the levels of diazinon, parathion, methyl parathion, and cypermethrin with the generation of some by-products such as paraoxon, diazaxon, oxons, methyl paraoxon, etc. Temperature also influences its activity, where treatment done at 14°C was better than treatment done at 24°C. In the preservation of fish, 2 ppm of OW and its ice flakes were used to analyze the freshness of hake (*Merluccius merluccius*). The fish samples were washed with OW and stored in cold storage with ice flakes made from OW at 2–3°C.

This result was proven by Lu et al. (2012), in which the combination of washing with ozone water and storing it with ozonized flake ice was performed on Japanese sea bass (*Lateolabrax japonicas*). The combination was the best in maintaining the sensory, microbiological, and chemical quality of fish for more than 12 days. Apples washed with ozone water showed a higher reduction of captan residues and a slight reduction of boscalid and pyraclostrobin residues. This method was effective because along with the reduction of residues it prevented further contamination due to the degeneration of active substances by ozone (Sadło et al., 2017). The application of OW (4 ppm for 1 or 2 mins) on the epidermis of postharvest Golden papaya fruit did not give any changes. The wax arrangement of the fruit could be the reason for the less reaction of ozone (Kechinski et al., 2012). Whereas in the case of the application of OW in controlling Anthracnose in papaya showed promising results. Ozone applied at 0.8 mg/L for 160 min and storing the papayas at 25°C reduced the severity of Anthracnose, minimized reduction in firmness, and lowered respiration rate (da Costa et al., 2021). Aflatoxin (AFB$_1$) content and its toxicity in corn were diminished by 90 mg/L of ozone. This was analyzed with the effect of ozone concentration, treatment time, and moisture content of corn. As the ozone concentration and treatment time were increased, the reduction was more; but the higher moisture content of corn affected the rate of degradation of aflatoxin due to formation of OH and other radicals (Luo et al., 2014).

OW in the preconditioning of corn seeds was performed. Hydropriming of corn seeds with aqueous ozone (AO) at different concentrations and varied periods of preconditioning showed that an ozone concentration of 20 mg/L and preconditioning time of 60 mins improved the physiological quality of corn seeds at the best. The cellular solutes which are responsible for the aging of corn seeds were reduced and a significant increase in plant height was observed (Monteiro et al., 2021). Ozone, in combination with sodium bicarbonate (SBC), was used for studying its synergistic effect of microbial load. The effect produced by a combination of both these substances on citrus *Clementina spp.* (stored clementine) (Strano et al., 2021) showed that there was a reduction in sour rots; reduction of total viable count by more than 1 log unit, reduced decay incidence were observed also the sensory parameters were not compromised compared to the control. Ozone applied at 2 mg/L along with organic manure enhanced tomato growth with good quality tomatoes (Tahamolkonan et al., 2022).

In liquid foods, the application of OW needs more study as the properties of the liquids can affect the efficiency of ozone treatment. Tender coconut water (TCW) treated by ozone and ultrasound treatment with Nisin showed that the browning index of ozone treated TCW is significantly less than ultrasound treatment stating that polyphenol oxidase and peroxidase enzymes were inactivated by ozone. Ozone treatment with Nisin extended shelf life for three weeks, good sensory attributes, and less effect on nutritional and physiochemical properties of TCW (Rajashri et al., 2019). The studies of Patil et al. (2010) show that with the treatment of 0.048 mg of OW at a flow rate of 0.12 L/min on cloudy apple juice for 5 mins, a 5-log reduction of *E.coli* was observed. Whereas other parameters such as phenolic content, color, and total phenolic content of treated apple juice are reduced. Therefore, a careful study of the properties of liquid foods with ozone is required before the treatment to prevent loss of physiological parameters.

Ozone gas at low concentrations is highly toxic, i.e., O$_3$ concentration at 0.2 ppm level and higher can cause damage to the pulmonary system, and the respiratory tract also may be fatal to humans. However, AO or OW is very suitable for surface treatment due to its nature of not leaving toxic by-products and residues. Even at very low concentrations of ozone, the sanitizing or oxidative effect is more and is an efficient process (Strano et al., 2021). Since it is a surface treatment, the alteration of internal properties of various commodities is limited. Therefore, use of OW enhances the properties of food and increases shelf life of food.

5.3.4 Plasma Activated Water

Plasma-activated water (PAW) is a novel technology that can be referred to as an extended application of cold plasma in food and agriculture. It is a promising technology for producing safer and higher-quality food products and increasing agricultural production (Guo et al., 2021). Specifically, the potent microbicidal properties of PAW in the inactivation of microbiota, preservation of biochemical properties of food during storage, germination of seeds, and the reduction of pesticide residues make it an exceptional technology in food processing (Ali et al., 2021).

PAW generation is associated with the interaction of gas phase plasma with water either in direct or indirect mode of application (Wang & Salvi, 2021). The reactive species in PAW induce oxidative stress and collapse the composition of the microbial cell by damaging their cell wall, proteins, lipids, and other internal organelles, ultimately leading to cell death. Gram-negative bacteria are more susceptible to death easily than gram-positive bacteria by PAW treatment (Mai-Prochnow et al., 2021). In the case of the virus, the generated plasma species damage their nucleic acids, DNA, RNA, and proteins leading to the death of the pathogenic viral particles (Xiang et al., 2018). The reactive species in gas phase plasma comprises of N_2, and O_2, H, OH, N, and NO, and the secondary species that can be generated in the liquid phase that have more shelf life are H_2O_2, O_3, HNO_2, and HNO_3 (Figure 5.3) (Zhou et al., 2020). Among these, reactive nitrogen species show enhanced

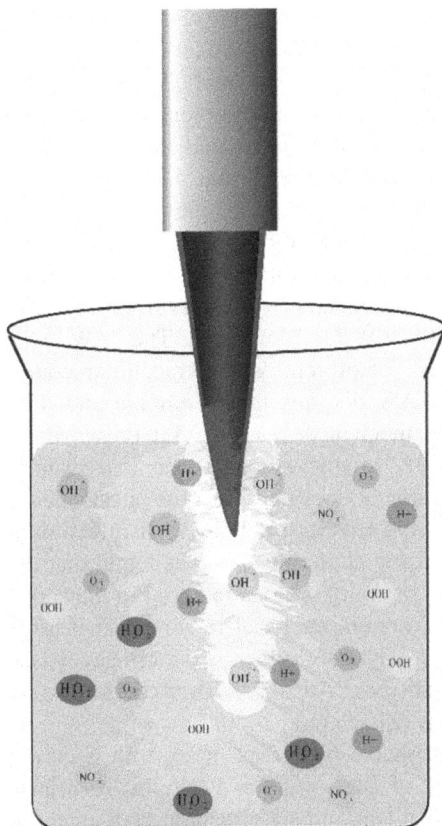

FIGURE 5.3   Plasma activated water generation.

antimicrobial action by penetrating into the cell wall and causing cell apoptosis. Atmospheric corona discharge, gliding arc discharge, dielectric barrier discharge, plasma jet, and surface micro discharge are some common plasma sources. PAW can be generated in various means such as direct discharge generation within liquids by supplying air bubbles, discharge of gas phase plasma jet over the liquid surface, or discharge of gas phase into the liquid in the form of bubbles (Dharini et al., 2023). The PAW generation is influenced by the following factors: voltage, generation mode, the gap between the plasma plume and liquid, the nature of the electrode, and the chemical environment (Al-Sharify et al., 2020). The physicochemical changes of generated PAW include electrical conductivity, OR potential, pH, and metal ions (Rahman et al., 2022).

PAW generation time and treatment time influence the disinfection and preservation of fruits and vegetables. For example, PAW generated in 10 and 20 minutes were treated for 10 and 15 minutes, respectively. It was found that PAW generated in 20 minutes had high ORP, and conductivity which in turn significantly reduced *S. aureus* in strawberries during treatment of 15 minutes. And biochemical characteristics such as color, TSS, and firmness of the fruit were also maintained by this treatment even at the end of 4 days (Ma et al., 2015). Similarly, PAW with high ORP and low pH reduced the Chinese bayberry's decay by 50% and maintained the TSS and firmness of berries to a greater extent even after 8 days of storage (Ma et al., 2016). PAW action on highly nutritious mung bean sprouts maintained its phenolic and flavonoid contents with no effect on its antioxidant potential. The reactive species produced during PAW generation, such as H_2O_2, nitrites, and nitrates, are correlated to disinfection efficiency by reducing the total aerobic bacteria, yeasts, and fungi load during refrigerator storage of 6 days (Xiang et al., 2019). PAW was found to have a promising preservation effect on fresh-cut fruits such as apples and kiwi fruits during storage periods of 12 days and 8 days, respectively (Liu et al., 2020; Zhao et al., 2019).

The presence of minimal quantity of carcinogenic and mutagenic pesticide and fungicide residues in fruits and vegetables can cause harmful effects to humans (Reeves et al., 2019). PAW treatment greatly reduced chlorpyrifos, carbaryl, and chlorothalonil, thiram residues in grapes, strawberries, and tomatoes, respectively. High OR potential and acidic environment of PAW and metastable reactive species contribute to pesticide and fungicide dissipation in fruits (Ali et al., 2021; Sarangapani et al., 2020). Therefore, PAW has a significant role in reducing postharvest losses by inactivating causative microorganisms and preserving the physiochemical properties of fruits and vegetables. The antibiofilm and antibacterial activities exhibited by PAW during preservation of food were studied by analyzing the membrane damage mechanism and sterilization ability. Herein, instead of meat sample, authors Xiang et al. (2018) treated isolated *Pseudomonas deceptionensis CM2* strains with PAW for 10 minutes. The PAW exposure time inactivated the microbe and induced the leakage of proteins and intracellular components through the holes created on cell surface. Thus, PAW disrupts the membrane and initiates the cellular components leakage. The above-studied mechanism was useful in preserving chicken breast using PAW. *Pseudomonas deceptionensis CM2* in chicken breast was inactivated with 12 minutes of PAW immersion. Additionally, the treatment does not affect the physiochemical and sensory attributes of the meat. However, the safety and efficacy of the PAW disinfection have to be studied for wide application in food processing (Kang et al., 2019). It was found that PAW inactivation efficiency is inversely correlated to concentration of organic material in the sample subjected to treatment. This may be due to presence of organic material in sample that protects the bacterial strains. Further, the phase of the sample may also influence the efficiency, such that the liquid organic material may affect the reactive species diffusion, whereas the solid particles do not have the ability to affect the diffusion process (Zhao et al., 2020).

The highly nutritious and perishable byproduct of soymilk is carrier of food spoiling microorganisms. However, immersion of bean curd in PAW for 30 minutes reduced the microbial load and increased the physiochemical and sensory properties of the plant-based product. PAW immersion time is directly related to microbial inactivation efficiency such that PAW activated for 90 seconds and immersion time of 30 minutes resulted in greater reduction of bacterial, yeast, and mold colonies. Color and textural properties contributing to sensory attributes were not significantly affected by the treatment. Additionally, it was found that PAW treatment did not destroy the functional compound of bean curd as no difference in the isoflavone content was observed with control and treated samples (Zhai et al., 2019). Similarly, immersing tofu in PAW for 15 minutes reduced microbial load without affecting the quality of tofu during storage for 28 days in refrigeration. About 80% of polyphenols in tofu was retained after treatment, in addition to minimal effect on color, texture, and water holding capacity (Frías et al., 2020). The effect of PAW on different applications on various food products is provided in Table 5.2.

Thus, the defensive mechanism of PAW can be used to inactivate pathogenic bacterial, fungal, and other viral species residing in food and preserve the biochemical properties of foods like firmness, color, texture, and others simultaneously. Indeed, high initial cost, high energy consumption during PAW generation, and critical control of process parameters are some challenges encountered during PAW treatment in food industry (Patra et al., 2022). More investigations on synergistic effect of PAW with novel technologies like ultrasound, high pressure processing would be welcoming. And, commercialization of PAW-treated foods, ahead step for USFDA regulatory approval would be future direction for PAW technology application in food industry (Han et al., 2023).

These are some of promising technologies among which some are commercially used and others in the nascent stage. Though these washing technologies can reduce the microbial load present on perishable products, it can't stop the contamination in the subsequent process. It was reported that a 10% of microbial load on the fresh commodities can't be removed irrespective of the washing methods used (Velderrain-Rodríguez et al., 2019). Thus, it is also necessary to make sure the washed commodities are processed.

5.4   SUMMARY

During minimal processing of perishable commodities, cellular contents get liberated which could promote the microbial spoilage. Thus, to stop it, several washing methods are carried out to reduce the growth of microbial load in perishables and retard its growth before reaching the next step of processing. As the use of commercial chlorinated water interacts with organic matter and forms carcinogen, the need for identifying a potential alternative processed water technology is high. Non-thermal and non-chemical technologies like electrolyzed water, OW, and plasma activated water were found efficient in preserving perishable foods without leaving any residues. However, one major limitation in using these technologies is the concentration limit that can be applicable for various perishable foods. Large concentration of active compounds in water could also oxidize the nutrients in the commodity; for example, higher quantities of reactive species in ozonized and plasma activated water could cause oxidation of lipids. Use of alkaline, slightly acidic, and neutral EW is preferable over acidic EW owing to the higher concentration of chlorine in acidic EW. Thus, alternate processed water technologies can be used at low concentrations of actives in the water based on the guidelines provided by regulatory bodies.

REFERENCES

Acoglu, B., & Omeroglu, P. Y. (2021). Effectiveness of different type of washing agents on reduction of pesticide residues in orange (*Citrus sinensis*). *LWT, 147*(February), 111690. https://doi.org/10.1016/j.lwt.2021.111690

Aider, M., Gnatko, E., Benali, M., Plutakhin, G., & Kastyuchik, A. (2012). Electro-activated aqueous solutions: Theory and application in the food industry and biotechnology. *Innovative Food Science and Emerging Technologies, 15*, 38–49. https://doi.org/10.1016/j.ifset.2012.02.002

Akther, S., Islam, R., Alam, M., & Alam, J. (2022). Impact of slightly acidic electrolyzed water in combination with ultrasound and mild heat on safety and quality of fresh cut cauliflower. *Postharvest Biology and Technology, 197*, 112189. https://doi.org/10.1016/j.postharvbio.2022.112189

Ali, M., Cheng, J. H., & Sun, D. W. (2021). Effect of plasma activated water and buffer solution on fungicide degradation from tomato (*Solanum lycopersicum*) fruit. *Food Chemistry, 350*. https://doi.org/10.1016/j.foodchem.2021.129195

Ali, A., Yeoh, W. K., Forney, C., & Siddiqui, M. W. (2018). Advances in postharvest technologies to extend the storage life of minimally processed fruits and vegetables. *Critical Reviews in Food Science and Nutrition, 58*(15), 2632–2649. https://doi.org/10.1080/10408398.2017.1339180

Almeida, M. De, Garruti, S., Machado, T. F., & Oliveira, D. (2021). The use of electrolyzed water as a disinfectant for fresh cut mango. *Scientia Horticulturae, 287*, 110227. https://doi.org/10.1016/j.scienta.2021.110227

Al-Sharify, Z. T., Al-Sharify, T. A., Al-Obaidy, B. W., & Al-Azawi, A. M. (2020). Investigative study on the interaction and applications of plasma activated water(PAW). *IOP Conference Series: Materials Science and Engineering, 870*(1). https://doi.org/10.1088/1757-899X/870/1/012042

Alvaro, J. E., Moreno, S., Dianez, F., Santos, M., Carrasco, G., & Urrestarazu, M. (2009). Effects of peracetic acid disinfectant on the postharvest of some fresh vegetables. *Journal of Food Engineering, 95*(1), 11–15. https://doi.org/10.1016/j.jfoodeng.2009.05.003

Belay, Z. A., Botes, W. J., & Caleb, O. J. (2021). Effects of alkaline electrolyzed water pretreatment on the physicochemical quality attributes of fresh nectarine during storage. *Journal of Food Processing and Preservation, 45*(10), 1–12. https://doi.org/10.1111/jfpp.15879

Betts, G., & Everis, L. (2005). Alternatives to hypochlorite washing systems for the decontamination of fresh fruit and Vegetables. In *Improving the Safety of Fresh Fruit and Vegetables*. Woodhead Publishing Limited. https://doi.org/10.1533/9781845690243.3.351

Bhilwadikar, T., Pounraj, S., Manivannan, S., Rastogi, N. K., & Negi, P. S. (2019). Decontamination of microorganisms and pesticides from fresh fruits and vegetables: A comprehensive review from common household processes to modern techniques. *Comprehensive Reviews in Food Science and Food Safety, 18*. https://doi.org/10.1111/1541-4337.12453

Bialoszewski, D., Bocian, E., Bukowska, B., Czajkowska, M., Sokół-Leszczyńska, B., & Tyski, S. (2010). Antimicrobial activity of ozonated water. *Case Reports and Clinical Practice Review, 16*(9), 71–75.

Bilek, S. E., & Turantaş, F. (2013). Decontamination efficiency of high power ultrasound in the fruit and vegetable industry, a review. *International Journal of Food Microbiology, 166*(1), 155–162. https://doi.org/10.1016/j.ijfoodmicro.2013.06.028

Botondi, R., Barone, M., & Grasso, C. (2021). A review into the effectiveness of ozone technology for improving the safety and preserving the quality of fresh-cut fruits and vegetables. *Foods, 10*(4). https://doi.org/10.3390/foods10040748

Chen, Xi, & Hung, Y.-C. (2018). Development of a chlorine dosing strategy for fresh produce washing process to maintain microbial food safety and minimize residual chlorine. *Journal of Food Science, 83*(6), 1701–1706.

Chen, Xiuqin, Tango, C. N., Daliri, E. B., & Oh, S. (2019). Disinfection efficacy of slightly acidic electrolyzed water combined with chemical treatments on fresh fruits at the industrial scale. *Foods, 8*(10), 1–18. https://doi.org/10.3390/foods8100497

da Costa, A. R., Faroni, L. R. D., Salomão, L. C. C., Cecon, P. R., & de Alencar, E. R. (2021). Use of ozonized water to control anthracnose in papaya (Carica papaya L.) and its effect on the quality of the fruits. *Ozone: Science and Engineering, 43*(4), 384–393. https://doi.org/10.1080/01919512.2020.1816450

da Silva, J. P., Costa, S. M., de Oliveira, L. M., Vieira, M. C. S., Vianello, F., & Lima, G. P. P. (2015). Does the use of ozonized water influence the chemical characteristics of organic cabbage (Brassica oleracea var. capitata)? *Journal of Food Science and Technology, 52*(11), 7026–7036. https://doi.org/10.1007/s13197-015-1817-0

Deng, L.-Z., Mujumdar, A. S., Pan, Z., Vidyarthi, S. K., Xu, J., Zielinska, M., & Xiao, H.-W. (2020). Emerging chemical and physical disinfection technologies of fruits and vegetables: A comprehensive review. *Critical Reviews in Food Science and Nutrition, 60*(15), 2481–2508. https://doi.org/10.1080/10408398.2019.1649633

Dharini, M., Jaspin, S., & Mahendran, R. (2023). Cold plasma reactive species: Generation, properties, and interaction with food biomolecules. *Food Chemistry, 405*(PA), 134746. https://doi.org/10.1016/j.foodchem.2022.134746

Dos Santos, L. M. C., da Silva, E. S., Oliveira, F. O., Rodrigues, L. de A. P., Neves, P. R. F., Meira, C. S., Moreira, G. A. F., Lobato, G. M., Nascimento, C., Gerhardt, M., Lessa, A. S., Mascarenhas, L. A. B., & Machado, B. A. S. (2021). Ozonized water in microbial control: Analysis of the stability, in vitro biocidal potential, and cytotoxicity. *Biology, 10*(6). https://doi.org/10.3390/biology10060525

Esua, O. J., Cheng, J. H., & Sun, D. W. (2021). Functionalization of water as a nonthermal approach for ensuring safety and quality of meat and seafood products. *Critical Reviews in Food Science and Nutrition*, *61*(3), 431–449. https://doi.org/10.1080/10408398.2020.1735297

Frías, E., Iglesias, Y., Alvarez-Ordóñez, A., Prieto, M., González-Raurich, M., & López, M. (2020). Evaluation of Cold Atmospheric Pressure Plasma (CAPP) and plasma-activated water (PAW) as alternative non-thermal decontamination technologies for tofu: Impact on microbiological, sensorial and functional quality attributes. *Food Research International*, *129*, 108859. https://doi.org/10.1016/j.foodres.2019.108859

Fu, T.-J., Li, Y., Awad, D., Zhou, T.-Y., & Liu, L. (2018). Factors affecting the performance and monitoring of a chlorine wash in preventing Escherichia coli O157:H7 cross-contamination during postharvest washing of cut lettuce. *Food Control*, *94*, 212–221.

Gonçalves, A. A., & Lira Santos, T. C. (2019). Improving quality and shelf-life of whole chilled Pacific white shrimp (*Litopenaeus vannamei*) by ozone technology combined with modified atmosphere packaging. *LWT*, *99*, 568–575. https://doi.org/10.1016/j.lwt.2018.09.083

Gu, G., Ottesen, A., Bolten, S., Ramachandran, P., Reed, E., Rideout, S., Luo, Y., Patel, J., Brown, E., & Nou, X. (2018). Shifts in spinach microbial communities after chlorine washing and storage at compliant and abusive temperatures. *Food Microbiology*, *73*, 73–84. https://doi.org/10.1016/j.fm.2018.01.002

Guo, D., Liu, H., Zhou, L., Xie, J., & He, C. (2021). Plasma-activated water production and its application in agriculture. *Journal of the Science of Food and Agriculture*, *101*(12), 4891–4899. https://doi.org/10.1002/jsfa.11258

Guo, Y., Zhu, Z., Zhao, Y., Zhou, T., Lan, B., & Song, L. (2021). Simultaneous annihilation of microorganisms and volatile organic compounds from municipal solid waste storage rooms with slightly acidic electrolyzed water. *Journal of Environmental Management*, *297*(July), 113414. https://doi.org/10.1016/j.jenvman.2021.113414

Haji, M., Kerbache, L., Muhammad, M., & Al-Ansari, T. (2020). Roles of technology in improving perishable food supply chains. *Logistics*, *4*(4), 33. https://doi.org/10.3390/logistics4040033

Han, Q.-Y., Wen, X., Gao, J.-Y., Zhong, C.-S., & Ni, Y.-Y. (2023). Application of plasma-activated water in the food industry: A review of recent research developments. *Food Chemistry*, *405*, 134797. https://doi.org/10.1016/j.foodchem.2022.134797

Horvitz, S., & Cantalejo, M. J. (2014). Application of ozone for the postharvest treatment of fruits and vegetables. *Critical Reviews in Food Science and Nutrition*, *54*(3), 312–339. https://doi.org/10.1080/10408398.2011.584353

Huang, X., Zhu, S., Zhou, X., He, J., Yu, Y., & Ye, Z. (2021). Preservative effects of the combined treatment of slightly acidic electrolyzed water and ice on pomfret. *International Journal of Agricultural and Biological Engineering*, *14*(1), 230–236. https://doi.org/10.25165/j.ijabe.20211401.5967

Iacumin, L., & Comi, G. (2019). Microbial quality of raw and ready-to-eat mung bean sprouts produced in Italy. *Journal of Food Microbiology*, *82*(July 2018), 371–377. https://doi.org/10.1016/j.fm.2019.03.014

Inpitak, P., & Udompijitkul, P. (2022). Effect of household sanitizing agents and electrolyzed water on Salmonella reduction and germination of sunflower and roselle seeds. *International Journal of Food Microbiology*, *370*(August 2021), 109668. https://doi.org/10.1016/j.ijfoodmicro.2022.109668

Iram, A., Wang, X., & Demirci, A. (2021). Electrolyzed oxidizing water and its applications as sanitation and cleaning agent. *Food Engineering Reviews*, 411–427. https://doi.org/10.1007/s12393-021-09278-9

Jee, D. Y., & Ha, J. W. (2021). Synergistic interaction of tap water-based neutral electrolyzed water combined with UVA irradiation to enhance microbial inactivation on stainless steel. *Food Research International*, *150*(PA), 110773. https://doi.org/10.1016/j.foodres.2021.110773

Kang, C., Xiang, Q., Zhao, D., Wang, W., Niu, L., & Bai, Y. (2019). Inactivation of Pseudomonas deceptionensis CM2 on chicken breasts using plasma-activated water. *Journal of Food Science and Technology*, *56*(11), 4938–4945. https://doi.org/10.1007/s13197-019-03964-7

Kechinski, C. P., Montero, C. R. S., Guimarães, P. V. R., Noreña, C. P. Z., Marczak, L. D. F., Tessaro, I. C., & Bender, R. J. (2012). Effects of ozonized water and heat treatment on the papaya fruit epidermis. *Food and Bioproducts Processing*, *90*(2), 118–122. https://doi.org/10.1016/j.fbp.2011.01.005

Kumar, A., Mangla, S. K., Kumar, P., & Karamperidis, S. (2020). Challenges in perishable food supply chains for sustainability management: A developing economy perspective. *Business Strategy and the Environment*, *29*(5), 1809–1831. https://doi.org/10.1002/bse.2470

Lee, M. Y., Kim, Y. K., Ryoo, K. K., Lee, Y. B., & Park, E. J. (2006). Electrolyzed-reduced water protects against oxidative damage to DNA, RNA, and protein. *Applied Biochemistry and Biotechnology*, *135*(2), 133–144. https://doi.org/10.1385/ABAB:135:2:133

Lin, Y., Simsek, S., & Bergholz, T. M. (2022). Impact of chlorinated water on pathogen inactivation during wheat tempering and resulting flour quality. *Journal of Food Protection*, *85*(8), 1210–1220.

Liu, C., Chen, C., Jiang, A., Sun, X., Guan, Q., & Hu, W. (2020). Effects of plasma-activated water on microbial growth and storage quality of fresh-cut apple. *Innovative Food Science and Emerging Technologies*, *59*, 102256. https://doi.org/10.1016/j.ifset.2019.102256

Liu, C., Chen, C., Jiang, A., Zhang, Y., Zhao, Q., & Hu, W. (2021). Effects of aqueous ozone treatment on microbial growth, quality, and pesticide residue of fresh-cut cabbage. *Food Science and Nutrition*, *9*(1), 52–61. https://doi.org/10.1002/fsn3.1870

Liu, Y., Wang, J., Zhu, X., Liu, Y., Cheng, M., & Xing, W. (2021). Effects of electrolyzed water treatment on pesticide removal and texture quality in fresh-cut cabbage, broccoli, and color pepper. *Food Chemistry*, *353*(February). https://doi.org/10.1016/j.foodchem.2021.129408

Liu, R., Yu, Z., Sun, Y., & Zhou, S. (2021). Food Bioscience The enzymatic browning reaction inhibition effect of strong acidic electrolyzed water on different parts of sweet potato slices. *Food Bioscience*, *43*(July), 101252. https://doi.org/10.1016/j.fbio.2021.101252

Lonigro, A., Montemurro, N., & Laera, G. (2017). Science of the total environment effects of residual disinfectant on soil and lettuce crop irrigated with chlorinated water. *Science of the Total Environment*. https://doi.org/10.1016/j.scitotenv.2017.01.083

Lu, F., Liu, S. L., Liu, R., Ding, Y. C., & Ding, Y. T. (2012). Combined effect of ozonized water pretreatment and ozonized flake ice on maintaining quality of Japanese sea bass (*Lateolabrax japonicus*). *Journal of Aquatic Food Product Technology*, *21*(2), 168–180. https://doi.org/10.1080/10498850.2011.589040

Luan, C., Zhang, M., Fan, K., & Devahastin, S. (2021). Effective pretreatment technologies for fresh foods aimed for use in central kitchen processing. *Journal of the Science of Food and Agriculture*, *101*(2), 347–363. https://doi.org/10.1002/jsfa.10602

Luo, X., Wang, R., Wang, L., Li, Y., Bian, Y., & Chen, Z. (2014). Effect of ozone treatment on aflatoxin B1 and safety evaluation of ozonized corn. *Food Control*, *37*(1), 171–176. https://doi.org/10.1016/j.foodcont.2013.09.043

Luu, P. (2019). *Efficacy of Aqueous and Gaseous Chlorine Dioxide in Reducing Salmonella, E. coli O157 :H7, and Listeria monocytogenes on Sweet Potatoes, Strawberries, and Blueberries* (Issue March). Louisiana State University and Agricultural and Mechanical College. https://repository.lsu.edu/gradschool_theses/4893

Lyu, F., Gao, F., Zhou, X., Zhang, J., & Ding, Y. (2018). Using acid and alkaline electrolyzed water to reduce deoxynivalenol and mycological contaminations in wheat grains. *Food Control*, *88*, 98–104. https://doi.org/10.1016/j.foodcont.2017.12.036

Ma, R., Wang, G., Tian, Y., Wang, K., Zhang, J., & Fang, J. (2015). Non-thermal plasma-activated water inactivation of food-borne pathogen on fresh produce. *Journal of Hazardous Materials*, *300*, 643–651. https://doi.org/10.1016/j.jhazmat.2015.07.061

Ma, R., Yu, S., Tian, Y., Wang, K., Sun, C., Li, X., Zhang, J., Chen, K., & Fang, J. (2016). Effect of non-thermal plasma-activated water on fruit decay and quality in postharvest Chinese bayberries. *Food and Bioprocess Technology*, *9*(11), 1825–1834. https://doi.org/10.1007/s11947-016-1761-7

Mai-Prochnow, A., Zhou, R., Zhang, T., Ostrikov, K. (Ken), Mugunthan, S., Rice, S. A., & Cullen, P. J. (2021). Interactions of plasma-activated water with biofilms: inactivation, dispersal effects and mechanisms of action. *npj Biofilms and Microbiomes*, *7*(1), 1–12. https://doi.org/10.1038/s41522-020-00180-6

Marín, A., Tudela, J. A., Garrido, Y., Albolafio, S., Hern, N., Andújar, S., Allende, A., & Gil, M. I. (2020). Chlorinated wash water and pH regulators affect chlorine gas emission and disinfection by-products. *Innovative Food Science & Emerging Technologies*, *66*(October). https://doi.org/10.1016/j.ifset.2020.102533

Mazhar, M. A., Khan, N. A., Ahmed, S., Khan, A. H., Hussain, A., Rahisuddin, Changani, F., Yousefi, M., Ahmadi, S., & Vambol, V. (2020). Chlorination disinfection by-products in municipal drinking water – a review. *Journal of Cleaner Production*, *273*. https://doi.org/10.1016/j.jclepro.2020.123159

Megahed, A., Aldridge, B., & Lowe, J. (2020). Antimicrobial efficacy of aqueous ozone and ozone–lactic acid blend on Salmonella-contaminated chicken drumsticks using multiple sequential soaking and spraying approaches. *Frontiers in Microbiology*, *11*(December), 1–11. https://doi.org/10.3389/fmicb.2020.593911

Moghassem Hamidi, R., Shekarforoush, S. S., Hosseinzadeh, S., & Basiri, S. (2021). Evaluation of the effect of neutral electrolyzed water and peroxyacetic acid alone and in combination on microbiological, chemical, and sensory characteristics of poultry meat during refrigeration storage. *Food Science and Technology International*, *27*(6), 499–507. https://doi.org/10.1177/1082013220968713

Monteiro, N. O. da C., de Alencar, E. R., Souza, N. O. S., & Leão, T. P. (2021). Ozonized water in the preconditioning of corn seeds: Physiological quality and field performance. *Ozone: Science and Engineering*, *43*(5), 436–450. https://doi.org/10.1080/01919512.2020.1836472

Nayak, S. L., Sethi, S., Sharma, R. R., Sharma, R. M., Singh, S., & Singh, D. (2020). Aqueous ozone controls decay and maintains quality attributes of strawberry (Fragaria × ananassa Duch.). *Journal of Food Science and Technology*, 57(1), 319–326. https://doi.org/10.1007/s13197-019-04063-3

Nyamende, N. E., Belay, Z. A., Keyser, Z., Oyenihi, A. B., & Caleb, O. J. (2022). Impacts of alkaline-electrolyzed water treatment on physicochemical, phytochemical, antioxidant properties and natural microbial load on "Granny Smith" apples during storage. *International Journal of Food Science\& Technology*, 57(1), 447–456.

Ogunniyi, A. D., Tenzin, S., Ferro, S., Venter, H., Pi, H., Amorico, T., Deo, P., & Trott, D. J. (2021). A pH-neutral electrolyzed oxidizing water significantly reduces microbial contamination of fresh spinach leaves. *Food Microbiology*, 93(August 2020), 103614. https://doi.org/10.1016/j.fm.2020.103614

Pandiselvam, R., Subhashini, S., Banuu Priya, E. P., Kothakota, A., Ramesh, S. V., & Shahir, S. (2019). Ozone based food preservation: A promising green technology for enhanced food safety. *Ozone: Science & Engineering*, 41(1), 17–34. https://doi.org/10.1080/01919512.2018.1490636

Patil, S., Torres, B., Tiwari, B. K., Wijngaard, H. H., Bourke, P., Cullen, P. J., O' Donnell, C. P., & Valdramidis, V. P. (2010). Safety and quality assessment during the ozonation of cloudy apple juice. *Journal of Food Science*, 75(7), 437–443. https://doi.org/10.1111/j.1750-3841.2010.01750.x

Patra, A., Prasath, V. A., Pandiselvam, R., Sutar, P. P., & Jeevarathinam, G. (2022). Effect of plasma activated water (PAW) on physicochemical and functional properties of foods. *Food Control*, 142, 109268. https://doi.org/10.1016/j.foodcont.2022.109268

Possas, A., Pérez-rodríguez, F., Tarlak, F., & García-gimeno, R. M. (2020). Quantifying and modelling the inactivation of Listeria monocytogenes by electrolyzed water on food contact surfaces. *Journal of Food Engineering*, 290, 110287. https://doi.org/10.1016/j.jfoodeng.2020.110287

Praeger, U., Herppich, W. B., & Hassenberg, K. (2016). Aqueous chlorine dioxide treatment of horticultural produce : Effects on microbial safety and produce quality: A review. *Critical Reviews in Food Science and Nutrition*, 58(2), 318–333. https://doi.org/10.1080/10408398.2016.1169157

Premjit, Y., Sruthi, N. U., Pandiselvam, R., & Kothakota, A. (2022). Aqueous ozone: Chemistry, physiochemical properties, microbial inactivation, factors influencing antimicrobial effectiveness, and application in food. *Comprehensive Reviews in Food Science and Food Safety*, 21(2), 1054–1085. https://doi.org/10.1111/1541-4337.12886

Rahman, M., Hasan, M. S., Islam, R., Rana, R., Sayem, A. S. M., Sad, M. A. A., Matin, A., Raposo, A., Zandonadi, R. P., Han, H., Ariza-Montes, A., Vega-Muñoz, A., & Sunny, A. R. (2022). Plasma- activated water for food safety and quality: A review of recent developments. *International Journal of Environmental Research and Public Health*, 19(11). https://doi.org/10.3390/ijerph19116630

Rahman, S. M. E., Khan, I., & Oh, D. (2016). Electrolyzed water as a novel sanitizer in the food industry : Current trends and future perspectives. *Comprehensive reviews in food science and food safety*, 15, 471–490. https://doi.org/10.1111/1541-4337.12200

Rajashri, K., Shivegowda, B., Pradeep, R., Negi, S., & Rastogi, N. K. (2019). Effect of ozone and ultrasound treatments on polyphenol content, browning enzyme activities, and shelf life of tender coconut water. *Journal of Food Processing and Preservation*, October 1–11. https://doi.org/10.1111/jfpp.14363

Reeves, W. R., McGuire, M. K., Stokes, M., & Vicini, J. L. (2019). Assessing the safety of pesticides in food: How current regulations protect human health. *Advances in Nutrition*, 10(1), 80–88. https://doi.org/10.1093/advances/nmy061

Roy, P. K., Mizan, M. F. R., Hossain, M. I., Han, N., Nahar, S., Ashrafudoulla, M., Toushik, S. H., Shim, W.-B., Kim, Y.-M., & Ha, S.-D. (2021). Elimination of Vibrio parahaemolyticus biofilms on crab and shrimp surfaces using ultraviolet C irradiation coupled with sodium hypochlorite and slightly acidic electrolyzed water. *Food Control*, 128, 108179.

Sadło, S., Szpyrka, E., Piechowicz, B., Antos, P., Józefczyk, R., & Balawejder, M. (2017). Reduction of captan, boscalid and pyraclostrobin residues on apples using water only, gaseous ozone, and ozone aqueous solution. *Ozone: Science and Engineering*, 39(2), 97–103. https://doi.org/10.1080/01919512.2016.1257931

Sarangapani, C., Scally, L., Gulan, M., & Cullen, P. J. (2020). Dissipation of pesticide residues on grapes and strawberries using plasma-activated water. *Food and Bioprocess Technology*, 13(10), 1728–1741. https://doi.org/10.1007/s11947-020-02515-9

Seo, J., Puligundla, P., & Mok, C. (2019). Decontamination of collards (Brassica oleracea var. acephala L.) using electrolyzed water and corona discharge plasma jet. *Food Science and Biotechnology*, 28(1), 147–153. https://doi.org/10.1007/s10068-018-0435-9

Shirahata, S., Hamasaki, T., & Teruya, K. (2012). Advanced research on the health benefit of reduced water. *Trends in Food Science and Technology*, *23*(2), 124–131. https://doi.org/10.1016/j.tifs.2011.10.009

Song, H., Lee, J. Y., Lee, H. W., & Ha, J. H. (2021). Inactivation of bacteria causing soft rot disease in fresh cut cabbage using slightly acidic electrolyzed water. *Food Control*, *128*(May), 108217. https://doi.org/10.1016/j.foodcont.2021.108217

Strano, M. C., Timpanaro, N., Allegra, M., Foti, P., Pangallo, S., & Romeo, F. V. (2021). Effect of ozonated water combined with sodium bicarbonate on microbial load and shelf life of cold stored clementine (*Citrus clementina* Hort. ex Tan.). *Scientia Horticulturae*, *276*(July 2020), 109775. https://doi.org/10.1016/j.scienta.2020.109775

Sun, J., Chen, H., Xie, H., Li, M., Chen, Y., Hung, Y. C., & Lin, H. (2022). Acidic electrolyzed water treatment retards softening and retains cell wall polysaccharides in pulp of postharvest fresh longans and its possible mechanism. *Food Chemistry: X*, *13*(September 2021), 100265. https://doi.org/10.1016/j.fochx.2022.100265

Tahamolkonan, M., Ghahsareh, A. M., Ashtari, M. K., & Honarjoo, N. (2022). Tomato (Solanum lycopersicum) growth and fruit quality affected by organic fertilization and ozonated water. *Protoplasma*, *259*(2), 291–299. https://doi.org/10.1007/s00709-021-01657-7

Tomasello, F., Pollesel, M., Mondo, E., Savini, F., Scarpellini, R., Giacometti, F., Lorito, L., Tassinari, M., Cuomo, S., Piva, S., & Serraino, A. (2021). Effectiveness of alkaline electrolyzed water in reducing bacterial load on surfaces intended to come into contact with food. *Italian Journal of Food Safety*, *10*(3), 39–42. https://doi.org/10.4081/ijfs.2021.9988

Velderrain-Rodríguez, G. R., López-Gámez, G. M., Domínguez-Avila, J. A., González-Aguilar, G. A., Soliva-Fortuny, R., & Ayala-Zavala, J. F. (2019). Minimal processing. *Postharvest Technology of Perishable Horticultural Commodities*, 353–374. https://doi.org/10.1016/B978-0-12-813276-0.00010-9

Wang, H., Duan, D., Wu, Z., Xue, S., Xu, X., & Zhou, G. (2019). Primary concerns regarding the application of electrolyzed water in the meat industry. *Food Control*, *95*(June 2018), 50–56. https://doi.org/10.1016/j.foodcont.2018.07.049

Wang, J., & Han, R. (2019). Removal of pesticide on food by electrolyzed water. In *Electrolyzed water in food: Fundamentals and applications* (pp. 39–65). Springer.

Wang, Q., & Salvi, D. (2021). Recent progress in the application of plasma-activated water (PAW) for food decontamination. *Current Opinion in Food Science*, *42*, 51–60. https://doi.org/10.1016/j.cofs.2021.04.012

Wu, J., Luan, T., Lan, C., Hung Lo, T. W., & Chan, G. Y. S. (2007). Removal of residual pesticides on vegetable using ozonated water. *Food Control*, *18*(5), 466–472. https://doi.org/10.1016/j.foodcont.2005.12.011

Xiang, Q., Kang, C., Niu, L., Zhao, D., Li, K., & Bai, Y. (2018). Antibacterial activity and a membrane damage mechanism of plasma-activated water against Pseudomonas deceptionensis CM2. *LWT*, *96*, 395–401. https://doi.org/10.1016/j.lwt.2018.05.059

Xiang, Q., Liu, X., Liu, S., Ma, Y., Xu, C., & Bai, Y. (2019). Effect of plasma-activated water on microbial quality and physicochemical characteristics of mung bean sprouts. *Innovative Food Science and Emerging Technologies*, *52*, 49–56. https://doi.org/10.1016/j.ifset.2018.11.012

Xinyu, L., Xiang, Q., Cullen, P. J., Su, Y., Chen, S., Ye, X., Liu, D., & Ding, T. (2020). Plasma-activated water (PAW) and slightly acidic electrolyzed water (SAEW) as beef thawing media for enhancing microbiological safety. *LWT – Food Science and Technology*, *117*. https://doi.org/10.1016/j.lwt.2019.108649

Xuan, X., & Ling, J. (2019). *Generation of electrolyzed water*. Springer Singapore. https://doi.org/10.1007/978-981-13-3807-6_1

Yan, P., Daliri, E. B., & Oh, D.-H. (2021). New clinical applications of electrolyzed water: A review. *Microorganisms*, *9*, 136.

Yan, J., & Xie, J. (2021). Removal of Shewanella putrefaciens biofilm by acidic electrolyzed water on food contact surfaces. *LWT*, *151*(January), 112044. https://doi.org/10.1016/j.lwt.2021.112044

Yang, H., Feirtag, J., & Diez-Gonzalez, F. (2013). Sanitizing effectiveness of commercial "active water" technologies on Escherichia coli O157:H7, Salmonella enterica and Listeria monocytogenes. *Food Control*, *33*(1), 232–238. https://doi.org/10.1016/j.foodcont.2013.03.007

Zhai, Y., Liu, S., Xiang, Q., Lyu, Y., & Shen, R. (2019). Effect of plasma-activated water on the microbial decontamination and food quality of thin sheets of bean curd. *Applied Sciences*, *9*(20), 4223.

Zhang, C., Zhao, Z., Yang, G., Shi, Y., Zhang, Y., Xia, X., & Shi, C. (2020). Effect of slightly acidic electrolyzed water on natural Salmonella reduction and seed germination in the production of alfalfa sprouts. *Food Microbiology*, *97*, 103414. https://doi.org/10.1016/j.fm.2020.103414

Zhao, Y., Chen, R., Liu, D., Wang, W., Niu, J., Xia, Y., Qi, Z., Zhao, Z., & Song, Y. (2019). Effect of nonthermal plasma-activated water on quality and antioxidant activity of fresh-cut kiwifruit. *IEEE Transactions on Plasma Science*, *47*(11), 4811–4817. https://doi.org/10.1109/TPS.2019.2904298

Zhao, Y. M., Ojha, S., Burgess, C. M., Sun, D. W., & Tiwari, B. K. (2020). Influence of various fish constituents on inactivation efficacy of plasma-activated water. *International Journal of Food Science and Technology*, *55*(6), 2630–2641. https://doi.org/10.1111/ijfs.14516

Zhou, R., Zhou, R., Wang, P., Xian, Y., Mai-Prochnow, A., Lu, X., Cullen, P. J., Ostrikov, K., Bazaka, K. (2020). Plasma activated water (PAW): Generation, origin of reactive species and biological applications. *Journal of Physics D: Applied Physics*. 53(30), 303001.

6 Ozone Technologies in Food Processing, Preservation

V. Monica, Anbarasan Rajan, and R. Mahendran

6.1 INTRODUCTION

Consumers have become increasingly concerned about their lifestyles, particularly their eating habits, in recent decades (Petrescu et al., 2019). A plate of contaminated food contains potentially harmful bacterial, viral, and parasitic organisms that can cause diseases ranging from mild diarrhoea to deadly cancer. It can also lead to a chain reaction of malnutrition and diseases in people of all ages (WHO, 2022). This could be possibly by environmental contaminants such as impure air, water, and soil, food processing contaminants such as unauthorized adulterants, food additives, and chemical or biologically contaminated packaging materials. So, it is important to preserve foods from exposure to these contaminating sources. For this, several non-thermal technologies, such as high-pressure processing, ultrasound, and cold plasma, are being applied (Hernández-Hernández et al., 2019). Ozone is one among them, and this chapter focuses on the properties of ozone, its mechanism of generation, factors influencing ozone processing, their application in different food processing industries, and others.

Ozone (O_3) is a triatomic oxygen molecule that is naturally present in the earth's atmosphere at very lower concentrations, 0.05 mg/L (A. Nath et al., 2010). In the presence of solar radiation, the diatomic oxygen molecule in the atmosphere undergoes photodissociation and forms a triatomic molecule, O3. These molecules form a layer in the stratosphere, commonly known as the O_3 layer, and it shields terrestrial living organisms from harmful UV radiation (Sarron et al., 2021). O_3 was first discovered by Schonbein in 1839, and it was not until the 1990s that it was used to sanitize water. Early in the 21st century, FDA approved the use of O_3 in food industries and labelled it as GRAS (Generally Recognized as Safe). Presently, O_3 is commercially used as a disinfectant in bottled water, swimming pools, cooling towers, and wastewater treatment. Specifically, in the food processing sector, O_3 has wide applications including seed germination, pesticide degradation, insect mortality, sanitation of food plant equipment, food preservation, etc. (Boopathy et al., 2022). Moreover, O_3 is a strong oxidant that is effective against various types of microorganisms like bacteria, viruses, and parasites invading cereals, fruits, vegetables, milk, meat, and their products (Chuwa et al., 2020). Despite the fact that O_3 is a gaseous compound, both aqueous and gaseous forms of O_3 can be used in commercial food applications.

6.2 PROPERTIES OF OZONE

The term ozone in Greek means 'smell' and comprises of three oxygen atoms arranged in a cyclic structure, each at a distance of 1.26 Å. O_3, in its pure form has soft and sky-blue colour with an acrid and pungent odour. At room temperature, O_3 is a colourless and has detectable limit of 0.01 ppm to 0.05 ppm level. At negative temperature ($-112°C$), O_3 condenses into dark blue liquid. Regarding its chemical properties, O_3 molecule has molecular weight of 48 g/mol with gas density of 2.14 kg/m³ at 0°C. The melting point and boiling point of O_3 molecules are $-192.5 \pm 0.4°C$ and $-111.9 \pm 0.3°C$, respectively. Half-life of the O_3 molecule is less, that gaseous O_3 takes 15–1524 minutes and aqueous O_3 takes 10–80 minutes to become half of its initial concentration and also it varies with

DOI: 10.1201/9781003359302-6

temperature. O_3 is 13 times more soluble than oxygen in water below 30°C (Epelle et al., 2023; Guzel-Seydim et al., 2004). O_3 is an inherently unstable molecule that it decomposes with an oxidative potential of 2.07 V. Aftermath of O_3 decomposition results in a diatomic oxygen molecule and a free singlet oxygen. The free O_2 molecule hinders the microbial growth and oxidizes the chemical compounds around it (Boopathy et al., 2022). The oxidative potential of different disinfectants are as follows O_3 (2.07 V) > hydrogen peroxide (1.80 V) > peracetic acid (1.76 V) > chlorine dioxide (1.50 V) > chlorine gas (1.36 V) (Ashby, 2012; Donohue, 2003; Xue et al., 2023; Zhang & Huang, 2020). This shows O_3 is a strong oxidizer and can kill broad spectrum of microbes including Gram-positive bacteria, Gram-negative bacteria, spores, viruses, and fungi (Joshi et al., 2013).

6.3   FACTORS AFFECTING OZONE EFFICACY

Either gaseous O_3 or aqueous O_3 can be used in food processing depending on the target of action. For example, gaseous O_3 is employed to process the harvested crops and alter the storage conditions of food products, especially grains. Here, O_3 serves as a fumigant and sterilizing agent. Whereas aqueous O_3 is employed in washing the food products via ozonized water washing, spraying, rinsing, and immersing after harvesting. Several factors, such as external and internal factors, affect the efficacy of O_3 action on food products (Tiwari & Muthukumarappan, 2012; Xue et al., 2023).

External (environmental) factors that affect O_3 efficacy include O_3 concentration, flow rate, treatment time, and temperature (Tiwari & Muthukumarappan, 2012). However, regardless of O_3 form, O_3 concentration and treatment time influence the O_3 efficacy. Aqueous ozonation at variable gas flow rate of 3 L/min, 6.5 L/min, 10 L/min, O_3 concentration of 10% wt/wt, 20% wt/wt, 30% wt/wt for 5 minutes, 12.5 minutes, 20 minutes was applied to sugar cane juice on the target of reducing the total plate count bacterial colonies and preserving biochemical properties of juices. About 3.72 log reduction of microbes and 67.8% of polyphenol oxidase enzyme inactivation were achieved at combination of gas flow rate, concentration, and contact time of 6.5 L/min, 26.4% wt/wt, and 20 minutes, respectively. The optimized O_3 parameters revealed that higher gas flow rate generated lesser number of large sized bubbles and resulted in poor dissolution of O_3 and ultimately lowered the inactivation rate and vice versa could be seen during lower gas flow rate. This shows that relatively higher O_3 concentration and longer O_3 contact time with optimized gas flow rate aids in obtaining desirable results of O_3 action on foods (Panigrahi et al., 2020).

Intrinsic (internal) factors such as pH and organic matter influence the O_3 efficacy. There is a crucial relationship between pH of the food products and O_3 stability. O_3 molecules are more stable at acidic conditions (pH 3–4) and gradually degrade with a decrease in acidic conditions due to formation of hydroxyl radicals. O_3 molecules become unstable and degrade at higher rate at alkaline conditions (pH 14) due to formation of peroxyl radicals and hydroxide ions. The unstable O_3 molecules have the inability of achieving higher inactivation rate of microbial load in foods (Sarron et al., 2021). When Jamil et al. (2017) applied O_3 at concentrations 1.5 mg/mL, 1.7 mg/mL, and 2 mg/mL to water of pH 6, 7, and 8 at different temperatures 15°C, 25°C, and 35°C for 5,10, 30, 45, and 60 seconds, this kind of relationship was observed. Water of pH 6 ozonated at 2 mg/L for 60 seconds had *E. coli* colonies of $5.8*10^1$ CFU/mL, of pH 7 had $4.2*10^3$ CFU/mL, and of pH 8 had $9.50*10^3$ CFU/mL. The microbial inactivation was found to be faster at acidic water than basic water because the rate constant K' values at pH 6, 7, and 8 were 0.06/s, 0.035/s, and 0.02/s, respectively. Furthermore, it was discovered that temperature had a marginal effect on *E. coli* inactivation in water. In terms of microbial inactivation, O_3 and temperature have an inverse relationship that higher microbial reduction could be achieved at lower temperatures as the solubility ratio of O_3 increases with temperature reduction (Pandiselvam et al., 2020). Presence of organic matter in food products during O_3 treatment can negatively affect its efficacy. O_3 treatment to filtered pummelo juice helped to achieve 2.3 log reduction of bacterial colonies while 1.2 log reduction was achieved in unfiltered fruit juice. Similar reduction trend was seen in yeast and mould count reduction of fruit juice. Also, O_3 treatment to filtered fruit juice showed better physiochemical properties such as total soluble solids (TSS), total phenolic content (TPC) than unfiltered fruit juice. This is because antibacterial activity of O_3 is attributed to its diffusivity through biological membranes of foods (Shah et al., 2019).

6.4 OZONE GENERATION MECHANISM

Besides natural formation, there are two other methods (Figure 6.1) of producing O_3. They are corona discharge method, and ultraviolet radiation source and electrolysis methods. Among these two, corona discharge method is common and commercial method of O_3 production in food processing industries. The instability of O_3 demands onsite production (Chuwa et al., 2020). The corona discharge method consists of two electrodes, separated by a dielectric medium and discharge gap. The discharge gap is filled with oxygen gas. When high voltage alternating current is applied across electrodes, the oxygen molecules in discharge gap get excited and split into oxygen atoms. The split oxygen atoms combine with diatomic oxygen molecules and result in formation of triatomic O_3 molecules (A. Nath et al., 2010). O_3 can even be produced by UV lamps. This method is analogous to the natural phenomenon of O_3 formation. The oxygen molecules between the electrodes are excited by the UV radiation with a wavelength of 188 nm that is emitted from the UV lamp. This excitation causes the bond between the oxygen molecules to break, resulting in the formation of oxygen atoms. Following this, coupling of an oxygen molecule and an oxygen atom result in O_3. After a certain period of time, the formed O_3 molecules get degraded to form oxygen molecules. Following are the reactions that happens during O_3 generation and destruction (Dubey et al., 2022).

Absorption of energy by oxygen molecules (Eq. 6.1)

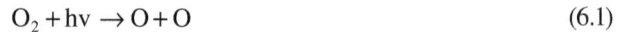

$$O_2 + hv \rightarrow O + O \tag{6.1}$$

Combination of oxygen atom and oxygen molecules (Eq. 6.2)

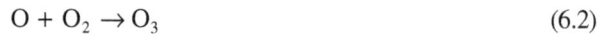

$$O + O_2 \rightarrow O_3 \tag{6.2}$$

Destruction of O_3 molecules (Eq. 6.3)

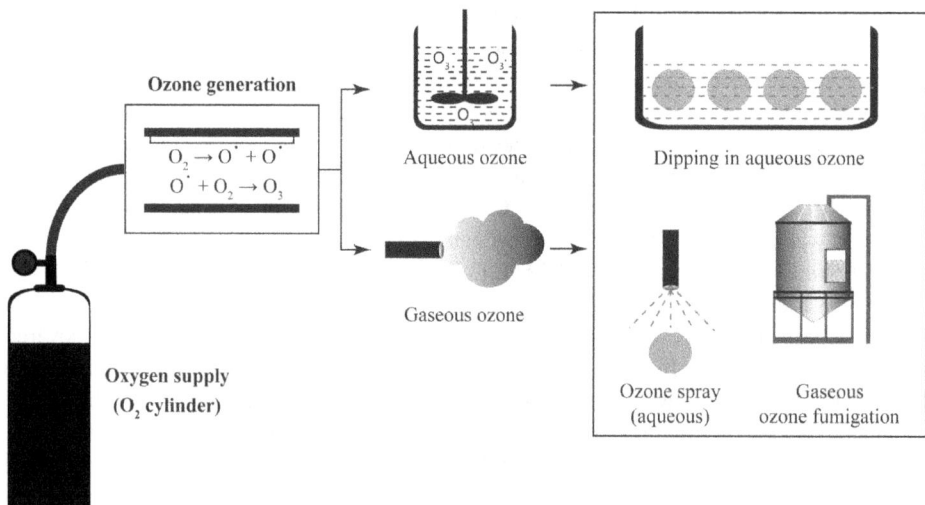

$$O_3 + hv \rightarrow O^- + O_2 \tag{6.3}$$

FIGURE 6.1 Overview on different modes of ozone application in food sector.

6.5 ANTIMICROBIAL MECHANISM

O_3 has been manifested for its strong antimicrobial activities against microbes such as Gram-positive, gram-negative bacteria, fungi, virus, yeast, and mould (Miller et al., 2021). The triatomic O_3 molecule targets the surface of the microbial cell and induces the cellular leakage through the pores created in the cell membrane of the microbial cell membrane via oxidation. The oxidational effect of O_3 breaks the double bonds of polyunsaturated fatty acids into acid peroxides. Further, O_3 also carries out proteolytic activity on protein structure and oxides the sulfhydryl group and forms short peptide chains from enzymes. Extensive degradation of protein in microbial cells causes heavy damage in the nucleic acids, leading to rapid death of the microbes. O_3 even has the potential to kill viruses by destroying the viral genetic material RNA by changing the polypeptide chains in protein coat of virus (Pandiselvam et al., 2020).

The mechanism of insecticidal action of O_3 is not widely discussed. However, it has been hypothesized O_3 initiates the discontinuous gas exchange with closure of the insect's spiracle region, where the respiration starts. Thus, the respiratory metabolism of insects gets affected by minimizing the inhalation of O_2 gas and exhalation of carbon dioxide gas. Eventually, the insect dies with improper gas exchange and decreasing respiration metabolism (Boopathy et al., 2022). Figure 6.2 shows the antimicrobial and insecticidal mechanism of O_3.

6.6 ROLE OF OZONE IN FOOD PROCESSING AND PRESERVATION

In food industries, O_3 takes up the role of fumigant, preservative, and antimicrobial agent during the processing of cereals, fruits, vegetables, milk, meat, and their products (Boopathy et al., 2022). Figure 6.3 shows the application of O_3 in various food processing industries.

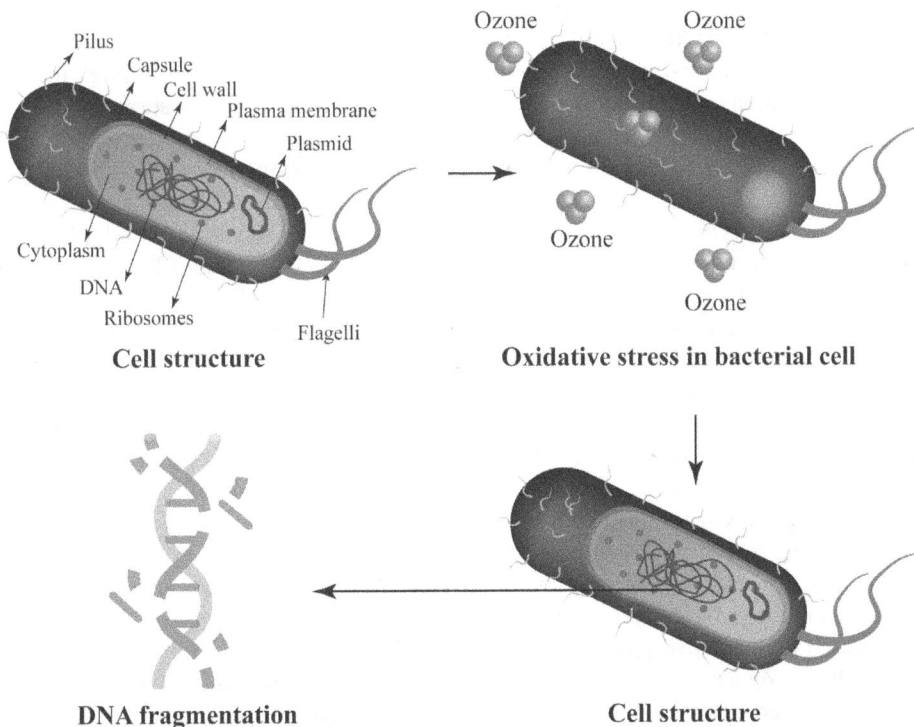

FIGURE 6.2 Antimicrobial and insecticidal mechanism of O_3.

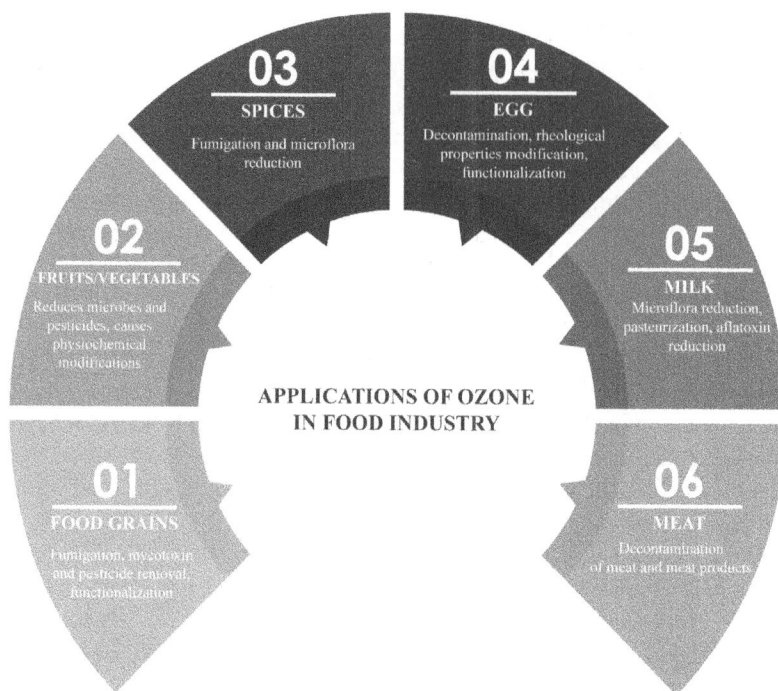

FIGURE 6.3 Application of O_3 in various food processing industries.

Insect infestation remains a problem in grain storage that may cause qualitative and quantitative losses in grains like rice, wheat, maize, barley, sorghum, etc. (Deshwal et al., 2020). About 40% of crops worth of $220 billion is lost every year due to insect pest infestation (FAO, 2021). O_3 has a vital role in disinfesting the grains and the insecticidal efficiency of O_3 in grains depends upon O_3 concentration, velocity of O_3, contact time, insect developmental stage, grain moisture content (MC), grain bed thickness, and other environmental conditions. Green gram infested with *Callosobruchus maculatus* was exposed to O_3 concentration varying from 500 to 1500 ppmv for different time periods. Hundred percent of mortality of *C. maculatus* adults, larvae, pupae, and eggs was achieved at its highest concentration (1500 ppmv) for varying time periods. Adult insects were least tolerant to O_3 exposure and O_3 at 500 ppmv for 274.40 minutes killed 90% of them. Whereas the insects at pupae stage were resistant to O_3 and required 500 ppmv for 1816.54 minutes to achieve 90% of mortality rate (Pandiselvam et al., 2019). Lemic et al. (2019) experimented to find out the effect of O_3 exposure directly on wheat weevils *Sitophilus granarius* and to wheat weevils *Sitophilus granarius* in maize. Hundred percent mortality was achieved when O_3 was directly exposed on wheat weevils. Nevertheless, not all weevils residing in maize were disinfested. In both cases, the walking response and velocity of the weevils were reduced after 120 minutes of O_3 exposure. In cowpea storage, gaseous O_3 treatment at concentration of 0.25, 0.5, 1.0, 1.5, and 2.0 g/m^3 was given to kill adults of *C. maculatus* and *C. chinesis* insects. The effectiveness of O_3 treatment on insect mortality was observed up to seven days of pea storage at intervals of two days. At the end of the seventh day, complete mortality of the two insects was observed at maximum concentrations of O_3 (1.5 g/m^3 and 2.0 g/m^3). Additionally, the oxidation of the O_3 during the treatment decreased the carbohydrates, protein, fat, phenolics, flavonoids, and tannins of the cowpea (Gad et al., 2021). Yet another problem in grain processing is mycotoxins, which is a complex toxic compound produced by fungi (Ensley & Radke, 2019). Trichothecenes, deoxynivalenol (DON), nivalenol (NIV), zearalenone (ZEN), T-2 and H-2 toxins are some common mycotoxins produced by Fusarium species that are associated with cereal grains (WHO, 2018). The oxidizing property of O_3 can be utilized to degrade the mycotoxins in cereal grains (Afsah-Hejri et al., 2020). ZEN and ochratoxin (OTA) are secondary metabolites that may contaminate the corn and cause serious health problems in humans. O_3 exposure of 10 mg/L for 5 seconds detoxified ZEN below the detection limit in corn. However, OTA was resistant to O_3 and only 65.7% of it was degraded after 120 seconds. It was interesting to note that MC of corn has an important role in the detoxification of ZEN and OTA. High MC can solubilize the O_3, which will increase the contact time between O_3 and mycotoxins. Moreover, in water, O_3 gets easily decomposed to oxygen molecules and hydroxyl ions, which may contribute to a higher detoxification rate. In the aspects of quality of the O_3 treated corn, the color and fatty acid content were negatively affected (Qi et al., 2016). O_3 not only limits its application to detoxification but also has a significant contribution in changing the structural and functional aspects of cereal and its products. Gaseous O_3 at the flow rate of 0.75 L/min reduced more than 50% of ZEN toxins in whole maize flour. Additionally, the viscosity of the flour was decreased as the O_3 exposure cleaved the glycosidic bonds of flour resulting in poor crosslinking effect. The peroxide value of the flour increased as the unsaturated bonds in the fatty acids were affected by O_3 exposure. In contrast to non-O_3 treated flour, O_3 treated flour demonstrated the presence of 10-pentadecenoic acid, heptadecenoic acid, and myristoleic acid which may be related to oxidizing effect of O_3 (Alexandre et al., 2019). As O_3 treatment could damage the wheat granules, the water absorption index (WAI) and water solubility index (WSI) of Korean wheat flour reached their maximum during the treatment time of 60 and 45 minutes, respectively. A remarkable increase in peak and final viscosities of the flour was seen with O_3 treatment. However, a significant reduction in wet gluten content of flour was seen because the longer exposure of O_3 avoids insoluble gluten aggregation and gluten structure formation. Also, ozonated flour failed to give distinct changes in protein bands of SDS-PAGE (Lee et al., 2017). O_3 treatment affects the oil quality and lipid content of oil crops. O_3 concentration of 50 ppm for 30 minutes increased the oil content of peanut grains (29.40 mg MA FM) and peanuts in pods (17.60 mg MA FM) to maximum with 2-thiobarbituric acid value (mg MA FM) of 276.20 and 222.22, respectively. However, significant changes in the peroxide values of the treated oils were not seen. With this, O_3 treatment also reduced the fungal colonies in the grains and podded grains (Cristina et al., 2019). Piechowiak et al. (2018) took an initiative to control enzymic activities in extracted wheat flour (fluid state) using O_3 technology. Controlling enzymic activity of the flour helps in obtaining good quality baked products. O_3 concentration of 5, 15, and 30 ppm for 5, 15, and 30 minutes. Amylolytic, proteolytic, lipolytic, and lipoxygenase activities of flour before and after O_3 treatment were determined. A declining trend in amylose, protease, and lipase activities were seen with increase in O_3 concentration. This may be because of oxidation of SH groups in enzymes and formation of disulfide bonds. However, the lipoxygenase activity of the ozonated flour increased, which may reduce the shelf life of the flour. Thus, O_3 remains an effective tool in controlling enzymatic activities of flour and aids in producing superior quality baked products. Interesting functional properties were exhibited by ozonated potato starch. WSI, gel strength up to 85°C, and in vitro digestibility of the potato starch at moderate gelatinization temperature increased after ozonation. On the other side, WAI and gel strength at 95°C decreased. Cereal products showing these kinds of functional properties would be helpful in formulating hydrocolloids, meat products, nursing care foods, and dysphagia foods (Castanha et al., 2019). The effect of addition of O_3 treated wheat bran (for 50 minutes) in noodles making was evaluated on the basis of rheological properties and microbiological properties. Almost 90% of microbial reduction was achieved during 30 days of storage. A proportion of 15% and 20% of treated bran were added to noodles as an additive. Tensile strength of the noodles increased as the starch and protein molecules got integrated. O_3 treatment encapsulates the starch molecules in the bran and strengthens the gluten network. This helps in increasing water absorption capacity and decreasing cooking loss (CL) of fortified noodles (Liu et al., 2020). Hence, O_3 has a constructive application in grain processing such as insect mortality, pesticide degradation, detoxification, modification of functional ingredients, etc., owing to give high quality food products. Table 6.1 summarizes the application of O_3 in cereal processing in food industries.

TABLE 6.1

Application of O$_3$ in Food Processing

Food Commodity	Objective	Processing Parameters	Result	Reference
Cowpea	To find the effect of ozone on adults of *Callosobruchus maculatus, Callosobruchus chinensis*	O$_3$ concentration: 0.25, 0.5, 1.0, 1.5, 2.0 g/m^3 for 24 hours	Mortality rate after seven days *Callosobruchus maculatus* at O$_3$ concentration 0.25 g/m^3: 91.9% 0.5 g/m^3: 96.2% 1.0 g/m^3: 100% 1.5 g/m^3: 100% 2.0 g/m^3: 100% *Callosobruchus chinensis* 100% mortality at all concentrations	(Gad et al., 2021)
Potato starch	To study the functional properties of ozonated starch	O$_3$ concentration: 47 mg/L Flow rate: 0.5 L/min Time: 0 to 60 minutes	60 minutes of Ozonated starch showed 10% of WSI at 85ºC 60 minutes of Ozonated starch showed 4% WAI at 70ºC 15 minutes of Ozonated starch at 65ºC, 70ºC showed gel strength energy value of 0.82 mJ, 0.80 mJ Ozonated starch showed 84% of rapid digestible starch	(Castanha et al., 2019)
Peanut grains Peanut grains in pods	To study the O$_3$ effect on fungal colonies and lipid content of peanuts	O$_3$ concentration: 10, 30, 50 ppm Time: 10, 30, 45, and 60 minutes	50 ppm for 60 minutes Fungal reduction: 75.79% in peanut grains Fungal reduction: 82.66% in podded peanut grains 2-thibarbituric acid value: 254.80 mg MA FM in grains 2-thiobarbituric acid value: 159.19 mg MA FM in podded peanut grains Oil content:12.30 mg MA FM in grains Oil content:12.30 mg MA FM in grains	(Cristina et al., 2019)
Semi-dried buckwheat noodles	To study impact of O$_3$ on physiochemical and microbiological properties of the noodles	Aqueous O$_3$ concentration: 0.43, 1.29, 1.88, 2.15, and 2.21 mg/L Flow rate: 5 g/h Time: 10, 20, 30, 40, and 50 minutes	Total plate count (TPC): 2.05 log CFU/g Cooking loss: <6.5% Water absorption: <1.1% Hardness: >4500 g Tension force: >16 g Free thiol (SH) group: 2.5 µmol/g	(Bai & Zhou, 2021)
Wheat bran	To investigate the addition of O$_3$ treated wheat bran in noodle fortification	Flow rate: L/min Time: 20, 30, 40, 50, and 60 minutes	15% of treated bran was added Addition of 15% of bran addition have Hardness: 5119.71 g WAI: 93.78% Cooking loss: 5.67%	(Liu et al., 2020)
Cantaloupe peels	To study the physiochemical properties of O$_3$ treated peels	Gaseous O$_3$ concentration: 31 g/L Time: 30 and 60 minutes	31 g/L for 30 minutes Colour value: L*: 39.15, a*: −12.71, b*: 28.66 pH: 5.64 TSS: 6.67ºBrix TPC: ≈190 µg/g Ascorbic acid content: ≈100 µg/g	(Miller et al., 2021)

(Continued)

TABLE 6.1 (*Continued*)
Application of O_3 in Food Processing

Food Commodity	Objective	Processing Parameters	Result	Reference
Watermelon juice	To study the impact of O_3 treatment on physiochemical properties of juice	O_3 flow rate: 1 L/ minute Time: 5, 10, 15, 20, and 25 minutes	TSS: 6.50°Brix pH: 5.883 TA: 0.061% of malic acid PME activity: 0.65 U/mL Ascorbic acid content: 0.525 µg/mL TPC: 192.606 µg GAE/mL Lycopene: 45.534 mg/kg	(B. J. Lee et al., 2022)
Whole milk and skim milk	To study the efficacy of O_3 in reducing microbial population of different kinds of milk	O_3 concentration: 25 and 28 mg/L Time: 5, 10, 15 minutes Gas flow rate: 1 L/minute	28 mg/L for 15 minutes Whole milk Pseudomonas spp: $>10^2$ CFU/mL Skim milk Pseudomonas spp: $\approx10^1$ CFU/mL	(Munhõs et al., 2019)
Butter	To study the effect of ozonated water on properties of butter	O_3 concentration: 0.15, 0.20, 0.25, and 0.30 mg/L	0.30 mg/L Dry matter: 79.67% Firmness: 347.46 N Particle size d (0.9): 164.50 µm Oxidative stability at 100°C: 21.12 hours	(Sert & Mercan, 2020)
Beef	To compare the efficiency of non-ozonated and ozonated water spray chilling in *E.coli* reduction	O_3 concentration: 12 ppm Time: 30 seconds Temperature: 3, 26, and 37°C	*E. coli*: 1.46 log reduction Aerobic bacteria: 0.99 log reduction	(Kalchayanand et al., 2019)
Turkey breast meat	To assess the physiochemical and microbiological properties of meat	O_3 concentration: 1 kg/m^3 Time: 2, 4, 6, and 8 hours	1 kg/m^3 for 8 hours TBARS: 0.37 mg of malonaldehyde per kg meat Carbonyl content: 5.24 nmol of carbonyl per mg of protein Water holding capacity: 83.32% Cooking yield: 73.73%	(Ayranci et al., 2020)
Egg yolk	To study the effect of O_3 mediated oxidation on flavour profile of egg yolk	O_3 concentration: 10 to 15 g/m^3 Time: 0, 10, 20, 30, and 40 minutes	10 to 20 g/m^3 for 30 minutes Total volatile substance Alkanes: – Terpenoids: 0.06, Aldehydes: 94.63, Ketones: 0.29, Alcohols: 4.92, Amines: 0.10	(B. Chen et al., 2022)
Egg yolk	To improve the functionality of egg yolk liquid by ozonation	Time: 0, 10, 20, 30, and 40 minutes	20 minutes of O_3 treatment Emulsifying activity: 0.5 to 0.6 Emulsifying stability: ≈15 Emulsifying capacity: 30 to 35 Creaming index: ≈20	(Li et al., 2022)

Some of common microorganisms that invade FaV include *Erwinia carotovora, Pectobacterium carotovorum, Xanthomonas campestris,* and *Bacillus spp* (Alegbeleye et al., 2022). Consumption of contaminated FaV causes food poisoning, as it contains toxins (Khashroom et al., 2019). Acting as a sanitizing agent, O_3 has a vital role in disinfecting FaV. Nevertheless, physical and chemical properties of FaV such as weight loss, firmness, colour and TSS, titratable acidity (TA), ascorbic acid content, pigments can also be maintained by O_3 treatment (Boopathy et al., 2022). Fresh-cut spinach leaves were dipped in aqueous O_3 for different time periods of less than 1 minute. Spinach leaves dipped in optimized ozonated water had longer shelf life up to 13 days at 8°C. During storage, no in loss of chroma, chlorophyll a, chlorophyll b, total chlorophyll pigments, and yellowing of leaves were observed. Flavonoid content of the treated leaves increased, whereas weight, pH, TSS, TPC, TA, and total antioxidant activity (TAC) of the leaves were not affected. Greater reduction in gram positive and Enterobacteriaceae population were achieved during first five days (Papachristodoulou et al., 2018). The oxidative properties exhibited by O_3 molecules can damage the tissues of FaV, leading to loss of its structural integrity. This mechanism of attack by O_3 reduced the microflora of fresh cut yellow onions with remarkable decrease in respiration rate, weight loss and thus extending the shelf life of onions. Besides, the interaction between O_3 and surface of onions removed 12.2% to 70.5% of pesticides such as chlorpyrifos, cypermethrin dimethyl dichlorovinyl phosphate (DDVP), methomyl, trichlorfon, and omethoate (Chen et al., 2020). Fruit juice is one of the major products from fruit processing industry that 50% of fruits are processed into juice (Kandemir et al., 2022).

Gaseous ozonation at flow rate of 0.2 L/min to citrus juices (orange, lemon, and lime) altered its physiochemical properties such as pH, colour, pectin methyl esterase, ascorbic acid content, TPC, and antioxidant activity. Gradual decrease in TSS of all the three juices was observed with increase in treatment time. The decomposition of O_3 molecules produced hydroxyl radicals that increased the pH of juice (pH <7). The pH of the orange juice increased gradually from 0 to 30 minutes and pH of lime and lemon juice increased drastically between 10 and 20 minutes and gradually from 20 to 30 minutes. Consequently, it decreased the activity of pectin methyl esterase (0.04–0.06 PME U/mL) as the enzyme activity is optimum at neutral pH. Additionally, the increase in pH changed the colour of the juice, with slight impairment on juice quality. Turbidity of the treated juices decreased with 11.9% loss in orange juice, 6.8% in lemon juice, and 6.47% in lime juice. Ascorbic content of the juices decreased drastically during initial 10 minutes due to instant reaction between the thermolabile ascorbic acid in juice and oxygen molecules and gradually decreased with continuous exposure to O_3 that 85% loss in orange juice, 60% in lime juice, and 21% in lemon juice were observed. TPC of the juices decreased with treatment time. However, the TPC of orange increased at 10 minutes and gradually decreased during 20 and 30 minutes. About 84.36%,78.7%, and 76.15% TPC was lost in O_3 treated lemon, lime, and orange juices, respectively. The authors found that the study of the effect of O_3 on juice needs further validation to determine the critical limits as the decrease in colour, TPC, ascorbic content, and pectin methyl esterase enzyme of the juices were not acceptable (Shah et al., 2019). However, optimized O_3 treatment conditions help in maintaining the physiochemical and microbiological properties of fruit juice. For example, 25 minutes of O_3 treatment inactivated the microbial colonies in clarified and unclarified watermelon juice and thus extended the shelf life of juice. In terms of physiochemical properties, the TSS and pectin methyl esterase enzyme activity were not altered with the O_3 treatment. In contrast to previous study, the colour of the fruit juice brightened after O_3 treatment. However, the attack of O_3 molecules on the double bonds of organic pigment reduced the redness and yellowness. Further ozonation reduced non-enzymatic browning (NEB) of unclarified and clarified juices attributed by lower NEB values (Lee et al., 2022). Thus, optimized O_3 concentration, gas flow rate, and treatment time have a positive impact on physiochemical properties of fruit juice. O_3 has favourable application in food waste valorization by aiding bioactive compound extraction. Combination of O_3 (40 μg/L) and convective drying (40°C) had slightly increased the essential oil yield of orange peels. O_3 concentration and treatment time had a notable contribution in increasing the extraction yield, antioxidant activity and reducing the fungal colonies. O_3 treatment weakened the plant tissues and disintegrated the cell membrane to facilitate oil extraction. However, the treated peels looked darker and pectin content was lesser than control peels (Bechlin et al., 2020). Increase in ascorbic acid content, TPC was observed in 30 minutes O_3-treated cantaloupe peels. Also, no significant effect on pH and chlorophyll were observed after 30 minutes of treatment. Intense oxidation during O_3 treatment produced stress in peels that reduced its antioxidant activity. Thus, O_3 remains as an effective method in retaining the physiochemical properties of FaV. The authors fixed an objective to produce antilisterial cantaloupe peels through O_3 treatment that could be used as an additive (Miller et al., 2021).

Yet another study was done to assess the potency of O_3 in increasing the antioxidant activity of the beer added with ozonated fruit pulp. The study initially involved ozonation of saskatoon fruit pulps from different cultivators followed by addition in barley beer on the seventh day of fermentation. It was found that the polyphenolic content and the antioxidant activity of the beer with ozonated saskatoon pulp of all the varieties increased. However, organic acids such as neochlorogenic acid, chlorogenic acid, and caffeic acid decreased in beer with treated pulps (Gorzelany et al., 2022). Hurdle technology involving O_3 and ultraviolet radiation (UV-C) substantially reduced the microbial load in persimmon fruits than individual treatments. UVC lamps at a distance of 24 cm with the radiance of 0.6034 mW/cm^3 for 0.5 hours and 9.81 mg/m^3 for 1 hour reduced the following microorganisms *Aspergillus niger, Curtobacterium pusillum, Luteibacter jiangsuensis, Penicillium griseofulvum* 0% in the fruits after drying for 36 and 72 hours. The combined treatment elevated the astringency taste while no significant change in colour, MC, water activity (a_w), or pectin content was observed (X. Chen et al., 2020). Table 6.1 lists the application of O_3 in fruit and vegetable processing in food industries.

Being perishable, milk and milk products are prone to microbial contamination. Microbes such as *Acinetobacter, Campylobacter,* Pseudomonas spp., *Listeria monocytogenes*, Salmonella spp., etc., are major pathogenic contaminants in milk and their products that are responsible for the deadliest disease outbreaks (Fusco et al., 2020). In addition to these microbes, biofilm formation in dairy processing equipment contaminates and reduces the quality of the processed products (Baskaran & Radhakrishnan, 2021). Therefore, it is very important to ensure microbial safety of milk and its products. O_3 has a promising role in decontaminating milk and its products, as well as processing equipment surface (Panebianco et al., 2022). Munhõs et al. (2019) found exposure time and milk composition influenced the efficacy of the ozonation in milk. Compared to control, greater logarithmic reduction in *Pseudomonas spp.* Present in whole and skim milks were achieved at increased treatment time (15 minutes). However, the ozonated skim milk had lesser population of *Pseudomonas spp* than ozonated whole milk. This was because the fat and protein in whole milk hindered the bactericidal effect of O_3 during decontamination. Younis et al. (2020) compared and contrasted the effectiveness of the thermal pasteurization and O_3 treatments in reducing pathogenic microbial population in raw milk. Thirty minutes of ozonation inactivated all microbes in milk including *Bacillus cereus, Staphylococcus aureus*, and *Salmonella. Typhimurium, Shigella flexneri*, and *E. coli* below 1 CFU/mL, whereas pasteurized milk had higher microbial colonies count. Ozonated milk showed no change in physiochemical properties such as TSS, TA, protein, fat, or lactose. Although it showed higher TSS, protein content, and lower fat content than raw and pasteurized milk. Butter churned from ozonated cream gave desirable results in microbial load reduction and showed some variations in its microstructural and physiochemical properties. Total plate count bacteria of treated butter was reduced to 1.59 log CFU/mL. Degradation of carotenoids during ozonation decreased the yellowness of butter, with increased lightness and redness values. The O_3 concentration, particle size, firmness, and spreadability of the butter are correlated in such a way that as the O_3 concentration increases, the particle size of butter decreases. As the particle size decreases, the firmness of the butter decreases which increases the spreadability of butter. In this study, when the O_3 concentration was increased from 0.15 mg/L to 0.30 mg/L, the particle size decreased and lied between the range of 105.50 μm and 117.50 μm. The firmness of the butter increased with O_3 concentration up to 347.46 N from 126.86 N. This decreased the spreadability nature of butter. In the matter of oxidative stability of the butter, it decreased with increased O_3 concentration. This may be due to interference of oxygen molecules with fat component of butter (Sert & Mercan, 2020). Similar results were obtained when ozonated cream was used in butter production (Sert et al., 2020). Aflatoxin M1 (AFM1) is one among the group1 carcinogens found in milk and milk products that are acute and chronic poisons to humans (Nguyen et al., 2020). O_3 has the ability to degrade these toxins to greater extent. For example,60 minutes of O_3 treatment degraded 18.9% and 9.9% of AFM1 in milk concentrate and whey concentrate, respectively. Additionally, the treatment degraded β glucan and tetracycline antibiotics in concentrates. O_3 reduced cohesivity of the concentrates that impacted on reduction of firmness and consistency (Sert & Mercan, 2021). Table 6.1 provides the application of O_3 in milk processing in food industries.

Due to their high pH and MC, meat and its products have a high risk of microbial contamination, enzymatic oxidation, and lipolytic oxidation (Pellissery et al., 2020). Herein, O_3 acts as a sanitizing agent in inactivating pathogenic microbes such as *Pseudomonas spp, Salmonella spp coliforms, E.coli*, etc., and also has a positive role in altering the biochemical properties of meat and their products. Exposing chicken fillets to 0.38 mg/L of O_3 for 120 minutes at 3°C inactivated 6.89 log CFU/g of aerobic mesophilic bacterial population to 6.10 log CFU/g. Increase in contact temperature (37°C) lowered the percentage of disinfected mesophilic bacterial population. O_3 exposure reduced the protein in chicken fillets by 1%, as the oxidation properties of O_3 denatured the proteolytic enzymes of meat. However, pH and water content of meat showed no significant change, even after the treatment (Fathul Karamah & Wajdi, 2018). Kalchayanand et al. (2019) suggested that ozonated water spray chilling would be better than traditional spray chilling to reduce *E. coli* colonies in fresh beef flanks. Low pressure spraying of the non-ozonated and ozonated water at concentration 12 ppm for 90 seconds at the interval of 30 minutes for 12 hours reduced *E. coli* colonies to 5.24 log CFU/cm² and 3.78 log CFU/cm², respectively. In addition to *E. coli* colonies inactivation, O_3 treatment caused sublethal injuries to the bacterial cells, which struggles to proliferate at low temperature and dies subsequently. The change in physiochemical properties of gaseous O_3 treated turkey breast meat was evaluated by placing the meat in O_3 gas chamber for 8 hours. The carbonyl content, thiobarbituric acid content (TBARS), increased by sixfold times due to the oxidation effect of O_3. Structural changes brought by the O_3 increased the water holding capacity (WHC) and cooking yield of meat. The colour of the meat became light due to bleaching effect of O_3 and partial denaturation of protein (Ayranci et al., 2020). Table 6.1 summarizes the application of O_3 in meat and its product processing in food industries.

In egg processing, O_3 has been applied to disinfect eggshells which is the major source of microorganisms. Further, the technology is also in egg pasteurization, without comprising the physiochemical, thermal, and dielectric properties of egg and their products (Bermudez-Aguirre & Niemira, 2023). O_3 treatment applied to egg yolk had influenced its physiochemical properties such as fatty acid, aldehydes, and volatile substances. Half an hour of ozonation increased the aldehyde content to 94.63% and decreased the fatty acid to 22.3%. Moreover, the dibutyl amine with strong ammonia taste had been reduced to 0% from 1.50%. These all could be attributed to oxidation products formed during treatment, which degraded the fatty acid of the yolk and also could have part in increasing volatile substance. However, some volatile compounds, alkanes, terpenoids, amines, alcohols, and ketones are reduced after the treatment. O_3 treatment had not significantly affected the proteins of yolk that it did not showed distinguishable changes in SDS-PAGE (B. Chen et al., 2022). Li et al. (Li et al., 2022) studied the effect of O_3 on the functional properties of egg yolk such as emulsifying activity, emulsifying capacity, emulsifying stability, and creaming index. The oxidational effect of O_3 makes the yolk proteins to unfold partially, and it became flexible with the exposure of hydrophobic groups to O_3. This eventually increased emulsifying activity and emulsifying stability of egg yolk by 1.42 and 1.75 times, respectively. The structural change brought by 20 minutes of O_3 is further supported by the zeta potential, which is increased by 16.08 times. Sun et al. (2021) studied the O_3-induced changes in properties of egg yolk protein gels by determining free sulfhydryl content (FSC), free carbonyl content (FCC), WHC, CL, and rheological properties.

During the study, it was found that O_3 undergoes redox reactions in solutions and generates hydroxyl radicals, hydrogenated O_3 radicals, superoxide anion radicals, etc., which attack disulfide bonds in proteins and oxidize it. This impacts on decrease in FSC of egg yolk with increase in treatment time. In contrast to reduction in FSC, the FCC of egg yolk increases with the formation of lipid hydroperoxides and conversion of free amino acids into active carbonyl groups during ozonation. In the case of WHC, treated gels exhibit a higher holding capacity, which can be attributed to two possible reasons. First, O_3 treatment enhances the gel network, reducing water loss. Second, O_3 improves the interaction between protein and water in the yolk gel, thus reducing water loss.. With increased WHC, the treated gels give lesser loss in cooking. Table 6.1 lists the application of O_3 in egg processing in food industries.

6.7 CHALLENGES

Despite having beneficial applications such as decontamination and preservation in food processing, O_3 exhibits some limitations at industrial scale. Large-scale O_3 fumigation is not feasible because O_3 is influenced by internal factors like grain bed thickness, MC, O_3 concentration, treatment time, half-life, and external factors like temperature and relative humidity (Pandiselvam et al., 2017). When it comes to degrading pesticide residues in the rough, hairy-structured FaV, agitation mechanism is required to bring the O_3 molecules into contact with the pesticide compounds there. Yet, only a certain percent of pesticides could be eliminated. Also, the oxidation property exhibited by O_3 molecules during treatment degrades proteins, lipids, colour, pH, pigments, and vitamins of FaV, comprising consumer acceptability (Pandiselvam, Kaavya, et al., 2020). To effectively use O_3, on-site production is required due to its instability, which, in turn, necessitates a high capital cost (Sivaranjani et al., 2021). O_3 has lethal effects on humans such that 0.1 to 0.5 ppm of O_3 exposure for 3 to 6 hours affects vision. About 5 to 10 ppm of exposure affects the human respiratory organs. At higher levels like 50 ppm exposure causes asthma, cardiovascular disease, and even death. Therefore, not only on microorganisms, O_3 has lethal effect on humans too (Boopathy et al., 2022). To mitigate this, USFDA (2023) has recommended 0.1 ppm of O_3 for 8 hours a day as the threshold limit for O_3 exposure.

6.8 SUMMARY

The contribution of O_3 technology in food processing is immense that it acts as fumigant in grains, disinfectant in FaV, milk and their products, meat, and their products. Added to this, O_3 has a part in maintaining the physiochemical properties of foods. However, to fully exploit the effectiveness of O_3 technology in food processing some research gaps have to be bridged. Comparing and contrasting the efficacy of gaseous and aqueous O_3 treatments would be fruitful to increase its efficacy. Studies on hurdle concept of O_3 with other non-thermal technologies such as ultrasound, cold plasma, and high-pressure processing would be welcoming. In depth research on large-scale application, influence on seed germination, and bioactive compound extraction would be a breakthrough in O_3 application.

REFERENCES

A. Nath, K. M., Swer, T., Dutta, D., Verma, N., Deka, B. C., & Gangwar, B. (2010). A review on the application of nanotechnology in food processing and packaging. *Food Engineering Progress*, *14*(4), 283–291.

Afsah-Hejri, L., Hajeb, P., & Ehsani, R. J. (2020). Application of ozone for degradation of mycotoxins in food: A review. *Comprehensive Reviews in Food Science and Food Safety*, *19*(4), 1777–1808. https://doi.org/10.1111/1541-4337.12594

Alegbeleye, O., Odeyemi, O. A., Strateva, M., & Stratev, D. (2022). Microbial spoilage of vegetables, fruits and cereals. *Applied Food Research*, *2*(1), 100122. https://doi.org/10.1016/j.afres.2022.100122

Alexandre, A. P. S., Castanha, N., Costa, N. S., Santos, A. S., Badiale-Furlong, E., Augusto, P. E. D., & Calori-Domingues, M. A. (2019). Ozone technology to reduce zearalenone contamination in whole maize flour: Degradation kinetics and impact on quality. *Journal of the Science of Food and Agriculture*, *99*(15), 6814–6821. https://doi.org/10.1002/jsfa.9966

Ashby, M. T. (2012). *Hypothiocyanite* (pp. 263–303). https://doi.org/10.1016/B978-0-12-396462-5.00008-8

Ayranci, U. G., Ozunlu, O., Ergezer, H., & Karaca, H. (2020). Effects of ozone treatment on microbiological quality and physicochemical properties of turkey breast meat. *Ozone: Science and Engineering*, *42*(1), 95–103. https://doi.org/10.1080/01919512.2019.1653168

Bai, Y. P., & Zhou, H. M. (2021). Impact of aqueous ozone mixing on microbiological, quality and physico-chemical characteristics of semi-dried buckwheat noodles. *Food Chemistry*, *336*(June 2020), 127709. https://doi.org/10.1016/j.foodchem.2020.127709

Baskaran, K., & Radhakrishnan, M. (2021). Surface Pasteurization and Disinfection of Dairy Processing Equipment Using Cold Plasma Techniques. In *Non-Thermal Processing Technologies for the Dairy Industry* (pp. 143–156). CRC Press. https://doi.org/10.1201/9781003138716-11

Bechlin, T. R., Granella, S. J., Christ, D., Coelho, S. R. M., & Paz, C. H. de O. (2020). Effects of ozone application and hot-air drying on orange peel: Moisture diffusion, oil yield, and antioxidant activity. *Food and Bioproducts Processing*, *123*, 80–89. https://doi.org/10.1016/j.fbp.2020.06.012

Bermudez-Aguirre, D., & Niemira, B. A. (2023). A review on egg pasteurization and disinfection: Traditional and novel processing technologies. *Comprehensive Reviews in Food Science and Food Safety*, *22*(2), 756–784. https://doi.org/10.1111/1541-4337.13088

Boopathy, B., Rajan, A., & Radhakrishnan, M. (2022). Ozone: An alternative fumigant in controlling the stored product insects and pests: A status report. *Ozone: Science & Engineering*, *44*(1), 79–95. https://doi.org/10.1080/01919512.2021.1933899

Castanha, N., Santos, D. N. e, Cunha, R. L., & Augusto, P. E. D. (2019). Properties and possible applications of ozone-modified potato starch. *Food Research International*, *116*, 1192–1201. https://doi.org/10.1016/j.foodres.2018.09.064

Chen, X., Liu, B., Chen, Q., Liu, Y., & Duan, X. (2020). Application of combining ozone and UV-C sterilizations in the artificial drying of persimmon fruits. *LWT*, *134*, 110205. https://doi.org/10.1016/j.lwt.2020.110205

Chen, C., Liu, C., Jiang, A., Zhao, Q., Liu, S., & Hu, W. (2020). Effects of ozonated water on microbial growth, quality retention and pesticide residue removal of fresh-cut onions. *Ozone: Science & Engineering*, *42*(5), 399–407. https://doi.org/10.1080/01919512.2019.1680527

Chen, B., Sun, Y., Jin, H., Wang, Q., Li, Z., Jin, Y., & Sheng, L. (2022). An efficient processing strategy to improve the flavor profile of egg yolk: Ozone-mediated oxidation. *Molecules*, *28*(1), 124. https://doi.org/10.3390/molecules28010124

Chuwa, C., &, Vaidya, D., Kathuria, D., Gautam, S., Sharma, S., Sharma B. (2020). Ozone O3 an emerging technology in the food industry. *Food & Nutrition Journal*. https://www.gavinpublishers.com/article/view/ozone-o3-an-emerging-technology-in-the-food-industry

Deshwal, R., Vaibhav, V., Kumar, N., Kumar, A., & Singh, R. (2020). Stored grain insect pests and their management: An overview. ~ 969 ~ *Journal of Entomology and Zoology Studies*, *8*(5), 969–974.

Donohue, J. M. (2003). Water Conditioning, Industrial. In *Encyclopedia of Physical Science and Technology* (pp. 671–697). Elsevier. https://doi.org/10.1016/B0-12-227410-5/00819-X

Dubey, P., Singh, A., & Yousuf, O. (2022). Ozonation: An evolving disinfectant technology for the food industry. *Food and Bioprocess Technology*, *15*(9), 2102–2113. https://doi.org/10.1007/s11947-022-02876-3

Ensley, S. M., & Radke, S. L. (2019). Mycotoxins in Grains and Feeds. In *Diseases of Swine* (pp. 1055–1071). Wiley. https://doi.org/10.1002/9781119350927.ch69

Epelle, E. I., Macfarlane, A., Cusack, M., Burns, A., Okolie, J. A., Mackay, W., Rateb, M., & Yaseen, M. (2023). Ozone application in different industries: A review of recent developments. *Chemical Engineering Journal*, *454*, 140188. https://doi.org/10.1016/j.cej.2022.140188

FAO. (2021). *Climate change fans spread of pests and threatens plants and crops, new FAO study*. Food and Agriculture Organisation of United Nations.

Fathul Karamah, E., & Wajdi, N. (2018). Application of ozonated water to maintain the quality of chicken meat: Effect of exposure time, temperature, and ozone concentration. *E3S Web of Conferences*, *67*, 04044. https://doi.org/10.1051/e3sconf/20186704044

Fusco, V., Chieffi, D., Fanelli, F., Logrieco, A. F., Cho, G., Kabisch, J., Böhnlein, C., & Franz, C. M. A. P. (2020). Microbial quality and safety of milk and milk products in the 21st century. *Comprehensive Reviews in Food Science and Food Safety*, *19*(4), 2013–2049. https://doi.org/10.1111/1541-4337.12568

Gad, H. A., Abo Laban, G. F., Metwaly, K. H., Al-Anany, F. S., & Abdelgaleil, S. A. M. (2021). Efficacy of ozone for *Callosobruchus maculatus* and *Callosobruchus chinensis* control in cowpea seeds and its impact on seed quality. *Journal of Stored Products Research*, 92, 101786. https://doi.org/10.1016/j.jspr.2021.101786

Gorzelany, J., Michałowska, D., Pluta, S., Kapusta, I., & Belcar, J. (2022). Effect of ozone-treated or untreated saskatoon fruits (*Amelanchier alnifolia* Nutt.) applied as an additive on the quality and antioxidant activity of fruit beers. *Molecules*, 27(6), 1976. https://doi.org/10.3390/molecules27061976

Guzel-Seydim, Z. B., Greene, A. K., & Seydim, A. C. (2004). Use of ozone in the food industry. *LWT – Food Science and Technology*, 37(4), 453–460. https://doi.org/10.1016/j.lwt.2003.10.014

Hernández-Hernández, H. M., Moreno-Vilet, L., & Villanueva-Rodríguez, S. J. (2019). Current status of emerging food processing technologies in Latin America: Novel non-thermal processing. *Innovative Food Science & Emerging Technologies*, 58. https://doi.org/10.1016/j.ifset.2019.102233

Jamil, A., Farooq, S., & Hashmi, I. (2017). Ozone disinfection efficiency for indicator microorganisms at different pH values and temperatures. *Ozone: Science & Engineering*, 39(6), 407–416. https://doi.org/10.1080/01919512.2017.1322489

Joshi, K., Mahendran, R., Alagusundaram, K., Norton, T., & Tiwari, B. K. (2013). Novel disinfectants for fresh produce. *Trends in Food Science and Technology*, 34(1), 54–61. https://doi.org/10.1016/j.tifs.2013.08.008

Kalchayanand, N., Worlie, D., & Wheeler, T. (2019). A novel aqueous ozone treatment as a spray chill intervention against *Escherichia coli* O157:H7 on surfaces of fresh beef. *Journal of Food Protection*, 82(11), 1874–1878. https://doi.org/10.4315/0362-028X.JFP-19-093

Kandemir, K., Piskin, E., Xiao, J., Tomas, M., & Capanoglu, E. (2022). Fruit juice industry wastes as a source of bioactives. *Journal of Agricultural and Food Chemistry*, 70(23), 6805–6832. https://doi.org/10.1021/acs.jafc.2c00756

Khashroom, A., Saewan, S., George, J., & Hamad, H. (2019). Chemical and microbial environmental contaminants in fruits and vegetables and their effects on health: A mini review. *Journal of Environmental Science, Toxicology and Food Technology*, 13(4), 70–76. http://www.class.jpu.edu.jo/juris/uploads/publication/sidr/20211121-0313258771.pdf

Laureth, J. C. U., Christ, D., Ganascini, D., & Coelho, S. R. M. (2019). Effect of ozone application on the fungal count and lipid quality of peanut grains. *Journal of Agricultural Science*. 11(5), 271–280. https://doi.org/10.5539/jas.v11n5p271

Lee, M. J., Kim, M. J., Kwak, H. S., Lim, S.-T., & Kim, S. S. (2017). Effects of ozone treatment on physicochemical properties of Korean wheat flour. *Food Science and Biotechnology*, 26(2), 435–440. https://doi.org/10.1007/s10068-017-0059-5

Lee, B. J., Ting, A. S. Y., & Thoo, Y. Y. (2022). Impact of ozone treatment on the physico-chemical properties, bioactive compounds, pectin methylesterase activity and microbiological properties of watermelon juice. *Journal of Food Science and Technology*, 59(3), 979–989. https://doi.org/10.1007/s13197-021-05102-8

Lemic, D., Jembrek, D., Bažok, R., & Živković, I. P. (2019). Ozone effectiveness on wheat weevil suppression: Preliminary research. *Insects*, 10(10). https://doi.org/10.3390/insects10100357

Li, Z., Sun, Y., Jin, H., Wang, Q., Jin, Y., Huang, X., & Sheng, L. (2022). Improvement and mechanism of emulsifying properties of liquid egg yolk by ozonation technology. *LWT*, 156, 113038. https://doi.org/10.1016/j.lwt.2021.113038

Liu, C., Zhang, Y., Li, H., Li, L., & Zheng, X. (2020). Effect of ozone treatment on processing properties of wheat bran and shelf life characteristics of noodles fortified with wheat bran. *Journal of Food Science and Technology*, 57(10), 3893–3902. https://doi.org/10.1007/s13197-020-04421-6

Miller, F. A., Fundo, J. F., Garcia, E., Silva, C. L. M., & Brandão, T. R. S. (2021). Effect of gaseous ozone process on cantaloupe melon peel: Assessment of quality and antilisterial indicators. *Foods*, 10(4), 727. https://doi.org/10.3390/foods10040727

Munhõs, M. C., Navarro, R. S., Nunez, S. C., Kozusny-Andreani, D. I., & Baptista, A. (2019). *Reduction of Pseudomonas Inoculated into Whole Milk and Skin Milk by Ozonation* (pp. 837–840). https://doi.org/10.1007/978-981-13-2119-1_130

Nguyen, T., Flint, S., & Palmer, J. (2020). Control of aflatoxin M1 in milk by novel methods: A review. *Food Chemistry*, 311, 125984. https://doi.org/10.1016/j.foodchem.2019.125984

Pandiselvam, R., Kaavya, R., Jayanath, Y., Veenuttranon, K., Lueprasitsakul, P., Divya, V., Kothakota, A., & Ramesh, S. V. (2020). Ozone as a novel emerging technology for the dissipation of pesticide residues in foods – a review. *Trends in Food Science and Technology*, 97, 38–54. https://doi.org/10.1016/j.tifs.2019.12.017

Pandiselvam, R., Kothakota, A., Thirupathi, V., Anandakumar, S., & Krishnakumar, P. (2017). Numerical simulation and validation of ozone concentration profile in green gram (Vigna radiate) bulks. *Ozone: Science & Engineering*, 39(1), 54–60. https://doi.org/10.1080/01919512.2016.1244641

Pandiselvam, R., Mayookha, V. P., Kothakota, A., Sharmila, L., Ramesh, S. V., Bharathi, C. P., Gomathy, K., & Srikanth, V. (2020). Impact of ozone treatment on seed germination – a systematic review. *Ozone: Science & Engineering, 42*(4), 331–346. https://doi.org/10.1080/01919512.2019.1673697

Pandiselvam, R., Thirupathi, V., Mohan, S., Vennila, P., Uma, D., Shahir, S., & Anandakumar, S. (2019). Gaseous ozone: A potent pest management strategy to control Callosobruchus maculatus (Coleoptera: Bruchidae) infesting green gram. *Journal of Applied Entomology, 143*(4), 451–459. https://doi.org/10.1111/jen.12618

Panebianco, F., Rubiola, S., & Di Ciccio, P. A. (2022). The use of ozone as an eco-friendly strategy against microbial biofilm in dairy manufacturing plants: A review. *Microorganisms, 10*(1), 162. https://doi.org/10.3390/microorganisms10010162

Panigrahi, C., Mishra, H. N., & De, S. (2020). Effect of ozonation parameters on nutritional and microbiological quality of sugarcane juice. *Journal of Food Process Engineering, 43*(11), 1–14. https://doi.org/10.1111/jfpe.13542

Papachristodoulou, M., Koukounaras, A., Siomos, A. S., Liakou, A., & Gerasopoulos, D. (2018). The effects of ozonated water on the microbial counts and the shelf life attributes of fresh-cut spinach. *Journal of Food Processing and Preservation, 42*(1), e13404. https://doi.org/10.1111/jfpp.13404

Pellissery, A. J., Vinayamohan, P. G., Amalaradjou, M. A. R., & Venkitanarayanan, K. (2020). Spoilage Bacteria and Meat Quality. In *Meat Quality Analysis* (pp. 307–334). Elsevier. https://doi.org/10.1016/B978-0-12-819233-7.00017-3

Petrescu, D. C., Vermeir, I., & Petrescu-Mag, R. M. (2019). Consumer understanding of food quality, healthiness, and environmental impact: A cross-national perspective. *International Journal of Environmental Research and Public Health, 17*(1), 169. https://doi.org/10.3390/ijerph17010169

Piechowiak, T., Józefczyk, R., & Balawejder, M. (2018). Impact of ozonation process of wheat flour on the activity of selected enzymes. *Journal of Cereal Science, 84*, 30–37. https://doi.org/10.1016/j.jcs.2018.09.010

Qi, L., Li, Y., Luo, X., Wang, R., Zheng, R., Wang, L., Li, Y., Yang, D., Fang, W., & Chen, Z. (2016). Detoxification of zearalenone and ochratoxin A by ozone and quality evaluation of ozonised corn. *Food Additives and Contaminants – Part A Chemistry, Analysis, Control, Exposure and Risk Assessment, 33*(11), 1700–1710. https://doi.org/10.1080/19440049.2016.1232863

Sarron, E., Gadonna-Widehem, P., & Aussenac, T. (2021). Ozone treatments for preserving fresh vegetables quality: A critical review. *Foods, 10*(3), 605. https://doi.org/10.3390/foods10030605

Sert, D., & Mercan, E. (2020). Effects of churning with different concentrations of ozonated water on particle size, texture, oxidation, melting and microbiological characteristics of butter. *International Dairy Journal, 111*, 104838. https://doi.org/10.1016/j.idairyj.2020.104838

Sert, D., & Mercan, E. (2021). Effects of ozone treatment to milk and whey concentrates on degradation of antibiotics and aflatoxin and physicochemical and microbiological characteristics. *LWT, 144*(February). https://doi.org/10.1016/j.lwt.2021.111226

Sert, D., Mercan, E., & Kara, Ü. (2020). Butter production from ozone-treated cream: Effects on characteristics of physicochemical, microbiological, thermal and oxidative stability. *LWT, 131*, 109722. https://doi.org/10.1016/j.lwt.2020.109722

Shah, N. N. A. K., Sulaiman, A., Sidek, N. S. M., & Supian, N. A. M. (2019). Quality assessment of ozone-treated citrus fruit juices. *International Food Research Journal, 26*(5), 1405–1415.

Shah, N. N. A. K., Supian, N. A. M., & Hussein, N. A. (2019). Disinfectant of pummelo (*Citrus grandis* L. Osbeck) fruit juice using gaseous ozone. *Journal of Food Science and Technology, 56*(1), 262–272. https://doi.org/10.1007/s13197-018-3486-2

Sivaranjani, S., Prasath, V. A., Pandiselvam, R., Kothakota, A., & Mousavi Khaneghah, A. (2021). Recent advances in applications of ozone in the cereal industry. *LWT, 146*(January), 111412. https://doi.org/10.1016/j.lwt.2021.111412

Sun, Y., Wang, Q., Jin, H., Li, Z., & Sheng, L. (2021). Impact of ozone-induced oxidation on the textural, moisture, micro-rheology and structural properties of egg yolk gels. *Food Chemistry, 361*, 130075. https://doi.org/10.1016/j.foodchem.2021.130075

Tiwari, B. K., & Muthukumarappan, K. (2012). Ozone in Fruit and Vegetable Processing. In *Ozone in Food Processing* (pp. 55–80). Wiley-Blackwell. https://doi.org/10.1002/9781118307472.ch5

USFDA. (2023). *Food and Drugs*. U.S. Food and Drug Administration.

WHO. (2018). *Mycotoxins*. World Health Organization.

WHO. (2022). *Food safety*. World Health Organization.

Xue, W., Macleod, J., & Blaxland, J. (2023). The use of ozone technology to control microorganism growth, enhance food safety and extend shelf life: A promising food decontamination technology. *Foods*, *12*(4), 814. https://doi.org/10.3390/foods12040814

Younis, F., Fayed, A., Elbatawy, O., & Elsisi, A. (2020). A comparison between ozonation and thermal process in relation to cow's milk attributes with emphasis on pathogens. *Arab Universities Journal of Agricultural Sciences*, *27*(5), 2593–2600. https://doi.org/10.21608/ajs.2020.20456.1131

Zhang, T., & Huang, C.-H. (2020). Simultaneous quantification of peracetic acid and hydrogen peroxide in different water matrices using HPLC-UV. *Chemosphere*, *257*, 127229. https://doi.org/10.1016/j.chemosphere.2020.127229

7 Electrospraying and Electrospinning

Applications in the Food Industry

S. K. Reshmi, L. Mahalakshmi, A. J. Moses,
and C. Anandharamakrishnan

7.1 INTRODUCTION

Since the start of time, fiber has been used as a fundamental component of human existence and is found abundant in nature both as continuous filaments or elongated forms. Natural processes such as silkworms making silk filaments to make coconuts have been a significant inspiration for the creation of man-made fibers (Heim, Keerl, & Scheibel, 2009). Synthetic fibers came into existence as polymer and chemistry improved. DuPont developed nylon in 1938, making it the first synthetic material to be commercially successful. In 1887, Charles V. Boys demonstrated that it was possible to extract fibers from viscoelastic solution in the presence of an applied external electric field by connecting the nozzle of the capillary tube containing the solution to an electrical supply. This process is now known as electrospinning and helps to develop ultrathin fiber on the nanometer scale. According to reports, the electrospinning method has produced electrospun fibers with diameters as small as 1 nm and even lower (Dzenis, 2004; Jian et al., 2018). A new life was breathed into electrospinning, eventually, it became the preferred way to produce fibers with nanometer diameters. Electrospinning began receiving increasing attention at the beginning of this century when new materials and formulations for fabricating composites and ceramic nanofibers became available (Dai et al., 2002; Li, Wang, & Xia, 2003). The electrospraying technique, also known as electrohydrodynamic atomization, was developed in the seventeenth century by William Gilbert. Vonnegut et al. developed the electrospinning technique for fabricating microparticles in 1952, which later evolved into electrospraying. Various researchers and engineers have been fascinated by electrospinning and electrospraying since 1980, and their studies have grown dramatically each year (Ding et al., 2019; Wu et al., 2020). Electrospinning and electrospraying are kind of physical techniques to produce one-dimensional nanofiber and three-dimensional spherical particles. Various researchers have manipulated the possibility of combining different properties associated with morphology, composition, porosity, and assembly of fibers and particles by controlling the major factors including processing conditions, solution parameters, instrument alignment methods, and atmospheric conditions in the electrospinning process (Duan et al., 2017; Qiao et al., 2019).

The advancement of electrospinning and electrospraying technology to the level of nanomaterials development makes it a versatile and viable technology that is suited to a variety of applications. Electrospraying and electrospinning processes have been used in a wide range of applications in different fields including electronics, material science, biomedical, pharmaceutical, nutraceutical, textile, food, etc. (Jayan et al., 2019). Electrospun fibers made up of photoactive material have been used for the treatment of drug-resistant bacteria and cancer therapy. Various drug delivery systems have developed by encapsulating the drug inside the electrosprayed or electrospun materials (Bhushani, 2014). Additionally, electrospun micro and nanostructures have been employed in energy storage technologies such as photovoltaic, solar, fuel, and battery systems, and electrosprayed structures based on catalysts have been used in catalytic applications (Tycova et al., 2021).

DOI: 10.1201/9781003359302-7

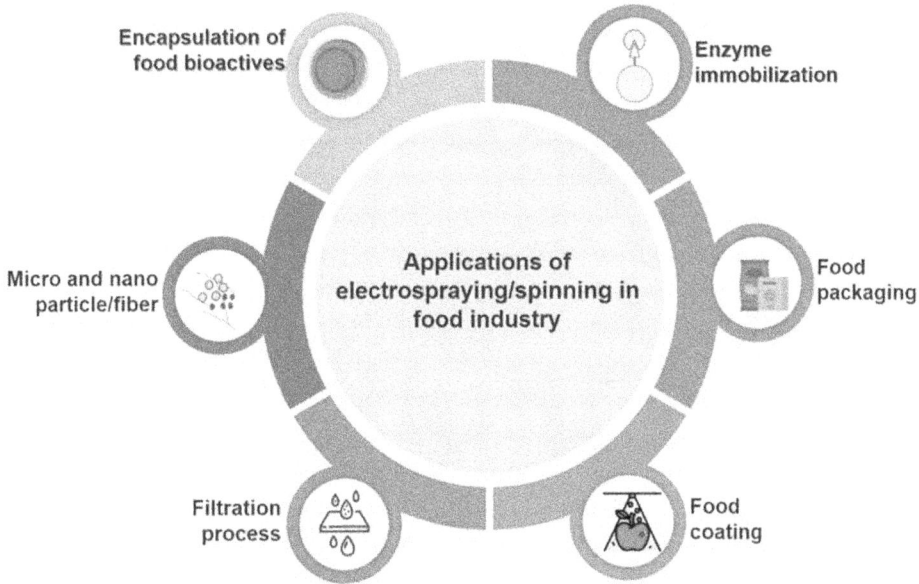

FIGURE 7.1 Applications of electrospraying/spinning in the food sector.

Various fibers electrospun with conductive composite, photoluminescence, and light-emitting can be used in optics, wearable, and nano-electronics. The hydrophobic surface coating can be done with electrospinning that can be utilized for cleaning applications. In the food industry, electrospraying and electrospinning processes are widely used for various applications including micro and nanoencapsulation of food bioactives or sensitive nutraceutical ingredients, filtration, food coating, food packaging, and enzyme immobilization (Figure 7.1). It possesses various functional advantages including non-thermal processed products, sustained and controlled release of active ingredients, limited degradation or denaturation, excellent encapsulation, and improved physicochemical stability using food-grade biopolymers (Bhushani, 2014). Science and engineering have devoted a great deal of attention to this emerging research area. We will discuss the electrospraying and spinning technique and its recent advancements in food application in this chapter, which will be useful to both newcomers and experts.

7.2 PRINCIPLE OF ELECTROSPRAYING AND ELECTROSPINNING PROCESSES

An electrohydrodynamic process uses electrohydrodynamic force to either draw or split up charged liquid into uniform ultrafine droplets. Electrospraying and electrospinning are two sister technologies of the electrohydrodynamic atomization process. Both processes employ the same technology and set up of instruments, which include a syringe pump to dispense the polymer solution at a regulated flowrate, a spinneret (nozzle), a ground collector plate, and a high-voltage source (Bock et al., 2011). The electrospraying principle is based on the concept of charged droplets, which states that the interface of a liquid droplet exit in a capillary tube can deform when an electric field is applied to it. At the capillary tube nozzle, two significant electrostatic forces are produced by the electric charge competing with the surface tension of the droplet, the charged droplet's hemispherical surface transforms into a conical shape known as the Taylor cone. When the electrostatic force produced by a high electric field exceeds the droplet surface tension, fine-charged droplets are pumped out of the tip of the nozzle at a constant rate controlled by a syringe pump on micro and nanosize and dissipate the excess charge, reducing the charge of the droplet without noticeably affecting its mass. The droplets scatter effectively without agglomeration or coalescence as they fly toward the

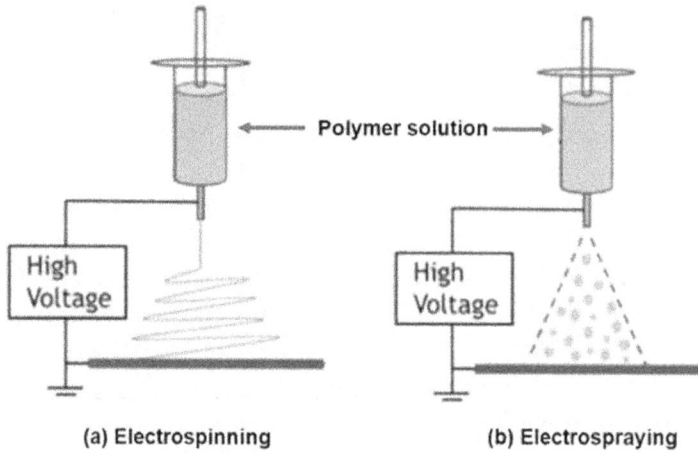

(a) Electrospinning   (b) Electrospraying

FIGURE 7.2 Process flow diagram for electrospinning (a) and electrospraying (b). (Adopted from Liu, Ramakrishna, & Liu, 2020.)

collector due to the Coulomb repulsion. Solvent gets evaporated during transit time from the tip to the collector, solidification of droplet takes place and fine spherical solid particles get deposited over the ground collector plate is known as the electrospraying process (Figure 7.2) (Behrouz Ghorani & Tucker, 2015).

If the feed solution viscosity is high, droplets scatter unevenly with the whipping and bending motion of the stream, elongation, and solvent evaporation will take place while traveling from tip to collector plate, and solid fibers get deposited over the ground collector plate known as electrospinning process. Also, the jet is subjected to drag force, gravitational force, and viscoelastic forces (Figure 7.3).

As the charged droplet or thread is propelled out by electrostatic force from the tip of the capillary tube to the collector, the high-velocity jet experiences drag force as a result of its contact with

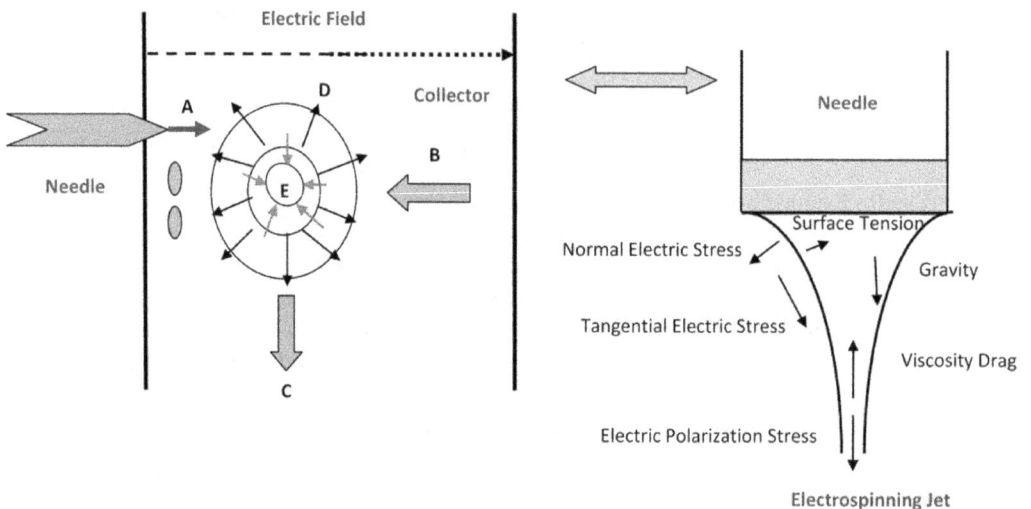

FIGURE 7.3 The electrospinning process's forces influenced the charged droplet, electrostatic force (A), drag force (B), gravitational force (C), Columbic repulsion force (D), and surface tension & viscoelastic force (E). (Adopted from Behrouz Ghorani & Tucker, 2015.)

the air around it. Surface tension and viscoelastic forces lead the droplet to compress, and columbic force causes it to expand. The electrostatic and columbic repulsion forces that are seen during the procedure directly related to the applied electric field. Charged droplets undergo distortion and expansion due to the applied electric field, changing their shape from spherical to spindle-like (Behrouz Ghorani & Tucker, 2015; Alehosseini et al., 2017).

7.3 PARAMETERS OF ELECTROSPINNING AND ELECTROSPRAYING

The production of particles or fiber with various morphologies depends heavily on the characteristics of the feed solution, the processing parameters, and the environmental factors.

7.3.1 SOLUTION PARAMETERS

Feed solution parameters such as molecular weight of the solute, concentration, surface tension, viscosity, surface charge, conductivity, strength, flexibility, degree of entanglement, and solubility of the solution have a great impact on the formation of the particles and fibers. In general, the solubility of the polymer increases with a decrease in molecular weight. As molecular weight increases, the degree and strength of entanglement of the polymer chain and viscosity increase. Electrospraying and bead formation were caused by low polymer concentration. When voltage is applied, electrostatic repulsion causes surface tension-induced Rayleigh instability, which is brought on by highly charged droplets, to self-disperse. The Taylor cone will emit a jet of solution as the polymer concentration increases (semi-dilute entangled regime), but instead of the solution breaking up, beads and fibers (beaded fibers) will form (Shenoy et al., 2005). Figure 7.4 shows the shape of the zein particles formed by electrospraying at various concentrations at a stream rate of 0.2 mL/hour with a voltage of 16 kV at a distance of 8 cm between the needle and the collector (Bhushani, Kurrey, & Anandharamakrishnan, 2017). Further increasing the concentration of polymer results in the development of uniform pure fiber due to whipping instability. The diameter of the fiber varies according to the solution viscosity (Gupta et al., 2005). The size of the fiber or particles is directly proportional to viscosity. Along with concentration, molecular weight and conductivity of the solution play a critical role in determining the morphology of the fiber or beads. Fibers or particles with small diameters can be obtained by increasing the conductivity of the solution and using a high molecular weight polymer.

When the voltage is applied, the elongation level of a jet depends on the electrical conductivity of the polymer solution which reflects the charged density of a jet. Also, the elongation of the jet increases along its axis when the solution conductivity is higher even under the same condition of applied voltage and distance between the tip and collector plate resulting in the formation of fiber with a smaller diameter. The applied electrostatic field and conductivity of the solution influence the net charge density of the jet.

The electrical conductivity of the solution reflects the charged density on a jet and determines the elongation level of the jet when the voltage is applied. Additionally, when the solution conductivity is higher even under the same conditions of applied voltage and the distance between the tip and collector plate, the production of fiber with a smaller diameter occurs. Thus, the conductivity of the solution and the applied electrostatic field have an impact on the net charge density of the jet.

When net charge density increases, the surface area also gets increased by the forces exerted on the jet, which induces the thin jet and prevents the formation of the bead (Tan et al., 2005). The morphology of the electrospinning and spraying structures also depends on the type of solvent used. In the case of spinning, 80–90% of the solution contains solvent, and the remaining 5–20% of the solution contains polymer concentration (Lu et al., 2006; Ghorani, Russell, & Goswami, 2013). Therefore, the choice of solvent greatly influences the density and morphology of bead or fiber formation. The drying of electrosprayed particles and electrospun fibers is attributed to the boiling point and density of the solvent. The drying of particles or fibers is indirectly proportional to the boiling point and density of the solvent (Wannatong, Sirivat, & Supaphol, 2004).

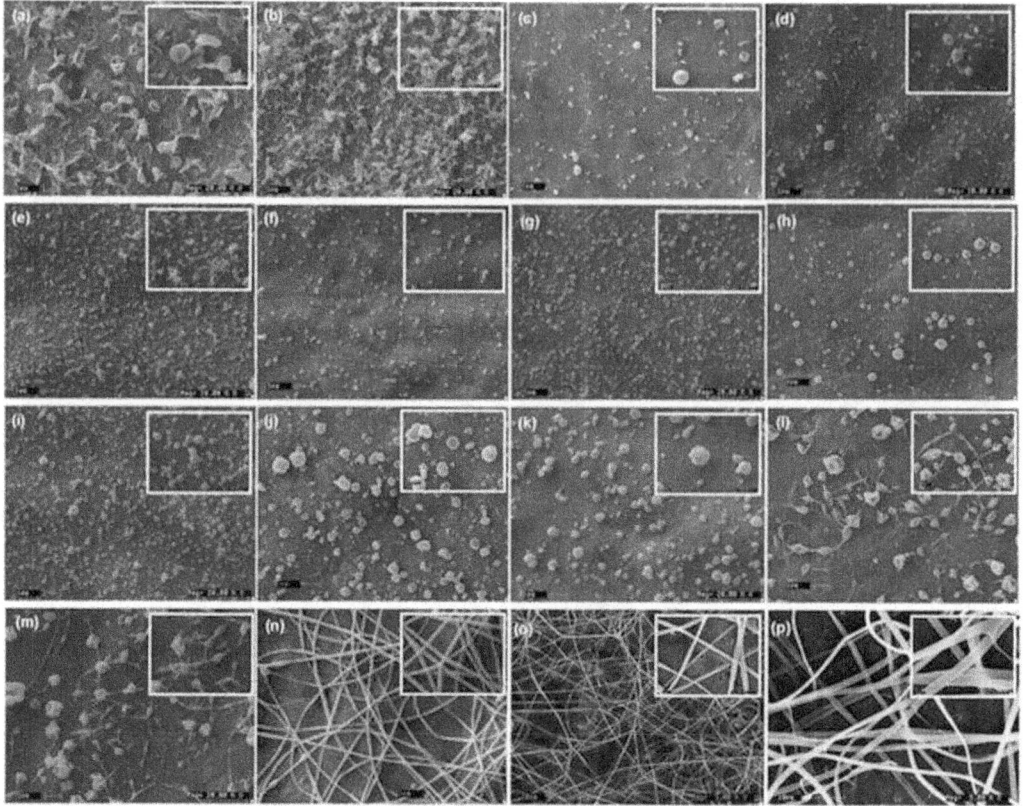

FIGURE 7.4 SEM image of zein particles produced by electrospraying process at different concentrations (a) 1.0%, (b) 2.0%, (c) 2.5%, (d) 3.0%, (e) 4.0%, (f) 5.0%, (g) 6.0%, (h) 7.5%, (i) 8.0%, (j) 10.0%, (k) 12.0%, (l) 15.0%, (m) 18.0%, (n) 20.0%, (o) 25.0%, and (p) 40.0%. (Adopted from Bhushani et al., 2017.)

7.3.2 PROCESSING PARAMETERS

Processing parameters include the kind of collector, flow rate, applied voltage, nozzle diameter, and distance from the nozzle tip to the ground collecting plate. Applied voltage plays a crucial role in the electrospraying and spinning process. Variations in the applied voltage manage the electric field strength and draw force between the spinneret and collector. When the voltage is applied to the solution, the solution gets charged. Spraying or spinning starts when the electrostatic forces are greater than the solution's surface tension. Taylor cone formation is more influenced by an applied electric field. The balance between electric force and surface tension decides the formation of the Taylor cone at the tip of the needle. Depending on feed flow rate is high, a higher or lower voltage is required to form a stable Taylor cone. When the applied voltage is more, more charges will accelerate the jet faster, so that more amount of solution will come out of the tip of the needle leading to the formation of a less stable and very small Taylor cone. When the viscosity of the feed solution is high, a higher voltage is required to overcome the surface tension of the feed solution to attenuate the jet and fiber formation. The threshold voltage fixation is required for each concentration of the solution to eject the solution out from the tip of the needle. The applied voltage and its electric field have an impact on the elongation and acceleration of the motion of the jet, which has a strong impact on the morphology of the particles or fibers. The diameter of the fiber or particle decreases with an increase in voltage (Ghorani & Tucker, 2015).

Feed flow rate has a strong impact on morphology (Jalili, Hosseini, & Morshed, 2005). Higher diameter particles or fibers are formed when the feed flow rate is higher and a low flow rate causes uniform stable particles or fibers (Zong et al., 2002). This is also due to the variations in feed velocity and drying time. The high flow rate causes droplets from the tip of the needle to become large, and more amount solvent to be evaporated during the travel time and the drying time will be very resulting in large bead formation and the junction between fibers (Yuan et al., 2004).

The inner diameter of the needle also influences the electrospraying and spinning process. If the diameter of the needle is large, clogging may happen due to surface tension (Mo et al., 2004). If the diameter of the needle is smaller, the surface will be more and the acceleration of the jet and velocity will be less. Therefore, the time for solvent evaporation, attenuation of jet, and splitting of droplet decreases. Thus, small and narrow size distributed particles or fibers can be formed (Zhao et al., 2004). If the diameter of the needle is very small, the ejection of highly concentrated solution is very tedious and related to surface tension.

The elongation, solvent evaporation, and drying take place between the distance of the tip to the collector. The distance between the tip to the collector influences the electric field strength and jet transit time. Improper solvent evaporation and drying occur when the distance between the tip to the collector is less due to inadequate time for the solvent to evaporate. It resulted in partially dried particles/fibers, agglomeration, a fusion of particles/fibers, and densely packed structures. Increasing the distance between the collector and the tip led to proper drying and complete solvent evaporation before reaching the collector. The distance between the tip and the collector also affects the size of the particle/fiber formation. Decreasing the distance also has the same impact as increasing the voltage on morphology. It has been asserted that increasing the collector-to-tip distance can result in a reduction in fiber diameter if the relative humidity is low because more solvent evaporation takes place across a greater area. However, in conditions of high relative humidity, the longer distance only weakens the field rather than enhancing evaporation (Barhate, Loong, & Ramakrishna, 2006; Yang et al., 2006).

The type of collector material used for the process has an impact on the degree of the surface charge arising. Highly conductive collector material increases the density of the fiber deposited. The amount of surface charge that builds up on the collection unit is indirectly proportional to the deposition of fiber/particles (Chowdhury & Stylios, 2010).

7.3.3 ENVIRONMENTAL FACTORS

Environmental factors include humidity, temperature, and pressure. Temperature and humidity have a direct impact on the morphology and production of particles/fibers. Variables like conductivity and solvent evaporation are directly correlated with the ambient temperature. The temperature has an impact on the solution's viscosity and surface tension as well. High humidity results in an increased diameter of particles/fibers (Casper et al., 2004; Chen & Yu, 2010).

Thus, identifying how each of these parameters affects particle dimensions and morphology is crucial to ensure proper process control.

7.4 DIFFERENT ELECTROSPRAYING/SPINNING CONFIGURATIONS

The electrospraying/spinning setup can be altered based on application and requirement by changing the configuration of the equipment. "Delft" type is one model that involves a nozzle ring (Figure 7.5a and b) kept inside the glass chamber and air/nitrogen is passed into it. A high voltage supply is given to the nozzle and a lower voltage is supplied to the ring. A potential difference is produced between the nozzle and the ring kept around the nozzle. The electrospraying pattern can be controlled effectively with the support of a ring. A grounded needle positioned in a vacuum produces corona discharge in the opposite direction of the charged nozzle to release the highly charged droplets. Dried particles can be gathered in two ways: first, by passing air or nitrogen through a filter-equipped chamber; second, by gathering particles around a grounded needle that has been placed in a petri dish. When employing water as the solvent, it was recommended that equipment with a ring model would be better suited for generating particles (Ding, Lee, & Wang, 2005; Ciach, 2006; Wu et al., 2009). But, the evaporation rate of the solvent is poor, and the production of small particles is not applicable in this configuration due to the usage of the chamber. Also, the yield is very less due to the deposition of dried particles around the glass chamber before it deposits over the filter. In the case of incorporation of drugs or bioactives using these techniques, the degradation and release are more due to reduced solvent evaporate rate and consequent polymer relaxation. As an alternative to the ring-nozzle setup, the beaker consists of immersed electrode or electrode placed around the beaker that was used as a collector unit (Figure 7.5c and d) (Xu & Hanna, 2006, 2007).

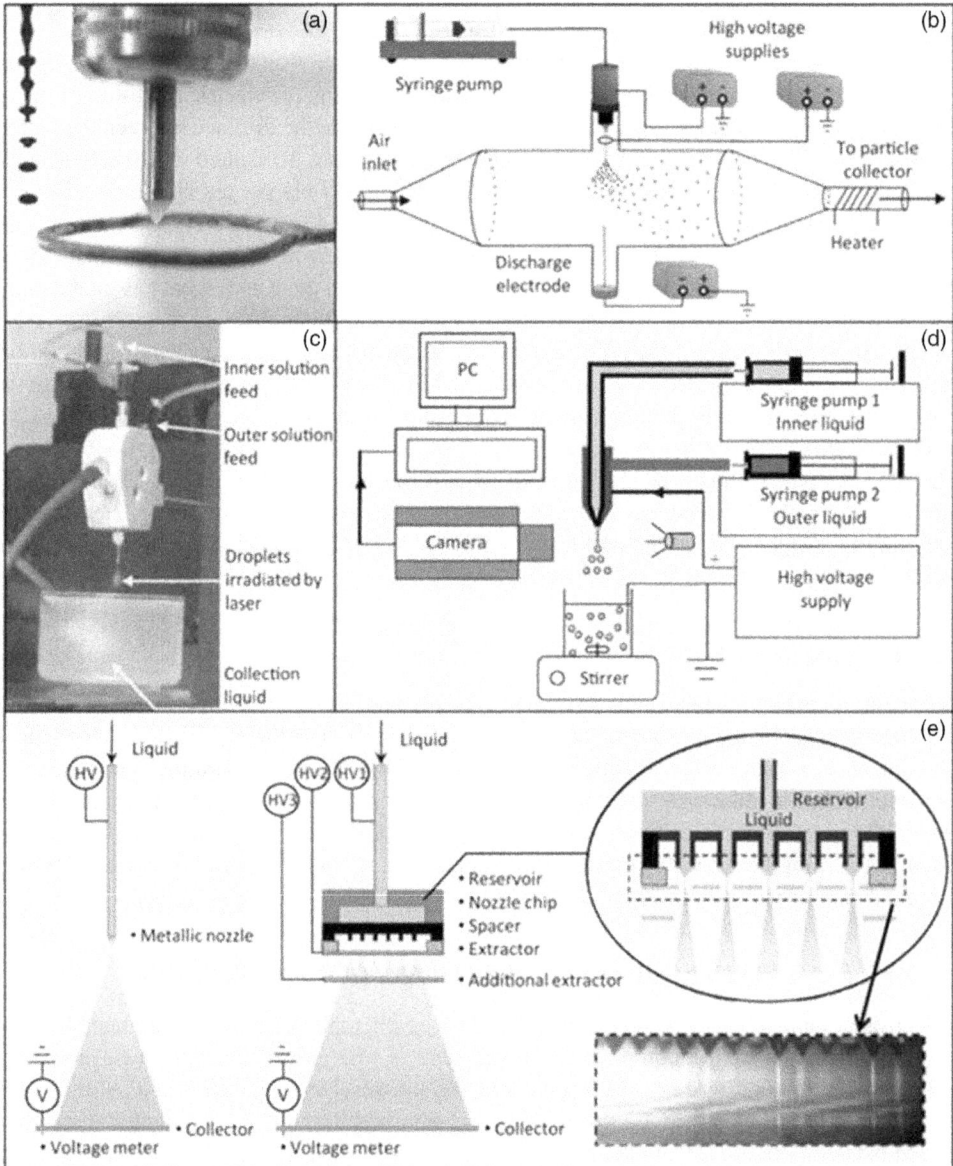

FIGURE 7.5 Different electrospraying device configurations. Nozzle-ring arrangement (a). (b) A glass chamber with a nozzle-ring configuration and an air or nitrogen flow. (c, d) Coaxial electrospraying in a fluid with laser optical spectrometer size measurement. (e) Ground collectors with single and multiplexed electrospray configurations. (Adopted from Bock, Dargaville, & Woodruff, 2012.)

The selection of solution for the collector beaker should have lower surface tension for prevention of particle aggregation or coalescence. This configuration requires a secondary step to filter and dry the particles. Also, there is more chance to release surface adsorbed molecules in the medium which leads to the loss of bioactives and low loading efficiency.

The most commonly used electrospraying system contains a conductive grounded collector unit (Figure 7.5e) (Arya et al., 2008; Hong et al., 2008). This conductive grounded collector unit prevents the deposition of particles to the charged area, loses and there is no requirement for further drying or filtration. It has control over the size, shape, and functional properties of the particles/fibers. Though this configuration provides better control over structural and functional advantages, low throughput and yield are major disadvantages of this technique. This can be overcome by using multiple needle sources to increase the yield (Figure 7.5e) (Almería, Fahmy, & Gomez, 2011).

Morphology and functional properties are the same as that of single needle configuration. Disturbance can occur between the sources and the electric field in multiple needle setups. Those interferences can be reduced by using an extractor between the sources. Also, core-shell or double-layered, or Janus particles/fibers can be produced by changing the nozzle configuration like twin fluid or three fluid nozzles. Additionally, to scale up electrospinning methods, researchers and engineers have already created numerous scaling-up systems using spinneret modifications (needleless stationary spinnerets and needleless rotating spinnerets) (Varesano, Carletto, & Mazzuchetti, 2009; Niu & Lin, 2012).

7.5 APPLICATIONS IN FOOD INDUSTRY

7.5.1 Enzyme Immobilization

Electrospun nanofibers, which feature an adjustable porosity and large surface area, are the most appropriate supports for enzyme immobilization. Today, enzyme immobilization using the electrospinning technique is a hot research topic, as it improves the performance, efficiency, and stability of immobilized biomolecules. The electrospinning process involves spinning polymeric solutions with an external electric field to produce fibers with nano or micron dimensions. The most frequently used electrospinning materials for enzyme immobilization are synthetic polymers since their excellent physiochemical properties allow fiber formation more easily than natural polymers. A synthetic polymer for immobilizing enzymes has also been found to exhibit excellent chemical and biological stability, as well as mechanical resistance. These advantages have prompted a lot of research in synthetic polymers. Most synthetic polymers such as polymethyl methacrylate (PMMA), polyethylene oxide (PEO), polyacrylonitrile (PAN), polystyrene, polyurethane, and polylactic acid (PLA) are used for electrospinning to produce fibers and utilized in enzyme immobilization. The polymers are normally not water soluble or biodegradable, but they can be electrospuned into fibers with a variety of organic solvents. Water-soluble polymers such as PEO and PVA can be electrospuned and work well with buffer solutions and enzymes. During electrospinning, these polymers can either be chemically derivatized or interacted directly with enzymes to produce nanofibers. Since the methacrylate groups in PMMA nanofibers have a low number of reactive functional groups, resulting in weak enzyme bonds, Kumar et al. (2013) functionalized and activated their electrospun PMMA nanofibers with glutaraldehyde and phenylenediamine, which introduced the aldehyde group on the surface of the fiber which gives appropriate site for covalent xylanase binding. Scanning electron microscope (SEM) and Fourier transform infrared (FTIR) analyses demonstrate covalently immobilized enzymes on nanofiber membranes, ensuring a shift in the texture of nanofiber surfaces.

To immobilize the enzyme lipase, Li, Chen, and Wu (2007) employed PAN nanofibers created by electrospinning in which the nitrile groups were activated from the amidation reaction process. A SEM scan showed clusters of enzyme molecules on the nanofiber's surface, and an FTIR analysis indicated that the enzyme molecules were covalently attached to the nanofibers. Using electrospun PVA nanofibers, Moreno-Cortez et al. (2015) successfully immobilized papain. For optimized stability and appropriate catalytic activity, Pinto et al. (2015) used water/oil emulsion electrospinning to prepare PCL (polycaprolactone) electrospun fibers to encapsulate trypsin. As part of the immobilization efforts, Temocin et al. (2018) constructed electrospun PVA and polyacrylamide (PAA) nanofibers for immobilizing horseradish peroxidase (HRP) enzyme. It has been shown that electrospun nanofibers can be an excellent technology for enzyme immobilization because of their smaller size in the nanoscale, higher surface area, numerous interaction sites, functional surfaces, and minimal mass transfer. There is still a challenge to produce nanofibers in batches though there are multiple spinnerets configurations available for the immobilization of enzymes. Additionally, electrospinning nanofibers with diameters lower than 10 nm, on which several enzymes can be immobilized, are challenging. Moreover, electrospinning is frequently utilized in industries; but as a result of the higher cost of manufacturing bigger-size fiber, it cannot compete with the old process when it comes to making fibers for filter applications. Large-scale production of nanofibers for enzyme immobilization is still a long way off; thus, current research is being conducted in the lab.

7.5.2   Food Packaging

Packaging serves a crucial function in safeguarding and containing food as it is processed, manufactured, delivered, handled, and stored. If food packaging has a benefit other than serving as an inert barrier against environmental factors, it might be regarded as active (Bhushani et al., 2017). Besides oxygen scavengers and antimicrobial agents, active packaging systems for food applications can also absorb moisture, release antioxidants, absorb flavors, or absorb odors (Rooney, 1995). Active packaging and nanostructured layers of the packaging material can be fabricated using the electrospinning technique. Active packaging by electrospinning in nanoscale has excellent functionalities because of its large area-to-volume ratio, submicron-to-nanoscale diameter, higher sensitivity toward atmospheric changes, and appropriateness for incorporating temperature-sensitive active ingredients (Vega-Lugo & Lim, 2009). Antimicrobial substances are used in active packaging systems to prevent the growth of bacteria by diffusing into the product while it is being stored. Considering this, consumers are willing to embrace natural, efficient, non-toxic, and flavorless food contact packaging materials (Neo et al., 2013). Recently, numerous naturally occurring plant extracts or active chemicals that are Generally Recognized as Safe (GRAS) materials have been used in active food packaging as antimicrobial agents due to their low health and environmental risks.

Optimization of the electrospinning process and solution parameters play an important role in the development of innovative environmentally friendly food packaging material along with its functional qualities. For example, the natural antimicrobial nanofibers (allyl isothiocyanate) were produced by electrospinning the protein from the soya source and PLA with β-cyclodextrin. Vega-Lugo and Lim (2009) showed that nanofibers, which release the antimicrobial agent by changes in humidity, can be used in active packaging. Also, the antibacterial characteristics of chitosan along with zein were used to produce the nanofibers which have excellent diffusion-resistant capacity and can be employed as food packaging material (Torres-Giner, Ocio, & Lagaron, 2009). In another study, the complex EGCG-Cu^{II}((−)-Epigallocatechin-3-gallate with PVA nanofibers have identified as a potential antibacterial agent (Sun et al., 2013). Also, the electrospun nanofibers were developed with the incorporation of antioxidant and antimicrobial using phenolic active material such as raspberry extract and gallic acid with soy protein and zein protein, respectively, after 60 days of storage at 21.5°C with 58% relative humidity, gallic acid incorporated zein nanofibers showed outstanding chemical and thermal stability (Neo et al., 2013; Wang et al., 2013). Fiber mats, the functional nanomaterial from both experiments, were demonstrated to be an efficient active packaging system for food preservation. Also demonstrated that it has potential antifungal activity due to its low water activity. Natural volatile compounds have antimicrobial activity and it is used in active food packaging. Nevertheless, safeguarding volatiles against deterioration and managing their release is a significant difficulty linked with their use in food packaging. Moreover, active food packaging systems utilize natural volatile substances with antimicrobial properties. However, protecting volatiles from deterioration and managing their release is a significant difficulty involved with their use in packaging systems (Mascheroni et al., 2013).

Chemical or biological sensors in the form of nanofibers are used in smart packaging or intelligent packaging to monitor food quality and safety as a result of changes in the interior or exterior surroundings (Kuswandi et al., 2011). Electrospun fibers can be used for not only active and smart packaging but also multi-layered structures. Electrospun nanofibers offer better mechanical, optical, and/or functional qualities than other multi-layered structures when used as packaging material's intermediate layers (Fabra et al., 2013). The ability of food packaging materials to control heat and act as an oxygen barrier is highly commendable.

Dodecane, a phase-changing material can be used to maintain the freezing temperature of foods that must remain frozen during storage and transportation. This chemical can be used to make bio-packaging materials by being entrapped in zein electrospun fibers (Perez-Masia et al., 2013). It is feasible to create transparent composite packaging materials with improved barrier characteristics and reduced oxygen permeability by combining these components into polymer matrixes (PLA/PEO). In polyhydroxy butyrate-co-valerate multilayers, it was also found that an electrospun zein nanofiber interlayer reduced oxygen and aroma permeability (Busolo et al., 2009). Future electrospinning food packaging material development must go from lab testing to commercial production. All of them have paved the way for potential electrospinning food packaging materials in the future.

7.5.3 FOOD COATING

Foods with edible coatings have a longer shelf life, which eliminates the need for packing materials and acts as a barrier material. Spray coating is a popular method for coating food, but electrospray enables accurate control of the deposition rate and film thickness. To reduce the amount and size of voids and cracks in the film, submicron or nanoscale monodisperse particles with a regulated size distribution should be sprayed on top of the film. By varying the voltage, flow rate, and concentration of the spraying solution used in the electrospraying system, the size of the jet may also be controlled. Starch sheets with a 40 mm diameter were produced by electrospraying and used as edible coatings (Pareta & Edirisinghe, 2006). Cut apple slices were coated with electrospun resveratrol-incorporated zein nanofibers, which effectively reduced moisture loss and preserved the color of the apple slices over 6 hours (Leena et al., 2020). As for oils, sunflower oil was used to produce thin and evenly distributed hydrophobic film coatings on both conductive and non-conductive surfaces. The features of ethanol droplet formation were examined using a single-nozzle and multiple-nozzle electrospraying system with soybean oil and medium-chain triglyceride oil. A droplet diameter of 2 mm was obtained for triglyceride oil droplets (Zhang et al., 2013). Also, electrosprayed the oil in water emulsion to create thin edible barrier films. The jets were stabilized by polyglycerol polyricinoleate, and the water vapor barrier qualities of the films were improved by whey protein isolate, respectively (Khan et al., 2012). The snack food and confectionery industry offers a fascinating range of food coating options, including chocolate coating. Since chocolate is a rather expensive ingredient in food processing, it is crucial to adopt effective techniques when coating items with chocolate because of how its rheological properties affect both the production process and the quality of the finished product. Electrospraying is a one-step, non-thermal, low-cost material and low-energy processing method for thin film coating that can create distinctive products. When film barrier qualities are crucial, the electrospraying method can be utilized instead of traditional spray coating. Additionally, electrospraying produces monodisperse particles that can help the confectionery industry save money. As compared to bulk chocolate, electrosprayed chocolate particles have slightly different organoleptic characteristics, particularly texture, and mouth feel. Electrospray systems offer an advantage over mechanical atomizers, so confectionery and related industries can use them as an alternative to non-electrohydrodynamic coating methods.

7.5.4 ENCAPSULATION OF BIOACTIVE COMPOUNDS

Various delivery systems were developed to encapsulate, protect, and enhance the controlled release of nutraceutical active ingredients to meet the consumer demand concern to improve nutritional intake and prevent various chronic diseases (Lesmes & McClements, 2009). Numerous micro and nanoscale nutraceutical delivery system has been developed using electrospraying and spinning processes (Leena et al., 2020; Mahalakshmi et al., 2022). Delivery systems have been developed in the form of matrices, core shells, or fibers with sizes ranging from micro to nanoscale. Morphology plays a major role in the delivery of bioactive substances. Electrospun nanofibers are more effective for encapsulation because of their porous structure, high surface area, increased gas penetration, and smaller interfibrous pores (Saeed & Park, 2010). Electrospraying/spinning can achieve ease of particle production, narrow particle-size distribution, and better loading efficiency by single-step processing, and these are the key advantages of this technique over other standard encapsulating technologies (Mahalakshmi et al., 2023). Electrospinning and electrospraying are commonly performed using polysaccharides, carbohydrates, proteins, and biopolymers for developing delivery systems. A very high encapsulation efficiency (>95%) has been reported for folic acid electrospun fibers using amaranth protein isolate-pullulan and did not show any degradation even after 2 hours of exposure to UV rays. It was found that in-vitro digestion of encapsulated bioactive compounds maintained a greater level of antioxidant capacity compared with those of unencapsulated compounds. The encapsulation of active compounds with proteins has also been the focus of many researchers due to biodegradability, excellent surface activity with the active ingredient, and amphiphilic nature to encapsulate both hydrophilic and lipophilic active compounds. Zein protein from corn grains has the advantage of being highly thermally resistant and oxygen-permeability. Electrospinning of zein may be used to stabilize light-sensitive and oxygen-sensitive food components in the future. In our previous study, nanoencapsulated β-carotene was produced within zein protein using electrospraying with enhanced encapsulation efficiency and improved bioaccessibility and intestinal permeability (Mahalakshmi et al., 2020). Also, curcumin nanoencapsulated particles were produced using electrospraying and had excellent encapsulation efficiency, and dissolution behavior over the spray drying technique (Mahalakshmi et al., 2023).

Furthermore, electrosprayed lycopene-incorporated whey protein capsules showed better protection against moisture and thermal degradation (Pérez-Masiá, Lagaron, & Lopez-Rubio, 2015). Another study showed that resveratrol-loaded zein nanoparticles produced an electrospraying process that had better encapsulation efficiency, bioaccessibility, and enhanced intestinal permeability of resveratrol (Jayan et al., 2019). Core-shell nanoparticles for the co-delivery of curcumin and resveratrol bioactives were developed using an electrospraying system (M. M. Leena et al., 2022). Humans and animals benefit from probiotics, which are live microorganisms. Food manufacturers and marketers are strongly motivated to develop food products with probiotics. The food industry and academia are very interested in encapsulating probiotics within food-grade materials to safeguard them throughout processing, storage, and digestion (Wang et al., 2022). Probiotics typically have a size of a few micrometers, and electrospraying is a trustworthy method of entrapping them. Electrosprayed particles or spun fibers should be larger than probiotics so that they completely envelop the probiotics. Probiotics were microencapsulated in an electrosprayed whey protein matrix and tested the probiotics' resistance in-vitro and during storage (Gomez-Mascaraque et al., 2016). It has been established that electrospraying technology can be utilized to produce probiotics for use in non-dairy goods instead of refrigeration. Zaeim et al. (2019) used electrospraying to test a core-shell microparticle for entrapment of probiotics. Probiotics and prebiotics were electrosprayed onto calcium alginate microparticles, which were subsequently dipped in chitosan solution to create a shell layer (Zaeim et al., 2019). The results of electrospraying as a means of co-encapsulating probiotics and high molecular weight prebiotics gave a new perspective for delivering living cells to the colon through meals.

7.5.5 Filtration or Clarification of Beverages

Several ways to apply electrospun nanofibers for beverage filtration are discussed below. Nanofiber films are now being tested in the beverage sector, such as in the production of fruit juices and beer. A preliminary study of the use of polyethylene terephthalate nanofibrous membrane (NFM) as an apple juice filtering membrane has shown twenty times more effective than ultrafiltration membrane in terms of juice flow and operates at seventy times lower pressure than ultrafiltration membrane. To preserve the nutritional value of the juice and improve shelf life and stability, the antioxidant qualities should be preserved. At the same time, the ability to remove phenols should also be enhanced during juice processing. NFM and oxygen radical absorption membranes have comparable filtering abilities to commercial polyamide membrane filters in terms of influencing apple juice's antioxidant properties. Research validated the effectiveness of electrospun NFM membranes for removing yeast cells and bacteria from beer using isothermal calorimetry.

Using naringinase and alginate both positively and negatively charged, self-assembly processes cover negatively charged nanofiber films. By removing the naringin and limonin in grapefruit juice, the (naringinase/alginate)10.5-CA process improves the sensory quality and commercial value of the juice. It is claimed that nanopores in these fibers are effective at capturing hydrocarbons from the air. Several applications of nanofibers in the beverage industry include pesticide detection, monitoring storage conditions, nanosensors for assessing quality, encapsulation matrices for protecting aromatic and volatile compounds in beverages, beverage packaging materials, and beverage filtration. Nanofibers were developed with an average porosity of 73.26%, diameter of 942 nm, and 150 μm mean thickness using the electrospinning process for neera filtration. Filtered neera has 2 log-reduction in yeast load and retained the nutritional content (Leena et al., 2021). As electrospun nanofibers have a high porosity and large surface area, they can be modified to meet specific requirements. These characteristics, along with their cost efficiency, make them ideal for beverage filtration.

7.6 OTHER APPLICATIONS

7.6.1 Drug Delivery

Due to their unique qualities, such as their larger specific surface area ratio, effective surface bioactivity, high intertwined porosity, tunable morphology, and structural similarity to the extracellular matrix (ECM), nanofibers have been used extensively in drug delivery applications (Suwantong, Ruktanonchai, & Supaphol, 2008; Yoo, Kim, & Park, 2009; Zou et al., 2012). Silk fibroin, chitosan, gelatin, collagen, PEO, PCL, and PLA have all been electrospun, along with other natural and synthetic polymers, to create scaffolds for use in tissue regeneration and drug delivery (Ma et al., 2011). A nanofibrous membrane that is loaded with the drugs can be applied topically to promote wound healing as well as the post-operative stage for the delivery of antibiotics, antifungals, antimicrobials, and anticancer drugs. During electrospinning, hydrophobic and hydrophilic drugs, as well as biomolecules such as protein and DNA, can be directly encapsulated into the fiber (Zong et al., 2004). The high surface area and three-dimensional open pores provide nanofibers with a more efficient drug release by reducing the constraint to drug diffusion (Luo et al., 2012). The use of electrospun fiber for local drug delivery reduces the required dosage, which reduces unwanted side effects and also results in less systemic absorption. Compared to other types of drug carriers, such as liposomes, hydrogels, and nano/microsphere, electrospun nanofiber greatly improves drug encapsulation efficiency and reduces burst release.

In contrast to a conventional film produced by solution-casting, an electrospun fibrous mat would allow drug molecules to diffuse into the surrounding environment (Taepaiboon, Rungsardthong, & Supaphol, 2006). Since electrospun membranes are highly porous, degradation products will not accumulate at the site of implantation. Additionally, electrospun fibers are trimmed to the required sizes, making them an excellent choice for clinical applications (Xie et al., 2010). Microparticles and nanoparticles made from polymeric material can be administered as oral, injectable, inhalable, topical, and local forms of drug-delivery systems. The most commonly used technique for manufacturing biodegradable polymeric micro- and nanoparticles includes solvent evaporation, single and double emulsion, spray-drying, porous glass membrane emulsification, and coacervation (Omi, 1996; Mu & Feng, 2001; Freitas, Merkle, & Gander, 2005). However, these techniques have some drawbacks, including scaling limitations, polydispersity in size, limited capacity to create small particles, and challenges incorporating hydrophilic medicines (Hans & Lowman, 2002; Enlow et al., 2011). Additional difficulties could arise if particle separation is required from dispersing solution or even from non-degradable surfactant. Furthermore, biomacromolecules are prone to degradation or inactivation due to exposure to high shear stress, high temperature, and organic solvent (Tamber et al., 2005; Lee et al., 2010). As a result, electrospraying has the advantage of overcoming the drawbacks of traditional particle production techniques.

7.6.2 TISSUE REGENERATION

This multidisciplinary field relates to the process of guiding cells to grow into new tissues using porous scaffolds made of biomaterials, first introduced by Langer & Vacanti (1993). In recent years, electrospinning research has been extensively explored for the fabrication of nanofibrous scaffolds, which replicate native ECM in terms of composition and architecture, whereas native ECM is tissue-specific, leading to specific applications. Due to their mechanical incompatibility with the native ECM, conventional electrospun scaffolds are not appropriate for tissue engineering applications, such as bone and cardiac regeneration. In tissue engineering and regenerative medicine, new electrospun scaffolds have the potential to address the aforementioned difficulties via a synergistic effect on stem cell differentiation. A scaffold design or configuration can generate topographical and biomechanical cues, whereas incorporating bioactive agents and biomolecules can provide biological cues, while material selection, fiber chemistry, thickness, and porosity control can provide electrochemical cues. A 3D network of fibrous proteins is formed in an ECM with a specific fiber orientation varying between tissues. 3D scaffolds are better at resembling natural ECM than 2D electrospun scaffolds. Electrospinning has been explored in tissue engineering applications as the potential solution to develop electrospun scaffolds for bone, cardiac, ligament, and cartilage regeneration (Liu et al., 2020).

Also, electrospraying/spinning has been used in garments, home finishing products, biomedical science, material science, electronics, and cosmetics.

7.7 FUTURE STUDIES

Although electrospun and electrosprayed materials have many benefits and potential uses, their fundamental drawback is low throughput which prevents their widespread commercial application. As a result, scaling up has taken on more importance. Therefore, working on structural aspects of the setup aimed at addressing the constraint is a vital area of research. Additionally, electrospun fibers and electrosprayed particles should be investigated in the food systems to bring them to life. Researchers need to look into studies to demonstrate the technique's efficacy in the area of food. In addition, the application of nanoparticles as functional ingredients in foods is dependent mainly on their ability to maintain their size at the nanoscale. Hence, it is imperative to handle and recover electrosprayed nanoparticle from collector material as the larger surface area of nanoparticle have a greater chance of aggregation resulting in size growth. Thus, future work in relation to the above-mentioned aspect is required to expand the end use of electrosprayed particles.

7.8 SUMMARY

As a promising field, nanotechnology boasts astounding results that could lead to the development of new food products that will benefit consumers and businesses. However, further studies are needed to determine how these nanostructures will interact long-term with products and human metabolism. It is important to evaluate the safety of nanomaterials and packaging before they are put on the market for food use. The electrospraying/spinning technique has various structural and functional advantages that can be explored more in various applications in the food sector. Electrospraying/spinning has configurable size, high surface area, and interwoven fiber structure as structural advantages. Functional advantages such as effective encapsulation, sustained release behavior, and improved stability of the active component present in food. However, for the food-related applications to be practical on an industrial scale, more sophisticated research on electrospinning/spraying processes is required.

REFERENCES

Alehosseini, A., Ghorani, B., Sarabi-Jamab, M., & Tucker, N. (2017) 'Principles of electrospraying: A new approach in protection of bioactive compounds in foods', Critical Reviews in Food Science and Nutrition, 8398(June), pp. 1–18. https://www.semanticscholar.org/paper/Principles-of-electrospraying%3A-A-new-approach-in-of-Alehosseini-Ghorani/c2fedff18124005f94e1b0feff8496982c3c0682.

Almería, B., Fahmy, T. M., & Gomez, A. (2011) 'A multiplexed electrospray process for single-step synthesis of stabilized polymer particles for drug delivery', Journal of Controlled Release, 154(2), pp. 203–210. doi: https://doi.org/10.1016/j.jconrel.2011.05.018.

Arya, N., Chakraborty, S., Dube, N., & Katti, D. S. (2008) 'Electrospraying: A facile technique for synthesis of chitosan-based micro/nanospheres for drug delivery applications', Journal of Biomedical Materials Research Part B: Applied Biomaterials, pp. 17–31. https://doi.org/10.1002/jbm.b.31085.

Barhate, R. S., Loong, C. K., & Ramakrishna, S. (2006) 'Preparation and characterization of nanofibrous filtering media', Journal of Membrane Science, 283(1), pp. 209–218. doi: https://doi.org/10.1016/j.memsci.2006.06.030.

Bhushani, J. A. (2014) 'Electrospinning and electrospraying techniques: Potential food based applications', Trends in Food Science & Technology. 38(1), pp. 21–33. https://www.semanticscholar.org/paper/Electrospinning-and-electrospraying-techniques%3A-Bhushani-Anandharamakrishnan/849d54961f4c8ce5f5a03a4da7519a34379bfab7.

Bhushani, J. A., Kurrey, N. K., & Anandharamakrishnan, C. (2017) 'Nanoencapsulation of green tea catechins by electrospraying technique and its effect on controlled release and in-vitro permeability', Journal of Food Engineering, 199, pp. 82–92. doi: https://doi.org/10.1016/j.jfoodeng.2016.12.010.

Bock, N., Woodruff, M. A., Hutmacher, D. W., & Dargaville, T. R. (2011) 'Electrospraying, a reproducible method for production of polymeric microspheres for biomedical applications', pp. 131–149. https://doi.org/10.3390/polym3010131.

Bock, N., Dargaville, T. R., & Woodruff, M. A. (2012) 'Electrospraying of polymers with therapeutic molecules: State of the art', Progress in Polymer Science., 37(11), pp. 1510–1551. https://www.semanticscholar.org/paper/Electrospraying-of-polymers-with-therapeutic-State-Bock-Dargaville/b5463805d4550c68d6ff327d79ec8754fbf535cb.

Busolo, M. A., Torres-Giner, S., & Lagaron, J. M. (2009). Enhancing the gas barrier properties of polylactic acid by means of electrospun ultrathin zein fibers. In ANTEC, proceedings of the 67th annual technical conference, Chicago, IL (pp. 2763–2768).

Casper, C. L. et al. (2004) 'Controlling surface morphology of electrospun polystyrene fibers: Effect of humidity and molecular weight in the electrospinning process', Macromolecules. American Chemical Society, 37(2), pp. 573–578. https://ui.adsabs.harvard.edu/abs/2004MaMol..37..573C/abstract.

Chen, H.-M., & Yu, D.-G. (2010) 'An elevated temperature electrospinning process for preparing acyclovir-loaded PAN ultrafine fibers', Journal of Materials Processing Technology, 210(12), pp. 1551–1555. doi: https://doi.org/10.1016/j.jmatprotec.2010.05.001.

Chowdhury, M., & Stylios, G. (2010) 'Effect of experimental parameters on the morphology of electrospun nylon 6 fibres', International Journal of Basic & Applied Sciences, 10(06), pp. 116–131.

Ciach, T. (2006) 'Microencapsulation of drugs by electro-hydro-dynamic atomization', International Journal of Pharmaceutics, 324(1), pp. 51–55. https://doi.org/10.1016/j.ijpharm.2006.06.035.

Dai, H., Gong, J., Kim, H., & Lee, D. (2002) 'A novel method for preparing ultra-fine alumina-borate oxide fibres via an electrospinning technique', Nanotechnology, 13(5), p. 674.

Ding, C., Fang, H., Duan, G., Zou, Y., Chen, S., & Hou, H. (2019) 'Investigating the draw ratio and velocity of an electrically charged liquid jet during electrospinning', RSC Advances, 9(24), pp. 13608–13613.

Ding, L., Lee, T., & Wang, C.-H. (2005) 'Fabrication of monodispersed Taxol-loaded particles using electro-hydrodynamic atomization', Journal of Controlled Release, 102(2), pp. 395–413. https://doi.org/10.1016/j.jconrel.2004.10.011.

Duan, G., Bagheri, A. R., Jiang, S., Golenser, J., Agarwal, S., & Greiner, A. (2017) 'Exploration of macroporous polymeric sponges as drug carriers', Biomacromolecules, 18(10), pp. 3215–3221.

Dzenis, Y. (2004) 'Spinning continuous fibers for nanotechnology', Science, 304(5679), pp. 1917–1919.

Enlow, E. M., Luft, J. C., Napier, M. E., & DeSimone, J. M. (2011) 'Potent engineered PLGA nanoparticles by virtue of exceptionally high chemotherapeutic loadings', Nano Letters, 11(2), pp. 808–813.

Fabra, M. J., Busolo, M. A., Lopez-Rubio, A., & Lagaron, J. M. (2013) 'Nanostructured biolayers in food packaging', Trends in Food Science & Technology, 31(1), pp. 79–87.

Freitas, S., Merkle, H. P., & Gander, B. (2005) 'Microencapsulation by solvent extraction/evaporation: Reviewing the state of the art of microsphere preparation process technology', Journal of Controlled Release, 102(2), pp. 313–332.

Ghorani, B., Russell, S. J., & Goswami, P. (2013) 'Controlled morphology and mechanical characterisation of electrospun cellulose acetate fibre webs', International Journal of Polymer Science. In W.-F. Lee (Ed.). Hindawi Publishing Corporation, 2013, p. 256161. https://www.readcube.com/articles/10.1155%2F2013%2F256161.

Ghorani, B., & Tucker, N. (2015) 'Fundamentals of electrospinning as a novel delivery vehicle for bioactive compounds in food nanotechnology', Food Hydrocolloids, 51, pp. 227–240. https://doi.org/10.1016/j.foodhyd.2015.05.024.

Gomez-Mascaraque, L. G., Morfin, R. C., Pérez-Masiá, R., Sanchez, G., & Lopez-Rubio, A. (2016) 'Optimization of electrospraying conditions for the microencapsulation of probiotics and evaluation of their resistance during storage and in-vitro digestion', LWT-Food Science and Technology, 69, pp. 438–446.

Gupta, P. et al. (2005) 'Electrospinning of linear homopolymers of poly(methyl methacrylate): Exploring relationships between fiber formation, viscosity, molecular weight and concentration in a good solvent', Polymer, 46(13), pp. 4799–4810. doi: https://doi.org/10.1016/j.polymer.2005.04.021.

Hans, M. L., & Lowman, A. M. (2002) 'Biodegradable nanoparticles for drug delivery and targeting', Current Opinion in Solid State and Materials Science, 6(4), pp. 319–327.

Heim, M., Keerl, D., & Scheibel, T. (2009) 'Spider silk: From soluble protein to extraordinary fiber', Angewandte Chemie International Edition, 48(20), pp. 3584–3596.

Hong, Y., Li, Y., Yin, Y., Li, D., & Zou, G. (2008) 'Electrohydrodynamic atomization of quasi-monodisperse drug-loaded spherical/wrinkled microparticles', Journal of Aerosol Science, 39(6), pp. 525–536.

Jalili, R., Hosseini, S. A., & Morshed, M. (2005) 'The effects of operating parameters on the morphology of electrospun polyacrilonitrile nanofibres', Iranian Polymer Journal (English Edition), 14(12), pp. 1074–1081.

Jayan, H., Leena, M. M., Sundari, S. S., Moses, J. A., & Anandharamakrishnan, C. (2019) 'Improvement of bioavailability for resveratrol through encapsulation in zein using electrospraying technique', Journal of Functional Foods, 57, pp. 417–424.

Jian, S., Zhu, J., Jiang, S., Chen, S., Fang, H., Song, Y., & Hou, H. (2018) 'Nanofibers with diameter below one nanometer from electrospinning', RSC Advances, 8(9), pp. 4794–4802.

Khan, M. K. I., Schutyser, M. A., Schroën, K., & Boom, R. (2012) 'The potential of electrospraying for hydrophobic film coating on foods', Journal of Food Engineering, 108(3), pp. 410–416.

Kumar, P., Gupta, A., Dhakate, S. R., Mathur, R. B., Nagar, S., & Gupta, V. K. (2013) 'Covalent immobilization of xylanase produced from Bacillus pumilus SV-85 s on electrospun polymethyl methacrylate nanofiber membrane', Biotechnology and Applied Biochemistry, 60(2), pp. 162–169.

Kuswandi, B., Wicaksono, Y., Abdullah, A., Heng, L. Y., & Ahmad, M. (2011) 'Smart packaging: Sensors for monitoring of food quality and safety', Sensing and Instrumentation for Food Quality and Safety, 5(3), pp. 137–146.

Langer, R., & Vacanti, J. P. (1993) 'Tissue engineering', Science, 260, pp. 920–926.

Lee, Y. H., Mei, F., Bai, M. Y., Zhao, S., & Chen, D. R. (2010) 'Release profile characteristics of biodegradable-polymer-coated drug particles fabricated by dual-capillary electrospray', Journal of Controlled Release, 145(1), pp. 58–65.

Leena, M. M., Anukiruthika, T., Moses, J. A., & Anandharamakrishnan, C. (2022) 'Co-delivery of curcumin and resveratrol through electrosprayed core-shell nanoparticles in 3D printed hydrogel', Food Hydrocolloids, 124, p. 107200.

Leena, M. M., Mahalakshmi, L., Moses, J. A., & Anandharamakrishnan, C. (2020). Nanoencapsulation of nutraceutical ingredients. In Biopolymer-based formulations (pp. 311–352). Elsevier.

Leena, M. M., Yoha, K. S., Moses, J. A., & Anandharamakrishnan, C. (2020) 'Edible coating with resveratrol loaded electrospun zein nanofibers with enhanced bioaccessibility', Food Bioscience, 36, p. 100669.

Leena, M. M., Yoha, K. S., Moses, J. A., & Anandharamakrishnan, C. (2021) 'Electrospun nanofibrous membrane for filtration of coconut neera', Nanotechnology for Environmental Engineering, 6(2), pp. 1–10.

Lesmes, U., & McClements, D. J. (2009) 'Structure–function relationships to guide rational design and fabrication of particulate food delivery systems', Trends in Food Science & Technology, 20(10), pp. 448–457.

Li, S. F., Chen, J. P., & Wu, W. T. (2007) 'Electrospun polyacrylonitrile nanofibrous membranes for lipase immobilization', Journal of Molecular Catalysis B: Enzymatic, 47(3–4), pp. 117–124.

Li, D., Wang, Y., & Xia, Y. (2003) 'Electrospinning of polymeric and ceramic nanofibers as uniaxially aligned arrays', Nano Letters, 3(8), pp. 1167–1171.

Liu, Z., Ramakrishna, S., & Liu, X. (2020) 'Electrospinning and emerging healthcare and medicine possibilities', APL Bioengineering, 4(3). https://pubmed.ncbi.nlm.nih.gov/32695956/.

Lu, C. et al. (2006) 'Computer simulation of electrospinning. Part i. Effect of solvent in electrospinning', Polymer, 47(3), pp. 915–921. https://doi.org/10.1016/j.polymer.2005.11.090.

Luo, X., Xie, C., Wang, H., Liu, C., Yan, S., & Li, X. (2012) 'Antitumor activities of emulsion electrospun fibers with core loading of hydroxycamptothecin via intratumoral implantation', International Journal of Pharmaceutics, 425(1–2), pp. 19–28.

Ma, G., Liu, Y., Peng, C., Fang, D., He, B., & Nie, J. (2011) 'Paclitaxel loaded electrospun porous nanofibers as mat potential application for chemotherapy against prostate cancer', Carbohydrate Polymers, 86(2), pp. 505–512.

Mahalakshmi, L., Choudhary, P., Moses, J. A., & Anandharamakrishnan, C. (2023). Emulsion electrospraying and spray drying of whey protein nano and microparticles with curcumin. Food Hydrocolloids for Health, 3, p.100122.

Mahalakshmi, L., Leena, M. M., Moses, J. A., & Anandharamakrishnan, C. (2020). 'Micro-and nano-encapsulation of β-carotene in zein protein: Size-dependent release and absorption behavior', Food & Function, 11(2), pp. 1647–1660.

Mahalakshmi, L., Yoha, K. S., Moses, J. A., & Anandharamakrishnan, C. (2022). Nano delivery systems for food bioactives. In Food, medical, and environmental applications of nanomaterials (pp. 205–230). Elsevier.

Mascheroni, E., Fuenmayor, C. A., Cosio, M. S., Di Silvestro, G., Piergiovanni, L., Mannino, S., & Schiraldi, A. (2013) 'Encapsulation of volatiles in nanofibrous polysaccharide membranes for humidity-triggered release', Carbohydrate Polymers, 98(1), pp. 17–25.

Mo, X. M. et al. (2004) 'Electrospun p(LLA-CL) nanofiber: A biomimetic extracellular matrix for smooth muscle cell and endothelial cell proliferation', Biomaterials, 25(10), pp. 1883–1890. https://doi.org/10.1016/j.biomaterials.2003.08.042.

Moreno-Cortez, I. E., Romero-García, J., González-González, V., García-Gutierrez, D. I., Garza-Navarro, M. A., & Cruz-Silva, R. (2015) 'Encapsulation and immobilization of papain in electrospun nanofibrous membranes of PVA cross-linked with glutaraldehyde vapor', Materials Science and Engineering: C, 52, pp. 306–314.

Mu, L., & Feng, S. S. (2001) 'Fabrication, characterization and in vitro release of paclitaxel (Taxol®) loaded poly (lactic-co-glycolic acid) microspheres prepared by spray drying technique with lipid/cholesterol emulsifiers', Journal of Controlled Release, 76(3), pp. 239–254.

Neo, Y. P., Swift, S., Ray, S., Gizdavic-Nikolaidis, M., Jin, J., & Perera, C. O. (2013) 'Evaluation of gallic acid loaded zein sub-micron electrospun fibre mats as novel active packaging materials', Food Chemistry, 141(3), pp. 3192–3200.

Niu, H., & Lin, T. (2012). Fiber generators in needleless electrospinning. Journal of Nanomaterials, 1–13.

Omi, S. (1996) 'Preparation of monodisperse microspheres using the Shirasu porous glass emulsification technique', Colloids and Surfaces A: Physicochemical and Engineering Aspects, 109, pp. 97–107.

Pareta, R., & Edirisinghe, M. J. (2006) 'A novel method for the preparation of starch films and coatings', Carbohydrate Polymers, 63(3), pp. 425–431.

Pérez-Masiá, R., López-Rubio, A., & Lagarón, J. M. (2013). Development of zein-based heat-management structures for smart food packaging. Food Hydrocolloids, 30(1), 182–191. https://doi.org/10.1016/j.foodhyd.2012.05.010

Pinto, S. C., Rodrigues, A. R., Saraiva, J. A., & Lopes-da-Silva, J. A. (2015) 'Catalytic activity of tryp-sin entrapped in electrospun poly (ε-caprolactone) nanofibers', Enzyme and Microbial Technology, 79, pp. 8–18.

Qiao, Y., Shi, C., Wang, X., Wang, P., Zhang, Y., Wang, D., & Zhong, J. (2019) 'Electrospun nanobelt-shaped polymer membranes for fast and high-sensitivity detection of metal ions', ACS Applied Materials & Interfaces, 11(5), pp. 5401–5413.

Rooney, M. L. (1995). Overview of active food packaging. In Active food packaging (pp. 1–37). Springer, 1-37..

Saeed, K., & Park, S. Y. (2010) 'Preparation and characterization of multiwalled carbon nanotubes/polyacry-lonitrile nanofibers', Journal of Polymer Research, 17(4), pp. 535–540.

Shenoy, S. L. et al. (2005) 'Role of chain entanglements on fiber formation during electrospinning of polymer solutions: Good solvent, non-specific polymer–polymer interaction limit', Polymer, 46(10), pp. 3372–3384. https://doi.org/10.1016/j.polymer.2005.03.011.

Sun, X. Z., Williams, G. R., Hou, X. X., & Zhu, L. M. (2013) 'Electrospun curcumin-loaded fibers with poten-tial biomedical applications', Carbohydrate Polymers, 94(1), pp. 147–153.

Suwantong, O., Ruktanonchai, U., & Supaphol, P. (2008) 'Electrospun cellulose acetate fiber mats containing asiaticoside or Centella asiatica crude extract and the release characteristics of asiaticoside', Polymer, 49(19), pp. 4239–4247.

Taepaiboon, P., Rungsardthong, U., & Supaphol, P. (2006) 'Drug-loaded electrospun mats of poly (vinyl alco-hol) fibres and their release characteristics of four model drugs', Nanotechnology, 17(9), p. 2317.

Tamber, H., Johansen, P., Merkle, H. P., & Gander, B. (2005) 'Formulation aspects of biodegradable polymeric microspheres for antigen delivery', Advanced Drug Delivery Reviews, 57(3), pp. 357–376.

Tan, S.-H. et al. (2005) 'Systematic parameter study for ultra-fine fiber fabrication via electrospinning pro-cess', Polymer, 46(16), pp. 6128–6134. https://doi.org/10.1016/j.polymer.2005.05.068.

Temoçin, Z., Inal, M., Gökgöz, M., & Yiğitoğlu, M. (2018) 'Immobilization of horseradish peroxidase on elec-trospun poly (vinyl alcohol)–polyacrylamide blend nanofiber membrane and its use in the conversion of phenol', Polymer Bulletin, 75(5), pp. 1843–1865.

Torres-Giner, S., Ocio, M. J., & Lagaron, J. M. (2009) 'Novel antimicrobial ultrathin structures of zein/chitosan blends obtained by electrospinning', Carbohydrate Polymers, 77(2), pp. 261–266.

Tycova, A., Prikryl, J., Kotzianova, A., Datinska, V., Velebny, V., & Foret, F. (2021) 'Electrospray: More than just an ionization source', Electrophoresis, 42(1–2), pp. 103–121. https://www.x-mol.net/paper/article/1298329998980255744.

Varesano, A., Carletto, R. A., & Mazzuchetti, G. (2009) 'Experimental investigations on the multi-jet electro-spinning process', Journal of Materials Processing Technology, 209(11), pp. 5178–5185.

Vega-Lugo, A. C., & Lim, L. T. (2009) 'Controlled release of allyl isothiocyanate using soy protein and poly (lactic acid) electrospun fibers', Food Research International, 42(8), pp. 933–940.

Wang, P., Ding, M., Zhang, T., Wu, T., Qiao, R., Zhang, F., & Zhong, J. J. (2022). 'Electrospraying technique and its recent application advances for biological macromolecule encapsulation of food bioactive sub-stances', Food Reviews International, 38(4), pp. 566–588.

Wang, S., Marcone, M. F., Barbut, S., & Lim, L. T. (2013) 'Electrospun soy protein isolate-based fiber fortified with anthocyanin-rich red raspberry (Rubus strigosus) extracts', Food Research International, 52(2), pp. 467–472.

Wannatong, L., Sirivat, A., & Supaphol, P. (2004) 'Effects of solvents on electrospun polymeric fibers: Preliminary study on polystyrene', Polymer International, 53(11), pp. 1851–1859. https://doi.org/10.1002/pi.1599.

Wu, Y. et al. (2009) 'Fabrication of elastin-like polypeptide nanoparticles for drug delivery by electrospray-ing', Biomacromolecules, 10(1), 19. https://pubmed.ncbi.nlm.nih.gov/19072041/

Wu, T., Ding, M., Shi, C., Qiao, Y., Wang, P., Qiao, R., & Zhong, J. (2020) 'Resorbable polymer electrospun nanofibers: History, shapes and application for tissue engineering', Chinese Chemical Letters, 31(3), pp. 617–625.

Xie, C., Li, X., Luo, X., Yang, Y., Cui, W., Zou, J., & Zhou, S. (2010) 'Release modulation and cytotoxicity of hydroxycamptothecin-loaded electrospun fibers with 2-hydroxypropyl-β-cyclodextrin inoculations', International Journal of Pharmaceutics, 391(1–2), pp. 55–64.

Xu, Y., & Hanna, M. A. (2006) 'Electrospray encapsulation of water-soluble protein with polylac-tide: Effects of formulations on morphology, encapsulation efficiency and release profile of particles', International Journal of Pharmaceutics, 320(1), pp. 30–36. https://doi.org/10.1016/j.ijpharm.2006.03.046.

Xu, Y., & Hanna, M. A. (2007) 'Electrosprayed bovine serum albumin-loaded tripolyphosphate cross-linked chitosan capsules: Synthesis and characterization', Journal of Microencapsulation. Taylor & Francis, 24(2), pp. 143–151. https://doi.org/10.1080/02652040601058434.

Yang, Y. et al. (2006) 'Experimental investigation of the governing parameters in the electrospinning of polyethylene oxide solution', IEEE Transactions on Dielectrics and Electrical Insulation, 13(3), pp. 580–585. https://www.semanticscholar.org/paper/Experimental-investigation-of-the-governing-in-the-Yang-Jia/289728751747f51f43019e7dd0f8f20d7081ec73.

Yoo, H. S., Kim, T. G., & Park, T. G. (2009) 'Surface-functionalized electrospun nanofibers for tissue engineering and drug delivery', Advanced Drug Delivery Reviews, 61(12), pp. 1033–1042.

Yuan, X. Y. et al. (2004) 'Morphology of ultrafine polysulfone fibers prepared by electrospinning', Polymer International, 53(11), pp. 1704–1710. https://www.researchgate.net/publication/243844682_Morphology_of_ultrafine_polysulfone_fibers_prepared_by_electrospinning.

Zaeim, D., Sarabi-Jamab, M., Ghorani, B., & Kadkhodaee, R. (2019) 'Double layer co-encapsulation of probiotics and prebiotics by electro-hydrodynamic atomization', LWT, 110, pp. 102–109.

Zhang, X., Kobayashi, I., Uemura, K., & Nakajima, M. (2013) 'Direct observation and characterization of the generation of organic solvent droplets with and without triglyceride oil by electrospraying', Colloids and Surfaces A: Physicochemical and Engineering Aspects, 436, pp. 937–943.

Zhao, S. et al. (2004) 'Electrospinning of ethyl-cyanoethyl cellulose/tetrahydrofuran solutions', Journal of Applied Polymer Science, 91(1), pp. 242–246. https://www.researchgate.net/publication/230055854_Electrospinning_of_ethyl-cyanoethyl_cellulosetetrahydrofuran_solutions.

Zong, X. et al. (2002) 'Structure and process relationship of electrospun bioabsorbable nanofiber membranes', Polymer, 43(16), pp. 4403–4412. https://doi.org/10.1016/S0032-3861(02)00275-6.

Zong, X., Li, S., Chen, E., Garlick, B., Kim, K. S., Fang, D., & Chu, B. (2004) 'Prevention of postsurgery-induced abdominal adhesions by electrospun bioabsorbable nanofibrous poly (lactide-co-glycolide)-based membranes', Annals of Surgery, 240(5), p. 910.

Zou, B., Liu, Y., Luo, X., Chen, F., Guo, X., & Li, X. (2012) 'Electrospun fibrous scaffolds with continuous gradations in mineral contents and biological cues for manipulating cellular behaviors', Acta Biomaterialia, 8(4), pp. 1576–1585.

8 Oscillating Magnetic Fields in Food Processing

S. Chittibabu and S. Shanmugasundaram

8.1 INTRODUCTION

In food industries, the food is undergoing several unit processes and unit operations. In each stage, temperature plays an important role in maintaining the quality of food materials. Conventionally, processing of food is carried out at high temperature to get the desired food products. Non-thermal processing methods are attractive to the food industry because they can preserve the nutritional and sensory properties of foods that may be degraded by heat. In addition, non-thermal processing methods can be used to achieve a range of effects on food, including microbial inactivation, enzyme inactivation, and modification of food structure and properties, without much increase in food processing temperatures. The high temperature processing techniques also affect the colour, texture, freshness, aroma, size, shape, and nutrient composition. Oscillating magnetic fields (OMFs) are categorized as non-thermal processing of foods because they do not rely on heat to process the foods (Ahmed & Ramaswamy, 2007).

An OMF is a type of magnetic field that changes direction and magnitude over time, typically in a regular or cyclical pattern. OMFs can be created by moving electrical charges or by changing the current in an electrical circuit. When an OMF interacts with a conductor or a magnetic material, it can induce an electrical current or a magnetic field in that material. OMFs generate an alternating magnetic field that induces electrical currents in the food. These electrical currents create an electric field that can affect the behaviour of charged particles such as ions and electrons in the food. The interaction of these charged particles with the electric field can result in several effects, including the inactivation of enzymes and microorganisms in the food. Because OMFs do not rely on heat to achieve these effects, they are considered a non-thermal processing method (Karel & Lund, 2003; Frankel, 2019).

OMFs are applied in the form of constant amplitude or decaying amplitude sinusoidal waves. The unit of magnetic field is newton-second per coulomb-metre (or newton per ampere-metre) and it is called tesla (T). Normally, the food to be processed by OMFs should be exposed to 1–100 pulses with 5–500 Hz frequency and intensity 5–50 T. The exposure time is 25–100 ms, whereas the temperature is within 50°C. It is important to maintain the frequency with 500 kHz otherwise, heat will be generated in the food system (Lipiec et al., 2004, 2005).

8.2 GENERATION OF OMF

OMFs are typically generated by passing an alternating electrical current through a coil of wire, which produces a changing magnetic field. The principle behind this process is known as Faraday's law of electromagnetic induction.

When an electrical current flows through a wire, it generates a magnetic field that encircles the wire. The strength and direction of the magnetic field depend on the amount and direction of the electrical current (Barbosa-Canovas et al., 2000). If the current is alternating (i.e., it changes direction periodically), the magnetic field will also change direction and strength accordingly.

By coiling the wire into a solenoid (i.e., a tightly wound coil), the magnetic field becomes more concentrated and uniform. When an alternating current (AC) is passed through the solenoid, it

112   DOI: 10.1201/9781003359302-8

produces a changing magnetic field that can be used for a variety of applications, including food processing. The magnetic force exerted depends on charged particle, intensity of magnetic field, and relative speed of the particle. The relationships between these parameters is expressed in Equation (1.1):

$$F = q(v \cdot B) \tag{1.1}$$

Where F – measure of magnetic force, q – no. of charged particles, B – intensity of magnetic field, v – relative speed in magnetic field.

The frequency and amplitude of the OMF can be controlled by adjusting the properties of the electrical current used to generate it. For example, the frequency can be adjusted by changing the frequency of the AC or by using a device called a frequency generator. The amplitude of the magnetic field can be adjusted by varying the voltage or current used to power the coil (Barnes, 2005).

There are several ways to produce an OMF, depending on the desired frequency and strength of the field. The common methods are given below:

Electromagnets: An electromagnet consists of a coil of wire with a current passing through it. By varying the current in the coil, an OMF can be generated. The frequency and strength of the field can be controlled by adjusting the current and the properties of the coil.

Alternating current (AC): AC is a type of electrical current that changes direction and magnitude at a specific frequency. By running AC through a coil of wire, an OMF can be generated. The frequency and strength of the field can be controlled by adjusting the frequency and voltage of the AC.

Permanent magnets: A permanent magnet produces a magnetic field that does not change over time. However, by moving a permanent magnet back and forth or rotating it, an OMF can be generated.

Induction: Induction is the process of generating an electrical current in a conductor by exposing it to a changing magnetic field. By using a coil of wire and a magnetic field that changes direction and magnitude, an OMF can be generated in the coil.

Ferrite cores: Ferrite cores are magnetic materials that can be used to create OMFs. By wrapping a coil of wire around a ferrite core and passing a current through the wire, an OMF can be generated.

The specific method used will depend on the intended application and the specific requirements for the magnetic field.

8.3   WORKING PRINCIPLE OF OMF GENERATORS

The working principle of an OMF generator involves the conversion of electrical energy into a magnetic field, which is then oscillated to produce the desired frequency (Figure 8.1). There are several types of OMF generators, but one common type is the pulsed electromagnetic field (PEMF) generator.

PEMF generators consist of a power source, a capacitor bank, a coil, and a control unit. The power source provides the initial electrical energy, which is then stored in the capacitor bank. When the control unit is activated, the capacitor bank discharges the stored energy into the coil, which generates a magnetic field.

The magnetic field produced by the coil is oscillated by switching the current direction in the coil at a specific frequency. This is typically accomplished using an electronic circuit that controls the current flow in the coil.

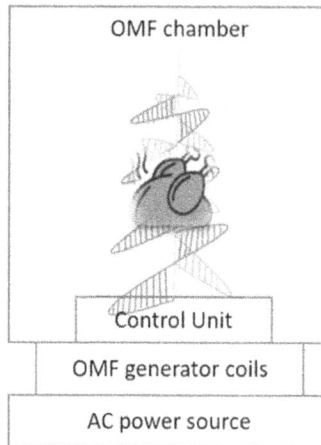

FIGURE 8.1 Schematic representation of OMF generator system.

8.4 CONVERSION EFFICIENCY OF AC TO MAGNETIC FIELD

Typically, the conversion efficiency of AC to magnetic field can range from a few percent to up to 90% or more in some high-performance systems. The conversion efficiency can be affected by factors such as the type of core material used, the coil design, the frequency of the AC, and the operating temperature (Barnes, 2005). Additionally, losses due to factors such as eddy currents and hysteresis can reduce conversion efficiency. The specific conversion efficiency of a given system would need to be measured or calculated based on the design and operating conditions.

8.5 CHARACTERISTICS OF OMF

OMFs have several characteristics that are important to understand when considering their potential applications in various fields, including food processing. Some of the important characteristics are listed below:

Frequency: OMFs have a specific frequency, which determines the rate at which the magnetic field oscillates. This frequency can range from a few hertz to several megahertz.

Amplitude: The amplitude of an OMF refers to the strength or intensity of the magnetic field. This can be controlled by adjusting the electrical current used to generate the field.

Direction: The direction of an OMF is determined by the orientation of the magnetic field relative to the direction of the electrical current used to generate it.

Uniformity: The uniformity of an OMF refers to how evenly the magnetic field is distributed over a given area.

Penetration depth: The penetration depth of an OMF refers to how deeply the magnetic field can penetrate into a material. This can be influenced by the frequency and amplitude of the field, as well as the magnetic properties of the material being exposed to the field.

Polarization: The polarization of an OMF refers to the orientation of the magnetic field relative to the direction of propagation. In some applications, such as magnetic resonance spectroscopy (MRS), the polarization of the magnetic field can be important in determining the information obtained from the material being studied.

Homogeneity: The homogeneity of an OMF refers to the uniformity of the magnetic field over a given volume. In food processing, it is important to have a homogeneous field to ensure that all parts of the food are exposed to the same field strength.

Directionality: The directionality of an OMF refers to the direction of the magnetic field with respect to the food being processed. OMF can be applied in different directions to achieve different effects on the food.

Duration: The duration of OMF exposure refers to the length of time that the food is exposed to the magnetic field. The duration of OMF exposure can range from a few seconds to several hours, depending on the application.

The characteristics of OMFs can vary depending on the specific application and the equipment used to generate the field (Barnes, 2005; Miñano et al., 2020). Understanding these characteristics is important for optimizing the use of OMFs in food processing.

8.6 MEASUREMENT THE INTENSITY OF OMFS

The intensity of an OMF can be measured using a device called a magnetometer. A magnetometer is an instrument that is used to measure magnetic fields. There are several types of magnetometers available for measuring different types of magnetic fields. For OMFs, one common type of magnetometer used is a fluxgate magnetometer. The fluxgate magnetometer works by measuring changes in the magnetic field intensity through a sensor that is composed of two coils of wire. The coils are wrapped around a core that is made of a highly magnetic material. When an OMF is applied to the sensor, it causes a change in the magnetic field in the core, which in turn induces a voltage in the coils. This voltage is proportional to the intensity of the OMF.

Another type of magnetometer that can be used is a search coil magnetometer. This type of magnetometer works by measuring the voltage that is induced in a coil of wire that is placed in the vicinity of the OMFs. The voltage in the coil is proportional to the rate of change of the magnetic field intensity.

Other types of magnetometers that can be used to measure OMFs include the proton precession magnetometer, the optically pumped magnetometer, and the superconducting quantum interference device (SQUID) magnetometer.

The intensity of the OMFs varies in the range of a few millitesla to several teslas, depending on the application and the food type being processed. For example, meat processing requires the field intensity between 0.1 and 0.5 T, fruit and vegetable processing requires from 1 to 5 millitesla. However, the exact intensity required for a specific application should be determined through experimentation and optimization.

8.7 MEASUREMENT THE FREQUENCY OF OMFS

The frequency of an OMF can be measured using a frequency counter or a spectrum analyzer. These instruments can measure the frequency of an electromagnetic wave accurately. To measure the frequency of an OMF, a sensor or a probe capable of detecting the magnetic field is needed. The sensor converts the magnetic field into an electrical signal, which can then be analyzed by the frequency counter or spectrum analyzer.

One common type of sensor used for measuring magnetic fields is the Hall effect sensor. This sensor works by measuring the voltage produced when a magnetic field is applied perpendicular to a current flowing through a thin film. The strength of the magnetic field can be calculated from the Hall voltage, and the frequency of the OMF can be determined using the frequency counter or spectrum analyzer.

Another type of sensor that can be used to measure the frequency of an OMF is the coil sensor. This sensor works by measuring the voltage induced in a coil by a changing magnetic field. The voltage waveform is used to calculate the frequency (Lednev, 1991; Miñano et al., 2020).

It is important to note that the accuracy of the frequency measurement depends on the quality of the sensor, as well as the accuracy of the frequency counter or spectrum analyzer used for the analysis.

8.8 EFFECT OF OMF ON FOODS

Exposure of food to magnetic fields affects the microstructure of the cells, increases the membrane permeability, protein receptors, induces movement of charges in molecules, and increases cell viability. These changes at the cellular level affect the biological reactions, molecular structure, signalling between cells and other connected cellular functions. OMFs can be used on foods in a variety of ways. Some common examples are:

Preservation: OMFs can be used to preserve food by inhibiting the growth of microorganisms, such as bacteria and fungi. The magnetic fields can disrupt the cell membranes of these microorganisms, leading to their death or reduced growth.

Extraction: OMFs can be used to extract certain compounds from foods, such as pigments or flavours. By exposing the food to an OMF, the compounds can be induced to move or vibrate, making them easier to extract.

Sorting: OMFs can be used to sort and separate different types of foods based on their magnetic properties. For example, magnetic fields can be used to remove metal fragments from foods or to sort different types of grains based on their magnetic properties.

Processing: OMFs can be used to process foods in various ways, such as drying, heating, or cooking. The magnetic fields can induce electrical currents in the food, leading to heating or other changes in the food's properties.

Pasteurization: OMFs can be used to pasteurize food by reducing the population of microorganisms, such as bacteria and yeasts, in the food. This can be particularly useful in reducing the risk of foodborne illness.

The effects of OMFs on foods will depend on the frequency, strength, and duration of the magnetic fields, as well as the specific properties of the food being treated (Grigelmo-Miguel et al., 2011; Hu et al., 2013; Terefe et al., 2015). Some research has shown promising results for the use of OMFs in food preservation and processing, although further research is needed to fully understand the effects and potential applications of this technology.

8.8.1 MECHANISM OF ENZYME INACTIVATION BY OMF

The mechanism of enzyme inactivation by OMFs during food processing is not fully understood, but it is believed to involve a combination of physical and chemical effects. One mechanism is that the OMF disrupts the structure of the enzyme molecules, causing them to lose their catalytic activity. The OMF can induce changes in the molecular conformation of the enzyme by generating forces that cause molecular vibrations and rotations. These changes can cause the enzyme to denature and lose its active site, which is essential for catalyzing chemical reactions. Another mechanism is that the OMF generates reactive oxygen species (ROS) that can react with the enzyme molecules, leading to their inactivation. The OMF can cause the formation of ROS, such as superoxide anions and hydrogen peroxide, which can react with the amino acid residues in the enzyme molecule and cause oxidative damage. This can lead to the loss of enzyme activity (Gerencser et al., 1962; Harte et al., 2001; Hu et al., 2013; Terefe et al., 2015; Veiga et al., 2020).

In conclusion, the mechanism of enzyme inactivation by OMF is likely to be complex and may involve multiple factors. The exact mechanism may depend on the specific enzyme and the processing conditions, such as the strength and frequency of the magnetic field, the duration of exposure, and the composition of the food matrix.

8.8.2 MECHANISM OF OMF IN MEAT PROCESSING

The mechanism of OMF on meat processing is believed to involve several physical and biochemical effects that can improve the quality, safety, and shelf life of meat products. One of the physical effects of OMF on meat is the induction of eddy currents, which generate localized heating and can help to increase the internal temperature of the meat. This can be particularly useful in meat products where thermal processing is required, such as sausages and hot dogs. OMF can also help to reduce cooking time and energy consumption, which can be beneficial in industrial meat processing. Another physical effect of OMF is the induction of mechanical stress and strain in the meat, which can help to tenderize the meat and improve its texture. OMF can also enhance the water-holding capacity of the meat, which can result in a juicier and more flavourful product. The biochemical effects of OMF on meat processing are due to the modulation of enzymatic activity and the reduction of microbial load. OMF can inhibit the activity of enzymes such as proteases, which can cause protein degradation and spoilage of meat. OMF can also reduce the population of microorganisms such as bacteria, yeasts, and Molds, which can cause foodborne illness and spoilage of meat (Kang et al., 2021).

The mechanism of OMF in meat tenderizing is complex and not yet fully understood. However, OMF works by inducing electrical currents within the muscle tissue, which generates heat and causes thermal and non-thermal effects. The thermal effect is due to the heat generated by the electrical currents, which causes the muscle tissue to expand and contract. This can lead to the breakdown of collagen and other connective tissue, resulting in increased meat tenderness. The non-thermal effect is due to the mechanical forces generated by the OMF. These forces can cause the muscle fibres to vibrate and move, resulting in further breakdown of collagen and other connective tissue. The mechanical forces can also disrupt the muscle fibres and increase the water-holding capacity of the meat.

The combination of thermal and non-thermal effects induced by OMF can result in significant improvements in meat tenderness without the need for traditional mechanical or chemical tenderization methods. However, the optimal processing parameters for OMF-based meat tenderization (such as frequency, intensity, and exposure time) may vary depending on the specific meat product and processing conditions (Mok et al., 2017a).

OMF has the potential to be a useful tool in meat processing by improving the quality, safety, and shelf life of meat products. However, further research is needed to optimize the processing parameters and to understand the long-term effects of OMF on the nutritional and sensory properties of meat.

8.9 APPLICATION OF OMF IN FOOD PROCESSING

8.9.1 OSCILLATION MAGNETIC FIELD APPLICATIONS IN FRUITS AND VEGETABLES PROCESSING

OMFs have shown promise in a variety of food processing applications, including the processing of fruits and vegetables. Here are some of the ways in which OMF can be used in fruit and vegetable processing:

Dehydration: OMF can be used to facilitate dehydration of fruits and vegetables by breaking down the cell walls and facilitating the release of water. This can reduce processing time and energy consumption compared to traditional drying methods.

Preservation: OMF has been shown to reduce the microbial load in fruits and vegetables, which can help to extend their shelf life and prevent spoilage. This can be particularly useful for fruits and vegetables that are prone to spoilage or that have a short shelf life. For example, OMF exposure of mango extended the shelf life without much loss in original quality attributed during the supercooling preservation of mangoes (Kang et al., 2021).

Texture modification: OMF can be used to modify the texture of fruits and vegetables by breaking down the cell walls and facilitating the release of pectin. This can help to soften fruits and vegetables and make them more palatable.

Flavour enhancement: OMF has been shown to enhance the flavour of fruits and vegetables by facilitating the release of volatile compounds. This can result in a more intense and complex flavour profile.

Enzyme inactivation: OMF can be used to inactivate enzymes that can cause browning and other undesirable changes in fruits and vegetables. This can help to maintain the colour, texture, and nutritional content of the products.

Disinfection: The influence of OMF pulses on pathogenic microorganisms of potatoes was studied and found that the bacterial load was reduced by three times, and for fungi, it was reduced by two times by the exposure to OMFs. Hence, the application of OMF pulses might be used in disinfecting agricultural products and food.

8.9.2   OMF Applications in Meat Processing

OMFs have a variety of potential applications in meat processing.

Curing: OMF has been shown to accelerate the curing process in meat products, which can reduce processing time and increase efficiency in meat processing. This can also help to ensure consistent quality in cured meat products.

Tenderization: OMF can be used to facilitate the tenderization of meat products by breaking down the connective tissue and muscle fibres. This can improve the texture and palatability of tougher cuts of meat.

Preservation: OMF has been shown to reduce the microbial load in meat products, which can help to extend their shelf life and prevent spoilage. This can be particularly useful for meat products that are prone to spoilage or that have a short shelf life.

Flavour enhancement: OMF can be used to enhance the flavour of meat products by facilitating the release of volatile compounds. This can result in a more intense and complex flavour profile.

Enzyme inactivation: OMF can be used to inactivate enzymes that can cause off-flavours and other undesirable changes in meat products. This can help to maintain the flavour, colour, and texture of the products.

Several studies have been carried out using both Pulsed electric field and OMF simultaneously on meat. Synergetic effect of Pulsed electric field and OMF was highly effective in the maintenance of original meat qualities without significant physical damages or chemical changes during long-term preservation of meat at sub-zero temperature (Kang et al., 2021; Mok et al., 2017). OMF shows promise as a non-thermal processing technique for meat products, especially in supercooling preservation stages.

8.9.3   OMF in Dairy Processing

In dairy processing, OMFs can be used at different stages in dairy processing.

Pasteurization: OMF can be used to pasteurize milk and other dairy products by reducing the microbial load. This can be particularly useful for reducing the risk of foodborne illness while maintaining the nutritional and sensory quality of the products.

Cheese production: OMF can be used to facilitate the coagulation and curd formation in cheese production. This can improve the yield, texture, and sensory characteristics of the final product.

Fermentation: OMF has been shown to enhance the fermentation process in dairy products such as yogurt and kefir which resulted in a more consistent and higher-quality products.

Shelf-life extension: OMF has been shown to extend the shelf-life of dairy products by reducing the microbial load and preventing spoilage. This can be particularly useful for dairy products that are prone to spoilage or that have a short shelf-life.

Enzyme inactivation: OMF can be used to inactivate enzymes that can produce unwanted flavours in dairy products. This can help to maintain the flavour, colour, and texture of the products (Tsuchiya et al., 1996).

8.9.4 OSCILLATION MAGNETIC FIELD APPLICATION IN GRAIN PROCESSING

OMFs have the potential to be used in the processing of grains, including cereals, pulses, and oilseeds. Here are some ways in which OMF can be used in grain processing:

Quality enhancement: OMF has been shown to improve the quality of grains by reducing insect infestation, mold growth, and mycotoxin contamination. This can help to maintain the nutritional and sensory quality of the grains and prevent spoilage.

Dehulling: OMF can be used to facilitate the dehulling process in grains such as rice, barley, and oats. This can improve the efficiency of the processing and the quality of the final product.

Milling: OMF can be used to enhance the efficiency of milling processes by reducing the energy requirements and improving the particle size distribution of the milled grains.

Fermentation: OMF has been shown to enhance the fermentation process in grains such as bread and beer.

Enzyme inactivation: OMF can be used to inactivate enzymes that can cause off-flavours and other undesirable changes in grain products (Barbosa-Cánovas et al., 1998; Li & Farid, 2016). This can help to maintain the flavour, colour, and texture of the products.

8.10 ADVANTAGES OF OMFS IN FOOD PROCESSING

Non-thermal processing: OMFs can be used to process food without relying on heat, which can help to preserve the nutritional content, texture, and flavour of the food.

Inactivation of microorganisms: OMFs can be used to inactivate microorganisms in food, including bacteria, yeasts, and molds. The magnetic fields disrupt the cell membranes and internal structures of microorganisms, leading to their inactivation. This can extend the shelf life of foods by reducing spoilage and preventing the growth of pathogenic microorganisms.

Enzyme inactivation: OMFs can also be used to inactivate enzymes in food. This can be useful in preventing undesirable reactions, such as browning and softening of fruits and vegetables, or inactivation of enzymes that cause off-flavours in dairy products.

Modification of food properties: OMFs can modify the physical and chemical properties of food, including texture, colour, and nutritional content. For example, OMFs can be used to modify the texture of fruits and vegetables by breaking down cell walls, or to increase the solubility and bioavailability of nutrients in dairy products.

Minimal processing: OMFs are considered a non-thermal processing technology, as they do not rely on heat to achieve their effects. This means that foods can be processed without significant changes to their nutritional or sensory properties, making them suitable for minimal processing.

Combination with other processing technologies: OMFs can be used in combination with other processing technologies, such as high-pressure processing or pulsed electric fields, to achieve synergistic effects on food preservation.

Reduced processing time: In some applications, OMFs can reduce the processing time required to achieve a desired result. For example, they have been shown to accelerate the curing process in meat products.

Improved food safety: OMFs have been shown to reduce the microbial load of food products, which can improve food safety and extend shelf life.

Enhanced quality: OMFs can be used to enhance the quality of certain food products by modifying their physical and chemical properties. For example, they have been shown to increase the hydration of bread dough, resulting in a softer and more elastic texture.

Reduced need for additives: The use of OMFs can reduce the need for certain additives, such as preservatives, by enhancing the natural preservation properties of the food.

Reduced environmental impact: Compared to thermal processing methods, the use of OMFs can reduce the energy consumption and environmental impact of food processing.

8.11 LIMITATIONS OF OMF IN FOOD PROCESSING APPLICATIONS

While OMFs have several potential applications in food processing, there are also some limitations and challenges associated with this technology. The major challenges are the inconsistent results and lack of reproducibility due to the design of experiments, time of exposure, and characteristics of magnetic fields. Some of the main limitations are given below:

Equipment cost: The equipment required to generate OMFs can be expensive, which may limit the accessibility of this technology for small-scale food processors.

Energy consumption: Generating OMFs requires energy, which can make this technology less environmentally friendly than some other food processing methods.

Equipment size: The size of the equipment required to generate OMFs can be quite large, which may limit its use in certain food processing settings.

Effect on sensory qualities: The use of OMFs can sometimes affect the sensory qualities of food, such as its texture, flavour, and aroma. This can be a limitation for certain types of food products.

Lack of standardized protocols: There is currently a lack of standardized protocols for the use of OMFs in food processing, which can make it difficult to compare results between studies and to optimize the technology for specific food products.

Limited understanding of mechanisms: While some research has been done on the effects of OMFs on food, there is still a limited understanding of the underlying mechanisms and the full range of potential effects on food quality and safety.

The OMFs show promise as a potentially useful food processing technology, further research is needed to fully understand the limitations and potential applications of this technology.

8.12 RESEARCH AND DEVELOPMENT IN OMF

Research and development in OMFs are a growing area of interest in various fields, including food processing.

Food preservation: Research is ongoing to understand how OMFs can be used to improve the preservation of food products. This includes investigations into how OMF can reduce the microbial load in food, extend shelf life, and prevent spoilage (Knorr et al., 2001).

Non-thermal processing: OMF is being studied as a non-thermal processing technique that can be used to process food products without relying on heat. This includes investigations into how OMF can be used to improve the texture, flavour, and nutritional content of foods (Huub Lelieveld et al., 2007).

Curing of meat products: OMF has been shown to accelerate the curing process in meat products, which could reduce processing time and increase efficiency in meat processing (Mok et al., 2017).

Chemical and physical changes: OMF can induce various chemical and physical changes in food products, including changes in texture, colour, and flavour. Researchers are studying how OMF can be used to modify the properties of food products to enhance their quality (Kayalvizhi et al., 2016).

Equipment design and optimization: Researchers are studying the design and optimization of equipment for generating OMFs. This includes investigations into how to maximize the efficiency and effectiveness of the OMF process.

Overall, research and development in OMFs is a growing area with a wide range of potential applications in food processing and other fields. As the technology continues to develop, we can expect to see new and innovative applications of OMF in the food industry and beyond.

8.13   RESEARCH GAP IN THE APPLICATION OF OMF IN FOOD PROCESSING

While OMF has shown potential as a novel technology for food processing, there are several research gaps that need to be addressed to fully understand its potential and limitations.

One major research gap is the lack of standardized processing protocols and equipment for OMF. There is currently no widely accepted standard for the strength, frequency, duration, and other processing parameters of OMF, which makes it difficult to compare results across different studies and to develop reliable processing equipment. Standardization is essential for ensuring the consistency and reproducibility of results, as well as for facilitating the adoption of OMF in industry (Gachovska et al., 2010; James et al., 2015).

Another research gap is the need for more fundamental research on the mechanisms of action of OMF. While there is some understanding of the physical and chemical effects of OMF on food, there is still much to be learned about the specific mechanisms of action that lead to the observed changes in food properties. This knowledge is essential for optimizing the processing parameters and for predicting the long-term effects of OMF on food quality and safety.

There is also a need for more research on the effects of OMF on the nutritional and sensory properties of food. While OMF has shown promise in improving food quality and safety, it is not yet clear how it affects the nutritional and sensory properties of food, which are critical for consumer acceptance. More research is needed to determine the effects of OMF on the nutrient content, flavour, texture, and other sensory attributes of food.

8.14   FUTURE TRENDS IN APPLICATION OF OMF IN FOOD PROCESSING

Combination with other processing technologies: OMFs can be combined with other processing technologies such as ultrasound, high-pressure processing, and pulsed electric fields to achieve synergistic effects on food. Such combinations can lead to improved microbial inactivation, enzyme inactivation, and modification of food properties.

Development of specific OMF frequencies and waveforms: There is a need to develop specific OMF frequencies and waveforms that can achieve the desired effects on food. Researchers may focus on optimizing OMF parameters for specific food types and applications.

Scale-up of OMF technology: Most of the research on OMFs in food processing has been conducted at the laboratory scale. There is a need to scale up the technology for commercial applications. This will involve the development of larger OMF generators and the design of processing equipment that can handle large volumes of food.

Application in new food products: There is a potential for the application of OMFs in the processing of new food products that require specific modifications of their properties. For example, OMFs can be used to modify the texture and sensory properties of fruits and vegetables, or to enhance the nutritional properties of dairy products.

Development of OMF sensors and monitoring systems: There is a need to develop sensors and monitoring systems that can accurately measure the OMF strength and uniformity in the food being processed. Such systems will be important for ensuring consistent processing conditions and for quality control (Cheng et al., 2017; Dalvi-Isfahan et al., 2017; Mok et al., 2017b; Priyadarshini et al., 2019; Valdramidis & Koutsoumanis, 2016).

The future trends in the application of OMFs in food processing will focus on optimizing the technology for specific food types and applications, scaling up the technology for commercial use, and developing new processing methods and products that can benefit from the unique effects of OMFs.

8.15   CHALLENGES OF USING OMF UNITS IN FOOD PROCESSING INDUSTRIES

There are several challenges associated with the use of OMF units in food processing industries. These challenges include:

Equipment cost: OMF units can be expensive to purchase and maintain, which can be a barrier to their adoption by smaller food processing companies.

Scale-up: While OMF technology has shown promise in lab-scale studies, scaling up the technology to industrial levels can be challenging. It can be difficult to maintain consistent magnetic fields over large volumes of food, which can limit the effectiveness of the technology.

Food heterogeneity: Different foods have different physical and chemical properties, which can affect the effectiveness of OMF treatment. For example, some foods may have a lower water content or higher fat content, which can affect the way magnetic fields interact with the food.

Regulatory issues: As a new technology, there may be regulatory barriers to the adoption of OMF units in food processing. Regulators may require additional data on the safety and effectiveness of the technology before approving its use.

Lack of understanding: There is still much to be learned about the mechanisms of OMF treatment and its effects on different foods. This lack of understanding can make it difficult to optimize the technology for different food types and processing conditions.

Energy requirements: OMF units require a significant amount of energy to operate, which can be a challenge for some food processing companies that are trying to reduce their energy consumption and environmental impact (Priyadarshini et al., 2019).

8.16   SUMMARY

More research is needed on the commercial feasibility and economic viability of OMF. While OMF has shown potential in laboratory and pilot-scale studies, its scalability and cost-effectiveness in industrial applications have not yet been fully demonstrated. The potential benefits of OMF technology for food processing are significant, but there are still many challenges to overcome to realize its full potential. It is important to determine the cost-benefit ratio of OMF compared to other food processing technologies and to identify the factors that may affect its adoption in industry. Continued research and development, as well as collaboration between industry, academia, and regulators, will be necessary to address these challenges and enable wider adoption of the technology.

REFERENCES

Ahmed, J. & Ramaswamy, H. S. (2007). *Applications of Magnetic Field in Food Preservation* (2nd ed.). CRC Press.

Barbosa-Cánovas, G. V., Góngora-Nieto, M. M., & Swanson, B. G. (1998). Nonthermal electrical methods in food preservation/Métodos eléctricos no térmicos para la conservación de alimentos. *Food Science and Technology International*, 4(5), 363–370. https://doi.org/10.1177/108201329800400508

Barbosa-Canovas, G. V., Schaffner, D. W., Pierson, M. D., & Zhang, Q. H. (2000). Oscillating magnetic fields. *Journal of Food Safety*, *65*, 86–89. https://doi.org/10.1111/j.1750-3841.2000.tb00622.x

Barnes, F. S. (2005). Mechanisms for electric and magnetic fields effects on biological cells. *IEEE Transactions on Magnetics*, *41*(11), 4219–4224. https://doi.org/10.1109/TMAG.2005.855480

Cheng, L., Sun, D.-W., Zhu, Z., & Zhang, Z. (2017). Emerging techniques for assisting and accelerating food freezing processes: A review of recent research progresses. *Critical Reviews in Food Science and Nutrition*, *57*(4), 769–781. https://doi.org/10.1080/10408398.2015.1004569

Dalvi-Isfahan, M., Hamdami, N., Xanthakis, E., & Le-Bail, A. (2017). Review on the control of ice nucleation by ultrasound waves, electric and magnetic fields. *Journal of Food Engineering*, *195*, 222–234. https://doi.org/10.1016/j.jfoodeng.2016.10.001

Frankel, R. B. (2019). *Biological Effects of Static Magnetic Fields* (Charles Polk, Ed.; 1st ed.). CRC Press. https://doi.org/10.1201/9781351071017

Gachovska, T., Cassada, D., Subbiah, J., Hanna, M., Thippareddi, H., & Snow, D. (2010). Enhanced anthocyanin extraction from red cabbage using pulsed electric field processing. *Journal of Food Science*, *75*(6), E323–E329. https://doi.org/10.1111/j.1750-3841.2010.01699.x

Gerencser, V. F., Barnothy, M. F., & Barnothy, J. M. (1962). Inhibition of bacterial growth by magnetic fields. *Nature*, *196* (4854), 539–541. https://doi.org/10.1038/196539a0

Grigelmo-Miguel, N., Soliva-fortuny, R., Barbosa-Cánovas, G. V., & Martín-Belloso, O. (2011). Use of Oscillating Magnetic Fields in Food Preservation. In *Nonthermal Processing Technologies for Food* (pp. 222–235). Wiley-Blackwell. https://doi.org/10.1002/9780470958360.ch16

Harte, F., Martin, M. F. S., Lacerda, A. H., Lelieveld, H. L. M., Swanson, B. G., & Barbosa-cánovas, G. V. (2001). Potential use of 18 tesla static and pulsed magnetic fields on *Escherichia coli* and *Saccharomyces cerevisiae*. *Journal of Food Processing and Preservation*, *25*(3), 223–235. https://doi.org/10.1111/j.1745-4549.2001.tb00456.x

Hu, W., Zhou, L., Xu, Z., Zhang, Y., & Liao, X. (2013). Enzyme inactivation in food processing using high pressure carbon dioxide technology. *Critical Reviews in Food Science and Nutrition*, *53*(2), 145–161. https://doi.org/10.1080/10408398.2010.526258

James, C., Reitz, B., & James, S. J. (2015). The freezing characteristics of garlic bulbs (Allium sativum L.) frozen conventionally or with the assistance of an oscillating weak magnetic field. *Food and Bioprocess Technology*, *8*(3), 702–708. https://doi.org/10.1007/s11947-014-1438-z

Kang, T., You, Y., Hoptowit, R., Wall, M. M., & Jun, S. (2021). Effect of an oscillating magnetic field on the inhibition of ice nucleation and its application for supercooling preservation of fresh-cut mango slices. *Journal of Food Engineering*, *300*, 110541. https://doi.org/10.1016/j.jfoodeng.2021.110541

Karel, M., & Lund, D. (2003). *Physical Principles of Food Preservation* (Marcus Karel & Daryl Lund, Eds.; 2nd ed.). CRC Press. https://doi.org/10.1201/9780203911792

Kayalvizhi, V., Pushpa, A. J. S., Sangeetha, G., & Antony, U. (2016). Effect of pulsed electric field (PEF) treatment on sugarcane juice. *Journal of Food Science and Technology*, *53*(3), 1371–1379. https://doi.org/10.1007/s13197-016-2172-5

Knorr, D., Angersbach, A., Eshtiaghi, M. N., Heinz, V., & Lee, D.-U. (2001). Processing concepts based on high intensity electric field pulses. *Trends in Food Science & Technology*, *12*(3–4), 129–135. https://doi.org/10.1016/S0924-2244(01)00069-3

Lednev, V. V. (1991). Possible mechanism for the influence of weak magnetic fields on biological systems. *Bioelectromagnetics*, *12*(2), 71–75. https://doi.org/10.1002/bem.2250120202

Lelieveld, H., Notermans, S. & de Haan, S. W. H. (2007). *Food Preservation by Pulsed Electric Fields: From Research to Application* (1st ed.). CRC Press.

Li, X., & Farid, M. (2016). A review on recent development in non-conventional food sterilization technologies. *Journal of Food Engineering*, *182*, 33–45. https://doi.org/10.1016/j.jfoodeng.2016.02.026

Lipiec, J., Janas, P., & Barabasz, W. (2004). Effect of oscillating magnetic field pulses on the survival of selected microorganisms. *International Agrophysics*, *18*(4), 325–328.

Lipiec, J., Janas, P., Barabasz, W., Pysz, M., & Pisulewicz, P. (2005). Effects of oscillating magnetic field pulses on selected oat sprouts used for food purposes. *Acta Agrophysica*, *5*(2), 357–365.

Miñano, H. L. A., Silva, A. C. de S., Souto, S., & Costa, E. J. X. (2020). Magnetic fields in food processing perspectives, applications and action models. *Processes*, *8*(7), 814. https://doi.org/10.3390/pr8070814

Mok, J. H., Her, J.-Y., Kang, T., Hoptowit, R., & Jun, S. (2017). Effects of pulsed electric field (PEF) and oscillating magnetic field (OMF) combination technology on the extension of supercooling for chicken breasts. *Journal of Food Engineering*, *196*, 27–35. https://doi.org/10.1016/j.jfoodeng.2016.10.002

Priyadarshini, A., Rajauria, G., O'Donnell, C. P., & Tiwari, B. K. (2019). Emerging food processing technologies and factors impacting their industrial adoption. *Critical Reviews in Food Science and Nutrition*, *59*(19), 3082–3101. https://doi.org/10.1080/10408398.2018.1483890

Terefe, N. S., Buckow, R., & Versteeg, C. (2015). Quality- related enzymes in plant-based products: Effects of novel food processing technologies part 2: Pulsed electric field processing. *Critical Reviews in Food Science and Nutrition*, *55*(1), 1–15. https://doi.org/10.1080/10408398.2012.701253

Tsuchiya, K., Nakamura, K., Okuno, K., Ano, T., & Shoda, M. (1996). Effect of homogeneous and inhomogeneous high magnetic fields on the growth of *Escherichia coli*. *Journal of Fermentation and Bioengineering*, *81*(4), 343–346. https://doi.org/10.1016/0922-338X(96)80588-5

Valdramidis, V. P., & Koutsoumanis, K. P. (2016). Challenges and perspectives of advanced technologies in processing, distribution and storage for improving food safety. *Current Opinion in Food Science*, *12*, 63–69. https://doi.org/10.1016/j.cofs.2016.08.008

Veiga, M. C., Fontoura, M. M., de Oliveira, M. G., Costa, J. A. V., & Santos, L. O. (2020). Magnetic fields: Biomass potential of Spirulina sp. for food supplement. *Bioprocess and Biosystems Engineering*, *43*(7), 1231–1240. https://doi.org/10.1007/s00449-020-02318-4

9 Ionizing Radiation Technologies in Food Preservation

Malini Buvaneswaran, Neeta S. Ukkunda,
V. R. Sinija, and R. Mahendran

9.1 INTRODUCTION

Ionizing radiation has about 100 years of history in scientific applications like medical, effluent treatment, food preservation, and disinfestation. The discovery of X-rays in 1895 and radioactivity in 1896 led to the studies of ionizing radiation on a living organism, and then the practical application was found in 1905. Radiation "refers to physical phenomena in which energy moves across space or matter". In the 1920s, ionizing radiation technology, also known as ionization or irradiation, was utilized to sterilize medical equipment at high doses ranging from 40 to 100 kGy (Munir & Federighi, 2020). Same period, agro-food industries have shown interest in ionizing radiation due to the advantages of these techniques in food preservation over heat treatments. Then a lot of studies concentrated on the application of ionizing radiation techniques to obtain microbial safety, storage stability, and to improve the shelf life of food products. Ionizing radiation refers to the process where ionizing energy is applied to materials like food to inhibit or destroy microorganisms, parasites, pests, insects, and chemical reactions. Ionizing radiation is a type of electromagnetic wave with enough energy to ionize atoms or molecules by removing electrons. In 1921, the Department of Agriculture proposed using X-rays to inactivate trichinae in pork, and in 1930, a French patent was awarded for the sterilization of packaged food (Diehl, 2002). On the other hand, the development and implementation of regulations and scientific frameworks for the application and marketability of food products treated by ionizing radiation further enhance the application of ionizing radiation. Food preservation generally denotes the prevention of food materials from a microorganism (bacteria, fungi, yeast) and chemical deterioration (rancidity, enzymatic browning, etc.). Recently, demand for high-nutritional food with less processing, minimally processed foods, ready-to-eat, and ready-to-cook foods are increased. This denotes the consumer concern about the nutritional aspects and processing techniques of food products. Based on these requirements and the advantage of ionizing radiation, it started to explore more food products. Food irradiation refers to the process of exposing food ingredients to a controlled amount of energy in the form of electromagnetic waves or high-speed particles (Farkas, 2006). The first half of the last century represents the age of inventors due to the non-practical application, in the middle of the 20th century, solid scientific findings and technical development enhances the commercial utilization of ionizing radiation techniques. In recent years, irradiation is permitted in more than 60 countries in the world and at least 10 countries using ionizing radiation to meet the phytosanitary requirements for trading fresh fruits and vegetables (Roberts, 2016). China placed first to produce the largest volume of irradiated foods. The Food and Agricultural Organization (FAO), the International Atomic Energy Agency (IAEA), and the World Health Organization (WHO) were the major partners in assisting the specific research programs and international cooperation in ionizing radiation application and implementation at the industrial level. Based on research works, the Joint FAO/IAEA/WHO Export Committee on the Wholesomeness of Irradiated Food (JECFI) reported that up to a dose of 10 kGy does not have any toxicological hazard, the dose level of 25–60 kGy (radappertization) is required in some applications of food preservation, while these foods are reported as safe to consume and nutritionally adequate by FAO/IAEA/WHO Study group of High Dose Irradiation (Farkas, 2006).

The main advantages of ionizing radiation in food preservation are (Ajibola, 2020):

   i. Sterilization or microbial inactivation is not dependent on temperature so the food can be processed at ambient temperature.
  ii. The irradiation effect does not depend on the product phase (gaseous, liquid, solid), dimensions (size, thickness), and density.
 iii. The food products can be processed with the package due to the higher penetration of ionizing radiation which eliminates food contamination after processing.
  iv. Green technology does not require any volatiles and chemical substances as well as not produce any waste or disposal.
   v. Application of single variable (exposure dosage/time).

This chapter provides an overview of the use of ionizing radiation in food preservation, including its benefits, commercial applications, and problems.

9.2   SOURCE AND MODE OF IONIZING RADIATION

9.2.1   SOURCE

The usage of ionizing technology for sterile materials is termed irradiation. Radiation is the transfer of energy from one source to another. Many forms of radiation energy are available from the electromagnetic spectrum, with ionizing radiation having the energy to ionize the materials by removing the bonded electrons from an atomic and molecular structure, which can be utilized for sterilizing or preserving food products. According to General Standard for Irradiated Foods (CODEX STAN 106-1983), three types of ionizing radiation potentially used in food industries are electromagnetic gamma (γ) rays, corpuscular beta (β) rays (E-beam), and X-rays (NAFDAC, 2021). Both γ and β radiations have different origins and modes of action. According to the International Agency for Atomic Energy (IAAE), the ionizing ratio for food treatment should be applied safely within the specific conditions of radiation sources, the energy of rays should not exceed 10 mega electron volts (MeV), X-ray should operate below 5 MeV energy level, similarly, the energy of rays should be less than 5 MeV. The gamma (γ) rays are usually produced from radioactive elements ^{60}Co and ^{137}Cs. Electrons in a cathodic tube are accelerated to form a stream of high-speed electrons driven to foods, beta (β) rays are formed. Hence, radioactive sources and electron accelerators are the primary sources of ionizing radiation and radiation, respectively (Munir & Federighi, 2020).

   i. Radioactive sources: Caesium-137 (^{137}Cs) and cobalt-60 (^{60}Co) are the major radioactive sources in industrial facilities. ^{60}Co, which is produced from Cobalt 59 in nuclear reactors, is mostly used. Due to radiation safety concerns, ^{137}Cs have been limited to laboratory facilities. Ration and two photons with an energy level of 1.17 and 1.33 MeV are emitted from ^{60}Co.
  ii. Electron accelerators: It is made up of an electron source (cathode), an accelerator tube, and a beam-shaping system. A cathode excited by electricity in a vacuum system emits electrons, which are then subjected to an electron gun to achieve the desired acceleration of electrons by applying an external voltage, and finally shaped to form a beam by the beam shaping system. The main advantage of this system is the energy, intensity, and geometry can be controlled in an electron accelerator. It is used to produce both X-ray and β radiations. China has 103 ^{60}Co facilities and six electron beam accelerators (Kume et al., 2009).

9.2.2 MODE OF ACTION OF IONIZING RADIATION

High-energy radiation causes ionization when radiation is absorbed by the material. Atoms and molecules in the material get excited by absorbed radiation, and they are highly reactive and cause chemical changes. If the absorbed energy is capable of removing an electron from the material, then the material is ionized and possesses a positive charge, which is unstable and highly reactive. In water molecules, the absorption of radiation causes the following reactions:

$$H_2O \rightarrow H_2O^+ + e^-$$

$$H_2O^+ \rightarrow OH^. + H^+$$

$$e^- + H_2O \rightarrow OH^- + H^.$$

$$H_2O \rightarrow H^+ + OH^-$$

Radiation produces radicals, atoms, or molecules that recombine to make hydrogen gas, hydrogen peroxide, or water molecules.

$$2H^. \rightarrow H_2$$

$$2OH^. \rightarrow H_2O_2$$

$$H^. + OH^. \rightarrow H_2O$$

The presence of oxygen further enhances the formation of hydrogen peroxide and hydroperoxyl radicals, which can act as oxidizing or reducing agents.

The biological effect, especially microbial inactivation by ionizing radiation, is achieved by both direct and indirect effects. (i) Direct effect: The radiation affects the DNA molecules directly by disturbing the molecular structure, which causes cell damage and cell death. The formation of cross-linking in DNA structure results in the distortion of a molecule and suppresses cell replication. It causes a chemical change in deoxy ribose sugar, purine, and pyrimidine bases, which causes the breakage of the phospho di-ester backbone. (ii) Indirect effect: In the indirect method, the free radicles produced by radiation cause cell damage. The water molecules in the microbial cell and other organic molecules produce hydroxyl ions (OH$^-$) and hydrogen atoms (H$^+$), which damage DNA molecules and cause molecule bonds and structural damage. This phenomenon causes the impairment of cell function and cell death. Both direct and indirect actions of ionizing radiation cause biological and physiological changes in cells.

9.3 FACTORS AFFECTING IONIZING RADIATION

The application of ionizing radiation in food products mainly varied based on the exposure time or dosage level. Accurate knowledge about the dose delivery and absorbed dose by food is necessary for food irradiation. In that case, dosimetry is used to monitor the dosage of irradiated foods.

i. Dosage: Dosage represents the amount of energy absorbed by the food material during radiation; the SI unit for dosage is Gy. Based on the field of application, the dose required to achieve the specific goal will vary, the dosage level and their requirement for specific food preservation of food products by electron beam are listed in Table 9.1. The average dosage of processed food should be lower than 10 kGy, also the maximum permissible dose should not exceed 150% of the minimum dose (NAFDAC, 2021). The free radicles produced

TABLE 9.1
Electron Beam Dose Range and Food Application

Approach	Dose Range	Application
Low energy	0.1–1 MeV	Surface sterilization, aseptic packaging, food packaging modification
Medium energy	1–5 MeV	Surface sterilization and food pasteurization
High energy	5–10 MeV	Sterilization, microbial inactivation, disinfestations, and waste treatment

Source: Lung et al. (2015); Munir & Federighi (2020).

during the indirect mode of action of ionizing radiation totally depend on the total dosage. DNA of viruses have the most resistance to irradiation followed by bacteria and fungi due to their hydrated cytoplasm. Yeast and molds have a wide dose range. Parasites and plant meristems (germ of bulb, tubers) are more sensitive to irradiation (Ravindran & Jaiswal, 2019). Also, the energy level, origin, and type of ionizing radiation affect the radioresistance of microorganisms.

   ii. Surrounding medium: The composition of the surrounding medium highly influences the dosage required to achieve microbial destruction. If the medium is protective against the free radicals produced during ionizing radiation, it protects the microbes. For example, the decimal reduction dosage of Salmonella in an egg is 0.632, meat is 0.558, and bone marrow is 0.91 (Munir & Federighi, 2020). The medium characteristics pH alters the sensitivity of microbes to ionizing radiation. Manganese and iron (Mn/Fe) concentrations are important in bacterial radioresistance because Mn^{2+} ions strengthen the enzymatic defense system against oxidative stress. The type and species of microbes, the initial population size, the growing condition of bacteria, the state of growth (vegetative or spore), and the mix of food all influence irradiation requirements.

  iii. Water content: The amount of moisture determines the effect of indirect radiation action. This is because the generation of free radicals at low water content is lower than at high water content, ultimately reducing damage to DNA structure and bonds.

  iv. Temperature: The treatment and product temperature alters the radiation resistance of microorganisms. For example in vegetative cells, a temperature above 45°C synergistically improves the microbial lethal rate. This is caused by increasing the temperature of the system above the ambient temperature where the microbial repair system normally obtains in ambient temperature. The lower temperature in system temperature enhances the microbial resistance to radiation, nearly two- to three-fold higher than ambient temperature. Prevention of diffusion and restricting the movement of radicals to frozen medium prevent the damage of DNA by free radicles is prevented (Andrews et al., 1998).

   v. Aeration: The availability of oxygen in the system causes higher efficiency if the system is in a vacuum; the resistance of the wet sample improved by 2–4 folds, whereas in dry products it was further increased to 8–17 folds (Ashraf et al., 2019).

  vi. Packaging materials: In some cases, the packaging material influences the required decimal reduction dose in irradiation. To avoid this packaging materials will be sterilized before filling the food materials, and then they undergo a radiation process, mostly E-beam is used to pre-sterilization of packaging material due to their low penetration. Polyethylene, polypropylene, and polystyrene are the generally used packaging material for irradiation.

 vii. Additives in food: In the hurdle technology concept, irradiation can be combined with other preservative techniques to reduce the irradiation treatment time or dosage level to ensure food safety. In that case, the addition of additives like nitrites can reduce the decimal reduction dosage for microbial inactivation by irradiation.

9.4 APPLICATION IN FOOD PRESERVATION

Based on the dosage level, the application of ionizing radiation in food preservation is classified into three categories such as radurization (<1 kGy), radicidation (1–10 kGy), and radappertization (>10 kGy) listed in Table 9.2. The ionizing radiation is used in food preservation for different reasons like pasteurization, sterilization, delayed senescence and maturation process, disinfestations, and inactivation of microorganisms. The ultimate goal of applying ionizing radiation in food preservation is to enhance the safety, quality, and shelf life of food products.

9.4.1 INACTIVATION OF MICROORGANISMS

Microbial contamination of food is a major concern of today's world because they are responsible for vivid health complications including foodborne illness and infections. The majority of alimentary diseases such as Campylobacteriosis, Listeriosis, Trichinellosis, and Salmonellosis caused by a variety of microbes. Meat and poultry, cereals, fruit, and vegetables are the carriers of these microbes. The pre- or post-harvest contamination or production of insanitary practices leads to an outbreak of foodborne illness or poisoning. Ionizing radiation's ability to cause chemical change during the inactivation of dangerous germs and insects is what makes it ideal for use in food. To destroy the molecular bonds in food, the radiation must have a high enough energy level. The energy of covalent chemical bonds ranges from 1 to 8 eV. All ionizing radiation can disrupt covalent bonds

TABLE 9.2
Gamma Irradiation Dose Range and Food Application

Approach	Dose Range (kGy)	Application	Food Products	Dose (kGy)
Radurization	<1 kGy	Inhibit sprouting	Potatoes, onions, yams, garlic, sugar beet, ginger	0.06–0.2
		Delay ripening	Strawberries, potatoes, mangoes, plantain	0.5–1.0
		Pest disinfestations and phytosanitation	Cereals, pulses, coffee beans, cocoa, spices, nuts, oil seeds, mangoes, papaya, dried fruits, fish	0.15–1.0
		Parasite inactivation (Trichina)	Meat and meat products, especially pig, fish, seafood	0.3–1.0
		Increase in yeast population	Soft cheese	
Radicidation	1–10 kGy	Inactivation of spoilage microorganisms	Meat, seafood, spices, and poultry (fresh and frozen)	1.0–7.0
		Extend shelf life	Raw and fresh fish, seafood, fresh produce, and frozen products	1.0–0.7
		Increased juice yield	In fruits	3.0–7.0
		Improve the quality	Decrease the cooking time in dehydrated foods	
Radappertization	>10 kGy	Enzymes (dehydrated)		10.0
		Sterilization	Packaging material	10.0–25.0
		Sterilization	Spices, dry vegetable seasoning	30.0
		Decontamination of food additives	Spices, natural gum	30.0
		Reduce pathogens to point of sterility	Space foods	44.0

Sources: Odueke et al. (2016); Yun et al. (2013).

since this energy is lower than that needed to ionize an orbital electron. As a response to ionizing radiation, OH radicals are created, which interact with the DNA of microbes. Due to DNA damage and genetic code alterations, these organisms no longer have the capacity to replicate. Thus, after being exposed to radiation, food is free of harmful microbes or insects, making it a safer product. Fruits, vegetables, dry goods, and spices no longer transfer live eggs or larvae into un-infested areas or lots.

Decontamination of sprouts (radish, mung bean, and alfalfa) by irradiation effectively reduces the foodborne pathogen content. The dosage of 3.3–5.3 kGy e-beam was successfully used for the elimination of L. monocytogenes on alfalfa sprout (Lung et al., 2015). Irradiation is a way to inactivate or reduce the high load of micro-flora on spices and herbs. Black pepper, when exposed to gamma rays at 10, 20, and 30 kGy, an increased dose of irradiation didn't show any effect on its (piperine) essential oil content. The energy of 5 MeV can be used to irradiate and the dosage range includes 7.5–15 kGy for spices with respect to sensory properties. Irradiation is able to achieve a 3–5 log reduction of pathogens like Salmonella spp., *Escherichia coli* O157:H7, *Listeria monocytogenes*, and *Staphylococcus aureus* without affecting the quality and nutritional value. Irradiation of spinach up to 4 kGy eliminates the *E. coli* O157:H7, likewise, it helps to prevent the outbreak of *E. coli* O104:H4 from fenugreek seeds in Europe (Prakash, 2016).

Most of the post-harvest losses are caused by fungal pathogens, nearly 25% in fruits and vegetables. The mechanical damage in fruits and vegetables during handling, transportation, and storage helps the growth of this mold and fungi. Any commodity contains a fungal growth considered to be decayed. The demand for organic products, the development of fungi resistance, and the cost of synthetic fungicides lead to the application of irradiation to inhibit fungal growth in food commodities (Hussain et al., 2012). The free radicals produced during ionization damage fungi cell molecules and chemical bonds, which cause changes in chemical, metabolic, and physiological changes. The application of gamma radiation exhibits the maximum efficiency of fungi inhibition in stored seeds and other dried products. The irradiation affects the mycelia growth, spore germination, and morphology of post-harvest fungal pathogens. The inactivation of individual fungal viability varies based on the species and other characteristics. For example, the inactivation radiation for Botrytis cinerea is 3–4 kGy, Penicillium expansum, and Rhizopus stolonifer is 1–2 kGy. This was mainly due to the radiosensitivity and radioresistance behavior of fungi (Jeong et al., 2015). The yeast antagonistic and fungistatic effect of irradiation on strawberries eliminates the fungal infection (Hussain et al., 2012). Dosages necessary for efficient control of fungal spoilage usually result in unwanted changes such as tissue or skin damage, changes in flavor, and texture impairing the appearance and natural ripening, because of this, post-harvest disease management appears to be restricted to only a few commodities (Munir & Federighi, 2020). Irradiation also can be used for the breakdown of fungal toxins, for example, maximum reduction of toxic compounds like aflatoxin B1 and ochratoxin A were obtained by 5–10 kGy of γ-irradiation and electron beam reduce the deoxynivalenol toxin by Fusarium in barley (Munir & Federighi, 2020). In some cases, the radiation source improves the mycotoxin and some fungal production, so necessary hurdle combinations should be used to ensure the desired effect. Other than bacteria and fungi, irradiation exhibits an effective application in the elimination of parasitic protozoa and helminths which cause major illness in children, immune-compromised patients, pregnant mothers, and the elderly (Li et al., 2017).

9.4.2 Pasteurization and Sterilization

In the past decade, the outbreak of human illness associated with food borne pathogens and the increase in the need to store food products for longer periods without microbial contamination increased the need for pasteurization and sterilization methods. A wide range of fruits and vegetables, including papayas and mangoes, cannot be suitable for cold storage, so irradiation plays a major role in preservation. In particular, for fruits, disinfestation and shelf life extension have been improved by ionizing irradiation. It has the ability to prevent both rotting and the growth of harmful bacteria, making it a viable approach for preserving the quality of fresh fruits and vegetables. Using irradiation at a pasteurization dosage (2–5 kGy) might prevent infections that affect fruit and vegetables and post-harvest spoiling without affecting the sensory attributes of the produce (Bisht et al., 2021). A pasteurization dosage of irradiation may cause quality changes in the product, such as softening. Significant changes and maybe even a drop in the vitamin content can be caused by doses of 0.25–1 kGy. Diseases associated with storage can be controlled with doses of more than 1.75 kGy. A dosage of between 1 and 3 kGy, when applied, it can cause the product to soften more quickly and develop unfavorable organoleptic characteristics (Sarkar & Mahato, 2020). When exposed to radiation at levels greater than 3 kGy, excessive ripening and the development of some unpleasant flavors occur. As a result, this technique should be applied along with other treatments. An absorbed dose of 1.5–2 kGy or as high as 3 kGy to treat strawberries which eliminates the causes of fruit rots, is the only exception (Hussain et al., 2012). The major spoilage microorganisms like L. monocytogenes and Salmonella are eliminated by irradiation at 3.5 and 4 kGy, respectively (Foley et al., 2002). In 1997, FDA approved food irradiation of refrigerated or frozen meat and meat products to enhance food safety. Application of low dosage helps to obtain the sterilization effect, which helps to maintain the quality and nutritional profile of food under storage conditions (Odueke et al., 2016).

9.4.3 Delayed Senescence and Maturation Process

Fruits and vegetables are living tissues that are actively respiring, and their quality must be preserved until consumption. To retain good quality, preservation processes should not destroy plant cells. Irradiation is successfully used to inhibit the ripening and sprouting most of agricultural products, especially fruits and vegetables. Sprouting of the tubers is an undesirable characteristic when they are meant to use in food products, sprouting reduces the weight and affects the appearance resulting in lower marketability of the products. The sprout growth was controlled by factors including temperature, humidity, light, and concentration of CO_2 and O_2. Compounds like Chlorpropham CIPC (Isopropyl Carbamate), alcohols, acetaldehydes, and ethylene are commonly used for sprout inhibition, but their usage comes with undesirable side effects. Ionizing radiation plays a major role in sprout inhibition without causing any adverse effect on the product and environment. It found that the irradiation of 0.1–0.15 kGy affects the mitotic activity and indole acidic acid synthesis in potato samples, which inhibits the sprouting for 5 months of storage period (Sarkar & Mahato, 2020). On the other hand, the wound healing process in potatoes involves the deposition of suberin below the wound surface and the formation of periderm, which requires mitotic activity and cell division. So, the exposure of potato >2 kGy affects the wound healing process and leads to weight loss under storage conditions. Food irradiation (<0.15 kGy) for sprout inhibition has been approved for commercial usage in many countries. In Japan, from 1973, irradiation was used to carry out the sprout inhibition in potatoes (Ehlermann, 2016). Irradiation is used in South American and Eastern European countries to prevent the sprouting of tuber crops such as onions, garlic, yams, and potatoes (Odueke et al., 2016).

Electron beam irradiation is used to prevent the germination of potatoes, ginger, garlic, and onions. Radiation doses of less than 1.0 kGy are suitable for inhibiting germination, delaying maturity, or reducing senescence in fresh fruits and vegetables, while an intermediate dose range of 1.0-3.0 kGy is required for increasing the shelf life of food products. Reduction in the germination of wheat grain was observed in 1 kGy. Imeh et al. (2012) reported the inhibition of sprouting up to 7 months in water yams. Likewise, a 20.4–58.8% reduction in the sprout length of soybean was obtained by irradiation (Yun et al., 2013). The commercial irradiation of potatoes was permitted in Hokkaido, Japan, for sprout inhibition in 1972, for more than 30 years it continued successfully (Kume et al., 2009). In most fruit and vegetable, irradiation causes alterations in the cell membrane (loss of turgor) and cell wall composition (pectin) and also affect the ripening process which leads to the off-flavor and aroma. The effect of irradiation on fruits and vegetables is highly dependent on dosage and treatment time. The dosage ranges of 0.25–1.0 kGy is used to inhibit the ripening of fruits and vegetables.

9.4.4 DISINFESTATION

Insects and pests are major concerns in the storage of agricultural commodities. They cause post-harvest losses and affect the quality of the products. Phytosanitary treatment prevents the introduction or spread of quarantine pests. Irradiation is one of the most effective phytosanitary treatments and has been adopted worldwide. The main goal of phytosanitary treatment is to prevent the development and reproduction of pests. The absence of pesticide residue, treatment in packaged form, and pallet loads further improve the application of this technology. In tobacco products, the first commercial insect control by irradiation was applied in 1910 (Hallman, 2011). Tribolium castaneum infestation in dried dates and raisins, Corcyra cephalonica infest dried apricots and figs can be prevented at doses 0.15–0.25 kGy or 0.25 kGy with storage at low temperatures can prevent infestation. Although most dry foods can be disinfested at lower levels than the 1 kGy dosage suggested by the Codex Alimentarius Commission (CAC), this amount is necessary for all agricultural foods and products. The resistance to radiation varies depending on the species of insect, with the Lepidoptera (moths) being the most resistant group and the Coleoptera (beetles) being the most vulnerable. The sterilizing dosages vary depending on the kind. A dose of 0.05 kGy is enough for beetles, and a dose of around 1 kGy is required for moths. An efficient method for reducing insect infestation is low-dose irradiation (McDonald et al., 2012). Irradiation, unlike fumigation, leaves no chemical traces behind, neither the environment nor the customer is harmed by this treatment. It has been suggested to use ionizing radiation at dosages of 0.15–0.75 kGy to eradicate insect infestations. At these levels, most insects are completely sterilized. However, it was shown that some moth species might not be sterilized by 1 kGy. Insect infestation was lessened by doses of 0.25 and 0.5 kGy, but after a year of storage at ambient room temperature, insect infestation was 100%. During a one-year storage period, the use of 0.25 kGy irradiation dosage and lower temperature fully prevented the growth of every type of insect. The Codex Alimentarius Commission advised that the dose of 1 kGy not be exceeded for treating rice, wheat, dates, legumes, and derivatives for the infestation in 1983 (Garbati Pegna et al., 2017). Several applications of ionizing radiation for food preservation are listed in Table 9.3.

9.4.5 OTHER APPLICATION

In pasteurized milk, the low-level gamma irradiation reduced the aflatoxin level up to 51.5% and 99% after four and eight days of exposure with a dose rate of 0.39 mGy/day, respectively. Also, the irradiation treatment is not suitable for milk due to its high-fat content leads to flavor by lipid oxidation (Hassanpour et al., 2019). In some cases, to extract the bioactive compounds more than 3.0 kGy of irradiation is required. The approved dosage level of ionizing radiation for various applications based on the classes of food products given by NAFDAC (2021) is shown in Table 9.4. Reduction or elimination of spoilage microorganisms helps to enhance the shelf life of irradiated food. The irradiation effectively improves the shelf life of products like sliced mushrooms, diced celery, and whole strawberries. The higher dosage of irradiation needs to eliminate the mold and yeast than bacteria. Delaying ripening in mangoes and papaya improves the shelf life by delaying the onset of senescence (Prakash, 2016).

9.5 IRRADIATION EFFECT ON FOOD COMPONENTS

Irradiation has an influence on various food components like water molecules, carbohydrates, protein, lipids, and vitamins. In water gamma radiation causes excitation, in which the water molecule enters a state of higher energy, which is another possibility. A variety of products, including atomic and molecular hydrogen, hydroxyl radicals, and hydrogen peroxide, may result from the interactions between the various ions that emerge and the water molecules. The wide variety of chemicals created is referred to as radio-induced products. Only molecular hydrogen and hydrogen peroxide are

TABLE 9.3

Application of Ionizing Radiation in Food Preservation

Food Product	Application	Dosage	Findings	References
Bibimbap (Korean cooked rice)	Sterilization	25 kGy	Decrease in the initial aerobic bacterial count of up to 6.3 log CFU/g	Park et al. (2012)
Strawberry	Retention of mold growth in refrigerated storage	2 kGy	Irradiation treatment combined with edible coating delayed the decay and mold growth for up to 18 days	Hussain et al. (2012)
Water yam	Sprout inhibition	3 Gy/minute	At 100 Gy only 10% of water yam sprouted in 3 months of storage No significant effect on lipid, protein, and carbohydrate content	Imeh et al. (2012)
Strawberry	Mold	2.0 kGy	The irradiation combined with carboxymethyl cellulose edible coating delays the decay process and inhibits mold growth for 18 days	Hussain et al. (2012)
Soybean	Sprout inhibition	1 kGy 3 kGy	A 20.4% reduction in sprout length A 58.8% reduction in sprout length	Yun et al. (2013)
Dairy products	Bacteria Flavor	0.5 kGy 40 kGy	Reduction in bacteria and mold (*L. monocytogenes*) in soft whey cheese Enhance sweetness flavor of the cheese and strawberry yogurt bar	Odueke et al. (2016)
Milk	Reduction of aflatoxin M1	0.39 mGy/day	A 51.5% reduction of aflatoxin in 4 days of exposure 99% reduction of aflatoxin in 8 days of exposure No change in the sensory and chemical quality of the milk	Hassanpour et al. (2019)
Brown rice	Storage stability	1–10 kGy	Improvement in the rehydration ratio and browning index, decrease in the total phenolic, antioxidant activity, β-carotene content, and microbiological stability increase	Jan et al. (2020)
Kimchi seasoning mixture	Shelf life and microbial safety	0–10 kGy	Reduction in microbial population along with the changes in pH, acidity, and sugar. No change in color	Jeong et al. (2020)
Pomegranate	Shelf life study	1–5 kGy	Increase in shelf life and reduction in polyphenol oxidase activity	Muhammad et al. (2014)
Wheat	Physical and chemical property	0–5 kGy	Decrease in rheological properties and dynamic module, increase in antioxidant activity	Bhat et al. (2020)
Beef loins	Microbiological and physicochemical quality	2.5 kGy	Enhanced microbiological safety and meat quality	Yim et al. (2016)
Peach	Insect disinfestation	0.66 kGy	Effective disinfestation. Altered the firmness and fresh color of peach	McDonald et al. (2012)

TABLE 9.4

NAFDC (National Agency for Food and Drug Administration and Control) Regulation for Food Irradiation (NAFDC, 2021)

Classes of Food	Purpose	Required Dose (kGy)
Class 1: Bulbs, roots, tubers (onions, yams, and potatoes)	To inhibit sprouting during storage	0.2
Class 2: Fresh fruits and vegetables (other than Class I). Plantains and mangoes	To delay ripening	1.0
	Insect disinfestations	1.0
	Shelf life extension	1.5
	Quarantine control	1.5
Class 3: Cereals and their milled products, nuts, oil seeds, pulses and dried fruits, beans, maize, millet, sorghum, cocoa, and kola nuts	Insect disinfestations	1.0
	Reduction of microbial load	5.0
Class 4: Fish, seafood, and their products (fresh and frozen)	Reduction of pathogenic microorganisms	5.0
	Shelf life extension	3.0
Class 5: Raw poultry and meat, and their products including (fresh and frozen), chicken, turkey, beef	Reduction of pathogenic microorganism	70
		3.0
	Shelf life extension	2.0
	Control of infections by parasites	
Class 6: Dry vegetables, spices and condiments, animal feeds, pepper, dry herbs, and herbal teas	Reduction of certain pathogenic	10.0
	Insect disinfections	1.0
Class 7: Dried food of animal origin, smoked fish, dried meat, stockfish	Insect disinfestations	1.0
	Control of mold	3.0
Class 8: Miscellaneous food including but not limited to honey, space foods, hospital foods, military rations, spices, liquid eggs, and thickeners	Reduction of microorganisms	<10
	Sterilization	<10
	Quarantine control	<10

stable molecules out of the rest of the group, despite the fact that they are consumed in the processes that happen after irradiation. The most reactive radio-induced products are the OH-radical and H_2O_2, which can combine with nearby molecules to produce new compounds. A potent oxidizing agent is a hydroxyl radical, while hydrogen and the aqueous electron are reducing agents. The unsaturated fatty acids and amino acids are highly susceptible to free radicals (Brewer, 2009). Aqueous electrons react to the hydrogen atoms in the C-H linkages of olefins or aromatic compounds and can be removed by the hydroxyl radical.

Due to the many radiolytic products generated under various circumstances and dosages, the impact of radiation on carbohydrates is complex over a broad dose range. High doses can result in hydrolysis, oxidation, degradation, and depolymerization. The formation of furan from fructose, glucose, sucrose, and organic acids was identified in irradiated fruits (Fan, 2005). Native protein can be denatured by irradiation, mostly by rupturing the hydrogen bonds and other links that are a part of secondary and tertiary structures. The ability of radiation to change proteins does not necessarily pose a problem when employing irradiation to treat food. The amino acids, which are crucial from a nutritional perspective, do quite well in the process. Radiation-induced alterations in meat products are minimal at the medium exposure level. Additionally, proteins that are irradiated produce radiolytic products, which include molecules like fatty acids, mercaptans, and other compounds. Even though they are insignificant, these compounds contribute to the sensory qualities of food. For example, the protein stability and antioxidant capacity of irradiated soybean increased by 10 kGy doses (Štajner et al., 2007). The majority of the lipids included in the diet are triglycerides.

The carbonyl group's oxygen atom is located in a fatty acid region with an electron shortage. Unsaturated fatty acids near the un-saturation core exhibit the other electron shortage. Cleavage happens at the oxygen atom or the double bond when it is exposed to radiation. Free radicals are produced as the major intermediates as a result ultimately; certain end products are produced as well, including CO_2, CO, H_2, and hydrocarbons, primarily alkanes and aldehydes. Additionally, unsaturated acids can create dimers and polymers, the production of which is enhanced by the presence of oxygen (Brewer, 2009). Production of free radicals in the presence of O_2 causes auto-oxidation of the lipids, and changes in the physical properties of lipids, such as melting point and viscosity.

Micronutrients in food called vitamins are radiation-sensitive, especially thiamine and vitamin C. This sensitivity is determined by the type of vitamin and the dose given to the food. Water-soluble vitamins should be assumed to be more susceptible to radiation damage due to the interactions of the hydroxyl radical created during the radiolysis of water. Thiamine (B1) and the vitamins A, E, C, and K are particularly radiation-sensitive. Riboflavin, niacin, pyridoxine, pantothenic acid, and vitamin D, in contrast, are significantly more stable (Woodside, 2015). Temperature and oxygen content have an impact on how quickly radiation-damaged vitamins deteriorate. Vitamin degradation can be slowed down by operating at low temperatures without oxygen, using vacuum packing, or working in a nitrogen-filled environment. Flavor and odor are derived from water-soluble components and volatile substances, respectively. Irradiation causes lipid oxidation which produces off odors and flavors. Irradiation enhances the sulfur-containing amino acids and causes cabbage, sulfur, or rotten vegetable-like odor. Most meat products contain high lipid content, and the low dosage of irradiation causes undesirable odor due to lipid oxidation (Brewer, 2009).

9.6   ADVANTAGE OF IONIZING RADIATION TECHNOLOGIES IN FOOD PRESERVATION

Food irradiation has various advantages over conventional methods of heat and chemical preservation. Some of the advantages are terminal food processing, chemical independence, temperature independence, flexibility, sterility assurance level, and ease of application. Ionizing radiation has a better penetration depth, so the products can be sealed and subjected to radiation after sealing or final packaging which greatly reduces the cross and post-sterilization contamination. No chemicals or toxic compounds or volatile substances are required; also, no end products are generated during processing which avoids disposal and environmental concerns. During irradiation, the temperature rise in the products is very minimal, and the process is effective at ambient temperature, so it does not require any additional cooling or heating system. All kinds of food products (gaseous, liquid, or solids) can be treated in irradiation; also, the efficiency of the process is not dependent on the product's physical property (size, thickness, and density) and product nature (homogeneous or heterogeneous systems). The irradiation process is highly effective for microbial inactivation which assures a high level of safety of this process, less than 106 microorganisms service in food products after the sterilization treatment (Ajibola, 2020). Only exposure dose/time controls the whole preservation process, thus simplifying the control system. Food irradiation has a very minimal influence on the flavor, texture, and nutritional quality of food. The primary advantage of the food irradiation process is the destruction of food microbes results in the production of safer foods, the extension of food shelf life through pest control, and delaying the of biological deterioration (Odueke et al., 2016). Irradiation is a versatile technology that can be used to achieve safety, security, and trade applications; a highly effective and efficient process where a broad spectrum of microbial inactivation, and insect and pest control can be achieved; and irradiated foods can be immediately distributed into the food supply chain without any specific environmental requirements during storage and transportation conditions (Yousefi & Razdari, 2014). In the United States, the Canadian and European community built a mobile irradiation facility, and a vessel with a portable ^{60}Co irradiator was equipped in the USA for coastal fisheries. Germany created the vessel carrying an X-ray facility for fisheries research, which exhibits the potential application of irradiation in food preservation (Ehlermann, 2016). These kinds of advancements in irradiation further improve the industrial-level application.

9.7 COMMERCIAL APPLICATION OF IONIZING RADIATION IN FOOD PRESERVATION

The first and foremost commercial application of irradiation was implemented in Germany in 1957 to improve the quality of the spices (Diehl, 2002), even though the new food law prohibited the usage of irradiation in 1959. In 2000, the first irradiated packaged beef went to the retail consumer market. The first irradiation company specially designed for food irradiation in 1994 was Food Technology Service, Inc. (FTS), Florida uses a cobalt-60 as a radiation source to process frozen poultry and food products. The survey conducted in 2005 indicated that approximately 4,05,000 tons of food are processed by irradiation worldwide. In the USA, nearly 4,00,000 tons of food were irradiated. Other Asian countries, especially China, have considerable progress in the application of irradiation in food preservation. In the USA, 80,000 tons of spices and 8,000 tons of meat are processed by irradiation. In China, spices, meat, and other products were preserved by irradiation. Decontamination of spices and herbs by irradiation was fairly used in the US, China, and other Asian countries. Spices are the main commodity preserved by irradiation for international trading. The USA, Mexico, Vietnam, Thailand, India, Australia, and New Zealand are constantly involved in the trading of irradiated fruits and vegetables, other than this, Pakistan, South Africa, Malaysia, and Indonesia play a role in export/import. Japan continues to treat potatoes for sprout inhibition, although volumes have steadily diminished since treatment first started in the early 1970s. Mr. Pranav Parekh, Agrosurg Irradiators (I) Pvt. Ltd, India, treated around 115 t of onion and 16 t of garlic in 2015 at a dose of 0.06 kGy. Australia exports mangoes, tomatoes, capsicums, and litchis to New Zealand (Roberts, 2016). Each year, the irradiated fresh products volume got increased to 20,000 tons. China ranked in the foremost position followed by the USA and Ukraine, these three countries contribute three-quarters of the total irradiated foods in the world. Ukraine, Vietnam, and Japan only treated the single commodities of grain, frozen seafood, and potatoes, respectively (Kume et al., 2009). The commercially available irradiated food products are listed in Table 9.5. In commodity, wise irradiation of spice and dry vegetables occupy half of the total amount, after that, sprout inhibition of garlic and potato, and disinfestations of grains and fruits come in data. Some of the EU member countries have national permission for the irradiation of food products, for example, France for frozen poultry meat; Belgium, France, and the Netherlands for frozen shrimp.

TABLE 9.5
Commercially Available Irradiated Food Products

Food Product	Company and Country	Application	Year
Potatoes	Canada	Inhibition of sprouting	1960
Spice herbs	Ion Beam Applications, Belgium	Microbial inactivation and shelf life extension	1986
Exotic fruit (papaya and lychees)	Hawaii (exported to the US mainland and Japan)	Pest control, especially fruit fly	1995
Frozen and fresh ground beef	Food Technology Service, Inc. (FTS), Florida	Microbial inactivation and shelf life extension	2000
Hamburgers	SureBeam Corporation, US	Microbial inactivation and shelf life extension	2000

9.8 CHALLENGES OF IONIZING RADIATION IN FOOD PRESERVATION

9.8.1 Safety Aspects

With the sponsorship of FAO, IAEA, and OECD, an international project in the field of food irra-diation was launched in 1970 with the explicit goal of carrying out a study program to examine the health safety of irradiated food. The program includes long-term animal feeding studies, short-term screening tests, and chemical changes in irradiated foodstuffs with a limited dosage range (up to 10 kGy). Based on the results of this program, JECFI concluded in 1980 that food treated with less than 10 kGy of irradiation dose level poses no toxicological hazard and poses no special nutritional or microbiological problems. In 1999, the JECFI stated that food treated with a high dosage level of up to 70 kGy does not have an adverse health effect. The concern about the changes caused by irradiation affects the marketability of irradiated foods, based on the serious research work, it was found that the occurrence of "photo nuclear activation" when the radiation energy is above 10 MeV. The induced radioactivity is very short-lived (only about a half hour), still further studies are required for the in depth studies of the free radicals' effect on different food molecules (Ehlermann, 2016). After several types of research work in this field, WHO and national and international export committees recognized food irradiation as a safe technology. The word "Wholesome" means that the eating of irradiated food does not cause any risk, and the irradiated food maintains all nutrition and other positive properties.

9.8.2 Public Acceptability of Irradiated Foods

According to recent research, irradiation foods are safe to consume and free of disease-causing germs, parasites, and pests. Irradiation does not render the food radioactive; the free radical created during radiation from water and other chemicals has a very short shelf life, thus, it is completely free of radioactive components. The outbreak of *E. coli* caused by the consumption of undercooked hamburgers sparked the first public interest in food irradiation (Munir & Federighi, 2020). Based on the product quality, improvement in shelf life and other parameters have a progressive acceptance rate in society. The understanding of the process by consumers and producers further enhanced the acceptance rate. The products like mangoes from Asian countries cannot go through phytos-anitary procedures like hot water treatment to maintain their quality, in that case, irradiation is the only choice to greatly eliminate the pest without any chemical usage (Munir & Federighi, 2020). Providing awareness and information about irradiated foods among the consumer, and encouraging them to choose irradiated food over conventionally treated foods will be the first step to enhancing the consumer acceptance rate, first, consumers have to try the products. Previous market research reported that 80–90% of consumers will buy irradiated foods after their first experience and know-ing the safety and benefits of the irradiated food. According to previous market research, 80–90% of customers will purchase irradiated foods following their first experience and learning about the safety and benefits of irradiation foods (Ravindran & Jaiswal, 2019). Other Pacific nations, such as Bangladesh and Thailand, have already agreed to use irradiation goods. China authorized the irra-diation of numerous agricultural goods, while other countries such as Sri Lanka, Pakistan, South Korea, and the Philippines established their irradiation facilities (Odueke et al., 2016).

9.8.3 Legislation Pertaining to Irradiated Foods

Several groups propose various national regulations on food irradiation; the European Council pro-posed the first draught of a European Directive on food irradiation in 1988. After ten years of delib-eration, a set of food irradiation guidelines was adopted and published in the EU's Official Gazette in 1999 (Diehl, 2002). Based on these, any food subjected to irradiation should be labeled as "irra-diated" or "treated with ionizing radiation" regardless of whether the whole food or part of the ingredients are subjected to irradiation. Irradiation (from any source, including gamma and electron X-rays) should be used in accordance with the International Atomic Energy Agency's safety regulations. The Codex code of practice should be followed by facilities that use irradiation for food commodities, and more information is available from the International Organization for Standardization (ISO). Commercial trade of irradiated foods necessitates the establishment of a system of international standards and national legislation. General Standard (CAC, 2003) claims that the maximum absorbed dose by the food material does not exceed 10 kGy except when it is needed for reasonable processing technique. Ionizing radiation-treated items should carry a written statement confirming treatment near the name of the food, and the use of the international food irradiation sign Radura is optional (Roberts, 2016). The usage of irradiated food as an ingredient in other products should be mentioned in the ingredient list. International Standards for Phytosanitary Measures (ISPM) provide technical guidance for the application of irradiation of pest control in fruits and vegetables (ISPM 18), Minimum dose level for pest regulation (ISPM 28) (Hernández-Hernández et al., 2019). ISPM 28 also acknowledges 150 Gy as the generic dose level for the prevention of adult Tephritid fruit flies (FAO, 2009). Analyzing the dosage absorbed by food and humans plays a major role in irradiation plants, severe international standards including IAEA (2002) and ISO/ASTM (2004, 2005, 2013, 2015) provide the guidelines for good dosimetry practice. ASTM has its standard guidelines for the packaging material used for food irradiation apart from that suitable packaging materials based on the USA research are updated on a website (USFDA, 2015). The irradiation of food products in different countries is listed in Table 9.6. In the US, irradiated meat products need to get approved by both FDA and the US Department of Agriculture's Food Safety and Inspection Service (USDA/FSIS).

TABLE 9.6
Countries and Their Irradiated Food Products

Country	Major Irradiated Food Commodity
Belgium	Spice, dehydrated vegetables, meat (frozen frog, fowl meat, and prawn), seafood, and egg
Brazil	Spices and fruits
Canada	Spices
China	Spices, dehydrated vegetables, cereals, garlic, health foods, and functional foods
France	Spice, dehydrated vegetables, meat (fowl meat, frog leg), seafood, gum Arabic, casein
Germany	Spices and dehydrated vegetables
Hungary	Spices and dehydrated vegetables
India	Spices (coriander, turmeric, red pepper), onion, mango
Indonesia	Spices, starch, seafood, cocoa powder
Israel	Spices
Japan	Potato
Malaysia	Spices, herbs, nutritional drinks
Netherland	Spices, dehydrated vegetables, meat (frozen frog leg, frozen meat), seafood (frozen or chilled prawns)
Poland	Spices, herbs, and dehydrated vegetables
South Africa	Spice and vegetables
South Korea	Spices (ginseng, pepper, chili powder), vegetables, food irradiation for allergy patients, military, and space foods
Thailand	Spices, fermented sausage, frozen seafood, fruits (mango, mangosteen, pineapple, rambutan, litchi, longan)
Ukraine	Grain and fruit
Ukraine	Grains (wheat, barley)
United States (US)	Spices, meat, poultry, fruits, and vegetables

Source: Kume et al. (2009).

The regulation proposal requested in 2002 for the irradiation of beef, poultry, shrimp, and prawns is currently being considered; however, detection measures are not necessary for Canada and the United States (Kume et al., 2009). In 2006, a law was passed authorizing the consumption of frozen irradiated frog legs from Belgium and France (Kume et al., 2009).

9.9 SUMMARY

Irradiation has an effective application for the preservation of food products and helps to enhance the quality and safety of the food. The application of irradiation is chemically safe and has less influence on the nutritional attributes of food. Initially, the application of food irradiation was limited to the decontamination of spices, herbs, and condiments, but recently, it has been utilized in the preservation of meat and fresh produce in several countries. Several advantages of this technology and no harmful effect on the environment and food products help the industrialization of this technology in the food sector. The major barrier to the commercialization of irradiated products in earlier days was consumer reluctance to accept the irradiated food. But recent studies denoted that the knowledge about irradiated foods get increased among consumers, and the rate of buying labeled irradiated food increased in several countries. But still, several manufacturers have persistent perceptions between food producers and other retail sellers that consumers will not buy irradiated food affecting the commercial production and distribution of irradiated foods. The cooperation between a processor and food trade, review, and harmonization of labeling requirements help to achieve a new goal in food irradiation.

REFERENCES

Ajibola, O. J. (2020). An overview of irradiation as a food preservation technique. Novel Research in Microbiology Journal, 4(3), 779–789. https://doi.org/10.21608/nrmj.2020.95321

Andrews, L. S., Ahmedna, M., Grodner, R. M., Liuzzo, J. A., Murano, P. S., Murano, E. A., Rao, R. M., Shane, S., & Wilson, P. W. (1998). Food preservation using ionizing radiation. Reviews of Environmental Contamination and Toxicology, 154, 1–53. https://doi.org/10.1007/978-1-4612-2208-8_1

Ashraf, S., Scholor, R., Sood, M., Professor, A., Bandral, J. D., Professor, A., Trilokia, M., Kashmir, J. &, Manzoor, I. M., & Manzoor, M. (2019). Food irradiation: A review. International Journal of Chemical Studies, 7(2), 131–136.

Bhat, N. A., Wani, I. A., & Hamdani, A. M. (2020). Effect of gamma-irradiation on the thermal, rheological and antioxidant properties of three wheat cultivars grown in temperate Indian climate. Radiation Physics and Chemistry, 176, 108953. https://doi.org/10.1016/j.radphyschem.2020.108953

Bisht, B., Bhatnagar, P., Gururani, P., Kumar, V., Tomar, M. S., Sinhmar, R., Rathi, N., & Kumar, S. (2021). Food irradiation: Effect of ionizing and non-ionizing radiations on preservation of fruits and vegetables – a review. Trends in Food Science and Technology, 114(June), 372–385. https://doi.org/10.1016/j.tifs.2021.06.002

Brewer, M. S. (2009). Irradiation effects on meat flavor: A review. Meat Science, 81(1), 1–14. https://doi.org/10.1016/j.meatsci.2008.07.011

CAC. (2003). Codex Alimentarius Commission. General Standard for Irradiated Foods (CODEX STAN 106-1983, Rev.1-2013). Codex Alimentarius, FAO/WHO, Rome.

Diehl, J. F. (2002). Food irradiation – Past, present and future. Radiation Physics and Chemistry, 63(3–6), 211–215. https://doi.org/10.1016/S0969-806X(01)00622-3

Ehlermann, D. A. E. (2016). The early history of food irradiation. Radiation Physics and Chemistry, 129(1953), 10–12. https://doi.org/10.1016/j.radphyschem.2016.07.024

Fan, X. (2005). Formation of furan from carbohydrates and ascorbic acid following exposure to ionizing radiation and thermal processing. Journal of Agricultural and Food Chemistry, 53(20), 7826–7831. https://doi.org/10.1021/jf051135x

FAO. (2009). Food and Agricultural Organisation of the United Nations. Phytosanitary Treatments for Regulated Pests. ISPM No. 28. FAO, Rome.

Farkas, J. (2006). Irradiation for better foods. Trends in Food Science and Technology, 17(4), 148–152. https://doi.org/10.1016/j.tifs.2005.12.003

Foley, D. M., Pickett, K., Varon, J., Lee, J., Min, D. B., Caporaso, F., & Prakash, A. (2002). Pasteurization of fresh orange juice using gamma irradiation : Microbiological, flavor, and sensory analyses. Food Microbiology and Safety, 67(4), 1495–1501.

Garbati Pegna, F., Sacchetti, P., Canuti, V., Trapani, S., Bergesio, C., Belcari, A., Zanoni, B., & Meggiolaro, F. (2017). Radio frequency irradiation treatment of dates in a single layer to control Carpophilus hemipterus. Biosystems Engineering, 155, 1–11. https://doi.org/10.1016/j.biosystemseng.2016.11.011

Hallman, G. J. (2011). Phytosanitary Applications of Irradiation. Comprehensive Reviews in Food Science and Food Safety, 10(2), 143–151. https://doi.org/10.1111/j.1541-4337.2010.00144.x

Hassanpour, M., Rezaie, M. R., & Baghizadeh, A. (2019). Practical analysis of aflatoxin M1 reduction in pasteurized Milk using low dose gamma irradiation. Journal of Environmental Health Science and Engineering, 17(2), 863–872. https://doi.org/10.1007/s40201-019-00403-9

Hernández-Hernández, H. M., Moreno-Vilet, L., & Villanueva-Rodríguez, S. J. (2019). Current status of emerging food processing technologies in Latin America: Novel non-thermal processing. Innovative Food Science and Emerging Technologies, 58, 102233. https://doi.org/10.1016/j.ifset.2019.102233

Hussain, P. R., Dar, M. A., & Wani, A. M. (2012). Effect of edible coating and gamma irradiation on inhibition of mould growth and quality retention of strawberry during refrigerated storage. International Journal of Food Science and Technology, 47(11), 2318–2324. https://doi.org/10.1111/j.1365-2621.2012.03105.x

Imeh, J., Y. Onimisi, M., & A. Jonah, S. (2012). Effect of irradiation on sprouting of water yam (Dioscorea Alata) using different doses of gamma radiation. American Journal of Chemistry, 2(3), 137–141. https://doi.org/10.5923/j.chemistry.20120203.07

Jan, A., Sood, M., Younis, K., & Islam, R. U. (2020). Brown rice based weaning food treated with gamma irradiation evaluated during storage. Radiation Physics and Chemistry, 177(April), 109158. https://doi.org/10.1016/j.radphyschem.2020.109158

Jeong, R. D., Shin, E. J., Chu, E. H., & Park, H. J. (2015). Effects of ionizing radiation on postharvest fungal pathogens. Plant Pathology Journal, 31(2), 176–180. https://doi.org/10.5423/PPJ.NT.03.2015.0040

Jeong, S.-G., Yang, J. E., Park, J. H., Ko, S. H., Choi, I. S., Kim, H. M., Chun, H. H., Kwon, M.-J., & Park, H. W. (2020). Gamma irradiation improves the microbiological safety and shelf-life of kimchi seasoning mixture. LWT, 134, 110144.

Kume, T., Furuta, M., Todoriki, S., Uenoyama, N., & Kobayashi, Y. (2009). Status of food irradiation in the world. Radiation Physics and Chemistry, 78(3), 222–226. https://doi.org/10.1016/j.radphyschem.2008.09.009

Li, C., He, L., Jin, G., Ma, S., Wu, W., & Gai, L. (2017). Effect of different irradiation dose treatment on the lipid oxidation, instrumental color and volatiles of fresh pork and their changes during storage. Meat Science, 128, 68–76. https://doi.org/10.1016/j.meatsci.2017.02.009

Lung, H. M., Cheng, Y. C., Chang, Y. H., Huang, H. W., Yang, B. B., & Wang, C. Y. (2015). Microbial decontamination of food by electron beam irradiation. Trends in Food Science and Technology, 44(1), 66–78. https://doi.org/10.1016/j.tifs.2015.03.005

McDonald, H., McCulloch, M., Caporaso, F., Winborne, I., Oubichon, M., Rakovski, C., & Prakash, A. (2012). Commercial scale irradiation for insect disinfestation preserves peach quality. Radiation Physics and Chemistry, 81(6), 697–704. https://doi.org/10.1016/j.radphyschem.2012.01.018

Muhammad, H., Ahn, J., Akram, K., Kim, H., Park, E., & Kwon, J. (2014). Chemical and sensory quality of fresh pomegranate fruits exposed to gamma radiation as quarantine treatment. Food Chemistry, 145, 312–318. https://doi.org/10.1016/j.foodchem.2013.08.052

Munir, M. T., & Federighi, M. (2020). Control of foodborne biological hazards by ionizing radiations. Foods, 9(7), 1–23. https://doi.org/10.3390/foods9070878

NAFDAC. (2021). National Agency for Food and Drug Administration and Control. "Food Irradiation Regulations". https://faolex.fao.org/docs/pdf/nig188327.pdf (accessed 15.10.2023).

Odueke, O. B., Farag, K. W., Baines, R. N., & Chadd, S. A. (2016). Irradiation applications in dairy products: A review. Food and Bioprocess Technology, 9(5), 751–767. https://doi.org/10.1007/s11947-016-1709-y

Park, J. N., Song, B. S., Kim, J. H., Choi, J. il, Sung, N. Y., Han, I. J., & Lee, J. W. (2012). Sterilization of ready-to-cook Bibimbap by combined treatment with gamma irradiation for space food. Radiation Physics and Chemistry, 81(8), 1125–1127. https://doi.org/10.1016/j.radphyschem.2012.02.042

Prakash, A. (2016). Particular applications of food irradiation fresh produce. Radiation Physics and Chemistry, 129, 50–52. https://doi.org/10.1016/j.radphyschem.2016.07.017

Ravindran, R., & Jaiswal, A. K. (2019). Wholesomeness and safety aspects of irradiated foods. Food Chemistry, 285(February), 363–368. https://doi.org/10.1016/j.foodchem.2019.02.002

Roberts, P. B. (2016). Food irradiation: Standards, regulations and world-wide trade. Radiation Physics and Chemistry, 129, 30–34. https://doi.org/10.1016/j.radphyschem.2016.06.005

Sarkar, P., & Mahato, S. K. (2020). Effect of gamma irradiation on sprout inhibition and physical properties of Kufri Jyoti variety of potato. International Journal of Current Microbiology and Applied Sciences, 9(7), 1066–1079. https://doi.org/10.20546/ijcmas.2020.907.125

Štajner, D., Milošević, M., & Popović, B. M. (2007). Irradiation effects on phenolic content, lipid and protein oxidation and scavenger ability of soybean seeds. International Journal of Molecular Sciences, 8(7), 618–627. https://doi.org/10.3390/i8070618

USFDA. (2015). United States Food and Drug Administration. Irradiated Food and Packaging. https://www.fda.gov/food/food-ingredients-packaging (accessed 07.03.2023).

Woodside, J. V. (2015). Nutritional aspects of irradiated food. Stewart Postharvest Review, 11(3), 1–6. https://doi.org/10.2212/spr.2015.3.2

Yim, D., Jo, C., Kim, H. C., Seo, K. S., & Nam, K. (2016). Application of electron-beam irradiation combined with aging for improvement of microbiological and physicochemical quality of beef loin. Korean Journal for Food Science of Animal Resources, 36(2), 215–222.

Yousefi, M. R., & Razdari, A. M. (2014). Irradiation' and its potential to food preservation. International Journal of Advanced Biological and Biomedical Research, 2(4), 477–481. https://www.ijabbr.com/article_12517_a9505a8b33aaceb87e168f3fdc2f1d22.pdf

Yun, J., Li, X., Fan, X., Li, W., & Jiang, Y. (2013). Growth and quality of soybean sprouts (*Glycine max* L. Merrill) as affected by gamma irradiation. Radiation Physics and Chemistry, 82(1), 106–111. https://doi.org/10.1016/j.radphyschem.2012.09.004

10 Emerging Applications of Power Ultrasound in Food Processing

Prakyath Shetty, Chaitanya Chivate,
S. Akalya, and Ashish Rawson

10.1 INTRODUCTION

The use of ultrasound technology in various fields has been well-established for several decades. The medical industry has been using ultrasound for diagnostic imaging and therapy, while the material science industry has been using it for the synthesis and processing of materials. In recent years, the food industry has also started to recognize the potential benefits of power ultrasound in enhancing various food processing techniques. Power ultrasound, also known as high-intensity ultrasound (HIU), is a non-thermal technology that utilizes high-frequency sound waves to create cavitation and mechanical agitation in liquid and solid food products.

Ultrasound is utilized in sonoprocessing to produce physical effects, as well as occasionally chemical ones. Because it is non-invasive and doesn't involve harmful radiation, the method offers a tool for modifying food ingredients in a way that customers will love. Moreover, it often functions at room temperature, so there may be less heat generated. The use of power ultrasound in food processing has shown promising results in enhancing the quality, safety, and shelf life of food products while also reducing processing time and energy consumption (Sengar et al., 2022). One of the key advantages of power ultrasound in food processing is its ability to improve the extraction of bioactive compounds from various food sources such as fruits, vegetables, and herbs. The increased extraction efficiency is attributed to the mechanical disruption of cell walls, resulting in the release of intracellular components. Traditional solid-liquid extraction methods take a lot of time, involve using a lot of hazardous organic solvents, and require using glassware. The desire in creating environmentally friendly (green) extraction technology has grown over the past 20 years because of this. Greening the extraction procedure entails reductions in energy, extraction time, cost, and consumption of organic solvents. The goal of the sustainable green extraction idea may be attained by using ultrasound-assisted extraction (UAE), which is a vital technology (Wang & Weller, 2006). UAE is highlighted as a major clean, environmentally friendly extraction method. It offers the essential theoretical context for the mechanism, suggestions for the best operational parameters to maximize the extraction yield, and the potential for linking UAE with some other analytical techniques. There have also been examples of recent applications of UAE in chemistry, biology, and technology (Rutkowska et al., 2017).

Moreover, power ultrasound has been proven effective in emulsification and homogenization of food products (Shanmugam & Ashokkumar, 2017). The mechanical agitation created by ultrasound waves facilitates the formation of stable emulsions, which can improve the texture and appearance of various food products. Additionally, ultrasound technology can also be used for sterilization and preservation of food products, reducing the need for thermal treatments that can negatively affect the nutritional and sensory properties of the food (Abrahamsen & Narvhus, 2022; Bermudez-Aguirre & Niemira, 2022).

142

DOI: 10.1201/9781003359302-10

Despite the potential benefits of power ultrasound in food processing, there are still limitations and challenges associated with its use. One of the major challenges is the lack of standardized protocols for the application of ultrasound in food processing, which can lead to inconsistent results. Furthermore, the cost of ultrasound equipment can be a significant barrier to its adoption in smaller food processing facilities.

10.1.1 CURRENT SCENARIO

Ultrasound technology has been widely used in the food industry in recent years for a variety of applications. Power ultrasound has been shown to enhance various food processing operations such as extraction, emulsification, mixing, homogenization, preservation, and cleaning. The use of power ultrasound has also been explored in the modification of food ingredients such as starch, protein, and lipids, which can improve their functionality and nutritional value. One of the main uses of ultrasound in food preservation is to extend the shelf life of fresh produce by inactivating microorganisms and enzymes that cause spoilage. This technique is known as ultrasonic pasteurization, and it has been shown to be effective in prolonging the shelf life of dairy products, fruits, and vegetable juices. Ultrasound technology has gained considerable attention in the food industry as an environmentally friendly and energy-efficient solution. Many studies have been conducted to explore the effects and applications of ultrasound in various food and environmental applications, and the results have been promising. Ultrasound can be applied in several food applications, including but not limited to extraction, mixing and homogenization, meat tenderization, microbial and enzyme inactivation, cutting, modification of structural and functional properties of biological macromolecules, aging of alcoholic beverages, fermentation, drying, freezing, cooking, foaming, and food waste valorization. The use of ultrasound in these applications has resulted in improved physicochemical and techno-functional properties of food products, increased shelf life during storage without significant effects on food quality attributes, and other notable advantages. One of the main advantages of ultrasound technology is its energy and solvent efficiency. Ultrasound processing requires less energy and solvent consumption compared to traditional methods, resulting in reduced operating costs and a more environmentally friendly process. Additionally, the process is relatively easy to operate, and there is less to zero-emission during the processing, making it a preferred technology in the food industry. As interest in green and sustainable technologies continues to grow, it is expected that ultrasound devices will be used extensively in the food industry. Ongoing research and development will undoubtedly lead to new applications and innovations in the use of ultrasound technology in food processing, ultimately benefiting both the industry and the environment (Taha et al., 2022).

10.2 MECHANISM OF ULTRASOUND PROCESSING

Ultrasound processing in food refers to the use of high frequency sound waves (typically 20–40 kHz) to alter the physical and chemical properties of food products. This can be achieved through a variety of mechanisms, including cavitation, mechanical deformation, and thermal effects. This technology has been used in a variety of food applications, including emulsion formation, homogenization, pasteurization, and preservation. The mechanism of ultrasound processing in food is based on the generation of microscopic bubbles within the food product. When the ultrasound waves interact with the food product, they create areas of high pressure and low pressure that cause the bubbles to rapidly expand and collapse. This process, known as cavitation, creates high shear forces and intense localized heating that can disrupt cell membranes, denature proteins, and break down large particles. One of the most well-established mechanisms of ultrasound processing in food is cavitation. Cavitation is the formation and collapse of bubbles in a liquid, which can generate intense local pressure and temperature fluctuations. These fluctuations can cause mechanical damage to food materials, leading to changes in texture, composition, and functional properties.

For example, cavitation has been shown to increase the solubility of proteins and polysaccharides in food materials, which can improve the nutritional value and functional properties of the food. In addition to cavitation, ultrasound processing can also cause other physical changes in food products. For example, it can increase the surface area of particles, promote the formation of emulsions, and enhance the diffusion of gases and liquids (Bhargava et al., 2021; Singla & Sit, 2021; Yao et al., 2020).

These effects can be used to improve the texture, stability, and nutritional properties of food products. Some examples of ultrasound processing in food include using it to increase the stability of emulsions in mayonnaise and salad dressings, homogenizing milk to make it more homogenous, pasteurizing juice to increase its shelf life, and preserving fruits and vegetables by reducing their bacterial and fungal populations. Overall, ultrasound processing in food is a promising technology that can improve the quality and safety of food products while reducing energy consumption and environmental impacts.

Another mechanism of ultrasound processing in food is mechanical deformation. High-frequency sound waves can cause mechanical vibrations in food materials, leading to changes in their structure and properties. For example, ultrasound has been shown to increase the size and number of pores in food materials, which can improve their water-holding and oil-absorbing capacity. Thermal effects also play a role in ultrasound processing in food. High-frequency sound waves can generate heat in food materials, leading to changes in their physical and chemical properties. For example, ultrasound has been shown to increase the gelation temperature of proteins, which can improve their functional properties (Dias et al., 2017).

10.2.1 METHODS OF GENERATION

As they take into consideration the energy entering the extraction system, ultrasound power, ultrasonic intensity, or acoustic energy density are frequently utilized as the main design parameters. Piezoelectric and magnetostrictive transducers are the two major kinds that are frequently employed in ultrasonic applications. Extraction yields are significantly influenced by where the transducer is placed. The effectiveness of extraction, the intensification of the process, and energy losses all depend on it. The extraction vessel's outer wall will allow ultrasonic waves to pass through it if transducers are positioned on each side of it. The development of a continuous extraction system may be possible with the use of several transducers within the flowing throughout systems for large-scale extraction. Extraction efficiency is improved by reducing acoustic energy losses when transducers are placed in close proximity to the sample and a suitable solvent (Tiwari, 2015).

The frequency of the waves is what really distinguishes sound from ultrasound. Three broad categories, each including a distinct frequency range, are used to classify sound waves: Infrasonic waves had frequencies just under the auditory range (less than 16 Hz), audible waves have frequencies between the hearing frequencies (more than 10 and less than 20 kHz) and microwaves frequencies (maximum 10 MHz), and ultrasonic waves have frequencies beyond the audible range (more than 20 kHz) (Tiwari, 2015).

10.2.2 TRANSDUCER IN ULTRASONIC PROCESS

The transducer of ultrasonic waves is a device that transforms desired electrical signals into mechanical energy. Magnetostrictive and piezoelectric transducers are the two most often used kinds in industrial applications. The operation of a magnetostrictive transducer follows the magnetostrictive principle, which describes interactions between the physical and magnetic fields in ferromagnetic materials including iron, nickel, and cobalt. If a ferromagnetic material is subjected to an external magnetic field, there is a size-changing phenomenon known as the "Joule effect" that causes the substance's length (or size) to vary. The "Wiedemann effect" refers to the shearing deformation that occurs in ferromagnetic materials under conditions of static magnetic field from

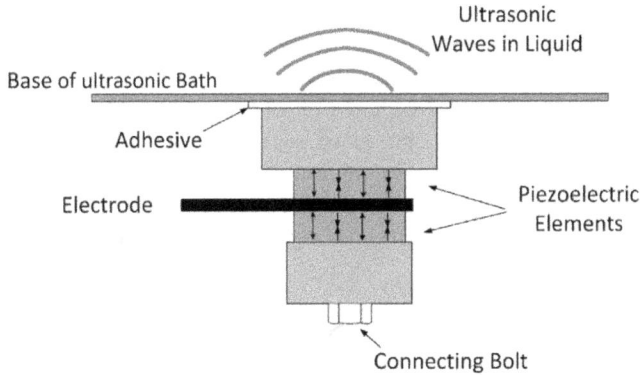

FIGURE 10.1 Langevin ultrasonic transducer fixed at the bottom of an ultrasonic bath. (Reproduced from Kentish, 2017.)

one direction and dynamic field opposite to the static field. Ultrasound bath type typically uses a Langevin piezoelectric transducer to generate sound vibrations. This device is composed of two ceramic components that respond to an electric field by changing their size in a precise and reproducible manner. Therefore, when an alternating electric field is applied, the ceramic components move up and down with high consistency, producing sound waves. Similarly, a loudspeaker converts electrical signals into audible sounds. In ultrasonic baths, typically four to six of these transducers are arranged in a regular pattern beneath the base of the bath, as shown in Figure 10.1 (Kentish, 2017). And generally, one or more piezoelectric discs wedged between two metal pieces make up a piezoelectric transducer. The length of the assembly, which is acoustically equivalent to one-half wavelength just at applied frequency, is resonant at a preset frequency. The piezoelectric ultrasonic transducer offers advantages over the magnetostrictive one, including a simple construction, the absence of electromagnetic radiation, high energy density, low cost, and more. Consequently, people have been concentrating more on the piezoelectric way of producing ultrasonic waves. For a variety of real-world uses, distinct vibrational modes can be induced in a variety of piezoelectric transducer types. There are, generally speaking, four basic vibration modes: longitudinal, radial, flexural (or bending), and torsional (Yao et al., 2020).

10.3 FACTORS AFFECTING EFFICACY

Several factors can affect the efficacy of ultrasound application in food processing. Understanding these factors is essential to optimize the use of ultrasound technology and ensure consistent and reproducible results. The following are some of the key factors that can impact the efficacy of ultrasound application in food processing.

10.3.1 ULTRASOUND FREQUENCY

The frequency of the ultrasound waves used in food processing can affect the efficacy of the treatment. Frequency has an effect on the pressure cycle. As the frequency increases, the pressure fluctuates rapidly, causing a lot of turbulent inside the liquid. Acoustic cavitation is dominated by the created phenomena called acoustic streaming at high frequencies and amplitudes. Low frequency ultrasound in the range of 20–40 kHz is generally best suited for most of the food processing applications. Increasing frequency alters the pressure cycle by causing abrupt pressure changes, creating high internal turbulence that can impede bubble growth and result in subpar cavitation (Pokhrel et al., 2016), and higher frequencies require more power to achieve the same cavitational effect due to lower bubble temperatures and insufficient time for solvent evaporation into the cavitation bubbles.

Furthermore, free radical formation, which can be detrimental to emulsion stability, increases with frequency, peaking at 200–600 kHz and then decreasing as frequency increases (Ashokkumar, 2011; Ashokkumar, 2015). In addition to the effect of streaming, the high-frequency experiment's reduced diffusion time inhibits the development of a cavitating bubble. Tiny bubbles who have a more homogeneous size distribution are produced with higher frequency (Singla & Sit, 2022). It is impractical to assume extreme situations like O_2 ionization at whichever frequency and power since the pace and variety of radical formation change with frequency and cause primary radicals as well as ions to only form within a specific frequency range (Merouani et al., 2013).

10.3.2 Ultrasound Power and Intensity

The power and intensity of the ultrasound waves can also affect the efficacy of the treatment. Generally, sonication power represented in nominal applied power (NAP), which is electrical power used to operate ultrasound equipment and actual power, i.e. power absorbed by the food matrix which is more appropriate. The actual power can be calculated by calorimetric method and recent equipments will measure the actual power by default. The calorimetric method usually entails taking a predetermined volume of water and treating with the ultrasound while recording the temperature variation over time for a given ultrasound power. The following equation can be used to determine the actual power delivered (Khan et al., 2022).

$$P = mC_p \frac{dT}{t}$$

where m = the mass of water (g); P = actual power (W); C_p = the specific heat capacity of water (4.18 $Jg^{-1}°C^{-1}$); dT = change in temperature (°C); t = treatment time (sec).

Ultrasound intensity (W/cm^2) can be expressed by sonication power emitting per unit area of the emitting surface (probe tip). So increasing probe diameter used for application will decrease ultrasound intensity. So probe used for specific application can influence efficiency of ultrasound treatment for that specific application (Rutkowska et al., 2017). Higher intensity waves can cause more damage to cells and microorganisms but can also be more effective in disrupting cell walls and disrupting microorganisms. However, sharply increasing power can change bubble behavior by allowing bubbles to develop irregularly during expansion, that can lead in less cavitation and material growth (Brotchie et al., 2009).

10.3.3 Treatment Time

The duration of the ultrasound treatment can also affect the efficacy of the treatment. Optimum treatment time for food process application will be varying by a large extent depending on ultrasonication power, frequency, probe type, food matrix, and type of application. Longer treatment times can result in more effective inactivation of microorganisms but can also effect negatively on food matrix.

10.3.4 Temperature

Temperature is a crucial factor in ultrasound application in food processing. Ultrasound can generate a significant amount of heat during treatment. So most of the cases, an external cooling system should be used to maintain optimum temperature for specific applications. Cavitation is influenced by two major factors: temperature and static pressure, both of which may be adjusted depending on the application. Higher temperatures can reduce the viscosity of the medium while increasing vapour pressure, resulting in bubble formation. However, when the temperature rises, the volume of vapour inside the bubbles rises as well, cushioning the collapse and reducing cavitation severity. As a result, there is an ideal temperature for acoustic cavitation. Increased pressure, on the other hand, prevents cavitation, but the energy produced during implosion is significantly greater (Astráin-Red et al., 2019). Higher temperatures can increase the efficacy of ultrasound application by reducing the viscosity of food products and increasing cavitation. However, excessively high temperatures can also denature proteins and other biomolecules, leading to changes in food quality.

10.3.5 Food Matrix

The properties of the food being treated can also affect the efficacy of the ultrasound treatment. Viscosity plays a major role in propagation of ultrasound to the food matrix, high viscosity of food matrix can restrict even distribution of ultrasound waves thought food matrix causing cavitation effect concentrated on specific areas of food matrix.

10.3.6 Ultrasound Equipment

The type and quality of ultrasound equipment used for food processing can also affect the efficacy of ultrasound application. Factors such as the power output, bath type, probe type, transducer size, and coupling medium can impact the penetration and distribution of ultrasound waves in the food matrix.

10.3.7 Processing Environment

The processing environment can also influence the efficacy of ultrasound application in food processing. Factors such as the presence of air bubbles, dissolved gases, or impurities can affect the formation and stability of cavitation bubbles, leading to inconsistent results.

10.4 ULTRASOUND EQUIPMENT FOR FOOD PROCESSING

10.4.1 Batch Type

Batch type ultrasound probe system (Figure 10.2) consists of ultrasound processor which has ultrasound generator and transducer, also different size probes which can be attached to the transducer to transfer generated ultrasound to the product. The ultrasonic generator receives energy from the

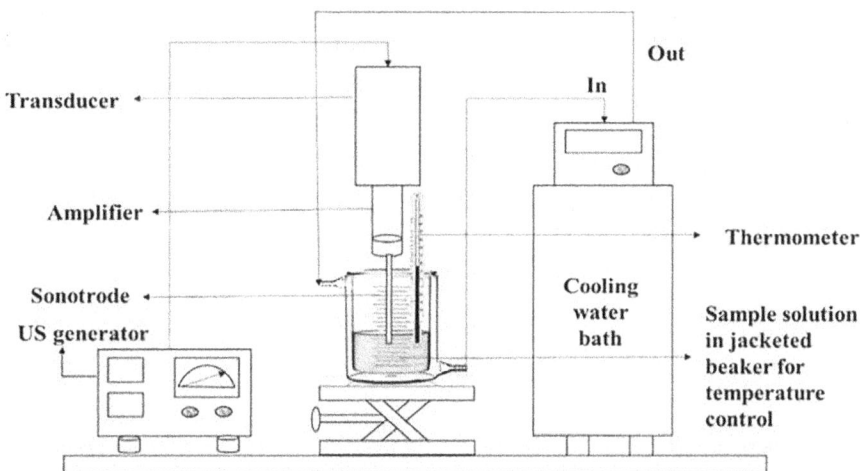

FIGURE 10.2   Batch type ultrasound probe system. (Reproduced from Suchintita Das et al., 2022.)

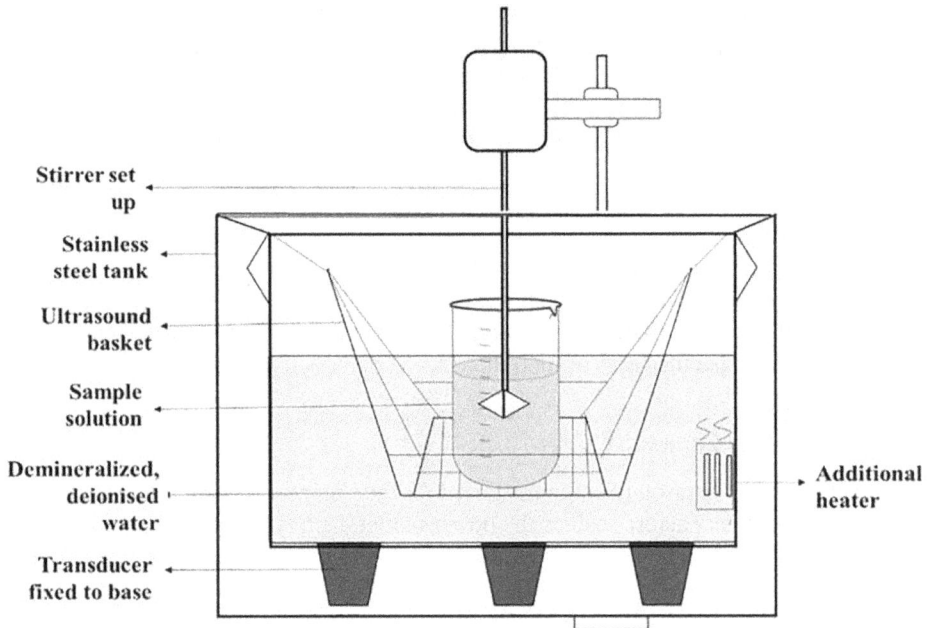

FIGURE 10.3 Batch type ultrasound bath system. (Reproduced from Suchintita Das et al., 2022.)

power source and converts it to the appropriate voltage, frequency, and amperage. Then, it is connected to the ultrasonic transducer which is a device that converts electrical signals into ultrasonic waves. Also, batch type ultrasonic bath system is used in many applications (Figure 10.3). Food matrices are treated in batch wise in batch type system. The energy is absorbed by the food products, causing physical and chemical changes to occur. One of the main advantages of batch type ultrasound is that it can be used to process a fixed volume of food matrix at once, also can achieve high specific energy density (J/mL) in less time compared to the continuous type, making it a highly efficient and cost-effective method. Additionally, ultrasound energy can be precisely controlled and adjusted, allowing for precise and consistent processing results. One of the most common applications of batch type ultrasound in food processing is the extraction of bioactive compounds from plant materials. This can be done by immersing plant materials in a liquid and applying ultrasound energy. The energy causes the cell walls to rupture, releasing the flavonoids and other compounds into the liquid. Another application of batch type ultrasound in food processing is the homogenization of liquids. This can be used to create emulsions, suspensions, and other types of mixtures. By applying ultrasound energy to the liquid, the panicles in the liquid are broken down, resulting in a homogenous mixture (Mason, 1998).

The batch type ultrasound system is widely used in laboratory settings for research and development purposes. However, when it comes to industrial applications, the system has certain limitations that make it less suitable for use in process flow. The system can only process a certain amount of material at one time, which makes it unsuitable for continuous production processes in industrial settings. This means that it is difficult to integrate the system into a larger manufacturing process without causing bottlenecks or slowing down production.

10.4.2 Continous Type

A batch-type ultrasound system can be converted into a continuous-type system by incorporating a flow cell, storage tank, and pump system to circulate the food matrix. The flow cell consists of inlet and outlet ports that allow for the continuous flow of the food matrix, with a probe fitted into the flow cell to transmit ultrasound energy to the food matrix. Flow cells are available in various volume capacities, and the flow rate (L/hr) significantly affects the treatment efficiency in continuous-type ultrasound systems. Therefore, optimizing the flow rate is crucial when scaling up ultrasound processing from batch-type. As the total volume processed increases, maintaining a similar specific energy density (J/mL) as in batch-type systems requires more treatment time. It is, therefore, essential to optimize ultrasonication power, intensity, time, flow rate, and volume of the flow cell when scaling up ultrasound processing. By optimizing these parameters, it is possible to convert the batch-type ultrasound system into a continuous-type system, resulting in improved processing efficiency and product quality in food processing applications.

The liquid whistle is one of the first methods for emulsification via cavitation. Process material is driven through an aperture under pressure created by a strong pump, where it exits and expanded into a mixing chamber. The system is robust and long-lasting because it has no moving parts other than a pump. Resonating tube reactors are the apparatuses most suited for widespread use in the food business. The liquid that has to be treated is essentially sent through a conduit with walls that vibrate ultrasonically. In this manner, the sound energy produced by transducers attached to the exterior of the tube is directly transmitted into the liquid that is flowing. Rectangular, pentagonal, hexagonal, and circular pipe cross-sections are available for commercial tube reactors, which are typically made of stainless steel. An alternate method would include coaxially inserting a radially emitted bar into the pipe that is carrying the moving liquid; this would only need minor modifications to the current pipework. A typical piezo transducer is used to drive a hollow tube in one of these systems, which is sealed off at one end. Another idea uses a cylindrical titanium rod with opposite piezoelectric transducers attached to either end. Both inserts are created so the ultrasonic frequency is radially emitted at half-wavelength intervals across their lengths (Moser et al., 2001; Sun, 2014).

10.5 ULTRASOUND PROCESSING IN FOOD SYSTEM

One of the main benefits of ultrasound processing is that it can inactivate microorganisms and enzymes, which can help to extend the shelf life of food products. This is done by creating high-pressure and high-temperature zones that disrupt the cell membranes of the microorganisms, causing them to die. This process is known as "cold pasteurization" as it does not require high heat treatment, which can cause thermal degradation of food products.

Ultrasound processing can also be used to improve the texture and release of bioactive compounds from food products. This is done by breaking down cells and tissues, which can make the food products more palatable and increase their nutritional value. In addition, ultrasound can be used for mixing, emulsifying, and dispersing ingredients, which can improve the homogeneity and stability of food products. This can be beneficial in many products like chocolate, cheese, sauces, and others. Overall, ultrasound processing is a versatile and non-thermal technology that can be used to improve the quality and safety of food products while preserving their nutritional value. However, it is important to note that more research is needed to fully understand the effects of ultrasound processing on food products and to ensure that the technology is used safely and effectively.

10.5.1 MICROBIAL INACTIVATION AND ACTIVATION

Ultrasound technology can be used for both microbial activation and inactivation. In microbial activation, ultrasound can be used to enhance the growth and reproduction of microorganisms by increasing the oxygen and nutrient supply to the cells and it is used to improve the efficiency of fermentation processes. In microbial inactivation, ultrasound can be used to disrupt the cell membrane and inactivate microorganisms by creating mechanical stress on the cells and it is used to inactivate microorganisms (Janghu et al., 2017). In food industry, microbial activation is done by fermentation process where as inactivation is done by pasteurization, sterilization, and other preservation methods (Meena et al., 2023).

Ultrasound treatment of goat milk at 881 W for 6 minutes resulted in the highest level of inactivation of *Staphylococcus aureus*, *Escherichia coli*, and mesophilic aerobic bacteria. A correlation was observed between increased power and time of treatment and a greater inactivation effect. The curds produced by the US-treated milk were found to be superior in terms of organoleptic properties compared to the control curd (Dumuta et al., 2022). Many studies didn't achieve 5 log reduction in microbial count which is standard for pasteurization, Red grape juice treated with ultrasound (750 W, 70%) for 10 min exhibited 3.4 log reduction in total plate count (Margean et al., 2020). Dhahir et al. achieved only 2 log reduction in pathogenic bacteria *E. coli* O157: H7 and Salmonella typhimurium in camel milk utilizing 900 W power. However, it didn't produce any negative effect on camel milk fatty acids, lipid peroxides, and protein fractions (Dhahir et al., 2020).

Ultrasound treatment alone found to be not effective to inactivate microorganisms at the pasteurization level (5 log reduction). So using combination of ultrasound with heat (thermosonication), pressure (manosonication), both heat and pressure (manothermosonication), and other techniques is necessary to achieve desirable results (Abrahamsen & Narvhus, 2022; Bermudez-Aguirre & Niemira, 2022).

Multiple studies have demonstrated the efficacy of utilizing a combination of treatments to achieve desired results in pathogenic inactivation. Specifically, Wang et al. (2020) demonstrated that the application of a combination of ultrasound and ultraviolet treatment at a power level of 600 W and a treatment time of 10 minutes resulted in 100% inactivation of pathogens (Jingyi Wang et al., 2020). Additionally, Das et al. (2020) reported that utilizing a combination of microwave (750 W) and ultrasound treatment (80%) on bottle gourd (Lagenaria siceraria) juice for 15 minutes at 70°C resulted in a 5 log reduction in L. monocytogenes count and an improvement in juice quality (Das et al., 2020). However, Yildiz et al. (2019) found that thermosonication alone (55°C, 120 μm, 24 kHz, 3 min) was sufficient for achieving a 5.69 log reduction in acid-adapted *E. coli* count (Yildiz et al., 2019).

The development of new ultrasonic reactors is essential to achieve desirable results in pathogenic inactivation. Baboli et al. (2020) reported on the efficacy of a new reactor in rapidly inactivating *E. coli* and *S. aureus* in apple juice. Specifically, the study found that the new reactor achieved a 5 log reduction of *E. coli* in 35 secs at 60°C and a 5 log reduction of *S. aureus* in 30 at 62°C. This research highlights the potential of new ultrasonic reactor designs in improving the efficiency and effectiveness of pathogenic inactivation in food and beverage products (Baboli et al., 2020).

Ultrasound technology can also be used to inhibit microbial growth in vegetables during storage periods. For example, a study by (Jian Wang & Fan, 2019) found that ultrasound treatment of green asparagus at a frequency of 40 KHz and a power of 360 W for 10 minutes was effective in retarding the growth of total number of colonies, molds, and yeasts during a storage period of 12 to 16 days. This research highlights the potential of ultrasound technology as a non-thermal preservation method for extending the shelf life of fresh produce and reducing food waste.

10.5.2 FERMENTATION

Fermentation is a biological process in which microorganisms such as yeast and bacteria convert sugars and other carbohydrates into various end products, such as alcohol, acids, and gases. Ultrasound can be used to enhance the fermentation process by increasing the rate of mass transfer, such as oxygen transfer, and by promoting the growth and activity of microorganisms. Ultrasound can be used to reduce fermentation time in dairy products like yogurt by using sound waves. Studies have shown that this can decrease production time cost, and improve consistency and texture. However, the effects of ultrasound on microorganisms can vary depending on the intensity and frequency used. Low intensity ultrasound can have positive effects on microbial growth, biofilm formation, and metabolism, but high intensity can be harmful. Factors such as the type of microorganism, medium composition, pH, and temperature also play a role. Using sub-lethal doses of ultrasonic irradiation can be beneficial for certain processes, but the intensity and duration must be carefully chosen to avoid negative effects and reduce energy costs (Akdeniz & Akalın, 2022).

Ultrasonication is a technique that can be used to reduce the fermentation time of dairy products like yogurt. It's been shown to decrease production time and cost, and improve consistency and texture. Additionally, it can reduce the lag phase of bacteria growth during fermentation without affecting the overall fermentation time. Overall, reducing fermentation time in dairy products can be beneficial in terms of production efficiency, cost, and product quality (Abesinghe et al., 2019). Similarly, another study found that ultrasound pretreatment has a significant impact on the curd-forming properties of milk. The addition of inoculants, the intensity of ultrasound applied, and the timing of inoculation all affect the curd-forming properties. The results showed that when inoculation was followed by ultrasound treatment, the time for curd gel formation was significantly reduced, providing a potential benefit for the dairy industry by reducing curd fermentation/gelation time and increasing production efficiency (Choudhary & Rawson, 2021).

10.5.3 EXTRACTION

UAE is a method of extracting bioactive compounds from a food matrix by using ultrasonic energy (Naik et al. 2022). This method is becoming increasingly popular in various industries, such as the food and beverage industry, the pharmaceutical industry, and the cosmetics industry because it offers several advantages over traditional extraction methods. One of the main advantages of UAE is that it is a relatively fast and efficient method of extraction. Traditional methods of extraction, such as Soxhlet extraction and maceration, can take several hours or even days to complete. In contrast, UAE can extract compounds in just a few minutes (Rawson & Radhakrishnan, 2023).

Traditional extraction methods, such as solvent extraction and maceration, have several drawbacks including low efficiency, high consumption of solvents, high energy requirements, and prolonged production time. UAE is a modern, environmentally friendly and rapidly growing technology that is suitable for increasing the yield and efficiency of extracting bioactive compounds. The process of UAE is primarily achieved through producing the cavitation bubbles in biological matrix. This method has been proven to achieve high yields and extraction rates of bioactive compounds. Additionally, UAE has significant economic and environmental benefits and holds immense potential for development and implementation (Wen et al., 2018; Sengar et al., 2020).

Ultrasound extraction works through various mechanisms such as fragmentation, erosion, capillarity, detexturation, and sonoporation (Chemat et al., 2017). The cavitation effect produces physical, chemical, and mechanical effects when applied to a liquid medium, which is the primary mechanism of UAE. Microturbulence created by violent collapse bubbles can generate a high temperature and shear forces in media leading to enhance the heat and mass transfer in food matrix that aid in the extraction process by various mechanisms such as cell disintegration, pitting and erosion of cell surfaces, increased solvent penetration, increased hydration and swelling of the cell, release of molecules from the cell surface and disruption of molecular aggregates to improve the efficiency of the extraction process. These mechanisms work together to increase the yield of bioactive compounds while also improving their functional properties (Rahman & Lamsal, 2021; Sengar et al. 2022).

Ultrasound is used to extract various bioactive components from plant/food matrix such as aromas, carotenoids, flavonoids, polyphenols, pigments, antioxidants, polysaccharides, proteins, etc. Raj and Dash conducted extraction of phytocompounds from the peel of dragon fruit using ultrasound with 100 W of power. The study found that betacyanin was successfully extracted with minimal degradation to the pigment, yielding high extraction efficiency, making it a viable source for a natural colorant. However, the results indicated that the extraction process was sensitive to several variables such as temperature, the ratio of liquid to solid, solvent concentration, and treatment duration (Raj & Dash, 2020).

10.5.4 ENZYME INACTIVATION AND ACTIVATION

Enzyme activation and inactivation are important processes in food production and other industries. Enzyme activation by ultrasound can increase the activity of enzymes and speed up the reaction, making the process more efficient, improving the texture, flavor, and nutritional value of food products, and helping to preserve food by breaking down potential spoilage agents (Thangaraju et al., 2023). On the other hand, enzyme inactivation by ultrasound can prevent spoilage and extend the shelf life of food products, eliminate harmful bacteria and other microorganisms in food products and be more efficient and cost-effective than other methods.

Ultrasound can both activate and inactivate enzymes, depending on the specific properties of the enzyme and the conditions of ultrasound application. Under mild ultrasonic conditions, enzyme activation can occur through the breakdown of enzyme aggregates, resulting in increased availability of active enzyme sites. However, intense ultrasound can cause extreme conditions such as significant shear and excessive free radicals, leading to damage of the enzyme's tertiary structure and polypeptide chains, resulting in enzyme inactivation. Studies have shown the effects of ultrasound on a variety of enzymes, including chitinase, β-D-glucosidase, papain, dextranase, cellulase, and lipase (Ma et al., 2020, Iqbal et al., 2019).

Ultrasound treatment can alter enzyme activity in milk through a combination of cavitation-induced physical effects and sonochemical products. Homogenization effects resulting from moderate US treatment, along with protein unfolding, can enhance mass transfer and enzyme-substrate interactions, leading to an increase in enzymatic activity. Conversely, extensive US pretreatment in combination with mild pressure or heat conditions can result in enzyme denaturation and subsequent impairment of enzymatic activity (Munir et al., 2019).

Ultrasound treatment did not result in an increase in the enzymatic activity of glucoamylase. Rather, when ultrasonic power exceeded 270 W, a decrease in enzymatic activity was observed, with complete inactivation occurring at 360 W with a treatment time greater than 40 min (Wang et al., 2020). Ultrasound treatment (33 kHz, 600 W) of ginger slices dipped in a beaker containing purified water at 30°C for 30 minutes reduced the activity of polyphenol oxidase and peroxidase by more than 70%, which can increase the shelf life of ginger (Osae et al., 2019). The activity of chitinase was evaluated under different treatment conditions. The results showed that when the chitinase was treated at an intensity of 25 W/mL for a duration of 20 minutes, there was a significant increase in enzyme activity compared to the untreated chitinase. Specifically, the chitinase activity was found to be increased by 19.17% (Hou et al., 2019). Enzyme activation and inactivation by ultrasound is a powerful tool in the food production industry, providing significant benefits such as increased efficiency, improved quality, extended shelf life, and cost savings.

10.5.5 MIXING AND HOMOGENIZATION

Homogenization, the process of reducing particle and globule size, is a crucial operation in food engineering. It is applied to products such as milk, fruit juice, tomato products, and sauces. The technique of emulsification, which involves blending two or more liquid phases, results in a stable suspension of droplets. This is an important aspect of the food and cosmetic industries, as it creates complex media with unique properties for specific uses.

Mixing and homogenization using ultrasound often leads to improved mixing performance. This is often observed visually, mixing is characterized by the degree of homogeneity of a single phase or a multi-phase medium and can be quantified at two scales: macromixing, which refers to the overall apparatus scale (such as mixing time in a batch condition or a dispersion of the residence time distribution in continuous operation), and micromixing, which refers to the lower scale (such as particles or droplets). The quality of micromixing is critical to many submicron-scale phenomena, including precipitation, emulsification, mass transfer at interfaces, reaction rate, and selectivity (Delmas & Barthe, 2015; Naik et al., 2023).

Emulsification is the process of blending two liquids where one liquid is dispersed in the other. There are different types of emulsions, such as oil-in-water and water-in-oil. Emulsions are widely used in industries such as food, cosmetics, pharmaceuticals, paints, and agrochemicals. For example, emulsions are used in functional foods as delivery systems for bioactive components like curcumin, β-carotene, polyphenols, and essential oils. The stability of these emulsions is important for their use in various applications, but instability mechanisms like flocculation, phase separation, coalescence, and Ostwald ripening can affect their effectiveness. Efficient emulsifiers are needed to produce stable emulsions (Taha et al., 2020). When compared to other techniques such as high-pressure and high-speed homogenization, ultrasound homogenization (300 W for 10 min) produced a stable and high-quality emulsion with a high emulsification activity of 81.05% and a minimum droplet size of 150 nm (Zhou et al., 2022). Ultrasonic treatment can reduce lipid and protein oxidation in food, leading to improved stability. A recent study utilized ultrasound (400 J/cm^3) to prepare a walnut oil and almond protein isolate emulsion, which resulted in improved physical stability of the emulsified lipids during storage. The study also found that sonication inhibited protein oxidation by limiting the increase in carbonyl formation, and preserving free sulfhydryl groups, intrinsic tryptophan fluorescence, and molecular weight (Zhang et al., 2021). Using ultrasound for emulsification can also enhance functional properties of food matrix and can increase bioaccessibility of bioactive components. Zhang et al. (2019) reported that ultrasound treatment (400 W) significantly increased lycopene content and bioaccessibility of lycopene (Zhang et al., 2019). So using ultrasound for emulsification can enhance the quality and nutrition of food matrix.

10.5.6 DRYING

Drying food products is a common method for preserving food by removing moisture. This helps to extend the shelf life of the food and prevent the growth of microorganisms that can cause spoilage. Ultrasonic technology is employed in the drying of food products, which encompasses both pretreatment and assisted drying techniques. The ultrasound pre-treatment can cause food products to lose or gain water, and higher ultrasonic parameters lead to more water loss or gain. The use of ultrasonic technology prior to drying has been observed to enhance the drying rate, although the results may be variable. In the case of ultrasonic assisted drying, an increase in ultrasound power has been found to positively impact the drying process, however, the extent of improvement is dependent on factors such as air velocity and temperature. The use of ultrasonic technology has also been shown to affect food quality, including reducing water activity, enhancing product color, and minimizing nutrient loss (Huang et al., 2020)

Generally, ultrasound is used as a pretreatment and combined with other drying methods. Application of ultrasound pretreatment (25.2–117.6 W/L for 5–15 min) to convective drying of mulberry leaves in a continuous operation enhanced convective drying kinetics and reduced total energy consumption having more influence on internal mass transfer resistance (Tao et al., 2016). Also, Tao et al. developed a hot-air convective dryer integrated with a contacting ultrasound system for the dehydration of garlic slices, resulting in a significantly accelerated drying process and an improvement in organosulfur compound retention and a reduction in browning (Tao et al., 2018). The use of ultrasound pretreatment in conjunction with pulsed electric field for drying carrots resulted in a shorter drying time and improved retention of color and carotenoids. It also displayed a reduced ability to absorb water vapor from the environment and an increased water diffusion rate of 63% (Wiktor et al., 2019). Combination of ultrasound with microwave and hot air drying reduced total drying time, energy consumption, and enhances drying rate and product quality of red beetroot. It serves as an alternative to hot air drying, producing a more porous, nicely colored, and crispy product (Szadzińska et al., 2020). And Cao et al. investigated the effect of air borne ultrasound and microwave assisted drying of broccoli floret showed that ultrasound treatment preserved phytochemicals in the sample better than microwave treated sample. However, drying efficiency was less than microwave treated sample (Cao et al., 2019). Also used for osmotic dehydration

(Li et al., 2021) and spray drying (Khaire & Gogate, 2020). The application of ultrasound prior to spray drying has a considerable impact on the particle arrangement, resulting in an enhancement of the mesopore volume. The use of a lower power of 45 W leads to the formation of spherical and symmetrically shaped particles, while a higher power of 90 W results in a less favorable outcome (Aghamohammadi et al., 2019).

10.5.7   MODIFICATION OF FOOD MACROMOLECULES: PROTEIN, CARBOHYDRATE, FAT

Ultrasound can alter the physical and biochemical properties of food macromolecules. Ultrasound is a type of energy which is produced by vibrating particles at a frequency above the range of human hearing. During this process, the ultrasound waves interact with the food macromolecules, causing them to change shape, size, and physical properties. Ultrasound modification of food macromolecules can be used to improve food texture, flavor, and stability, as well as to modify the functional properties of proteins, lipids, and carbohydrates (Sengar et al., 2022).

Proteins can be modified by ultrasound to enhance functional properties such as solubility, textural properties, emulsifying properties, absorption of essential amino acids and digestibility, and reduce the extent of denaturation, polymerization, and aggregation of protein molecules. Additionally, power ultrasound can be used to modify proteins and reduce allergens by decreasing the disulfide bonds from their quaternary structure and improving the glycation reaction (Sengar et al., 2022). This can be beneficial for patients with digestive disorders who have difficulty digesting proteins. Ultrasound treatment affected the structure and molecular flexibility of soybean protein isolate (SPI). The treatment results in increased β-sheet content, flexibility, and sulfhydryl content, and unfolds its tertiary structure. SPI shows the best solubility, turbidity, surface hydrophobicity, emulsifying activity, and stability at 400 W ultrasound power (Yan et al., 2021). Ultrasound affects starch structures positively based on ultrasound parameters such as amplitude, time, temperature, and energy, causing structural changes like pores, cracks, ruptures, and granule deformation. The effect of ultrasound on starch granules varies with power and application method. Generally, original crystalline types of starches are retained after ultrasonication, with a reduction in crystallinity varying based on processing conditions. Ultrasonication causes degradation of molecular chains, damage to crystalline structure, and reduction in short-range order, resulting in higher proportion of amorphous starch and lower crystallinity. Exceptionally, retrograded high-amylose maize starch forms a new short-range ordered structure after ultrasonication, all these contribute in enhancing hydration properties, thermal properties, rheological properties, and digestibility (Wu et al., 2022).

The impact of dual-frequency ultrasound on the conformation of polysaccharide extracted from Lentinus edodes was studied at 20/40 and 20/60 kHz frequency modes. Results showed a transformation of the rigid triple-helix chain into a loose and flexible triple-helix chain, with the strongest immune activity observed. These results suggest that dual-frequency ultrasound has a significant effect on the conformation of lentinan, and the spiral number, stiffness, flexibility, and side-chain characteristics of the polysaccharide chain play a crucial role in determining its immune activity (Hua et al., 2022). Additionally, ultrasound modification can be used to increase the bioavailability of essential fatty acids. HIU may be a useful tool for creating improved organogels for food applications (Sharifi et al., 2019). Effect of HIU on physical properties of oleogels made from monoglycerides (MG) and high oleic sunflower oil showed that slow cooling rate and high MG concentration improved the oil binding capacity and hardness of the network. HIU application was also efficient in increasing the oil binding capacity and hardness, especially at high MG concentration. Microscopy images showed that HIU and cooling rate changed the microstructure of the oleogels, resulting in smaller crystals (Giacomozzi et al., 2019).

Overall, ultrasound modification of food macromolecules can be used to increase the nutritional value, texture, taste, and shelf-life of food products. This process can help to improve the health of individuals who have difficulty digesting certain macromolecules or who need to manage their blood sugar levels.

10.5.8 Tenderization of Meat

Tenderization is the process of making meat more tender by breaking down its muscle fibers and connective tissue. This can be achieved through various methods such as marination, mechanical methods, enzymes, or aging. The goal of tenderization is to make the meat easier to chew and digest, resulting in a more enjoyable eating experience. Ultrasound is a highly efficient technique for tenderizing meat in a short amount of time, and can be tailored to target specific muscles for improved taste, appearance, and storage life of meat products.

Ultrasound technology can be used to improve the quality of meat before it is stored, or after it has undergone aging. The application of ultrasound waves to meat prior to storage (pre-rigor meat) helps to reduce bacterial growth and accelerate tenderization by accelerating the glycolysis process by increasing the release of calcium in the cytosol by damaging the membranes, while ultrasound applied after aging (post-rigor meat) can help to tenderize the meat and enhance its flavor with probable mechanisms of physical disruption in the muscle microstructure by cavitation and/or release and activation of proteolytic enzymes (Bhat et al., 2018). The effect of ultrasound treatments on tenderizing meat was investigated in several studies. In all studies, ultrasound was found to be an effective method for improving meat tenderness. In one study on goose meat, the ultrasound treatment (600 W for 30 min) resulted in the lowest shear force and cooking loss, and increased the fragmentation of myofibrillar fraction(Li et al., 2018). Finally, a study on chicken gizzard found that ultrasound treatment (500 W for 30 min) decreased shear force, muscle fiber diameter, and increased myofibril fragmentation, leading to improved thermal stability and overall tenderization of the gizzard (Du et al., 2021).

10.5.9 Germination

Germination with ultrasound involves exposing seeds to ultrasonic waves to enhance and improve the germination process. The mechanical energy generated by the waves creates stress on the seed coat, breaking seed dormancy and promoting germination. Furthermore, ultrasound boosts seed absorption of water and nutrients, leading to quicker and more effective germination. Research has shown that ultrasound improves germination rates and seedling growth in various seeds. Application of HIU on the germination of wholegrain brown rice that ultrasound significantly improved germination rates and led to reduction in starch contents and size of starch granules. Also, increase in reducing sugar. Additionally, it increased the accumulation of gamma-aminobutyric acid, antioxidants, and proline, and improved the nutritional index of free amino acids, as well as the bioaccessibility of calcium and iron (Xia et al., 2019). Hydration of wheat grains used in the food industry for malt production using ultrasound found that higher temperature and acoustic density improved hydration kinetics, the maximum germination was observed at 5 days with 86% of the grains germinated. The results suggest that ultrasound could be a clean alternative for intensifying the malting process and reducing processing time (Guimarães et al., 2020). Similarly, modulation of the hydration process during paddy germination by ultrasound resulted in sono-hydro priming had a higher hydration rate compared to hydropriming. The activation energy required was lower than hydropriming, and sono-hydro priming resulted in higher germination rate and reduced the soaking time by 50% compared to hydropriming (Kalita et al., 2021). Zhang et al. found that the use of HIU prior to soaking could be a potential strategy to improve the germination and nutritional characteristics of pre-germinated seeds.

10.5.10 Aging of Alcoholic Beverages

The aging process of alcoholic beverages affects their color and flavor, traditionally done by storing in wooden barrels. Alternative methods such as using wood fragments, ultrasound, micro-oxygenation, pulsed electric field, high hydrostatic pressure, and microwave and gamma irradiation are being studied to optimize the aging process, reduce production costs, and increase competitiveness (Krüger et al., 2022). Ultrasonic treatment (380 W) enhanced the maturation and increase the aroma components. The variety of flavor components increased with longer storage time and a dynamic balance and harmony of esters, alcohols, acids, and aldehydes was achieved through ultrasonic aging, resulting in a soft and mellow kiwi wine (Zhang et al., 2023). Also, ultrasound improved color characteristic of wine by by enhancing the co-pigmentation between caffeic acid and monomeric anthocyanins in the wine, resulting in the improvement of wine quality (Xue et al., 2022). Similarly, ultrasound treatment (100 W for 20 min) enhanced color stability by reduction in anthocyanin degradation, and increased antioxidative ability by inactivation of polyphenol oxidase during Sanhua plum wine aging (Zhiqian Wu et al., 2022).

High power ultrasound (HPU) is a promising technology in winemaking and has been approved for grape treatment. Polyphenols, particularly anthocyanins, have been shown to have an impact on red wine sensory qualities and evolution during storage (Celotti et al., 2020). Similarly, ultrasound treatment (180 W, 20 min) of blueberry wine increased anthocyanins without adverse effects on anti-oxidant activity during the early stage of aging (Xusheng Li et al., 2020). The use of ultrasound was found to be an effective method for shortening the aging time of vinegar and improving its taste. This was demonstrated in a study of crabapple vinegar, where ultrasound treatment increased the levels of volatile components such as esters, aldehydes, and heterocycles, and made the taste of the vinegar softer (Zhai et al., 2021).

10.6   CHALLENGES AND RESEARCH NEEDS

The utilization of ultrasound in the food processing industry has been demonstrated to significantly enhance the quality of food products. The integration of ultrasound with other techniques has been shown to produce even better outcomes, making it an attractive area for further exploration and improvement. However, the optimization of parameters, safety considerations, energy efficiency, and the use of advanced techniques such as Barbell Horn Ultrasonic Technology and manothermosonication must be considered in order for ultrasound to be developed for use in a commercial setting. Research has consistently demonstrated the benefits of using ultrasound in place of, or in conjunction with, conventional food processing techniques. The integration of ultrasound with other techniques has been shown to lead to a higher overall quality of the final product, which could become the focus of future research endeavors. In order for the application of ultrasound in the food processing industry to be optimized, research must be conducted at a ground-level to assess the impact of ultrasound on bulk food production. Additionally, consideration must be given to the safety and adverse effects of ultrasound on human health. Energy requirements play a significant role in the commercialization and industrialization of ultrasound, and as such, much energy will be required for this technology to be implemented on a large scale. Therefore, research into ultrasound should be specifically aimed at industrial-level implementation in order to effectively overcome the obstacles associated with its application in the food industry.

1. Energy uniformity – The viscosity of the treating sample can affect the energy dissipation of ultrasound through the sample volume. In some food matrices, viscosity increases as treatment time increases (as does temperature), which can affect the uniformity of energy dissipation. If the viscosity reaches a certain point, the sample can become more viscous around the probe, causing damage to the probes due to power overload. (This problem can be solved by using a continuous system or a constant mixing mechanism.)
2. Wear and tear of probes – Probes can wear after a long period of use, but when we use an inappropriate power level based on the food matrix, they can wear much faster than normal. To extend the life of the probe, all parameters must be optimized.
3. There is a need of development of high durable probe and optimum ultrasonic reactors to increase the efficiency and durability of probes.

4. The majority of ultrasound applications used batch type systems, when converting the same operation into continuous operation, we must use a different optimization strategy because using the same conditions might not produce the same results. Therefore, it may be beneficial to develop a proper protocol for ultrasound treatment for various applications.
5. Scaling up of operation can be challenging because lack of researches on large scale operations which will have different sets of challenges regarding type, physical and chemical properties of food matrix, requirement of more energy (or proper utilization of energy), etc.

10.7 CONCLUSION

Ultrasound technology is being applied in the fruit, juice, and dairy industries for analysis, control, and processing. It has the potential to improve food processing by destroying microorganisms and inactivating enzymes at low to moderate temperatures without altering organoleptic and nutritional properties, modification of food macromolecules to improve its physical and functional properties, activation of microorganisms and enzymes to accelerate various processes like fermentation, efficient extraction of bioactive components from food in less time and resources, etc. While they may not completely replace traditional processing methods, they can enhance or complement them, leading to improved sensory and physicochemical properties in food products. Integrating ultrasound with other techniques such as emulsification, drying, and extraction can also increase process efficiency, though initial investment may be required due to high capital expenditure. With continued progress in commercialization and research, the cost of equipment is expected to decrease, making nutritious and safe food products available to consumers at an affordable cost.

REFERENCES

Abesinghe, A. M. N. L., Islam, N., Vidanarachchi, J. K., Prakash, S., Silva, K. F. S. T., & Karim, M. A. (2019). Effects of ultrasound on the fermentation profile of fermented milk products incorporated with lactic acid bacteria. *International Dairy Journal*, *90*, 1–14. https://doi.org/10.1016/j.idairyj.2018.10.006

Abrahamsen, R. K., & Narvhus, J. A. (2022). Can ultrasound treatment replace conventional high temperature short time pasteurization of milk? A critical review. *International Dairy Journal*, *131*, 105375. https://doi.org/10.1016/j.idairyj.2022.105375

Aghamohammadi, S., Haghighi, M., & Ebrahimi, A. (2019). Pathways in particle assembly by ultrasound-assisted spray-drying of kaolin/SAPO-34 as a fluidized bed catalyst for methanol to light olefins. *Ultrasonics Sonochemistry*, *53*, 237–251. https://doi.org/10.1016/j.ultsonch.2019.01.009

Akdeniz, V., & Akalın, A. S. (2022). Recent advances in dual effect of power ultrasound to microorganisms in dairy industry: Activation or inactivation. *Critical Reviews in Food Science and Nutrition*, *62*(4), 889–904. https://doi.org/10.1080/10408398.2020.1830027

Ashokkumar, M. (2011). The characterization of acoustic cavitation bubbles – An overview. *Ultrasonics Sonochemistry*, *18*(4), 864–872. https://doi.org/10.1016/J.ULTSONCH.2010.11.016

Ashokkumar, M. (2015). Applications of ultrasound in food and bioprocessing. *Ultrasonics Sonochemistry*, *25*(1), 17–23. https://doi.org/10.1016/j.ultsonch.2014.08.012

Astráin-Redin, L., Ciudad-Hidalgo, S., Raso, J., Condón, S., Cebrián, G., & Álvarez, I. (2019). Application of High-Power Ultrasound in the Food Industry. *Sonochemical Reactions*. https://doi.org/10.5772/intechopen.90444

Baboli, Z. M., Williams, L., & Chen, G. (2020). Rapid pasteurization of apple juice using a new ultrasonic reactor. *Foods*, *9*(6). https://doi.org/10.3390/foods9060801

Bermudez-Aguirre, D., & Niemira, B. A. (2022). Pasteurization of foods with ultrasound: The present and the future. *Applied Sciences (Switzerland)*, *12*(20). https://doi.org/10.3390/app122010416

Bhargava, N., Mor, R. S., Kumar, K., & Sharanagat, V. S. (2021). Advances in application of ultrasound in food processing: A review. *Ultrasonics Sonochemistry*, *70*(June 2020), 105293. https://doi.org/10.1016/j.ultsonch.2020.105293

Bhat, Z. F., Morton, J. D., Mason, S. L., & Bekhit, A. E. D. A. (2018). Applied and emerging methods for meat tenderization: A comparative perspective. *Comprehensive Reviews in Food Science and Food Safety*, *17*(4), 841–859. https://doi.org/10.1111/1541-4337.12356

Brotchie, A., Grieser, F., & Ashokkumar, M. (2009). Effect of power and frequency on bubble-size distributions in acoustic cavitation. *Physical Review Letters*, *102*(8), 1–4. https://doi.org/10.1103/PhysRevLett.102.084302

Cao, Y., Tao, Y., Zhu, X., Han, Y., Li, D., & Liu, C. (2019). Effect of microwave and air-borne ultrasound-assisted air drying on drying kinetics and phytochemical properties of broccoli floret. *Drying Technology*, *38*(13), 1–16. https://doi.org/10.1080/07373937.2019.1662437

Celotti, E., Stante, S., Ferraretto, P., Román, T., Nicolini, G., & Natolino, A. (2020). High power ultrasound treatments of red young wines: Effect on anthocyanins and phenolic stability indices. *Foods*, *9*(10). https://doi.org/10.3390/foods9101344

Chemat, F., Rombaut, N., Sicaire, A. G., Meullemiestre, A., Fabiano-Tixier, A. S., & Abert-Vian, M. (2017). Ultrasound assisted extraction of food and natural products. Mechanisms, techniques, combinations, protocols and applications. A review. *Ultrasonics Sonochemistry*, *34*, 540–560. https://doi.org/10.1016/j.ultsonch.2016.06.035

Choudhary, P., & Rawson, A. (2021). Impact of power ultrasound on the quality attributes of curd and its fermentation/gelation kinetics. *Journal of Food Process Engineering*, *44*(7), 1–12. https://doi.org/10.1111/jfpe.13698

Das, M. J., Das, A. J., Chakraborty, S., Baishya, P., Ramteke, A., & Deka, S. C. (2020). Effects of microwave combined with ultrasound treatment on the pasteurization and nutritional properties of bottle gourd (*Lagenaria siceraria*) juice. *Journal of Food Processing and Preservation*, *44*(12). https://doi.org/10.1111/jfpp.14904

Delmas, H., & Barthe, L. (2015). Ultrasonic Mixing, homogenization, and emulsification in food processing and other applications. In *Power Ultrasonics: Applications of High-Intensity Ultrasound*. Elsevier. https://doi.org/10.1016/B978-1-78242-028-6.00025-9

Dhahir, N., Feugang, J., Witrick, K., Park, S., & AbuGhazaleh, A. (2020). Impact of ultrasound processing on some milk-borne microorganisms and the components of camel milk. *Emirates Journal of Food and Agriculture*, *32*(4), 245–254. https://doi.org/10.9755/ejfa.2020.v32.i4.2088

Dias, A. L. B., Arroio Sergio, C. S., Santos, P., Barbero, G. F., Rezende, C. A., & Martínez, J. (2017). Ultrasound-assisted extraction of bioactive compounds from dedo de moça pepper (*Capsicum baccatum* L.): Effects on the vegetable matrix and mathematical modeling. *Journal of Food Engineering*, *198*, 36–44. https://doi.org/10.1016/j.jfoodeng.2016.11.020

Du, X., Li, H., Nuerjiang, M., Shi, S., Kong, B., Liu, Q., & Xia, X. (2021). Application of ultrasound treatment in chicken gizzards tenderization: Effects on muscle fiber and connective tissue. *Ultrasonics Sonochemistry*, *79*, 105786. https://doi.org/10.1016/j.ultsonch.2021.105786

Dumuta, A., Vosgan, Z., Mihali, C., Giurgiulescu, L., Kovacs, M., Sugar, R., & Mihalescu, L. (2022). The influence of unconventional ultrasonic pasteurization on the characteristics of curds obtained from goat milk with the low cholesterol content. *Ultrasonics Sonochemistry*, *89*(June), 106155. https://doi.org/10.1016/j.ultsonch.2022.106155

Giacomozzi, A. S., Palla, C. A., Carrín, M. E., & Martini, S. (2019). Physical properties of monoglycerides Oleogels modified by concentration, cooling rate, and high-intensity ultrasound. *Journal of Food Science*, *84*(9), 2549–2561. https://doi.org/10.1111/1750-3841.14762

Guimarães, B., Polachini, T. C., Augusto, P. E. D., & Telis-Romero, J. (2020). Ultrasound-assisted hydration of wheat grains at different temperatures and power applied: Effect on acoustic field, water absorption and germination. *Chemical Engineering and Processing – Process Intensification*, *155*(May). https://doi.org/10.1016/j.cep.2020.108045

Hou, F., Ma, X., Fan, L., Wang, D., Wang, W., Ding, T., Ye, X., & Liu, D. (2019). Activation and conformational changes of chitinase induced by ultrasound. In *Food Chemistry* (Vol. 285). Elsevier Ltd. https://doi.org/10.1016/j.foodchem.2019.01.180

Hua, Y., Zhang, H., Fu, Q., Feng, Y., & Duan, Y. (2022). Effects of Ultrasound Modification with Different Frequency Modes on the Structure, Chain Conformation, and Immune Activity of Polysaccharides from *Lentinus edodes*. *Foods*, *11*(16), 2470. https://doi.org/10.3390/foods11162470

Huang, D., Men, K., Li, D., Wen, T., Gong, Z., Sunden, B., & Wu, Z. (2020). Application of ultrasound technology in the drying of food products. *Ultrasonics Sonochemistry*, *63*, 104950. https://doi.org/10.1016/j.ultsonch.2019.104950

Iqbal, A., Murtaza, A., Hu, W., Ahmad, I., Ahmed, A., & Xu, X. (2019). Activation and inactivation mechanisms of polyphenol oxidase during thermal and non-thermal methods of food processing. *Food and Bioproducts Processing*, *117*, 170–182. https://doi.org/10.1016/j.fbp.2019.07.006

Janghu, S., Bera, M. B., Nanda, V., & Rawson, A. (2017). Study on power ultrasound optimization and its comparison with conventional thermal processing for treatment of raw honey. *Food Technology and Biotechnology*, *55*(4), 570–579. https://doi.org/10.17113/ftb.55.04.17.5263

Kalita, D., Jain, S., Srivastava, B., & Goud, V. V. (2021). Ultrasonics – sonochemistry Sono-hydro priming process (ultrasound modulated hydration): Modelling hydration kinetic during paddy germination. *Ultrasonics – Sonochemistry*, *70*(September 2020), 105321. https://doi.org/10.1016/j.ultsonch.2020.105321

Kentish, S. E. (2017). Engineering Principles of Ultrasound Technology. In *Ultrasound: Advances in Food Processing and Preservation*. Elsevier. https://doi.org/10.1016/B978-0-12-804581-7.00001-4

Khaire, R. A., & Gogate, P. R. (2020). Novel approaches based on ultrasound for spray drying of food and bioactive compounds. *Drying Technology*, *39*(12), 1–22. https://doi.org/10.1080/07373937.2020.1804926

Khan, S. A., Dar, A. H., Bhat, S. A., Fayaz, J., Makroo, H. A., & Dwivedi, M. (2022). High intensity ultrasound processing in liquid foods. *Food Reviews International*, *38*(6), 1123–1148. https://doi.org/10.1080/87559129.2020.1768404

Krüger, R. T., Alberti, A., & Nogueira, A. (2022). Current technologies to accelerate the aging process of alcoholic beverages: A review. *Beverages*, *8*(4), 65. https://doi.org/10.3390/beverages8040065

Li, X., Wang, Y., Sun, Y. Y., Pan, D. D., & Cao, J. X. (2018). The effect of ultrasound treatments on the tenderizing pathway of goose meat during conditioning. *Poultry Science*, *97*(8), 2957–2965. https://doi.org/10.3382/ps/pey143

Li, L., Yu, Y., Xu, Y., Wu, J., Yu, Y., Peng, J., An, K., Zou, B., & Yang, W. (2021). Effect of ultrasound-assisted osmotic dehydration pretreatment on the drying characteristics and quality properties of Sanhua plum (Prunus salicina L.). *LWT*, *138*(September 2020). https://doi.org/10.1016/j.lwt.2020.110653

Li, Xusheng, Zhang, L., Peng, Z., Zhao, Y., Wu, K., Zhou, N., Yan, Y., Ramaswamy, H. S., Sun, J., & Bai, W. (2020). The impact of ultrasonic treatment on blueberry wine anthocyanin color and its in-vitro antioxidant capacity. *Food Chemistry*, 127455. https://doi.org/10.1016/j.foodchem.2020.127455

Ma, X., Cai, J., & Liu, D. (2020). Ultrasound for pectinase modification: An investigation into potential mechanisms. *Journal of the Science of Food and Agriculture*, *100*(12), 4636–4642. https://doi.org/10.1002/jsfa.10472

Margean, A., Lupu, M. I., Alexa, E., Padureanu, V., Canja, C. M., Cocan, I., Negrea, M., Calefariu, G., & Poiana, M. A. (2020). An overview of effects induced by pasteurization and high-power ultrasound treatment on the quality of red grape juice. *Molecules*, *25*(7), 1–16. https://doi.org/10.3390/molecules25071669

Mahendran, R., & Rawson, A. (2023). Advances in thermal and nonthermal food engineering interventions in millet processing and its value addition. *Journal of Food Process Engineering*. https://doi.org/10.1111/jfpe.14354

Mason, T. (1998). Power ultrasound in food processing—The way forward. In: Povey, M. J. W. and Mason, T. J. (Eds.), *Ultrasound in Food Processing*, Blackie Academic and Professional, London, 105–126.

Meena, L., Buvaneswaran, M., Byresh, T. S., Sunil, C. K., Rawson, A., & Venkatachalapathy, N. (2023). Effect of ultrasound treatment on white finger millet-based probiotic beverage. *Measurement: Food*, *10*, 100090. https://doi.org/10.1016/j.meafoo.2023.100090

Merouani, S., Hamdaoui, O., Rezgui, Y., & Guemini, M. (2013). Effects of ultrasound frequency and acoustic amplitude on the size of sonochemically active bubbles-theoretical study. *Ultrasonics Sonochemistry*, *20*(3), 815–819. https://doi.org/10.1016/j.ultsonch.2012.10.015

Moser, W. R., Find, J., Emerson, S. C., & Krausz, I. M. (2001). Engineered synthesis of nanostructured materials and catalysts. *Advances in Chemical Engineering*, *27*, 1–48. https://doi.org/10.1016/S0065-2377(01)27002-1

Munir, M., Nadeem, M., Qureshi, T. M., Leong, T. S. H., Gamlath, C. J., Martin, G. J. O., & Ashokkumar, M. (2019). Effects of high pressure, microwave and ultrasound processing on proteins and enzyme activity in dairy systems – a review. *Innovative Food Science and Emerging Technologies*, *57*, 102192. https://doi.org/10.1016/j.ifset.2019.102192

Naik, M., Natarajan, V., Thangaraju, S., Modupalli, N., & Rawson, A. (2023). Assessment of storage stability and quality characteristics of thermo-sonication assisted blended bitter gourd seed oil and sunflower oil. *Journal of Food Process Engineering*, *46*(6). https://doi.org/10.1111/jfpe.14070

Osae, R., Zhou, C., Xu, B., Tchabo, W., Tahir, H. E., Mustapha, A. T., & Ma, H. (2019). Effects of ultrasound, osmotic dehydration, and osmosonication pretreatments on bioactive compounds, chemical characterization, enzyme inactivation, color, and antioxidant activity of dried ginger slices. *Journal of Food Biochemistry*, *43*(5), 1–14. https://doi.org/10.1111/jfbc.12832

Pokhrel, N., Vabbina, P. K., & Pala, N. (2016). Sonochemistry: Science and engineering. *Ultrasonics Sonochemistry*, *29*(July), 104–128. https://doi.org/10.1016/j.ultsonch.2015.07.023

Rahman, M. M., & Lamsal, B. P. (2021). Ultrasound-assisted extraction and modification of plant-based proteins: Impact on physicochemical, functional, and nutritional properties. *Comprehensive Reviews in Food Science and Food Safety*, *20*(2), 1457–1480. https://doi.org/10.1111/1541-4337.12709

Raj, G. V. S. B., & Dash, K. K. (2020). Ultrasound-assisted extraction of phytocompounds from dragon fruit peel: Optimization, kinetics and thermodynamic studies. *Ultrasonics Sonochemistry*, 105180. https://doi.org/10.1016/j.ultsonch.2020.105180

Rutkowska, M., Namieśnik, J., & Konieczka, P. (2017). Ultrasound-assisted extraction. *The Application of Green Solvents in Separation Processes*, 301–324. https://doi.org/10.1016/B978-0-12-805297-6.00010-3

Sengar, A. S., Rawson, A., Muthiah, M., & Kalakandan, S. K. (2020). Comparison of different ultrasound assisted extraction techniques for pectin from tomato processing waste. *Ultrasonics Sonochemistry*, *61*, 104812. https://doi.org/10.1016/j.ultsonch.2019.104812

Sengar, A. S., Thirunavookarasu, N., Choudhary, P., Naik, M., Surekha, A., Sunil, C. K., & Rawson, A. (2022). Application of power ultrasound for plant protein extraction, modification and allergen reduction – A review. *Applied Food Research*, *2*(2), 100219. https://doi.org/10.1016/j.afres.2022.100219

Shanmugam, A., & Ashokkumar, M. (2017). Ultrasonic preparation of food emulsions. *Ultrasound in Food Processing*, 287–310. https://doi.org/10.1002/9781118964156.ch10

Sharifi, M., Goli, S. A. H., & Fayaz, G. (2019). Exploitation of high-intensity ultrasound to modify the structure of olive oil organogel containing propolis wax. *International Journal of Food Science and Technology*, *54*(2), 509–515. https://doi.org/10.1111/ijfs.13965

Singla, M., & Sit, N. (2021). Application of ultrasound in combination with other technologies in food processing: A review. *Ultrasonics Sonochemistry*, *73*, 105506. https://doi.org/10.1016/j.ultsonch.2021.105506

Singla, M., & Sit, N. (2022). *Raising challenges of ultrasound-assisted processes and sonochemistry in industrial applications based on energy efficiency* (pp. 349–373). https://doi.org/10.1016/B978-0-323-91937-1.00017-7

Suchintita Das, R., Tiwari, B. K., Chemat, F., & Garcia-Vaquero, M. (2022). Impact of ultrasound processing on alternative protein systems: Protein extraction, nutritional effects and associated challenges. *Ultrasonics Sonochemistry*, *91*(November), 106234. https://doi.org/10.1016/j.ultsonch.2022.106234

Sun, D. W. (2014). Emerging technologies for food processing. *Emerging Technologies for Food Processing*, 1–635. https://doi.org/10.1016/B978-0-12-411479-1.01001-9

Szadzińska, J., Mierzwa, D., Pawłowski, A., Musielak, G., Pashminehazar, R., & Kharaghani, A. (2020). Ultrasound- and microwave-assisted intermittent drying of red beetroot. *Drying Technology*, *38*(1–2), 93–107. https://doi.org/10.1080/07373937.2019.1624565

Taha, A., Ahmed, E., Ismaiel, A., Ashokkumar, M., Xu, X., Pan, S., & Hu, H. (2020). Ultrasonic emulsification: An overview on the preparation of different emulsifiers-stabilized emulsions. In *Trends in Food Science and Technology* (Vol. 105, pp. 363–377). Elsevier Ltd. https://doi.org/10.1016/j.tifs.2020.09.024

Taha, A., Mehany, T., Pandiselvam, R., Anusha Siddiqui, S., Mir, N. A., Malik, M. A., Sujayasree, O. J., Alamuru, K. C., Khanashyam, A. C., Casanova, F., Xu, X., Pan, S., & Hu, H. (2022). Sonoprocessing: Mechanisms and recent applications of power ultrasound in food. *Critical Reviews in Food Science and Nutrition*, 1–39. https://doi.org/10.1080/10408398.2022.2161464

Tao, Y., Wang, P., Wang, Y., Kadam, S. U., Han, Y., Wang, J., & Zhou, J. (2016). Power ultrasound as a pretreatment to convective drying of mulberry (*Morus alba* L.) leaves: Impact on drying kinetics and selected quality properties. *Ultrasonics Sonochemistry*, *31*, 310–318. https://doi.org/10.1016/j.ultsonch.2016.01.012

Tao, Y., Zhang, J., Jiang, S., Xu, Y., Show, P. L., Han, Y., Ye, X., & Ye, M. (2018). Contacting ultrasound enhanced hot-air convective drying of garlic slices: Mass transfer modeling and quality evaluation. *Journal of Food Engineering*, *235*, 79–88. https://doi.org/10.1016/j.jfoodeng.2018.04.028

Thangaraju, S., Modupalli, N., Naik, M., Rawson, A., & Natarajan, V. (2023). Changes in physicochemical characteristics of rice bran oil during mechanical-stirring and ultrasonic-assisted enzymatic degumming. *Journal of Food Process Engineering*, *46*(6). https://doi.org/10.1111/jfpe.14123

Tiwari, B. K. (2015). Ultrasound: A clean, green extraction technology. *TrAC – Trends in Analytical Chemistry*, *71*, 100–109. https://doi.org/10.1016/j.trac.2015.04.013

Wang, Jian, & Fan, L. (2019). Effect of ultrasound treatment on microbial inhibition and quality maintenance of green asparagus during cold storage. *Ultrasonics Sonochemistry*, *58*, 104631. https://doi.org/10.1016/j.ultsonch.2019.104631

Wang, D., Hou, F., Ma, X., Chen, W., Yan, L., Ding, T., Ye, X., & Liu, D. (2020). Study on the mechanism of ultrasound-accelerated enzymatic hydrolysis of starch: Analysis of ultrasound effect on different objects. *International Journal of Biological Macromolecules*, *148*, 493–500. https://doi.org/10.1016/j.ijbiomac.2020.01.064

Wang, Jingyi, Liu, Q., Xie, B., & Sun, Z. (2020). Effect of ultrasound combined with ultraviolet treatment on microbial inactivation and quality properties of mango juice. *Ultrasonics Sonochemistry*, *64*, 105000. https://doi.org/10.1016/j.ultsonch.2020.105000

Wang, L., & Weller, C. L. (2006). Recent advances in extraction of nutraceuticals from plants. *Trends in Food Science and Technology, 17*(6), 300–312. https://doi.org/10.1016/j.tifs.2005.12.004

Wen, C., Zhang, J., Zhang, H., Dzah, C. S., Zandile, M., Duan, Y., Ma, H., & Luo, X. (2018). Advances in ultrasound assisted extraction of bioactive compounds from cash crops – A review. *Ultrasonics Sonochemistry, 48*, 538–549. https://doi.org/10.1016/j.ultsonch.2018.07.018

Wiktor, A., Dadan, M., Nowacka, M., Rybak, K., & Witrowa-Rajchert, D. (2019). The impact of combination of pulsed electric field and ultrasound treatment on air drying kinetics and quality of carrot tissue. *LWT, 110*, 71–79. https://doi.org/10.1016/j.lwt.2019.04.060

Wu, Zhiqian, Li, X., Zeng, Y., Cai, D., Teng, Z., Wu, Q., Sun, J., & Bai, W. (2022). Color stability enhancement and antioxidation improvement of Sanhua plum wine under circulating ultrasound. *Foods, 11*(16). https://doi.org/10.3390/foods11162435

Wu, Zhuoting, Qiao, D., Zhao, S., Lin, Q., Zhang, B., & Xie, F. (2022). Nonthermal physical modification of starch: An overview of recent research into structure and property alterations. *International Journal of Biological Macromolecules, 203*(November 2021), 153–175. https://doi.org/10.1016/j.ijbiomac.2022.01.103

Xia, Q., Tao, H., Li, Y., Pan, D., Cao, J., & Liu, L. (2019). Characterizing physicochemical, nutritional and quality attributes of wholegrain *Oryza sativa* L. subjected to high intensity ultrasound-stimulated pre-germination. *Food Control, 106827*. https://doi.org/10.1016/j.foodcont.2019.106827

Xue, Z., Wang, T., & Zhang, Q. (2022). Effects of ultrasound irradiation on the co-pigmentation of the added caffeic acid and the coloration of the cabernet sauvignon wine during storage. *Foods, 11*(9). https://doi.org/10.3390/foods11091206

Yan, S., Xu, J., Zhang, S., & Li, Y. (2021). Effects of flexibility and surface hydrophobicity on emulsifying properties: Ultrasound-treated soybean protein isolate. *LWT, 142*(December 2020), 110881. https://doi.org/10.1016/j.lwt.2021.110881

Yao, Y., Pan, Y., & Liu, S. (2020). Power ultrasound and its applications: A state-of-the-art review. *Ultrasonics Sonochemistry, 62*, 104722. https://doi.org/10.1016/j.ultsonch.2019.104722

Yildiz, S., Pokhrel, P. R., Unluturk, S., & Barbosa-Cánovas, G. V. (2019). Identification of equivalent processing conditions for pasteurization of strawberry juice by high pressure, ultrasound, and pulsed electric fields processing. In *Innovative Food Science and Emerging Technologies* (Vol. 57). Elsevier Ltd. https://doi.org/10.1016/j.ifset.2019.102195

Zhai, X., Wang, X., Wang, X., Zhang, H., Ji, Y., Ren, D., & Lu, J. (2021). An efficient method using ultrasound to accelerate aging in crabapple (*Malus asiatica*) vinegar produced from fresh fruit and its influencing mechanism investigation. *Ultrasonics Sonochemistry, 72*, 105464. https://doi.org/10.1016/j.ultsonch.2021.105464

Zhang, H., Mo, W., Liao, S., Jia, Z., Zhang, W., Zhang, S., & Liu, Z. (2023). Ultrasonics sonochemistry ultrasound promotes germination of aging Pinus tabuliformis seeds is associated with altered lipid metabolism. *Ultrasonics Sonochemistry, 93*(October 2022), 106310. https://doi.org/10.1016/j.ultsonch.2023.106310

Zhang, Y., Qiu, Q., Xu, Y., Zhu, J., Yuan, M., & Chen, M. (2023). Fast aging technology of novel kiwifruit wine and dynamic changes of aroma components during storage. *Food Science and Technology (Brazil), 43*, 1–15. https://doi.org/10.1590/fst.98422

Zhang, S., Tian, L., Yi, J., Zhu, Z., Dong, X., & Decker, E. A. (2021). Impact of high-intensity ultrasound on the chemical and physical stability of oil-in-water emulsions stabilized by almond protein isolate. *LWT, 149*(June), 111972. https://doi.org/10.1016/j.lwt.2021.111972

Zhang, W., Yu, Y., Xie, F., Gu, X., Wu, J., & Wang, Z. (2019). High pressure homogenization versus ultrasound treatment of tomato juice: Effects on stability and in vitro bioaccessibility of carotenoids. *LWT, 116*(800), 108597. https://doi.org/10.1016/j.lwt.2019.108597

Zhou, L., Zhang, W., Wang, J., Zhang, R., & Zhang, J. (2022). Comparison of oil-in-water emulsions prepared by ultrasound, high-pressure homogenization and high-speed homogenization. *Ultrasonics Sonochemistry, 82*, 105885. https://doi.org/10.1016/j.ultsonch.2021.105885

11 Ultrasound – An Emulsification Tool in Food Processing

Addanki Mounika, Prakyath Shetty,
S. Akalya, and Ashish Rawson

11.1 INTRODUCTION

11.1.1 EMULSION, EMULSIFIERS, ULTRASOUND

An emulsion is a mixture of two liquids, one of which behaves as a continuous phase and the other as a scattered phase; the two liquids are immiscible. A liquid medium is used in the ongoing phase (Zychowski, 2015). The use and manufacturing of these emulsions are required in the food business and other industries such as cosmetics, detergents, medicines, agriculture, petroleum, and so on. The emulsions' standard size ranges from 0.2 to 100 μm. Usually, oil and water are immiscible liquids, and to attain stability emulsifiers have been used for decades; some of the traditional emulsifiers are calcium carbonate, carrageenan, xanthan gum, and casein. Emulsions are classified according to their size, such as nanoemulsions, microemulsions, Pickering emulsions, and macroemulsions. They are excellent for numerous applications because of their stability, availability, and changes in rheological parameters such as yield stress, viscosity, and storage modulus (Karakuş, 2018). To produce or create various forms of emulsions, homogenization or combining of two immiscible liquids with the aid of an emulsifier to achieve stability aids in blending two beverages without losing their composure and helps to sustain strength for a more extended period. Some techniques used to prepare emulsions are high-shear homogenizer, piston homogenizer, high-pressure processing, microfluidization, colloidal mills, high-intensity ultrasound, rotary stator, and high-pressure homogenization (Kale & Deore, 2016).

Ultrasonic (US) is one such unique non-thermal technology works on the principle of acoustic cavitation, in which ultrasound waves create cavitation bubbles with a frequency of 20 kHz. These waves travel through the liquid and grow until they explode at 5000°C. Thus, it causes the cells to burst, breaking the cell walls and releasing many bioactive compounds or changes in the properties of emulsions; there are two types of ultrasound frequency: one is high intensity. Another one is low intensity (Zhou et al., 2021). In a study, the preparation of soy protein isolates (SPI) – pectin emulsion was stabilized by high-intensity ultrasound. Soybean protein isolates and complex pectin mixture were mixed with soybean oil by homogenizer and were stabilized using high-intensity ultrasound. At the power of 450 Watts, which showed the best emulsion stability index of 25.18 ± 1.24 minutes and particle size of 559.82 ± 3.17 nm (Wang et al., 2022).

Emulsifiers are the materials that will absorb the oil-water interface and lower the interfacial tension, and help maintain the stability of the oil droplets without undergoing flocculation, phase inversion, coalescence, or destabilization mechanism after homogenization or mixing. Emulsifiers such as protein, polysaccharides, xanthan gum, and sodium alginate help maintain the stability around the oil droplets like a membrane because of their binding properties such as amphiphilic and resistance to change in ionic strength and temperature thus, prevent from collapsing (Taha et al., 2020). Different types of emulsions using ultrasound, emulsifier, and ratios for preparing emulsions are given in Table 11.1.

DOI: 10.1201/9781003359302-11

TABLE 11.1

Different Types of Emulsions Compositions by Using Ultrasound

Type of Emulsion	Technique Used	Emulsifiers or Surfactants and Oils Used	Surfactant:Oil:Water or Percentage or Milliliters	References
Oil-in-water nanoemulsion	Complex titration ultrasound homogenization	CAPB and C.B. (biodegradable amphoteric surfactants), linoleic acid, and oleic acid	3:1:96	(Waglewska & Bazylińska, 2021)
Oil-in-water macroemulsion	Ultrasound	Propylene-glycol and water; sucrose and Habo Monoester P90	2.50 to 2.75:1; 18.50 to 20.00:1; 0.8 g	(Espinosa-álvarez et al., 2019)
Microemulsion	Ultrasound-assisted extraction	Tween 20, 80, soy lecithin, deionized water	1:50; 2:1; water: 40 mL	(Jalali-Jivan et al., 2019)
Microemulsion	Ultrasound	Nickel solution, molybdenum solution	0.36–0.38	(Ahmadi Khoshooei et al., 2021)
Oil-water emulsions, Pickering emulsion	Ultrasound and high-pressure homogenization	Deionized water and NaCl/ NaF stock solutions	20 wt % oil, 0.4 wt % of NaCl/NaF	(Benetti et al., 2019)
Pickering emulsions	Ultrasonic homogenizer	Corn oil, quinoa, water	0.2 and 0.7 corn oil; 0.25–6% w/v of oil and water	(Qin et al., 2018)
Water-in oil	Ultrasonic treatment	Silicone oil, water, TX-100	Silicone and water – 100 mL, TX-100 5–25%	(Luo et al., 2019)
Water-in-oil nanoemulsion	Ultrasonic standing waves	Palm oil, NaCl, water(10 wt%), PGPR, MCT, lecithin	Palm oil – 61.25 wt% MCT – 26.25, PGPR – 2.5 wt% and lecithin to PGPR, 2:1	(Raviadaran et al., 2019)

11.2 PRINCIPLE, TYPES OF EMULSION, AND MECHANISM OF EMULSIFICATION

Emulsification is mixing two or more non-diluted liquids with food surfactants to gain stability (Schroën et al., 2020). Of those two liquids, one is oil, and another is water. The emulsion consists of two or more liquid phases dispersed into one another because of their immiscible nature.

11.2.1 TYPES OF EMULSIONS

These emulsions are derived into four types. Oil-in-water (O/W), water-in-oil (W/O), water-in-oil-in-water (W/O/W), and oil-in-water-in-oil (O/W/O). The word emulsion is derived from emulgeo, which means "to milk." Each type of emulsion has two phases: one is a continuous phase, and another one is a dispersed phase. Generally, in oil-in-water emulsions, the dispersed phase is oil, and the continuous phase is water. These oil droplets are embedded inside the water and are represented in Figure 11.1(a), and vice-versa for the water-in-oil emulsion in Figure 11.1(b) (Akbari & Nour, 2018).

In a water-in-oil-water emulsion, the water is trapped inside the oil droplets and suspended in the continuous phase water, and vice versa for the oil-in-water-in-oil. These are also called multiple emulsions. The general diagram for multiple emulsions is presented in Figure 11.1(c) (Chang & Nickerson, 2018).

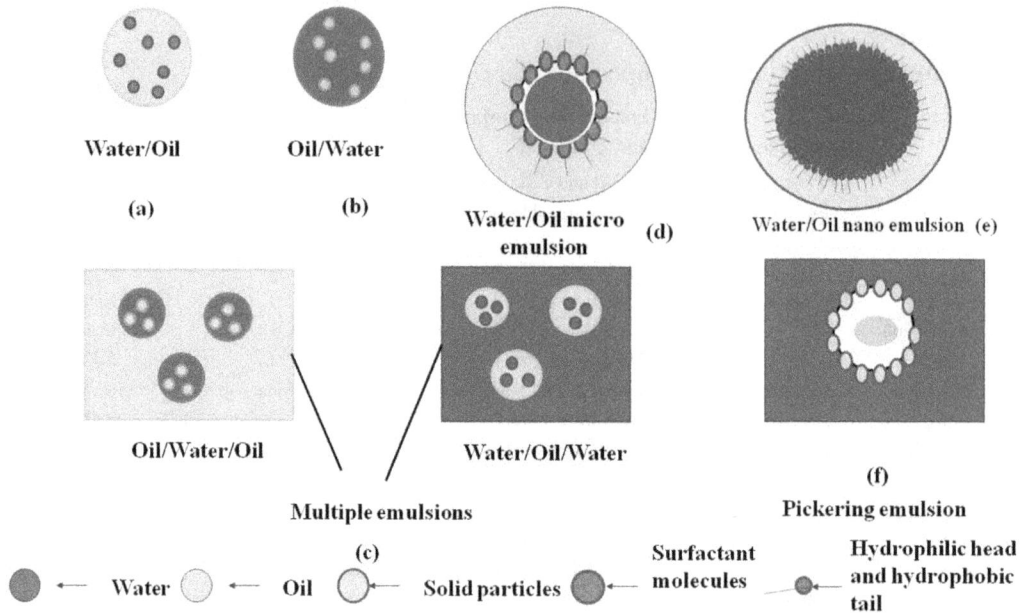

FIGURE 11.1 Various types of emulsions and their appearance. (Adapted and modified from Aswathanarayan & Vittal, 2019.)

11.2.2 MICROEMULSIONS

Schulman and Hoar (1943) described microemulsions more than 75 years ago; it is the mixing of oil, water, and surfactant by titrating milky o/w emulsion. The word "microemulsions" was invented by Schulman and colleagues. According to Danielsson and Lindman, a microemulsion is a thermodynamically stable liquid solution composed of water, oil, and amphiphile. Microemulsion has a diameter range of 100–600 with a size spectrum of 10–300 nm, represented as shown in Figure 11.1(d) (Roohinejad et al., 2018).

The thermodynamic theory for microemulsion generation is based on the equation given below:

$$\Delta G_f = \gamma \Delta A - T \Delta S$$

Where ΔG_f is the free energy, $\gamma \Delta A$ is the interfacial tension of oil and water and change in the interfacial area, ΔS is the entropy, and T is the temperature. Based on the free energy mixing, the droplet radius generally depends on the free energy. However, in the microemulsion, it is negative; this happens when there is more surface area reduction.

The preparation of microemulsions can be done with the help of high-intensity ultrasound using essential oil. The droplet size diameters of the microemulsions ranged from 1.98 to 5.46 µm, with high encapsulation efficiency (79.91–81.97%) and low separation rates (2.50–6.67%). When compared to non-encapsulated essential oils, there is a 20–75% reduction in minimum inhibitory concentration (MIC) and minimum bacterial concentration (MBC) for both bacteria *Listeria monocytogenes* and *Escherichia coli* (9%). Essential oils used in this study are cinnamon, oregano, and rosemary at concentrations of 10, 45, and 30%, respectively. Microemulsions were prepared by using 5% of inulin and 3% Tween 80 (w/w). The ultrasound parameters are ultrasound wave amplitude of 84% for 15 minutes for rosemary and cinnamon essential oils and 30 minutes for oregano essential oil (Dávila-Rodríguez et al., 2020).

11.2.3 Nanoemulsions

Nano emulsions are a transparent system with droplets ranging in size from 20 to 500 nm; the Brownian motion of the oil/water droplets is responsible for the nanoemulsions' stability. Nanosized particles overcome gravitational forces in nano emulsions, rendering them kinetically stable and resistant to particle aggregation. Brownian motion surpasses the emulsions' gravitational forces to establish the kinetic stability of the nano emulsions, resulting in the termination of particle accumulation (Ashaolu, 2021). These nanoemulsions are more miniature than microemulsions and have a larger surface area, as shown in Figure 11.1(e). In nanoemulsion, destabilization occurs because of the Ostwald ripening process. High-pressure homogenization, ultrasonication, and microfluidizers are examples of high-energy emulsification techniques (Mungure et al., 2018).

In a study, the preparation of thyme oil nanoemulsion with the help of natural emulsifiers such as sodium caseinate and soybean lecithin was used as an antimicrobial emulsifier to prevent microorganisms (*E. coli, L. monocytogenes,* and *Salmonella enteritidis*), in the preparation of thyme oil-based nanoemulsions. These nanoemulsions are tiny droplets produced by high-shear mixing. A certain ratio of gum Arabic to lecithin in preparing antimicrobial eugenol nanoemulsion acts effectively against *L. monocytogenes* (McClements & Jafari, 2018).

11.2.4 Pickering Emulsions

Pickering emulsions can be stabilized with the help of solid particles (micro or nanoparticles) as an emulsifier and contain good stability due to the higher coalescence resistance; Ramsden first discovered it in 1903. As soon as the usage of solid particles as an emulsifier, these emulsions prevent the formation of the coagulants of oil droplets when they are attached to adsorbed to the oil-water interface to form a film as represented in Figure 11.1(f) (Albert et al., 2019). To use solid particles as an emulsifier, these should contain the degree of wettability; particle size should be smaller than the emulsion droplet; these two points, along with the concentration, affect the stability of the Pickering emulsions. Pickering emulsions are more advisable than conventional or macro emulsions (Chen et al., 2020). Some solid particles for the Pickering emulsion stability are rice starch, Protein polysaccharide complex, Janus particles, and colloidal particles.

11.2.5 Mechanisms of Destabilization

There are different mechanisms for the de-emulsification process or breakdown of droplets: creaming and sedimentation, flocculation, phase inversion, coalescence, and Ostwald ripening. In the creaming and sedimentation mechanism, the droplets. Flocculation happens when droplets attract each other, which results in the development of loosely clumped particles. Ostwald ripening process aims for the mass diffusion from more minor to larger droplets due to the inside pressure difference between the tiny and larger droplets. In step inversion, the dispersed phase becomes the ongoing phase, and the continuous phase becomes the dispersed phase. There are two forms of phase inversion in emulsions: catastrophic and transitional. In emulsions, the destructive phase inversion is caused by modifying the oil-to-water ratio (Bains & Pal, 2020; Grammatikopoulos et al., 2019; Costa et al., 2019).

All these mechanisms are represented in Figure 11.2. These are unstable individually and reversible, but when combined and result in the formation of droplet coalescence, which is the irreversible stage of two droplets fusing due to the rupture of the regulating layer, which causes the formation of distinct oil and water phases.

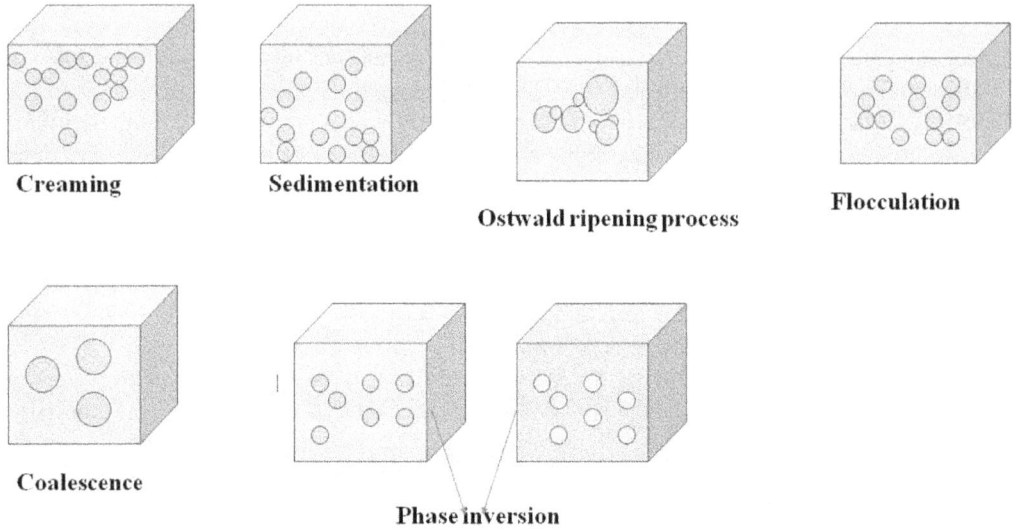

FIGURE 11.2   Destabilization mechanism of emulsions. (Adapted and modified Piacentini, 2014.)

11.2.6   FACTORS AFFECTING EMULSIFICATION

Several factors affect the stability of emulsions; it is essential to know the strength of the emulsion to avoid unstable emulsification. According to Shao et al. (2020), the two main factors affecting stability are steric and electrostatic interactions, which are closely related to the interfacial properties.

11.2.6.1   Droplet Thickness

When an emulsifier adsorbs the oil-water interface of the droplets to produce a stable emulsion, the interfacial layer thickness varies, which has a more significant impact on the strength of the steric interactions between the droplets. The droplet's thickness is insufficient, resulting in instability (Shao et al., 2020).

11.2.6.2   Electrostatic Interactions

In this interfacial strength plays a significant role in the stability of emulsions; when interfacial availability of droplets is poor, the potency of the electrostatic repulsion's weakness due to the change in the variations results in the instability of the emulsions and the mechanisms involved are flocculation and coalescence, and the reason is the Van der Waals interactions (Shao et al., 2020).

11.2.6.3   Depletion of Surfactant Solution

This phenomenon is due to the high increase of the demanded interfacial area correlated to the fragmentation of the dispersed phase; the resulting surfactant adsorption may result in severe matrix phase impoverishment, limiting the adsorption process and the accomplishment of the high coverage degree required to stabilize droplets against coalescence (Ravera et al., 2021).

11.3   EMULSION STABILIZATION

Stability can be achieved by an emulsifier or stabilizer once the emulsion is produced. Usually, these emulsifiers are high molecular weight polymers capable of maintaining the oil-water interface without destroying the droplets from destabilization mechanisms. Similarly, stabilizers are low molecular weight, and fast diffusing molecules contain rapid adsorption on newly formed interfaces, thus arrest the droplet movement (Costa et al., 2019).

11.3.1 Stabilization by Biomolecules

Synthetic, animal-based surfactants are commonly used biomolecules to stabilize the emulsions, but natural biomolecules from plants, such as phospholipids, glycosides, proteins, and polysaccharides, are replacing and are presently in demand to reduce the usage or replace the animal based or synthetic biomolecules because of their low toxicity, high selectivity, necessary activity at high temperature and pH for better sustainability (Costa et al., 2019).

11.3.2 Lecithin

Lecithin contains various phospholipids and anionics with charges depending on pH. The best example of a natural emulsion to produce oil-water emulsion is sunflower lecithin, with the droplet sizes recorded from 30 to 160 μm. Lecithin prevents the aggregation under neutral acidity circumstances, this is because the negatively charged contact exhibits substantial electrostatic repulsion. Recently, lysolecithin has become widely employed in the food business for nutritious (Costa et al., 2019).

11.3.3 Proteins

Greater stability can be achieved by the protein-protein intermolecular interactions from viscoelasticity around the droplets; the primary mechanism to maintain the strength of the emulsions for the droplets is electrostatic repulsion to prevent flocculation; this happens when the proteins are more exposed to the hydrophobic groups toward oil phase. Proteins with high molecular weights and elongated structures have higher steric repulsion and better resist aggregation. Because of intermolecular cross-linking, globular proteins tend to form thicker and viscoelastic gel-like surfaces (Costa et al., 2019).

11.3.4 Cellulose

Getting cellulose in sufficient amounts without high prices can be done, which can be used in the food industry as an alternative to synthetic cellulose. It is an amphiphilic polymer that adsorbs on the oil-water interface and prevents coalescence formation because it is a good stabilizer. The Pickering mechanism for cellulose is a self-assembly mechanism in which the solid particles self-assemble themselves at oil-water interfaces and forms a protective layer around the droplets (Costa et al., 2019).

11.4 COMMON AND COMMERCIAL EMULSIFICATION TOOLS

11.4.1 High-Pressure Homogenizer

High-pressure homogenizers (Figure 11.3(a)) are the current most popular technique for creating fine emulsions in the food industry. This system primarily utilizes a high-pressure pump and valves to produce homogenized products. A high-pressure pump applies high pressure to the fluid system, and a coarse emulsion is forced through a narrow homogenizer valve, where the oil droplets are disrupted by disruptive forces, resulting in fine emulsions with droplet sizes of up to 0.1 μm (Modarres-Gheisari et al., 2019). Mainly, two mechanisms impact the particle size breakdown of oil droplets. Firstly, droplet breakdown is caused by a combination of shear force generated by food matrix passing through the valve opening, cavitation forces caused by pressure drop when passing through the valve, and turbulent flow generated by the passage of food matrix through valve opening. Secondly, a collision between droplets causes coalescence and droplet expansion (Perrin et al., 2022; Yadav & Kale, 2020). Recent design advances, such as intensifiers, homogenization chamber

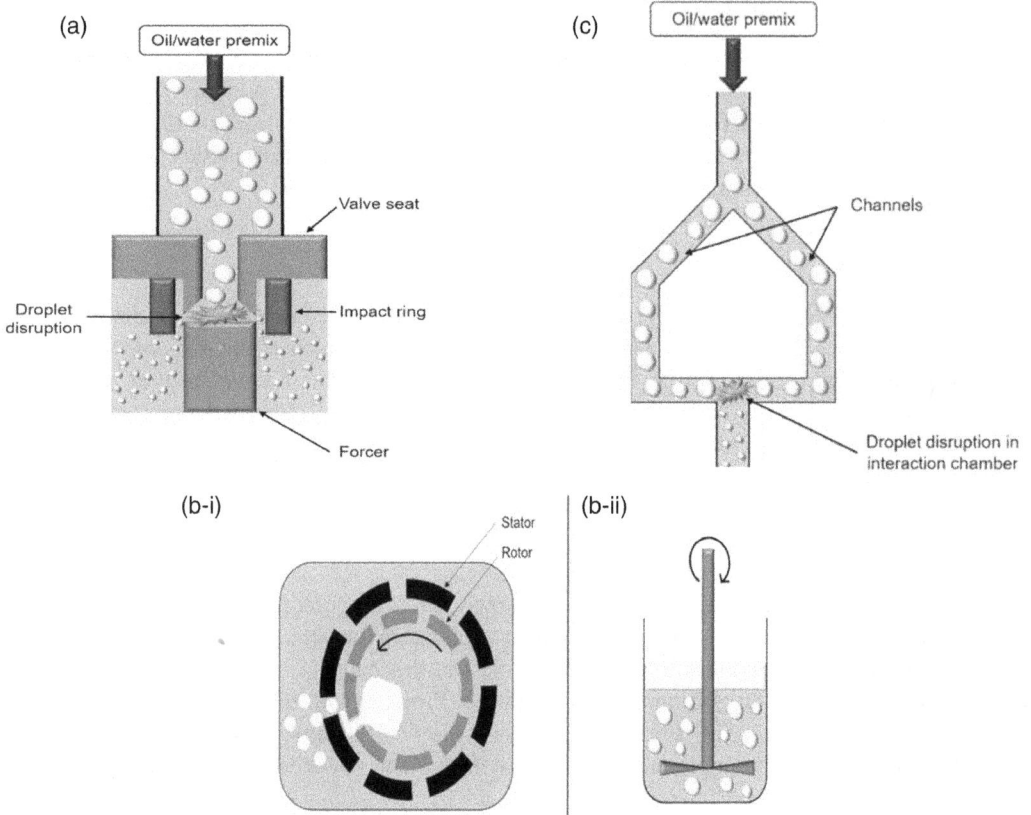

FIGURE 11.3 (a) High-pressure homogenizer. (Reproduced from Perrin et al., 2022.) (b) High-shear homogenizers system (i) rotor stator system (ii) disc system. (Reproduced from Perrin et al., 2022.) (c) Schematic representation of microfluidizer. (Reproduced from Perrin et al., 2022.)

geometries, and the development of new homogenization valves with ceramic seats and needles, as well as, more recently, diamond coating, allow for operating pressure levels of up to 400 MPa (Levy et al., 2021). Although it is the most commonly used commercial technique, it is energy and cost intensive, and cleaning the homogenizer is time consuming and tedious, as it requires the use of delicate mechanical parts. It is also only suitable for medium and low viscosity emulsions and cannot be used to treat shear sensitive compounds.

11.4.2 High-Shear Homogenizers

High-shear homogenizers, also known as rotor/stator mixers, are commercially used to produce macroemulsions, pre-emulsions, or premix prior to using other emulsification methods. The mixing head consists of specially designed impeller rotating at a wide range of speeds starting from 6000 rpm to 25000 rpm inside a stationary ring with slots. Food matrix is drawn into the rotating part and expelled through the slots, resulting in extremely high shear in a small volume (Shanmugam & Ashokkumar, 2017). High-shear mixing is a technique that involves rapidly rotating a mixing head (Figure 11.3(b)) to disrupt the interface between oil and water phases, break down larger droplets into smaller ones, and distribute the liquids evenly throughout the vessel with the help of baffles. Different mixing head designs, such as blades, propellers, and turbines, are available for various applications. The strong centrifugal force created by the rotor as it spins quickly creates a vacuum effect, drawing food matrix into the processing chamber. The food matrix is subjected to a variety of mechanical forces within this chamber, including shearing, compression, impact, and turbulence, which combine to break it down into smaller droplets. The pressure rises again after the liquid exits the stator holes, causing a cavitation effect, as shown in Figure 11.3(b). This technique is used mainly to produce coarse emulsions and results in good product stability and the homogenization effects of these devices can be improved by increasing the rotation speed and the homogenization time (McClements, 2016).

11.4.3 MICROFLUIDIZER

Microfluidizer is a high energy emulsification method that produces very small droplets less than 0.1 μm. A microfluidizer is made up of a food matrix inlet, a high-pressure pump, and an interaction chamber with a Y-type, as shown in Figure 11.3(c) or Z-type narrow channel through which food matrix flow and interact with one another. A coarse pre-emulsion is fed into the system, then a high-pressure pump accelerates the premix to high velocity within channels, sometimes water and oil phases are fed in separate channels depending on channel design, then it is led to collide on each other on a solid surface inside the interaction chamber and disruption of larger droplets caused by intense disruptive forces generated when the two fluid streams collide (McClements, 2016). Commercially available microfluidizers have capacities ranging from 10 mL to 12,000 L h^{-1} and can operate at pressures of up to 270 MPa, producing very fine droplets. By raising the homogenization pressure, the droplet size can be decreased, number of passes, and emulsifier concentration. This method is considered the most efficient high energy method for producing very small droplet size, and it is best suited for low and intermediate viscosity fluids.

11.4.4 MEMBRANE AND MICROCHANNEL HOMOGENIZERS

Membrane homogenizers consist of a membrane which is an essential part of the system where disruption of droplet takes place. Membranes are made up of various organic and inorganic materials and porous media such as glass and ceramic, polymeric membranes, and metallic membranes. The type of membrane used depends on the type of emulsion, with hydrophilic membranes used in O/W emulsions and hydrophobic membranes used in W/O emulsions to avoid membrane wetting by dispersed phase. The coarse emulsion can be pressurized and passed through membrane having a small pore size up to 1μm to break larger droplets. The emulsification can be done in two ways, i.e. direct homogenization is the process of forming an emulsion directly by passing oil phase through a membrane to water phase to reduce oil droplet size using suitable emulsifier and premix homogenization, it involves the preparation of coarse emulsion by using magnetic bar or impeller in low speed (400–600 rpm) then passed through membrane to obtain finer emulsion. Membrane homogenizers are available in batch and continuous type. In the batch type, droplets are formed by passing the dispersed phase through a cylindrical membrane dipped in a vessel containing the continuous phase. In the continuous type, the homogenizer consists of a cylindrical membrane that the continuous phase flows through, and it is located within a tube that the dispersed phase flows through, as shown in Figure 11.4(a). The pressurized dispersed phase is forced through the membrane, forming small droplets in the continuous phase (McClements, 2016; Nazir & Vladisavljević, 2021). Microchannel/ Microfluidic homogenizers consist of a microchannel with well-defined geometry to break down droplets. It will work similarly to membrane homogenization, but instead of the membrane, the droplets are formed from hundreds of microchannels that are fabricated by photolithography. Microstructured devices provide better control over the droplet generation process and require less energy than conventional techniques. However, because of the low dispersed phase volumes, which are typically less than 1 mL/h for droplet sizes smaller than 20 μm, and the high cost of microfabrication processes, their use is primarily limited to advanced applications (Nazir & Vladisavljević, 2021; Vladisavljević et al., 2013).

(a)

(b)

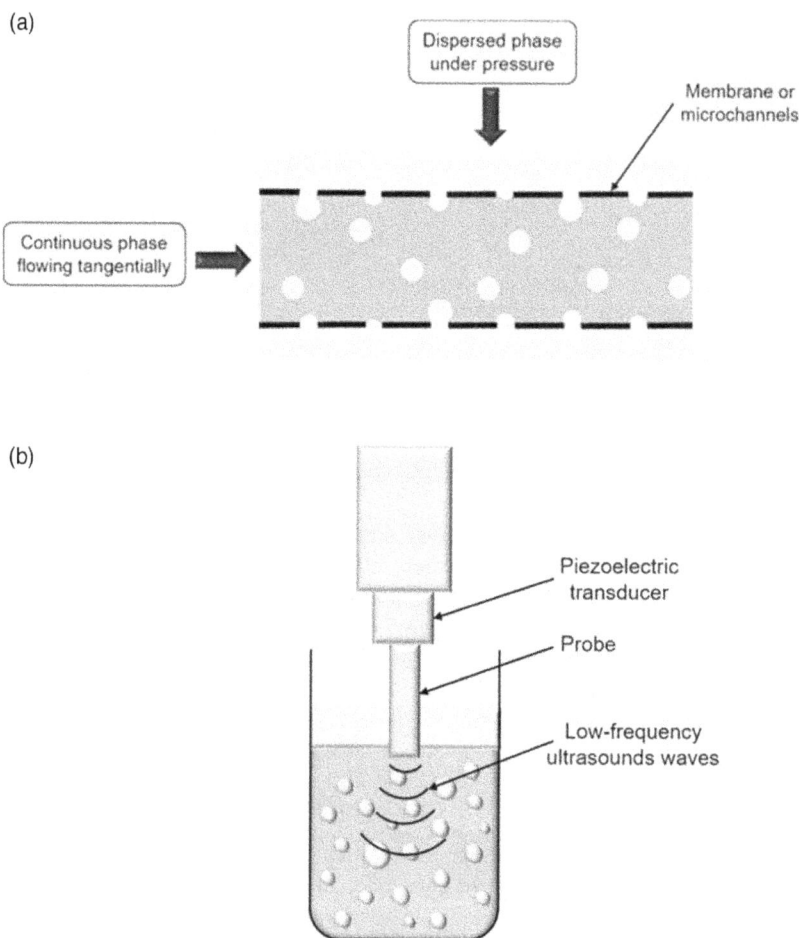

FIGURE 11.4 (a) Schematic representation of membrane and microchannel homogenizers; (b) schematic representation of probe type ultrasound homogenizer. (Reproduced from Perrin et al., 2022.)

11.5 EMULSIFICATION BY ULTRASOUND

Ultrasound homogenizers are a popular green technology for producing emulsions and nanoemulsions which comes under the high energy method of emulsification. Ultrasound is a mechanical wave that has a frequency greater than 20 kHz and typically classified into two categories, i.e. high-energy low-frequency (>1 W cm^{-2}, 20–100 kHz) and low-energy high-frequency (<1 W cm^{-2}, 100 kHz–1 MHz) (Bhangu & Ashokkumar, 2016).

11.5.1 EQUIPMENT DESIGN AND OPERATION

Generally, ultrasound homogenizer (Figure 11.4(b)) consists of ultrasound processor which has ultrasound generator and transducer, also different size probes which can be attached to the transducer to transfer generated ultrasound to the product. The ultrasonic generator receives energy from the power source and converts it to the appropriate voltage, frequency, and amperage. Then it is connected to the ultrasonic transducer which is a device that converts electrical signals into ultrasonic waves. There are a number of methods available to generate ultrasound waves, but most commonly used is piezoelectric transducer.

A piezoelectric transducer consists of one or more piezoelectric discs/crystals sandwiched between two metal sections in a protective casing. The entire assembly is tuned to a specific frequency. An electrical wave from generator applied to the transducer which causes the piezoelectric disc/crystal inside it to rapidly oscillate and generate an ultrasonic wave. The generated wave passes through probe to reach food matrix (Yao et al., 2020; McClements, 2016). The most significant effect of ultrasound on emulsions is acoustic cavitation produced by ultrasound. Cavitation generates physical forces primarily as a result of the violent burst of the cavitation bubble, which causes shock waves, shear forces, high pressures, high speed microjets, and turbulence. Emulsions primarily form through three mechanisms: the breakdown of oil droplets, transportation of emulsifier to the surface of oil droplets for absorption, and the merging and colloid of emulsifier-coated droplets. Also, cavitation can generate free radicals, which can oxidize the oil phase and natural emulsifiers, as well as heat, which can affect the protein, polysaccharides, and oil in the food matrix. So optimization of ultrasonic parameters are very important to achieve desire result (Zhou, Zhang, Xing, et al., 2021).

11.5.2 Factors Affecting Sono-Emulsification

The emulsification procedure may be impacted by the cavitation efficiency during sonication. Many variables, including ultrasound frequency, ultrasonic power, temperature, time, and the medium's surface tension or viscosity, can affect cavitation efficiency.

11.5.2.1 Ultrasound Frequency

Low frequency ultrasound in the range of 20–40 kHz is generally best suited for emulsification. Increasing frequency alters the pressure cycle by causing abrupt pressure changes, creating high internal turbulence that can impede bubble growth and result in subpar cavitation (Pokhrel et al., 2016), and higher frequencies need more power for the cavitational effect due to lower bubble temperatures and insufficient solvent evaporation time. Furthermore, free radical formation, which can be detrimental to emulsion stability, increases with frequency, peaking at 200–600 kHz and then decreasing as frequency increases (Ashokkumar, 2011; Ashokkumar, 2015). The power available within the ultrasound range varies inversely with frequency, and mostly frequency between 16 and 100 kHz (Powerful ultrasound) interacts with matter, causing physical and chemical changes (Kaci et al., 2014).

11.5.2.2 Sonication Power

Sonication power is a major factor that influences emulsification efficiency. Generally, sonication power represented in nominal applied power (NAP), which is electrical power used to operate ultrasound equipment and actual power, i.e. power absorbed by the food matrix which is more appropriate in the emulsion preparation. The actual power can be calculated by calorimetric method and recent equipments will measure the actual power by default. The calorimetric method usually entails taking a predetermined volume of water and treating with the ultrasound while recording the temperature variation over time for a given ultrasound power. The following equation can be used to determine the actual power delivered (Khan et al., 2022).

$$P = mC_p \frac{dT}{t}$$

Where m = the mass of water, g; P = actual power, W; Cp = the specific heat capacity of water, 4.18 $Jg^{-1}{}^{\circ}C^{-1}$; dT = change in temperature, °C; t = treatment time, seconds.

Increased sonication power results in increased sonochemical effects, bubble size, and population. As a result, the collapse of a bubble can generate high-shear force, allowing for more efficient disruption of oil droplets. As a result, the size of the oil droplets will decrease, increasing emulsion stability (Ashokkumar et al., 2016; Shanmugam et al., 2012; Shanmugam & Ashokkumar, 2014a, 2014b, 2015). Many researchers are concluded that increasing ultrasound power will increase emulsion stability and quality. However, higher power can facilitate more cavitational intensity which can adversely affect the food matrix, emulsifiers, and other bioactive components. Ultrasonic power higher than 400 W produced adverse effects on emulsion stability by increasing particle size and decreasing zeta-potential values (Sui et al., 2017). Finding the optimum ultrasonic condition during ultrasound homogenization is critical for achieving a stable and high-quality emulsion. Another parameter that can be used to represent power is energy or power density. It is also useful for comparing various emulsification techniques. The following equation can be used to find energy density (Shanmugam & Ashokkumar, 2017).

$$\text{Energy density } (\text{JmL}^{-1}) = (\text{Actual power delivered, } W) \times \text{Time, s} / \text{Volume, mL}$$

For continuous system, it can be obtained by dividing the measured energy input (Js^{-1}) by the volumetric flow rate (mLs^{-1}). Similar energy density at different amplitude showed significant difference in droplet size. This implies that using higher amplitudes results in the formation of smaller droplets with similar energy densities. In other words, higher amplitudes with shorter residence times are more effective than lower amplitudes with longer residence times (Santos et al., 2019).

11.5.2.3   Temperature

The temperature of the solution will rise during ultrasound emulsification due to various physical and chemical effects. Cavitational activity decreases as temperature rises due to cushioned collapse and the formation of vaporous cavities. The properties of emulsions may be significantly influenced by the elevated temperature and the instantaneous high temperature produced at a local point. The effects of temperature on the food matrix include the unfolding of proteins, the characteristics of polysaccharides, and the oxidation of oils (Zhou, Zhang, Xing, et al., 2021). So, maintaining optimum operation temperature should be followed to avoid unwanted change in food matrix. Some changes can positively affect on the emulsion stability, like increased viscosity of gelatin solution at temperature of more than 30°C can increase stability of emulsion during storage (Zhu et al., 2018).

11.5.2.4   Sonication Time

The duration of sonication has a significant impact on emulsion stability. In general, increasing the sonication time at constant power and frequency increases the applied energy density to the solution, which reduces emulsion droplet size. Many studies indicated that increasing sonication time had a positive effect on emulsion stability (Belgheisi & Motamedzadegan, 2020; Mehmood et al., 2019; Sharma et al., 2021). However, overprocessing can have a negative impact on emulsion stability. Some authors observed that at low amplitudes (<40%), increasing time had an adverse influence on droplet size in nanoemulsions (Espinosa-Andrews & Páez-Hernández, 2020; Páez-Hernández et al., 2019; Sharma et al., 2021). Increasing sonication time up to 24 minutes at lower power (150 W) decreased particle size; however, increasing sonication time up to 24 minutes at high power (300 and 450 W) showed significant increases in particle size, indicating a potential to recoalescence (Sui et al., 2017). So, it is important to correlate with all the parameters to achieve stable emulsion.

11.6   SONO-EMULSIONS OF DIFFERENT FOOD MATRICES

Different food matrices require different combinations of emulsification parameters because the nature of the food matrix can be affected differently by various emulsification parameters. In the section that follows, sonoemulsions of protein/milk, polysaccharides, protein-polysaccharide mixture, and surfactant-based emulsions are all explored.

11.6.1 PROTEIN/MILK-BASED EMULSIONS

Protein molecules can function as emulsifiers to lessen the interfacial tension at the oil-water interface because some amino acid residues are hydrophilic and others are hydrophobic. The development of protein-coated oil droplets after emulsification significantly boosts the emulsion's stability. However, some proteins' limited solubility, high molecular weight, and weak electrostatic repulsions may have an adverse effect on their emulsifying properties. As a result, ultrasound is used to overcome these limitations and obtain a stable emulsion (Zhou, Zhang, Xing, et al., 2021). Proteins from different sources are used to stabilize emulsion, as shown in Table 11.2. Milk derived proteins are used in majority of cases followed by plant and animal derived proteins. Basically, ultrasound treatment will improve emulsion stability by increasing emulsification properties of food proteins by improving solubility and adsorption at interface and changing protein structure which can enhance protein flexibility by forming protein layer between oil droplet and water interface. The decrease in protein particle size caused by ultrasound treatment, as well as the structural changes that cause protein unfolding, appeared to be the primary causes of the increase in protein solubility. Also, it enhanced the compactness of the interface film and the adsorption of protein at the oil and water interface. The improved interfacial properties and intermolecular forces were primarily responsible for the resistance to droplet coalescence in protein emulsion (K. Li et al., 2020; M. Silva et al., 2020, 2021; J. Zhang et al., 2022). Sonication energy density and amount of protein play an important role in the physical and chemical stabilities of emulsion. However, high energy density exhibited

TABLE 11.2
Protein/Milk-Based Emulsion

Emulsifiers and Oil Phase	Emulsification Condition	Key Findings	References
Pea protein isolate (87.31%), soybean oil	200 W to 600 W, 20 kHz, 5 seconds ON, 2 seconds OFF, 10 minutes, below 20°C, 6 mm probe	• Treatment at 500 W significantly increased protein solubility and decreased protein particle size.	(J. Zhang et al., 2022)
Casein and whey protein (ratio: 80:20 to 40:60), grape seed oil	500 W, 60%, 20 kHz, 7 and 10 minutes, below 20°C, energy density of 81.9 and 117.0 J mL^{-1}	• Emulsion treated with 81.9 J mL^{-1} showed greater emulsion stability than higher energy density. Casein and whey protein ratios of 60:40, 50:50, and 40:60 showed greater results.	(M. Silva et al., 2021)
Antarctic krill protein (0.5%, 1.0%, 1.5%, and 2.0% w/v), soybean oil (25%)	Ultrasound intensity of 6690 W/cm^2 and 2-minute treatment	• The emulsion gel that was prepared had an elastic network structure without any chemical modifications.	(Zhou et al., 2021)
Sodium caseinate (1.5 g in 66 g of phosphate buffer), soybean oil (55%)	412.5 W to 712.5 W, 20 kHz, 4 seconds ON, 2 seconds OFF, 0–9 minutes, ice bath used, 13 mm probe	• Ultrasound treatments for 3, 6, and 9 minutes produced droplet size peaks of 615.1, 396.1, and 295.3 208 nm, respectively. But 6 minutes treatment showed better stability.	(K. Li et al., 2020)
Whey PI, soy PI, bovine gelatin, peanut PI, and corn zein (1%, w/v), rapeseed oil	Intensity of 50–55 W cm^{-2}, 20 kHz, 2 seconds ON, 2 seconds OFF, 6 minutes, 63.6 mm probe, 540 J/mL	• Stable emulsion formed using SPI, WPI, and gelatin as emulsifiers. Unable to form stable emulsions using PPI and zein at any ionic strength. It was discovered that emulsions made from gelatin had smaller droplet sizes.	(Taha et al., 2019)

physical instability due to the depletion flocculation mechanism and induced oxidation reactions at the critical casein concentration (Silva et al., 2021). Although under optimum conditions, ultrasound treatment will improve oxidation stability (by lowering peroxide value and thiobarbituric acid reactive substance) during storage period (Li et al., 2020). In the case of milk proteins, the casein/whey protein ratio has a significant effect on the minimum energy required to produce a stable emulsion (Silva et al., 2020).

11.6.2 POLYSACCHARIDE-BASED EMULSIONS

Polysaccharides are polymerized carbohydrates made up of at least ten monosaccharide chains linked together by glycosidic bonds. Polysaccharides that are commonly used include homopolysaccharides like starch and cellulose, as well as heteropolysaccharides like gum Arabica, xanthan gum, pectin, and guar gum. During the emulsification process, ultrasonic treatment could reduce the molecular weight and rheology of polysaccharide molecules in the water phase. These modifications could hasten the adsorption of polysaccharide molecules at the oil/water interface, lowering interfacial tension and thus improving emulsion stability via steric repulsion (Taha et al., 2020; Zhou, Zhang, Xing et al., 2021). Different types of polysaccharides are used to stabilize emulsion, as shown in Table 11.3. Ultrasound treatment can depolymerize polysaccharides, lowering the

TABLE 11.3
Polysaccharide-Based Emulsion

Emulsifiers and Oil Phase	Emulsification Condition	Key Findings	References
Cellulose micro/ nanocrystals (2.5–5% w/w), flaxseed oil (10–20% w/w)	525 W, 20 kHz, 4 minutes, 10°C **Probe:** 13 mm **Immersion depth:** 3 mm	• Flaxseed oil droplets were encased by smaller crystals, cellulose crystal can stabilize O/W emulsion through the Pickering mechanism.	(Meirelles et al., 2020b)
OSA modified starch (8–15% w/w), cinnamon essential oil (5%)	450 W, 3-second interval, 2.5, 5, and 10 minutes	• Ultrasound used as an emulsification tool to incorporate cinnamon essential oil into film.	(Feng et al., 2020)
Cellulose nanocrystal (0.5% and 1.0% w/w), flaxseed oil (2.5–7.5% w/w)	525 W, 20 kHz, 4 minutes, 10°C, 13 mm probe **Immersion depth:** 3 mm	• The cellulose particle concentration and flaxseed oil proportion impacted stabilizing properties. US improved stability, yielding consistent droplet sizes and a slightly thicker consistency.	(Meirelles et al., 2020a)
Potato starch (3% w/w), alginate (0.5% w/w), gelatin (0–1.5% w/w), sunflower oil (20% w/w)	150 W to 375 W, 20 kHz, 5 minutes, below 45°C, 13 mm probe	• Sonication produced stable emulsions with smaller droplet sizes. Power levels at 225 and 375. W showed improved emulsion characteristics.	(K. C. G. Silva & Sato, 2019)
Modified starch (2% w/v), Copaiba oleoresin	**Power:** 160–640 W **Time:** 2–8 minutes	• Ultrasound treatment at 480 W for 6 minutes exhibited smallest droplet diameter.	(Pasquel et al., 2017)
Modified starch (16% w/w), Annatto seed oil (4%)	160–640 W, 19 kHz, 0.5–5 minutes, 41.6°C, **Probe:** 13 mm	• Sonication at 480 W for 5 minutes produced kinetically stable emulsion with smallest droplet size. Modified starch emulsion showed less uniform droplet size than whey protein isolate at the same power and time level.	(E. K. Silva et al., 2016)

molecular weight which can reduce interfacial tension between the dispersed and continuous phases which can promote the formation of small droplet emulsions while also improving emulsion stability (Cui & Zhu, 2021). Emulsion stabilized by cellulose crystals (CC) demonstrated good emulsion stability, which can be attributed to the large crystals dispersed between oil droplets, which prevent flocculation, while oil droplets were completely surrounded by cellulose nanocrystals, indicating Pickering type stabilization, which is not associated with a decrease in interfacial tension (Meirelles et al., 2020b). Lower proportion of CC showed better stability than high concentration of CC (Meirelles et al., 2020a).

A polysaccharide mixture will generally have better functional properties than a single polysaccharide. The preparation of alginate, potato starch, and gelatin stabilized emulsions by varying gelatin concentration revealed that ultrasound treatment decreased viscosity, surface charge, and improved the emulsion's interfacial properties, increasing power reduced droplet size and increased flocculation, and gelatin concentrations greater than 1% produced stable emulsions (K. C. G. Silva & Sato, 2019). Using modified starch as a food matrix, ultrasound treatment produced a stable emulsion, but increasing the intensity of the ultrasonication process had no effect on reducing droplet size (E. K. Silva et al., 2016).

11.6.3 PROTEIN-POLYSACCHARIDE-BASED EMULSIONS

The combination of protein and polysaccharide (Table 11.4) can result in emulsions with small droplets and excellent stability against environmental stresses such as ionic strength, pH, and temperature. Protein and polysaccharide complexes can be formed via covalent and non-covalent interactions, which can improve the interfacial properties of adsorbed layers and the dispersion's stability.

TABLE 11.4
Protein and Polysaccharide-Based Emulsion

Emulsifiers and Oil Phase	Emulsification Condition	Key Findings	References
Soy protein isolate (1%) and pectin (0.3%), Soybean oil (10% v/v)	150 W to 600 W, 20 kHz, 4 seconds ON, 2 seconds OFF, 15 minutes, below 20°C, 6 mm probe	• The best results were found when the emulsion was treated at 450 W with droplet size of 559.82 nm.	(Wang et al., 2022)
Whey protein concentrate (1.83% w/w) and pectin (3% w/w), model vegetable oil (18 %)	700 W, 20 kHz, 75%, 100%, 2 minutes, below 35°C, 12.5 mm probe	• The emulsion stabilized by WPC-pectin had the smallest droplet size among all the emulsions, measuring 1.02 μm.	(Vélez-Erazo et al., 2021)
Soy protein isolate and high methoxyl pectin (1:1 and 4:1), Soybean oil (5–15%)	120 W, 20 kHz, 1 minutes, 12.7 mm probe and immersion depth of 1 cm	• SPI and pectin ratio of 1:1 produced stable emulsion without phase separation. 1:1 ratio and 5% oil combination produced lowest droplet size of 3.36 μm.	(Albano et al., 2020)
Whey protein isolate (2.0% w/w) and soluble soybean polysaccharides (0.5% w/w), sunflower oil (30% w/w)	750 W, 20 kHz, 70%, 2 minutes, icebath used, 28 mm probe	• Emulsion prepared with the protein-polysaccharide mixture exhibited higher stability to freeze-thawing than WPI alone.	(Cabezas et al., 2019)
Sodium caseinate and corn starch conjugate (30% w/w). Palm oil (4.5%)	Pre-emulsion at 4000 rpm for 2 minutes. 700 W, 20 kHz, 100%, 7 minutes, below 35°C 12.5 mm probe	• Parameters used to prepare conjugate such as heating time significantly affected the emulsion properties and droplet size varied from 0.72 to 1.03 μm.	(Consoli et al., 2018)

Sonication improves the emulsion stability and functional properties of mixtures that can be used in the low fat system, allowing consumers to consume healthier products (Taha et al., 2020, Albano & Nicoletti, 2018). Ultrasound emulsification was found to be an effective method for enhancing the emulsifying properties of food matrixes based on a complex of SPI and propylene glycol alginate (PGA). This method led to a reduction in interfacial tension and apparent viscosity, as well as an improvement in thermal and storage stability of the emulsion (Wang et al., 2022). However, it was also observed that the ratio of SPI and PC in the emulsion and the percentage of pectin utilized had a significant impact on the stability of the emulsion (Albano et al., 2020). The particle size, interfacial activity, and non-sedimentable protein content all increased as a result of the sono-emulsion of the 30% sunflower oil phase with SPI and soluble soybean polysaccharide mixture. However, soluble soybean polysaccharide combination and SPI that had not been heated showed greater freeze-thawing stability than cooked mixtures (Cabezas et al., 2019). Also, formation of proper protein and polysaccharide is important to achieve a stable emulsion with a smaller droplet size, many studies showed that the droplet size of mixture was more than a single component emulsion (Consoli et al., 2018).

11.6.4 SURFACTANT-BASED EMULSION

A surfactant is a substance that can greatly alter the interface of a solution system by adding a small quantity. In the food industry, Tween 20, Tween 40, Tween 80, Span 80, polyglycerol polyricinoleate, and lecithin are commonly used surfactants, as shown in Table 11.5. These surfactants are small molecule active ingredients and their molecular structure is amphiphilic: one end has hydrophilic

TABLE 11.5
Surfactant-Based Emulsion

Emulsifiers and Oil Phase	Emulsification Condition	Key Findings	References
Tween 20, sesame oil (oil to surfactant ratio, 1:1 to 1:5)	Pre-emulsion at 540 rpm for 3 minutes. 20 kHz, 10, 20, and 30 minutes	• Mean particle size of 189.5 nm achieved at 1:3 oil to surfactant	(Namratha Vinaya et al., 2021)
Chitosan and span 80 (span 80 fraction is 1 to 50%), oil (5–35%)	950 W, 25 kHz, amplitude of 6–64%, 20 seconds ON, 40 seconds OFF, 0–30 minutes, 30 ± 5°C, 6 mm probe	• At amplitude of 32% for 15 minutes produced lowest droplet size of 198.5 nm in 35% span 80 and 8% oil, molecular weight of chitosan had significant effect	(Zhang et al., 2020)
Tween 80 (3% w/w) and soya lecithin (1% w/w), canola oil (8% w/w)	**Frequency:** 20 kHz **Time:** 5 minutes	• Produce stable nanoemulsion against different environmental conditions with droplet size of 140.15 nm	(Mehmood & Ahmed, 2020)
Tween 80 (1–3 % w/w) Cinnamon essential oil	400 W, 24 kHz, 90%, 180.23–351.77 seconds, 4.82–45.18°C, 22 mm dia. Probe	• Minimum droplet size 65.98 nm was observed at 266 seconds, 4.82°C, and 3% w/w of Tween 80. Exhibited good stability storing at 4–45°C for 90 days	(Pongsumpun et al., 2020)
Tween 80 (1%) and modified starch (30 mg/mL) Laurel essential oil (6, 9, and 12 mg/mL)	500 W, 20 kHz, 70% amplitude at 2, 4, and 6 minutes	• Droplet size of emulsion varied from 239.5 to 357 nm. At oil concentration of 6 mg/mL, ultrasound treatment for 2 minutes produced best emulsion with droplet size of 297 nm	(Reis et al., 2019)
Tween 80 and lecithin (36:64 weight ratio) Crude oil (20%)	Bath sonication, time: 30 minutes	• Lecithin and tween 80 added to deionized water and sonicated to produce micelle solution	(Rocchio et al., 2017)

groups, while the other end has hydrophobic groups. As a small molecule, the surfactant can quickly spread and rearrange at the oil and water interface during emulsification to form a stable emulsion (Zhou, Zhang, Xing et al., 2021). Namratha Vinaya et al. conducted a study on the ultrasonic preparation of sesame oil in water nanoemulsion using Tween 20 as surfactant. The results indicated that the nanoemulsion exhibited good stability, and the oil to surfactant ratio and treatment time were found to be crucial factors in determining the droplet size and physical stability of the nanoemulsion. (Namratha Vinaya et al., 2021). The study by Pongsumpun et al. aimed to investigate the potential of using Tween 80 as a surfactant to produce cinnamon essential oil (CEO) nanoemulsions via ultrasound bath and probe. The study revealed that nanoemulsions created through the probe ultrasound method have smaller droplet sizes, a lower degree of droplet size variation, and higher viscosity when compared to those created through bath ultrasonication. The emulsions were stable up to 90 days at varying temperatures. Furthermore, the study highlighted that the concentration of Tween 80, sonication temperature, and emulsification duration played a significant role in determining the physicochemical properties of the CEO nanoemulsions (Pongsumpun et al., 2020). Combination of Tween 80 and soya lecithin is used as an emulsifier to form nanoemulsions (Mehmood & Ahmed, 2020; Rocchio et al., 2017). combination of Tween 80 and soya lecithin as emulsifiers resulted in better emulsion stability compared to lecithin alone, and comparable emulsion stability to Tween 80 alone. These results suggest that the primary mechanism for stabilization in this system is through the steric stabilization provided by Tween 80 (Rocchio et al., 2017). In another study, chitosan and Span 80 were used to produce water-in-oil emulsions through ultrasound treatment. The findings showed that the molecular weight of chitosan played a crucial role in determining the mean droplet diameter and polydispersity index of the emulsion. It was observed that higher molecular weight of chitosan resulted in smaller droplet size. Additionally, the position of the ultrasonic probe was found to have a significant impact on the droplet size (K. Zhang et al., 2020).

11.7 ADVANTAGE OF SONO-EMULSIONS OVER OTHER EMULSIONS

Ultrasound mainly it is a green technology used to produce different food emulsions. Sono-emulsions offer several advantages over other emulsions as follows.

- Sonication is a highly efficient method of emulsification, and it can often achieve stable emulsions with a high droplet size uniformity in a relatively short period of time.
- Unlike other techniques such as high-pressure homogenization, sonication does not involve the use of high pressure or mechanical shear, which can potentially damage or degrade sensitive components in the emulsion.
- Sonication can be easily scaled up for industrial applications, and the equipment required for sonication is relatively simple and inexpensive.
- The intensity, frequency, and duration of ultrasonic treatment can be easily controlled, allowing for precise adjustments to the emulsion's properties.
- Sonication can be used to emulsify a wide range of liquids, including oils, water, and other polar and non-polar liquids, making it suitable for a wide range of emulsion systems.
- Sonication requires less energy than other emulsification methods, making it an environmentally friendly option. The power required to produce the same volume of emulsion and same droplet size is less than that of a high-pressure homogenizer. Making it more cost-effective.
- They exhibit improved stability, greater solubilization of hydrophobic compounds, increased bioavailability of active ingredients, and more efficient production processes.
- Sono-emulsions are more thermally and mechanically stable than other emulsions due to their smaller droplet size. This makes them better suited for use in products exposed to extreme temperatures or mechanical stress.
- Ultrasound homogenizer don't have any moving parts which makes cleaning is lot easier and safe to operate.

It's important to note that sonoemulsion has also some limitations such as the viscosity of the liquids, the size and chemical nature of the droplets, and the stability of the emulsion. Therefore, the choice of the technique should be based on the desired properties of the final emulsion and the specific application. Many studies concluded that ultrasound-assisted emulsification produces more stable, monodisperse emulsion using less energy compared to high-pressure homogenization (Jiang et al., 2021; Leong et al., 2018; Y. Li & Xiang, 2019; Zhou et al., 2022), mechanical stirring (Sumitomo et al., 2019), microfluidizer (Alliod et al., 2019; Santos et al., 2019). Zhou et al. (2022) compared preparation of oil in water emulsion by ultrasound with high-pressure homogenizer (HPH) and high speed homogenizer. Ultrasound-treated sample exhibited high emulsifying activity and concluded that ultrasound emulsification is more efficient method than other to form stable myofibrillar protein – soybean oil emulsion with lowest droplet size of 150 nm (Zhou et al., 2022). High-intensity ultrasound prepared peanut protein stabilized emulsion exhibited high resistance for flocculation and separation due to gravitation during storage period. Although it produced larger particle size than HPH (Jiang et al., 2021). Also, the emulsion prepared by ultrasound showed higher stability, while HPH prepared emulsion showed aggregation during 30-day storage (Li & Xiang, 2019). Another study used ultrasound, microfluidizer, and premix membrane emulsification (PME) to prepare nanoemulsion resulted in ultrasound producing smaller droplet size compared to other techniques, but PME produced monodispersed droplets at higher droplet size (Alliod et al., 2019). Emulsification using ultrasound produces finer emulsions at high energy densities compared to mechanical stirring. However, mechanical stirring is more effective in creating coarser emulsions at lower energy densities (Sumitomo et al., 2019). Santos et al. (2019) compared sonication with microfluidization to prepare lemongrass essential oil and Appyclean 6552 emulsion, both techniques produced similar droplet size, but emulsion produced by microfluidization at high pressure underwent recoalescence. However, emulsion prepared at lower pressure produced stable emulsion. Also, at similar energy input levels, both techniques produce similar results, but after reaching optimum energy level, further increased energy did not show any change in sono emulsion, while microfluidization technique demonstrated recoalescence (Santos et al., 2019). Also, ultrasonication produced more stable and better monodisperse double emulsions than HPH at similar specific energy density of 20 J/g (Leong et al., 2018).

11.8   CHALLENGES AND RESEARCH NEEDS

Ultrasound-assisted emulsification is a powerful technique for the formation of stable and homogeneous emulsions, but there are several challenges and research needs that must be addressed in order to fully understand and optimize the process. Some of the key challenges include:

- One of the main challenges in ultrasound-assisted emulsification techniques is the scalability of the process. Ultrasound-assisted emulsification has primarily been studied at laboratory scale, and more research is needed to develop scalable and practical methods for industrial implementation. Mainly, focus on continuous operation is needed.
- Currently, the ultrasound probes used in ultrasound-assisted emulsification techniques are not very efficient. Therefore, there is a need to develop improved ultrasound probes with better efficiency.
- There is a need to develop new techniques and strategies for ultrasound-assisted emulsification, such as using high intensity focused ultrasound, multi-frequency ultrasound, and ultrasound-induced cavitation.

11.9   CONCLUSION

Ultrasound-assisted emulsification technology is a promising and environmentally friendly method that involves the application of ultrasonic energy to a liquid mixture in order to generate mechanical forces that break down the liquid into small droplets, resulting in stable and homogeneous emulsions. This method has gained significant traction in the food industry because of the increased focus on developing sustainable food processing systems. This chapter focuses mainly on the mechanism, different techniques used for emulsification, sonoemulsion of different food matrices, and future challenges of ultrasound-assisted emulsification. Studies have shown that ultrasound technology is more energy-efficient than other methods, such as HPH and microfluidizers. However, there is still room for improvement in terms of the stability of ultrasound-induced emulsions under different conditions, such as changes in pH, temperature, and ionic strength.

REFERENCES

Ahmadi Khoshooei, M., Scott, C. E., Carbognani, L., & Pereira-Almao, P. (2021). Ultrasound-assisted bimetallic NiMo nanocatalyst preparation using microemulsions for in-situ upgrading application: Impact on particle size. *Catalysis Today*, *365*(January), 132–141. https://doi.org/10.1016/j.cattod.2020.06.078

Akbari, S., & Nour, A. H. (2018). Emulsion types, stability mechanisms and rheology: A review. *International Journal of Innovative Research and Scientific Studies*, *1*(1), 14–21. https://papers.ssrn.com/sol3/papers.cfm?abstract_id=3324905

Albano, K. M., Cavallieri, Â. L. F., & Nicoletti, V. R. (2020). Electrostatic interaction between soy proteins and pectin in O/W emulsions stabilization by ultrasound application. *Food Biophysics*, *15*(3), 297–312. https://doi.org/10.1007/s11483-020-09625-z

Albano, K. M., & Nicoletti, V. R. (2018). Ultrasound impact on whey protein concentrate-pectin complexes and in the O/W emulsions with low oil soybean content stabilization. *Ultrasonics Sonochemistry*, *41*, 562–571. https://doi.org/10.1016/j.ultsonch.2017.10.018

Albert, C., Beladjine, M., Tsapis, N., Fattal, E., Agnely, F., & Huang, N. (2019). Pickering emulsions: Preparation processes, key parameters governing their properties and potential for pharmaceutical applications. *Journal of Controlled Release*, *309*(April), 302–332. https://doi.org/10.1016/j.jconrel.2019.07.003

Alliod, O., Almouazen, E., Nemer, G., Fessi, H., & Charcosset, C. (2019). Comparison of three processes for parenteral nanoemulsion production: Ultrasounds, microfluidizer, and premix membrane emulsification. *Journal of Pharmaceutical Sciences*, *108*(8), 2708–2717. https://doi.org/10.1016/j.xphs.2019.03.026

Ashaolu, T. J. (2021). Nanoemulsions for health, food, and cosmetics: A review. *Environmental Chemistry Letters*, *19*(4), 3381–3395.

Ashokkumar, M. (2011). The characterization of acoustic cavitation bubbles – An overview. *Ultrasonics Sonochemistry*, *18*(4), 864–872. https://doi.org/10.1016/j.ultsonch.2010.11.016

Ashokkumar, M. (2015). Applications of ultrasound in food and bioprocessing. *Ultrasonics Sonochemistry*, *25*(1), 17–23. https://doi.org/10.1016/j.ultsonch.2014.08.012

Ashokkumar, M., Cavalieri, F., Chemat, F., Okitsu, K., Sambandam, A., Yasui, K., & Zisu, B. (2016). Handbook of ultrasonics and sonochemistry. *Handbook of Ultrasonics and Sonochemistry*, 1–1487. https://doi.org/10.1007/978-981-287-278-4

Aswathanarayan, J. B., & Vittal, R. R. (2019). Nanoemulsions and their potential applications in food industry. *Frontiers in Sustainable Food Systems*, *3*(November), 1–21. https://doi.org/10.3389/fsufs.2019.00095

Bains, U., & Pal, R. (2020). Rheology and catastrophic phase inversion of emulsions in the presence of starch nanoparticles. *ChemEngineering*, *4*(4), 1–15.

Belgheisi, S., & Motamedzadegan, A. (2020). Impact of ultrasound processing parameters on physical characteristics of lycopene emulsion. *Journal of Food Science and Technology*. 58, 484–493. https://doi.org/10.1007/s13197-020-04557-5

Benetti, J. V. M., do Prado Silva, J. T., & Nicoletti, V. R. (2019). SPI microgels applied to Pickering stabilization of O/W emulsions by ultrasound and high-pressure homogenization: rheology and spray drying. *Food Research International*, *122*(April), 383–391. https://doi.org/10.1016/j.foodres.2019.04.020

Bhangu, S. K., & Ashokkumar, M. (2016). Theory of sonochemistry. *Topics in Current Chemistry*, *374*(4). https://doi.org/10.1007/s41061-016-0054-y

Cabezas, D. M., Pascual, G. N., Wagner, J. R., & Palazolo, G. G. (2019). Nanoparticles assembled from mixtures of whey protein isolate and soluble soybean polysaccharides. Structure, interfacial behavior and application on emulsions subjected to freeze-thawing. *Food Hydrocolloids*, *95*, 445–453. https://doi.org/10.1016/j.foodhyd.2019.04.040

Chang, C., & Nickerson, M. T. (2018). Encapsulation of omega 3-6-9 fatty acids-rich oils using protein-based emulsions with spray drying. *Journal of Food Science and Technology*, *55*(8), 2850–2861. https://doi.org/10.1007/s13197-018-3257-0

Chen, L., Ao, F., Ge, X., & Shen, W. (2020). Food-grade Pickering emulsions: Preparation, stabilization and applications. *Molecules*, *25*(14), 3202.

Consoli, L., Dias, R. A. O., Rabelo, R. S., Furtado, G. F., Sussulini, A., Cunha, R. L., & Hubinger, M. D. (2018). Sodium caseinate-corn starch hydrolysates conjugates obtained through the Maillard reaction as stabilizing agents in resveratrol-loaded emulsions. *Food Hydrocolloids*, *84*, 458–472. https://doi.org/10.1016/j.foodhyd.2018.06.017

Costa, C., Medronho, B., Filipe, A., Mira, I., Lindman, B., Edlund, H., & Norgren, M. (2019). Emulsion formation and stabilization by biomolecules: The leading role of cellulose. *Polymers*, *11*(10). https://doi.org/10.3390/polym11101570

Cui, R., & Zhu, F. (2021). Ultrasound modified polysaccharides: A review of structure, physicochemical properties, biological activities and food applications. *Trends in Food Science and Technology*, *107*(November 2020), 491–508.

Dávila-Rodríguez, M., López-Malo, A., Palou, E., Ramírez-Corona, N., & Jiménez-Munguía, M. T. (2020). Essential oils microemulsions prepared with high-frequency ultrasound: Physical properties and antimicrobial activity. *Journal of Food Science and Technology*, *57*(11), 4133–4142. https://doi.org/10.1007/s13197-020-04449-8

Espinosa-Álvarez, C., Jaime-Matus, C., & Cerezal-Mezquita, P. (2019). Some physical characteristics of the O/W macroemulsion of oleoresin of astaxanthin obtained from biomass of Haematococcus pluvialis Algunas características físicas de la macroemulsión O/W de oleorresina de astaxantina obtenida a partir de biomasa de H. *Dyna*, *86*(208), 136–142.

Espinosa-Andrews, H., & Páez-Hernández, G. (2020). Optimization of ultrasonication curcumin-hydroxylated lecithin nanoemulsions using response surface methodology. *Journal of Food Science and Technology*, *57*(2), 549–556.

Feng, X., Wang, W., Chu, Y., Gao, C., Liu, Q., & Tang, X. (2020). Effect of cinnamon essential oil nanoemulsion emulsified by OSA modified starch on the structure and properties of pullulan based films. *LWT*, *134*(September), 110123.

Grammatikopoulos, P., Sowwan, M., & Kioseoglou, J. (2019). Computational modeling of nanoparticle coalescence. *Advanced Theory and Simulations*, *2*(6), 1–26.

Jalali-Jivan, M., Abbasi, S., & Scanlon, M. G. (2019). Microemulsion as nanoreactor for lutein extraction: Optimization for ultrasound pretreatment. *Journal of Food Biochemistry*, *43*(8), 1–12. https://doi.org/10.1111/jfbc.12929

Jiang, Y. S., Zhang, S. B., Zhang, S. Y., & Peng, Y. X. (2021). Comparative study of high-intensity ultrasound and high-pressure homogenization on physicochemical properties of peanut protein-stabilized emulsions and emulsion gels. *Journal of Food Process Engineering*, *44*(6), 1–13. https://doi.org/10.1111/jfpe.13710

Kaci, M., Meziani, S., Arab-Tehrany, E., Gillet, G., Desjardins-Lavisse, I., & Desobry, S. (2014). Emulsification by high frequency ultrasound using piezoelectric transducer: Formation and stability of emulsifier free emulsion. *Ultrasonics Sonochemistry*, *21*(3), 1010–1017. https://doi.org/10.1016/j.ultsonch.2013.11.006

Kale, S. N., & Deore, S. L. (2016). Emulsion micro emulsion and nano emulsion: A review. *Systematic Reviews in Pharmacy*, *8*(1), 39–47. https://www.sysrevpharm.org/articles/emulsion-micro-emulsion-and-nano-emulsion-a-review.pdf

Karakuş, S. (2018). Introductory chapter: The perspective of emulsion systems. *Science and Technology Behind Nanoemulsions*, 1–8. https://doi.org/10.5772/intechopen.75727

Khan, S. A., Dar, A. H., Bhat, S. A., Fayaz, J., Makroo, H. A., & Dwivedi, M. (2022). High intensity ultrasound processing in liquid foods. *Food Reviews International*, *38*(6), 1123–1148. https://doi.org/10.1080/87559129.2020.1768404

Leong, T. S. H., Zhou, M., Zhou, D., Ashokkumar, M., & Martin, G. J. O. (2018). The formation of double emulsions in skim milk using minimal food-grade emulsifiers – a comparison between ultrasonic and high pressure homogenisation efficiencies. *Journal of Food Engineering*, *219*, 81–92. https://doi.org/10.1016/j.jfoodeng.2017.09.018

Levy, R., Okun, Z., & Shpigelman, A. (2021). High-pressure homogenization: Principles and applications beyond microbial inactivation. *Food Engineering Reviews*, *13*(3), 490–508. https://doi.org/10.1007/s12393-020-09239-8

Li, K., Li, Y., Liu, C. L., Fu, L., Zhao, Y. Y., Zhang, Y. Y., Wang, Y. T., & Bai, Y. H. (2020). Improving interfacial properties, structure and oxidative stability by ultrasound application to sodium caseinate prepared pre-emulsified soybean oil. *LWT*, *131*(136). https://doi.org/10.1016/j.lwt.2020.109755

Li, Y., & Xiang, D. (2019). Stability of oil-in-water emulsions performed by ultrasound power or high-pressure homogenization. *PLoS ONE*, *14*(3), 1–14.

Luo, X., Gong, H., Cao, J., Yin, H., Yan, Y., & He, L. (2019). Enhanced separation of water-in-oil emulsions using ultrasonic standing waves. *Chemical Engineering Science*, *203*, 285–292. https://doi.org/10.1016/j.ces.2019.04.002

Ma, W., Wang, J., Xu, X., Qin, L., Wu, C., & Du, M. (2019). Ultrasound treatment improved the physicochemical characteristics of cod protein and enhanced the stability of oil-in-water emulsion. *Food Research International*, *121*(March), 247–256.

McClements, D. J. (2016). *Food emulsions: Principles, Practices, and Techniques*. CRC Press.

McClements, D. J., & Jafari, S. M. (2018). Improving emulsion formation, stability and performance using mixed emulsifiers: A review. *Advances in Colloid and Interface Science*, *251*, 55–79. https://doi.org/10.1016/j.cis.2017.12.001

Mehmood, T., & Ahmed, A. (2020). Tween 80 and soya-lecithin-based food-grade nanoemulsions for the effective delivery of vitamin D. *Langmuir*, *36*(11), 2886–2892.

Mehmood, T., Ahmed, A., Ahmed, Z., & Ahmad, M. S. (2019). Optimization of soya lecithin and Tween 80 based novel vitamin D nanoemulsions prepared by ultrasonication using response surface methodology. *Food Chemistry*, *289*, 664–670.

Meirelles, A. A. D., Costa, A. L. R., & Cunha, R. L. (2020a). Cellulose nanocrystals from ultrasound process stabilizing O/W Pickering emulsion. *International Journal of Biological Macromolecules*, *158*, 75–84. https://doi.org/10.1016/j.ijbiomac.2020.04.185

Meirelles, A. A. D., Costa, A. L. R., & Cunha, R. L. (2020b). The stabilizing effect of cellulose crystals in O/W emulsions obtained by ultrasound process. *Food Research International*, *128*(May 2019), 108746. https://doi.org/10.1016/j.foodres.2019.108746

Modarres-Gheisari, S. M. M., Gavagsaz-Ghoachani, R., Malaki, M., Safarpour, P., & Zandi, M. (2019). Ultrasonic nano-emulsification – a review. *Ultrasonics Sonochemistry*, *52*, 88–105. https://doi.org/10.1016/j.ultsonch.2018.11.005

Mungure, T. E., Roohinejad, S., Bekhit, A. E. D., Greiner, R., & Mallikarjunan, K. (2018). Potential application of pectin for the stabilization of nanoemulsions. *Current Opinion in Food Science*, *19*, 72–76. https://doi.org/10.1016/j.cofs.2018.01.011

Namratha Vinaya, K., Mary John, A., Mangsatabam, M., & Anna Philip, M. (2021). Ultrasound-assisted synthesis and characterization of sesame oil based nanoemulsion. *IOP Conference Series: Materials Science and Engineering*, *1114*(1), 012085.

Nazir, A., & Vladisavljević, G. T. (2021). Droplet breakup mechanisms in premix membrane emulsification and related microfluidic channels. *Advances in Colloid and Interface Science*, *290*. https://doi.org/10.1016/j.cis.2021.102393

Páez-Hernández, G., Mondragón-Cortez, P., & Espinosa-Andrews, H. (2019). Developing curcumin nanoemulsions by high-intensity methods: Impact of ultrasonication and microfluidization parameters. *LWT*, *111*, 291–300.

Pasquel Reátegui, J. L., Barrales, F. M., Rezende, C. A., Queiroga, C. L., & Martínez, J. (2017). Production of Copaiba oleoresin particles from emulsions stabilized with modified starches. *Industrial Crops and Products*, *108*(March), 128–139.

Perrin, L., Desobry-Banon, S., Gillet, G., & Desobry, S. (2022). Review of high-frequency ultrasounds emulsification methods and oil/water interfacial organization in absence of any kind of stabilizer. *Foods*, *11*(15), 2194.

Piacentini, E. (2014). Emulsion. *Encyclopedia of Membranes*, 1–4. https://doi.org/10.1007/978-3-642-40872-4_1066-1

Pokhrel, N., Vabbina, P. K., & Pala, N. (2016). Sonochemistry: Science and engineering. *Ultrasonics Sonochemistry*, *29*(July), 104–128.

Pongsumpun, P., Iwamoto, S., & Siripatrawan, U. (2020). Response surface methodology for optimization of cinnamon essential oil nanoemulsion with improved stability and antifungal activity. *Ultrasonics Sonochemistry*, *60*, 104604.

Qin, X. S., Luo, Z. G., & Peng, X. C. (2018). Fabrication and characterization of quinoa protein nanoparticle-stabilized food-grade Pickering emulsions with ultrasound treatment: Interfacial adsorption/arrangement properties. *Journal of Agricultural and Food Chemistry*, *66*(17), 4449–4457. https://doi.org/10.1021/acs.jafc.8b00225

Ravera, F., Dziza, K., Santini, E., Cristofolini, L., & Liggieri, L. (2021). Emulsification and emulsion stability: The role of the interfacial properties. *Advances in Colloid and Interface Science*, *288*, 102344. https://doi.org/10.1016/j.cis.2020.102344

Raviadaran, R., Ng, M. H., Manickam, S., & Chandran, D. (2019). Ultrasound-assisted water-in-palm oil nano-emulsion: Influence of polyglycerol polyricinoleate and NaCl on its stability. *Ultrasonics Sonochemistry*, *52*(August), 353–363.

Reis, P. M. C. L., Mezzomo, N., Aguiar, G. P. S., Senna, E. M. T. L., Hense, H., & Ferreira, S. R. S. (2019). Ultrasound-assisted emulsion of laurel leaves essential oil (*Laurus nobilis* L.) encapsulated by SFEE. *Journal of Supercritical Fluids*, *147*, 284–292.

Rocchio, J., Neilsen, J., Everett, K., & Bothun, G. D. (2017). A solvent-free lecithin-Tween 80 system for oil dispersion. *Colloids and Surfaces A: Physicochemical and Engineering Aspects*, *533*, 218–223. https://doi.org/10.1016/j.colsurfa.2017.08.038

Roohinejad, S., Greiner, R., Oey, I., & Wen, J. (2018). *Emulsion-Based Systems for Delivery of Food Active Compounds: Formation, Application, Health and Safety*, 1–294. https://doi.org/10.1002/9781119247159

Santos, J., Jiménez, M., Calero, N., Undabeytia, T., & Muñoz, J. (2019). A comparison of microfluidization and sonication to obtain lemongrass submicron emulsions. Effect of diutan gum concentration as stabilizer. *LWT*, *114*(February), 108424.

Schroën, K., de Ruiter, J., & Berton-Carabin, C. (2020). The importance of interfacial tension in emulsification: Connecting scaling relations used in large scale preparation with microfluidic measurement methods. *ChemEngineering*, *4*(4), 1–22.

Schulman, J. H., & Hoar, T. P. (1943). Transparent Water-in-Oil Dispersions: The Oleopathic Hydro-Micelle. *Nature*, *152*, 102–103.

Shanmugam, A., & Ashokkumar, M. (2014a). Functional properties of ultrasonically generated flaxseed oil-dairy emulsions. *Ultrasonics Sonochemistry*, *21*(5), 1649–1657.

Shanmugam, A., & Ashokkumar, M. (2014b). Ultrasonic preparation of stable flax seed oil emulsions in dairy systems – Physicochemical characterization. *Food Hydrocolloids*, *39*, 151–162. https://doi.org/10.1016/j.foodhyd.2014.01.006

Shanmugam, A., & Ashokkumar, M. (2015). Characterization of ultrasonically prepared flaxseed oil enriched beverage/carrot juice emulsions and process-induced changes to the functional properties of carrot juice. *Food and Bioprocess Technology*, *8*(6), 1258–1266. https://doi.org/10.1007/s11947-015-1492-1

Shanmugam, A., & Ashokkumar, M. (2017). Ultrasonic preparation of food emulsions. *Ultrasound in Food Processing*, 287–310. https://doi.org/10.1002/9781118964156.ch10

Shanmugam, A., Chandrapala, J., & Ashokkumar, M. (2012). The effect of ultrasound on the physical and functional properties of skim milk. *Innovative Food Science and Emerging Technologies*, *16*, 251–258. https://doi.org/10.1016/j.ifset.2012.06.005

Shao, P., Feng, J., Sun, P., Xiang, N., Lu, B., & Qiu, D. (2020). Recent advances in improving stability of food emulsion by plant polysaccharides. *Food Research International*, *137*(January), 109376. https://doi.org/10.1016/j.foodres.2020.109376

Sharma, N., Kaur, G., & Kumar, S. (2021). Optimization of emulsification conditions for designing ultrasound assisted curcumin loaded nanoemulsion: Characterization, antioxidant assay and release kinetics. *LWT*, *141*(January), 110962.

Silva, E. K., Azevedo, V. M., Cunha, R. L., Hubinger, M. D., & Meireles, M. A. A. (2016). Ultrasound-assisted encapsulation of annatto seed oil: Whey protein isolate versus modified starch. *Food Hydrocolloids*, *56*, 71–83.

Silva, K. C. G., & Sato, A. C. K. (2019). Sonication technique to produce emulsions: The impact of ultrasonic power and gelatin concentration. *Ultrasonics Sonochemistry*, *52*, 286–293. https://doi.org/10.1016/j.ultsonch.2018.12.001

Silva, M., Zisu, B., & Chandrapala, J. (2020). Interfacial and emulsification properties of sono-emulsified grape seed oil emulsions stabilized with milk proteins. *Food Chemistry*, *309*, 125758. https://doi.org/10.1016/j.foodchem.2019.125758

Silva, M., Zisu, B., & Chandrapala, J. (2021). Stability of oil–water primary emulsions stabilised with varying levels of casein and whey proteins affected by high-intensity ultrasound. *International Journal of Food Science and Technology*, *56*(2), 897–908.

Sui, X., Bi, S., Qi, B., Wang, Z., Zhang, M., Li, Y., & Jiang, L. (2017). Impact of ultrasonic treatment on an emulsion system stabilized with soybean protein isolate and lecithin: Its emulsifying property and emulsion stability. *Food Hydrocolloids*, *63*, 727–734.

Sumitomo, S., Ueta, M., Uddin, M. A., & Kato, Y. (2019). Comparison of oil-in-water emulsion between ultrasonic irradiation and mechanical stirring. *Chemical Engineering and Technology*, *42*(2), 381–387. https://doi.org/10.1002/ceat.201800139

Taha, A., Ahmed, E., Hu, T., Xu, X., Pan, S., & Hu, H. (2019). Effects of different ionic strengths on the physicochemical properties of plant and animal proteins-stabilized emulsions fabricated using ultrasound emulsification. *Ultrasonics Sonochemistry*, *58*, 104627. https://doi.org/10.1016/j.ultsonch.2019.104627

Taha, A., Ahmed, E., Ismaiel, A., Ashokkumar, M., Xu, X., Pan, S., & Hu, H. (2020). Ultrasonic emulsification: An overview on the preparation of different emulsifiers-stabilized emulsions. In *Trends in Food Science and Technology* (Vol. 105, pp. 363–377). Elsevier Ltd. https://doi.org/10.1016/j.tifs.2020.09.024

Vélez-Erazo, E. M., Consoli, L., & Hubinger, M. D. (2021). Spray drying of mono- and double-layer emulsions of PUFA-rich vegetable oil homogenized by ultrasound. *Drying Technology*, *39*(7), 868–881. https://doi.org/10.1080/07373937.2020.1728305

Vladisavljević, G. T., Kobayashi, I., & Nakajima, M. (2013). Emulsion formation in membrane and microfluidic devices. *Emulsion Formation and Stability*, 77–98.

Waglewska, E., & Bazylińska, U. (2021). Biodegradable amphoteric surfactants in titration-ultrasound formulation of oil-in-water nanoemulsions: Rational design, development, and kinetic stability. *International Journal of Molecular Sciences*, *22*(21), 11776.

Wang, T., Wang, N., Li, N., Ji, X., Zhang, H., Yu, D., & Wang, L. (2022). Effect of high-intensity ultrasound on the physicochemical properties, microstructure, and stability of soy protein isolate-pectin emulsion. *Ultrasonics Sonochemistry*, *82*, 1–9.

Yadav, K. S., & Kale, K. (2020). High pressure homogenizer in pharmaceuticals: Understanding its critical processing parameters and applications. *Journal of Pharmaceutical Innovation*, *15*(4), 690–701.

Yao, Y., Pan, Y., & Liu, S. (2020). Power ultrasound and its applications: A state-of-the-art review. *Ultrasonics Sonochemistry*, *62*, 104722.

Zhang, J., Liu, Q., Chen, Q., Sun, F., Liu, H., & Kong, B. (2022). Ultrasonics sonochemistry synergistic modification of pea protein structure using high-intensity ultrasound and pH-shifting technology to improve solubility and emulsification. *Ultrasonics Sonochemistry*, *88*(June), 106099. https://doi.org/10.1016/j.ultsonch.2022.106099

Zhang, K., Mao, Z., Huang, Y., Xu, Y., Huang, C., Guo, Y., Ren, X., & Liu, C. (2020). Ultrasonic assisted water-in-oil emulsions encapsulating macro-molecular polysaccharide chitosan: Influence of molecular properties, emulsion viscosity and their stability. *Ultrasonics Sonochemistry*, *64*, 105018.

Zhou, L., Zhang, J., Lorenzo, J. M., & Zhang, W. (2021). Effects of ultrasound emulsification on the properties of pork myofibrillar protein-fat mixed gel. *Food Chemistry*, *345*, 128751. https://doi.org/10.1016/j.foodchem.2020.128751

Zhou, L., Zhang, W., Wang, J., Zhang, R., & Zhang, J. (2022). Comparison of oil-in-water emulsions prepared by ultrasound, high-pressure homogenization and high-speed homogenization. *Ultrasonics Sonochemistry*, *82*, 105885.

Zhou, L., Zhang, J., Xing, L., & Zhang, W. (2021). Applications and effects of ultrasound assisted emulsification in the production of food emulsions: A review. *Trends in Food Science and Technology*, *110*(1), 493–512. https://doi.org/10.1016/j.tifs.2021.02.008

Zhu, Q., Qiu, S., Zhang, H., Cheng, Y., & Yin, L. (2018). Physical stability, microstructure and micro-rheological properties of water-in-oil-in-water (W/O/W) emulsions stabilized by porcine gelatin. *Food Chemistry*, *253*(June 2017), 63–70.

Zychowski, L. (2015). Introduction to Emulsion Technology, Emulsifiers and Stability. https://cybercolloids.net/wp-content/uploads/2022/12/Introduction-to-Emulsion-Technology-Emulsifiers-and-Stability.pdf

12 Hydrodynamic Cavitation and Its Applications in the Food Industries

V. Athira, U. K. Aruna Nair, N. Bhanu Prakash Reddy, Yuvraj Bhosale Khasherao, and V. R. Sinija

12.1 INTRODUCTION

In the current global scenario, the demand for nutritious, health-promoting foods with greater nutraceutical values, products with close to natural taste, greater shelf-life, additive-free, and available in a convenient form, i.e., most of the times ready-to-eat (RTE) or ready-to-serve (RTS) form has been increasing (Martynenko et al., 2015). According to a report on functional foods and drinks – global market trajectory and analytics – 2022, the global functional foods and beverages market was valued at US $149.4 billion in 2020 and is predicted to reach a size of US $229.1 billion by 2027. Beverages, one of the key market segments, is expected to reach US $83.2 billion by the end of 2027 (Research and Markets, 2022).

This trend requires developing, adopting, and commercialising a wide range of technologies for processing raw food materials and reducing waste while processing. Most traditional food processing techniques are based on contact heating. They involve multiple processing steps that introduce the fruit into the atmosphere to promote oxidation, leading to the spoilage of the food product or degradation of processed products' quality. To address this issue alongside adopting concepts like "pure natural" and "no additives and preservatives", it was indeed necessary to explore the search for and development of new novel, non-thermal, and minimal processing technologies yet feasible in terms of cost-effectiveness and those technologies that keep food safe and fresh with minimal thermal processing (Martynenko et al., 2015). Consumers generally perceive minimally processed foods as close to naturality, more nutritious, healthier, and more satisfying to consume than thermally (rather conventionally) processed foods. This perception may be because of the experience gained from culinary practices where heating destroys taste, nutrients, and the natural form of the food product. However, novel food processing technologies require a different approach to achieve these consumer benefits (i.e., natural taste, microbial-free, nutrient-rich) (Arya et al., 2020). Any new food processing technology must undergo a thorough fundamental study to determine its efficacy.

Additionally, the industry requires the processing activities to be carried out most efficiently, i.e., in terms of product quality, energy, time, or cost. Alternative new technologies are continually being investigated to lower the overall processing cost while preserving or improving product quality in an ecologically friendly way. Moreover, it is important to examine whether using these technologies can destroy bacteria and enzymes, keep sensory qualities, and not create new toxic and harmful compounds that are undesirable while processing (Gogate, 2011a).

Some technologies that offer abovesaid advantages have been investigated for application in food processing; each work based on different microbe and enzyme inactivation mechanisms. They are high-pressure processing (HPP) or high hydrostatic pressure (HHP), which uses pressure force as a method of inactivation; high to moderate intensity pulsed electric field (PEF), which uses electrical energy to kill microbes and inactivate enzymes; ultraviolet (UV) which uses UV-C light (100–280 nm) to do the job; pulsed-light (PL), which uses high-intensity light energy in the pulsed form to

DOI: 10.1201/9781003359302-12

pasteurise the foods; ultrasonication or ultrasound (US), which uses sound energy for processing, and irradiation, which uses radiant energy to inactivate enzymes and microbes by disrupting chemical bonds, and many more alternative techniques that do the job at the molecular level (Hernández-Hernández et al., 2019; Knorr & Augustin, 2021; Zhang et al., 2019).

Unsurprisingly, one of those novel food processing technologies underexplored is hydrodynamic cavitation (HC) (Arya et al., 2020), often may be due to the perception in peoples' minds about cavitation as a destructive one. Cavitation is the formation, growth, and collapse of bubbles within a fraction of a second in a liquid flow system. Cavitation releases energy. It increases temperature and pressure up to 5,000 K and 100 to 5,000 bars, respectively. The energy released is much higher than in conventional processes.

Depending on the generation mode, cavitation can be classified into four types: acoustic cavitation, HC, optic cavitation, and particle cavitation. In acoustic cavitation, pressure variations are created by sound waves like ultrasound waves of 16 kHz to 100 MHz. HC can be produced by creating velocity variation using the system's geometry. For example, bypassing a liquid through a venturi or an orifice, a sudden change in velocity and pressure can create this HC. The cavitation phenomenon caused by laser optics is known as optic cavitation. Particle cavitation can be produced by using high-energy particles like a proton. In general, some of the mechanical (physical) effects of cavitation effect in liquids are high-intensity local turbulence, shock waves, microjet streaming, and greater shear stress. Thermal effects by cavitation result in bubble implosion, leading to vapour compression, which further leads to local hotspot formation. These hotspots are extremely localised heat changes (heating and cooling) in 10^{10} K/s that occur within micro to milli seconds (Sun, Liu, et al., 2020). These effects are primarily accountable for strengthening physical processes like nanoemulsion synthesis, formation of nanoparticles, microbial inactivation by disrupting cell walls, and disinfection. But, in contrast, highly reactive free radicals are generated in the aqueous medium accountable for the synthesis of various undesirable chemical products, degradation of nutrients, etc., among some of the unwanted chemical effects of the process (Carpenter, George, et al., 2017). Among these, acoustic cavitation and HC can be used in food process applications because of the effects produced by these two techniques and the easiness of producing cavitations (Asaithambi et al., 2019).

Initially, studies were conducted to reduce the HC cavitation effects in the system because of the damage and corrosion produced by this. Nowadays, scientists are trying to utilise these effects in the food and beverage industries. Applications of HC include disruption of microbial cells, food sterilisation, encapsulation, increased antioxidant activity, ethanol production, protein extraction, brewing, protein accessibility, waste management, etc.

12.2 THE MECHANISM IN HYDRODYNAMIC CAVITATION

HC can be induced by passing the food through a very small constriction like orifices, venturi holes (Figure 12.1), etc. It increases the velocity and kinetic energy. Millions of cavities are formed if the throttling pressure is below the threshold. Here, threshold pressure is usually the vapour pressure of the medium at the operating temperature. Later, when the liquid expands after the constriction, the cavities break and energy is released. Cavitation number (C_v) is an important term in this technology which can be expressed as in Equation 12.1.

$$C_v = P_2 - P_v / \left(1/2 \; \rho v_{th}^2 \right) \tag{12.1}$$

Where

P_2 is the "downstream pressure",
P_v is the "vapour pressure of the liquid", and
v_{th} is the "constriction velocity" (Carpenter, George et al., 2017).

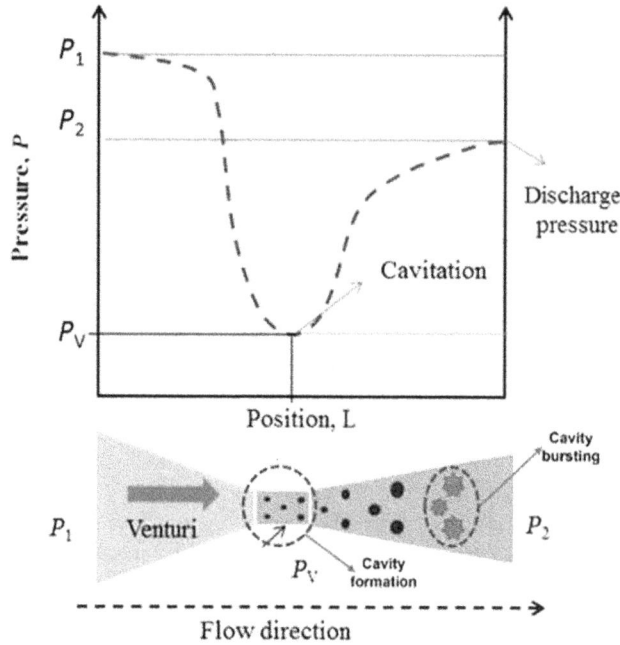

FIGURE 12.1 Pressure variation concerning the flow conditions in the venturi.

The initiation of the cavitation starts at the cavitation inception number. When the inlet pressure of the liquid stream increases, volumetric flow and velocity also increase. This velocity increase results in a low C_v; if the C_v is too low, super cavitation occurs, and no cavitational collapse occurs. In the region where $0.1 < C_v < 1$, cavitation formed is strong, and the collapse of the bubbles is more. Due to dissolved gases and other solid particles in the fluid(s), cavitation can be possible if $C_v > 1$ (Panda et al., 2020). Normal cavitation is absent in the case of $C_v > 4$ (Albanese & Meneguzzo, 2019). If the operating pressure is higher, then velocity decreases, and the Cv decreases. A small C_v indicates more cavities yielding more hydroxyl ('OH) radicals because of more cavity formation. These 'OH radicals cause oxidation or degradation of pollutants present in the water (Dhanke & Wagh, 2020). The operational temperature has an important role in HC; if more operating temperatures, then vaporous cavities get formed, which results in lower cavitation (Joshi & Gogate, 2012). Cavitation is generally more acidic (Carpenter, Badve, et al., 2017). In HC, the collapse intensity would be less compared to ultrasound cavitation. Intensity can be increased by changing the geometry and operating conditions of the system.

12.2.1 Effects of HC

Due to the cavitation phenomenon occurring with sudden variations in pressure and velocity simultaneously, air bubbles or cavities form, enlarge, and collapse, inducing some chemical and physical effects in the medium, further leading to desired changes. Moreover, the required changes or transformation in the medium depends on the method of cavity collapse. Collapse can be in two forms: symmetric collapse and asymmetric collapse. In symmetric collapse, the cavity remains spherical until the point of collapse, resulting in reactive free radical formation and organic molecule thermal pyrolysis. But in contrast, during the asymmetric collapse, bubbles stay non-spherical, and effects like intense local turbulence and high-velocity liquid jets are produced that are beneficial in most cases of desired physical transformation. Though both forms of cavity collapse obtain similar energy from the pressure and velocity fluctuations, the energy is dissipated mainly in two different forms, i.e., intense local turbulences or localised hot spot formation with high temperature and pressure.

With the cavitation and bubble collapse, a huge amount of localised energy is released into the surrounding environment in three forms: mechanical, thermal, and chemical (Carpenter, Badve et al., 2017; Sun, Liu et al., 2020).

12.2.1.1 Mechanical Effect

One of the most devastating HC effects involves mechanical or physical effects. This effect is attributed to the formation of high shear stresses by microjets, shock waves, and fluid collapse. A rapid change near the gas-liquid interface can cause adiabatic compression of bubbles and generate shock waves. In areas of maximum compression, the bladder wall stops expanding and contracts rapidly. At the onset of the ejection, a high-pressure shock wave with strong non-linear propagation properties is generated by contracting the liquid reflected from the bubble interface. Shock waves can travel through water at nearly three to four times the speed of sound in water (approximately 1,500 m/s). This means that at pressures up to 6,000 to 7,150 GPa, the smallest bubble radius is up to 4,000 m/s. Asymmetries can arise from free surfaces, solid boundaries, gravity, and acoustic pressure and can give rise to high-velocity re-entry jets (microjets) during rapid bubble contraction (Kuppa and Moholkar, 2010). The potential energy of the expanding bubble is translated during this process into the kinetic energy of the liquid around the collapsing bubble.

As the top surface accelerates inward faster than the bottom surface when it begins to deflate, the bubble will self-fold before being punctured by a microscopic jet of liquid. Shock waves are also generated when the water jet hits the bottom surface before the ring collapses. On the other hand, the opposing jet forms a rigid surface and moves in its direction. In addition, the interaction effect between the bubble and the pressure pulses caused by the collapse of other bubbles and the propagation of the shock wave can produce more powerful water jets. In such cases, the speed may exceed 370 or 400 m/s.

For example, the water hammer pressure can exceed 200 MPa when the water jet hits the aluminium material at a speed of 151 m/s. In addition, due to the conical or circular shape of the tip, the effective impact can be three times the water hammer pressure. Bubble collapse also produces high shear stresses. Strong, unstable boundary layer flows with higher shear rates are produced by secondary Bjerknes forces, causing bubbles to expand and contract near hard boundaries to move in the direction of the boundary (Carpenter, Badve, et al., 2017).

12.2.1.2 Thermal Effect

The hotspot theory is coherent with the experimental results and can explain the nature of the thermal effect caused by cavitation. Gases and vapours in foaming bubbles are compressed during bursting or collapse. It generates a large amount of heat instantaneously, only increasing the temperature of the liquid around the bubble and forming a small hot spot. A few thousand Kelvin local hot spots increase heating and cooling rates above 1010 K/s within microseconds to milliseconds. The distance from the centre significantly impacts the heating effect that is the source of homogeneous electrochemistry. There are three zones around the hotspot area: (Zone-I): inside the core of a bubble or cavity, where the gas phase can reach a peak temperature of 5,000 K to 10,000 K and a pressure of 1,000 atmospheres; inside this central region, trapped molecules split into smaller particles that generate reactive free radicals. These free radicals disrupt target molecules and trigger further dissociation. For example, water molecules are dissociated into •H and •OH radicals; (Zone-II): at the interface, where the thin layer of liquid surrounding the bubble's collapse can reach peak temperatures of 1,900 K to 2,000 K near atmospheric pressure. At this point, cavity oscillation and its subsequent collapse lead to high microjet and shear turbulence. Reactive free radicals will react with organic molecules and thermally decompose the following molecules; (Zone-III): is in the bulk medium where the temperature and pressure reach values close to normal atmospheric temperature and pressure (300 K and 1 atm). However, the bubble's collapse has no direct effect to volume temperature. The radicals generated here diffuse into the bulk of the liquid and interact with the desired molecules (Adewuyi, 2001; Sun, Chen, et al., 2020).

12.2.1.3 Chemical Effect

•OH has a high standard reduction or oxidation capacity (2.8 V) but slightly less than fluorine (3.03 V). •The OH radical can remove an electron from other compounds with an unpaired electron. It is believed to aid in the breakdown of chemicals (organic pollutants) and the inactivation of microorganisms and to play an essential role in acoustic chemistry. The water (H_2O) and dissolved oxygen (O_2) molecules in the cavities can be decomposed into various types of hydrogen atoms (•H), oxygen atoms (•O), hydroperoxyl radicals (HO_2•), and other atoms. Other particles, in extreme cases, are produced by the collapse of bubbles in the gas phase and gas-liquid interface. Without solute, these primary radicals can form H_2O, •O, and O_2. Outside the cooling interface or around the heated bubbles, the recombination of •OH and HO_2• produces hydrogen peroxide (H_2O_2). •H and •OH types can also react with H_2O_2 to produce OH and HO_2 (Ashokkumar, 2011; Sun, Chen, et al., 2020).

12.3 HC REACTORS

The important parts of the HC reactor are a feed tank, a plunger pump, a pressure gauge, and a cavitation chamber. HC can be classified into the orifice, venturi tube, or throttling according to the geometry or type of constriction used for creating the effect. According to the operating ways, HC can be classified into pulsating HC, continuous HC, and shear-induced HC (Gogate & Pandit, 2001).

12.3.1 Liquid Whistle Cavitation Reactor

Mixing and homogenisation are the two important functions of this reactor. A "jet edge tone" process creates cavitation energy within the liquid stream. When process liquid passes through a specially designed lip-shaped orifice, it will be subjected to extreme pressure and liquid jet vortices perpendicular to the original flow vector and create oscillations similar to those created by ultrasound (Gogate, 2011b). This device is very flexible, and maintenance problems are fewer. The cavitation intensity is less for the liquid whistle reactor. So, there are no reports about this reactor in chemical processing (Gogate & Pandit, 2001).

12.3.2 High-Pressure Homogenizer

"High-pressure discharge technique" is the basic principle of high-pressure homogenizer (HPH). The main components of this device are a high-pressure positive displacement pump and a throttling device. This system consists of two stages (first stage and second stage). The pressure in the first stage is 1,000 psi and increases to 10,000 psi in the second stage when the valve is used. The liquid from the tank is pumped to the first stage by a pump. Various pressure relief valves ensure fluid circulation to the supply tank. Cold water is used as the temperature regulator in this system. The critical discharge pressure for cavitation depends on the geometry of the throttling device. Applying the liquid to the device is also important to achieve the critical discharge pressure (apart from this, foaming the critical discharge pressure would be more efficient and appropriate). HPH can foam and emulsify in pharmaceutical, bio-processing, and food industries. The main disadvantage of this HPH is that there is little control over the intensity and volume of foaming action (Gogate, 2011b).

12.3.3 High-Speed Homogeniser (HSH)

Rotating elements can also produce HC. When the rotating element reaches its critical speed, the pressure on the outer (peripheral) part of the rotating element, such as the impeller, will decrease and reach the vapour pressure of the liquid. This pressure drop creates vapours. These bubbles move away from the impeller, so the pressure also increases, and this increase in pressure causes the

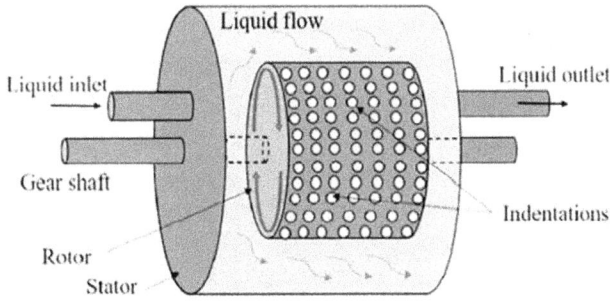

FIGURE 12.2   High-speed homogeniser having a stator rotor assembly.

bubble to collapse near the wall of the tank. HSH consists of a rotor and a stator (Figure 12.2). This kind of assembly is mainly made of stainless steel. There is a plate with holes in the stator to eliminate vortex formation and surface aeration. The power supply can control the speed of this device, and cavitation is initiated beyond the critical rotational speed. The main disadvantages of this type of device are higher power consumption and less flexibility with respect to design parameters. The suitable rpm for this device is 4,000 to 20,000 (Gogate, 2011b).

12.3.4   Low-Pressure Cavitation Reactor

The flow passes through a narrow portion and causes a restriction to the flow so that velocity increases (because of the reduction in area) and pressure decreases. Sometimes this pressure may decrease to the vapour pressure of the liquid, and bubble formation occurs. In many studies, venturi tubes and orifices (single and multiple) can be used as constrictions. A holding tank, a control valve, a centrifugal pump, and flanges are the main parts of a single orifice plate device. A multiple-hole orifice reactor has many advantages over a single-hole orifice device. The generation of different cavitation intensities is possible using a multiple-hole orifice reactor. Flexibility over operating and geometric conditions is another advantage of a multiple-hole orifice reactor. This flexibility helps select operating and geometric parameters for a particular application with minimum energy utilisation and maximum cavitational effects. A novel HC reactor by (Sampath Kumar & Moholkar, 2007) uses a converging-diverging nozzle for creating cavitation. Bubbles are introduced externally to the water flow. Different gases can be used to create cavitation. The gas distributor's size and flow rate are other important factors in controlling the cavity's bubble size and gas composition. If the nuclei are small, the cavitation intensity will be higher. The studies by Sampath Kumar and Moholkar (2007) indicated that the design could destroy microbial cells and sterilise food.

12.3.5   Microfluidisers

Microfluidisers (Figure 12.3) produce a jet velocity of up to 200 m/s and can produce a pressure drop of about 2 kbar. Cell disruption and emulsification are the main applications of microfluidics. A large pneumatic pump is used to push the reaction solution into the interaction chamber (Mozafari et al., 2008). Two streams with a pressure of 190 m/s are passed through the orifice of the jewel employing a back pressure regulator. The reactor and its accessories are immersed in a thermal equilibrium bath to control the temperature. For example, the water hammer pressure can exceed 200 MPa when the jet hits the aluminium material at 151 m/s. In addition, due to the conical or circular shape of the tip, the effective impact can be nearly three times the water hammer pressure. Bubble collapse also produces high shear stresses.

The secondary Bjerknes force produces a strong and unstable boundary layer flow with high shear rates, which causes bubbles that are expanding and contracting close to rigid boundaries to

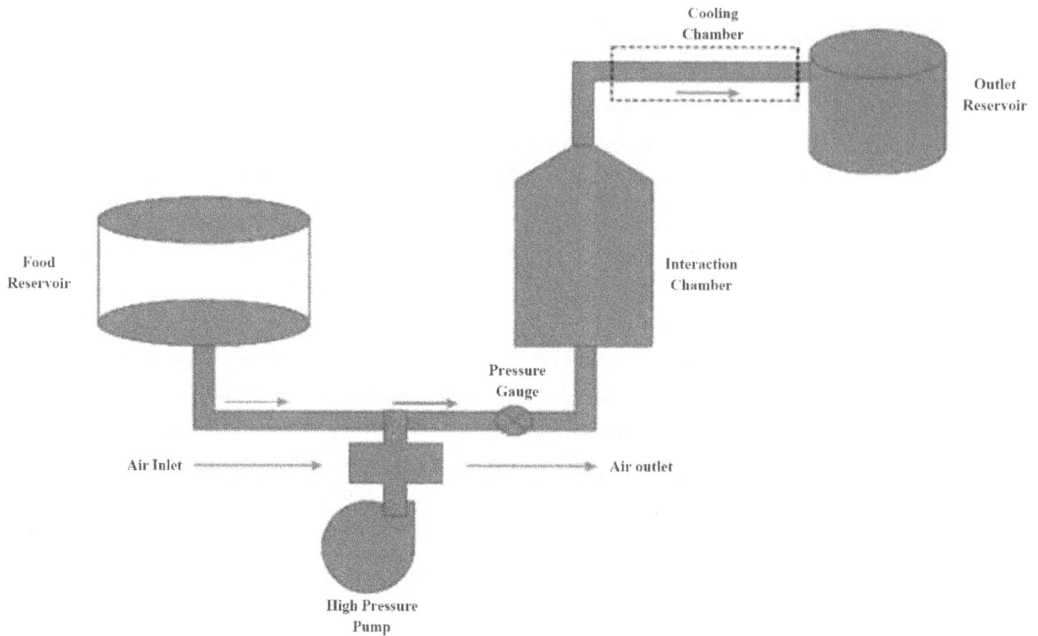

FIGURE 12.3 Schematic diagram of microfludisation assembly.

move in the direction of the boundary. High shear stresses result from the radial flow's interaction with a jetting flow that accelerates the liquid's impact on and spread along the wall and rises in the liquid medium. The studies by Suslick et al. (1997) exhibited that the Weissler reaction rate is very high in the acoustic cavitation reactor than in HC reactors.

12.3.6 SHOCK WAVE REACTOR

Studies by Milly et al. (2007, 2008) established that a shock wave reactor could be used for pasteurisation or sterilising fluid foods. The same reactor consists of an inner rotating cylinder and an outer stationary cylinder. The liquid is passed through the annular space between the cylinders. The inner cylinder has multiple holes on the outer surface. When this inner cylinder rotates at high speed, cavitation occurs in the liquid in contact with the inner cylinder. Bypass configuration in the reactor helps control the fluid's inlet pressure and flow rate. This bypass configuration is also helpful for recirculating the fluid to the storage tank without subjecting it to cavitational effects. A high-speed rotating cylinder causes the pressurisation of the fluid through the holes, and depressurisation occurs when it reaches a high area, which causes the liquid to vaporise. The dissolved gases present in liquid incorporate with the vapours to produce bubbles. Parameters like the size of holes, flow rate, etc., can be controlled in this reactor, but there is no control over cavitation intensity. The main disadvantage of this shock wave reactor is the lower intensity of cavitation. The energy generated due to the cavity or bubble collapse is less than cavities with voids or gaseous cavities.

12.4 APPLICATION OF HYDRODYNAMIC CAVITATION

12.4.1 MICROBIAL CELL DESTRUCTION

Microbial cell disruption is an important application of HC. Cell disruption is helpful for the release of intracellular components from microorganisms. The studies done by Save et al. (1994) showed that among various technologies used for microbial cell disruption, HC was the most energy-efficient technology. Save et al. (1997) studied the effect of HC on baker's yeast and showed that cell ruptur- ing and enzyme release increase with the number of passes and treatment time. One of the previous studies on HC showed the disruption of yeast in large-quantity and growing cells, i.e., cells in the exponential phase were seen as more susceptible to destruction than those in the stationary phase. At low discharge pressures, increasing cell concentration causes a significant increase in the quan- tity of cell breakdown. The intensity of cavitation increases as discharge pressure increases. As a result, nearly every cell may come into direct contact with the collapsing cavities. Thus, the extent of cell breakage is greater at higher discharge pressures. However, with larger cell concentrations, the likelihood of a cell colliding with a cavity is significant even at lower discharge pressures (Save et al., 1997). Azuma et al. (2007) made a high-pressure cavitation device. It combines two cavitating orifices in series with a plunger pump capable of producing pressure up to 1,050 bar. The variables in this experiment are the Cv (0.037–0.487) and the upstream nozzle velocity (176–384 m/s). During the second stage of the experiment, the sterilisation rate of gram-positive and gram-negative bacte- ria (*E. Coli, Pseudomonas putida*) was studied. At C_v 0.154, complete disinfection of water and *E. coli* in three successive treatments is achieved.

Gram-positive bacteria are more resistant than gram-negative bacteria to cavitation because of the strong cell wall at two conditions (C_v 0.104 and 0.037). Therefore, we can conclude that HC is an efficient method for cell destruction. Lee and Han (2015) showed an increased yield of lipids in wet microalgae when treated using HC. They identified a highly energy-efficient method for extracting oil from algae. Dong and Qin (2018) reported that cavitation created by microjets and shockwaves disrupts pathogenic microorganisms. The studies by Mevada et al. (2019) showed that energy effi- ciency increased up to 63.19 times and selective release of intracellular enzymes increased up to 4.79 times with HC. The studies on brewer's yeast (Balasundaram & Harrison, 2006) show that a C_v of 0.13 is the maximum for optimum cell disruption among C_v 0.09 to 0.99. They also studied the effect of HC in *E. coli* cell disruption by evaluating the periplasmic and cytoplasmic protein release. The maximum extent of protein release was at C_v 0.17 among the C_v 0.13 to 0.92. In 2011, they stud- ied the effect of the system's geometry and orifice number on periplasmic protein selective release. They selected square, rectangular, and circular holes. Results showed that the total soluble protein released was the same for the equal hole area. The release of acid phosphate was more in circular holes. The effect of flow rate also studied after 1,000 passes, showed that the release percentage is higher for the high flow rate.

12.4.2 FOOD STERILISATION

HC creates cellular inactivation by developing physical stresses in cells. HC can be used either individually or in combination with thermal methods to sterilise food items (Table 12.1). The main advantages of some methods using HC have reduced energy consumption, achieved microbial inac- tivation at low temperatures, and kept the freshness of the food products. Studies conducted by Milly et al. (2007) in liquid foods (tomato juice, apple juice, and skim milk) showed that HC could destroy bacteria, yeast, yeast ascospores, and heat-resistant bacterial spores. Commonly occurring spoilage-causing organisms like yeast and lactic acid bacteria can be inactivated at low tempera- tures with the combined effect of temperature and HC. This study also indicates the factors affect- ing the mortality rate of microorganisms by using HC. Those factors include moderate processing temperature (normally 65.6°C), elevated product exit temperature with maximal SPR residence time, and SPR rotor speed. HC helps to reduce sterilisation temperature so that it can give products of superior quality. In 2008, Milly et al. (2008) conducted another study based on applying a shock wave reactor in apple juice to inactivate *Saccharomyces cerevisiae*. The results showed that an increased reduction in *Saccharomyces cerevisiae,* up to 6.27-log cycle reduction, can be achieved by providing sufficient HC at low processing temperatures (65.6 and 76.7°C). HC is an energy-efficient method compared to PEF technology, so the efficiency of fruit juice industries can be enhanced using the HC effects. This study also indicated that scale-up of the equipment is very easy and can

TABLE 12.1

Some of the Past Studies Related to Microbial Destruction and Sterilisation Using HC

Sl. No.	Food Product	Treatment and Reactor	Observations	Reference
1	Skim milk	HC reactor	• *Clostridium sporogenes* were reduced by 2.84 log cycles when processed at 3000 to 3600 rpm reaching temperatures up to 115.6°C.	(Milly et al., 2007)
2	Apple juice	The reactor consists of an annulus with a rotating pock-marked inner cylinder	• *S. cerevisiae* was reduced by 6.57 log CFU/mL. • Energy consumption was less than thermal processing done at 65 and 76.7°C.	(Milly et al., 2008)
3	Liquid foods with *S. cerevisiae*	HC reactor with hydraulic circuit	• 90% yeast strains reduced at a far lower temperature (6.3–9.5°C) were achieved. • Energy input taken was about 20% lower.	(Albanese et al., 2015)
4	Raw water with pathogenic microorganisms	Venturi-type reactor with different throat lengths	• *E. Coli* and total plate count were reduced significantly at lower cavitation numbers. • Increasing throat length and cavitation time enhanced the inactivation of microbes.	(Dong & Qin, 2018)
5	Raw milk	Cavitator with a spinning rotor with 4 rows of 40 holes each	• A greater than 1-log reduction in total bacterial count was observed. • Milk fat globule size was brought down to <1.50 μm.	(Gregersen et al., 2020)
6	Milk	Continuous hydrodynamic cavitation	• 5.89, 5.53, and 2.99 log reductions of *E. coli*, *S. aureus*, and *B. cereus* were achieved, respectively.	(Sun et al., 2021)

be used in large-scale industries with improved energy efficiency. Another work done by De Bonis and Ruocco (2010) reported that direct steam injection could also be a reason for the formation of HC if the injecting parameters are controlled accordingly.

12.4.3 Water Disinfection

There are many physical and chemical methods for water disinfection. Chlorination and floc-culation using some physical methods, like UV treatment, etc., are some of the methods used for water treatment. But these methods are not 100% efficient. The disadvantages of these methods make people think about alternative methods that are more efficient than present ones. The research was done in bore well water disinfection by using HC reactors such as high-speed homogeniser (alone), HPH and orifice plate set up, and comparison of the effect with acoustic cavitation by using ultrasonic horn-type reactor which is operating at 22 kHz and power rating of 240 W showed that disinfection increases with increase in supplied energy to the system. More pressure on the orifice plate reactor with 75 L capacity increases the disinfection rate. This study showed that cavitation is effective not only in disinfection but also in a less time-consuming process. Comparing disinfection per unit of energy consumed in all the equipment showed that a high operating pressure orifice plate setup was more efficient than other cavitation reactors. The extent of disinfection for orifice plate setup, ultrasonic horn, and HPH is 310 CFU/J, 45 CFU/J, and 55 CFU/J, respectively. The disinfection rate is higher for HPHs, but energy efficiency is less and only 5 CFU/J (Jyoti & Pandit, 2001).

Studies conducted by Arrojo and Benito (2008) showed the effect of cavitation designs and different operational parameters on the inactivation rate of *E. coli*. Three different configurations with an equal overall free cross-section and different hole numbers. One hole with a 5-mm diameter, 6 holes with a 2-mm diameter, and 25 holes with a 1-mm diameter was the configuration used for the study. A venture tube is also used in this study. Mechanical disruption of bacteria happened with slow pressure oscillations that mean with low frequency. Disinfection rates were maximum in the configuration and condition, promoting large bubble formation, extended pressure oscillations, and increased cavitations. These conditions are shown by venture tubes, so more disinfection was possible in the venturi tube.

Jyoti and Pandit (2001) investigated the effect of different cavitation methods like hydrodynamic and acoustic, with some chemical methods like H_2O_2 and O_3 for bore well water treatment. HPC bacteria will be destroyed more when chemical methods are combined with ultrasonication. The disinfection of HPC was more when HC was combined with chemical methods rather than treatment with HC alone. The rate of cell disruption can be further improved when combining the HC reactor with hydrogen peroxide and ultrasonic flow cells. Free radical generation and individual effects of hydrogen peroxide, ultrasonic cavitation, and HC help to improve the result. Studies were done by Sawant et al. (2008) in water treatment using liquid whistle HC combined with ozone treatment. A suspension of *E. coli* is used to study the effect of treatment. HC causes 22% disinfection, and a combination of ozone causes 75% disinfection. HC has a low extent of disinfection; at the same time, ozone treatment is costly. A combination method is a good alternative to individual methods.

Mezule et al. (2009) studied the disinfection of *E. coli* on a laboratory scale using HC. In this study, a rotor generated cavitation. A simple milling cutter was used for driving the rotor. The CTC (5-cyano-2,3-ditolyl tetrazolium chloride) method is used to measure the respiratory condition, which is a key element in evaluating the disinfection power. A direct viable count (DVC) method is used to measure the reproduction or multiplication. Results showed that an energy input of 490 W/L for 3 minutes prevented the division of 75% *E. coli* cells. However, there is no defect for respiration, concluding that bacteria are in ABNC (active but not in the culturable state). The results indicate that HC can be an effective water disinfection method if we increase the cavitation intensity so that the destruction of the cells is possible. Optimisation of the cavitation chamber is the best method to increase the cavitation intensity; we can attain complete cell disruption.

Cost is very important in selecting a new technique for any process. The technique should be cheap compared to normally using conventional methods. Jyoti and Pandit (2001) estimated the cost of treatments using different cavitation reactors and compared them with ozone and chlorine treatment. The results indicate that cavitation produced by orifice plate setup and high-speed homogeniser is the most cost-effective method. The cost of the treatment using a high-speed homogeniser is US $0.81/m^3$, and for the orifice plate technique, it is US $1.4/m^3$. The techniques using sonochemical reactors and HPHs cost US $14.88 and US $6.55/m^3$, respectively. But these costs are very high when compared with chlorination and ozone treatment. The cost of chlorination treatment is US $0.0071/m^3$; for ozonation, the cost is almost US $0.024/m^3$. These cost estimations are based on small-scale applications. So, cavitation is more suitable when a bulk quantity of water is used, like large-scale municipal corporation wastewater treatment plants and ballast water treatment. There are no hazardous by-products in the case of HC compared to traditional chlorination and ozone treatment. We can modify the reactors to get high cavitation intensity, and combining HC with traditional techniques will reduce the cost of the techniques.

12.4.4 PROTEIN SOLUBILISATION

Proteins are essential nutrients for the human body. They are one of the building blocks of body tissue and can also serve as a fuel source. HC of animal- and plant-based proteins has been experimented with so far. Studies were done by Chen et al. (2016) to solubilise myofibrillar proteins in chicken breast using an HPH. 0.1, 69, 103, and 168 MPa pressure for two passes in an HPH creates cavitation. Myofibrillar protein's solubility, particle property, and microstructure were tested after treatment. Results showed that high-pressure homogenization (HPH) reduced the particle size and produced an aqueous suspension of myofibrillar protein with high absolute zeta potential. The reduced size and intermolecular repulsion cause high chicken breast myofibrillar protein solubility. So, HPH is an acceptable method for myofibrillar solubilisation. Chen et al. (2017) continued researching myofibrillar protein powder to improve its functional properties. Structural change can improve the myofibrillar protein powder's functional properties. An HPH and hydrogen peroxide can increase the thermal stability of the myofibrillar protein (Chen et al., 2017). The research was done to study the effect of high-intensity ultrasound (HIU), HPP, and HPH on myofibrillar protein suspension's physicochemical and functional properties and compared it with an untreated control sample. The results showed that the sample treated with HPP had increased solubility. The sample treated with HPH showed less solubility. Rheological studies of the sample water treatment showed that HPH had poor gel formation ability compared to HPP and HIU. This study concluded that HPP is a good and effective approach for manufacturing gel-based meat products (Xue et al., 2018).

Studies were done in 2017 to improve the functional properties of freeze-dried myofibrillar protein powder by modifying the structure using HPH. The intensity of homogenisation used for this study ranges from 0 to 20000 psi. The dissociation of myosin and the myofibrillar polymer was detected in water-soluble myofibrillar protein powder (WSMP-P) after HPH. HPH also caused the dissociation of myofibril and myosin and actin trimer formation by partial unfolding and rearrangement, thereby modifying the structure of WSMP-P. This dissociation and actin trimer formation created an amorphous protein structure with more thermal stability. They showed that HPH could increase the water-solubility and emulsifying properties of WSMP-P by reducing particle size to unfolded structure to submicron level and creating a high surface net charge in aqueous suspension (Chen et al., 2017). This study gives hope to using WSMP-P as a protein supplement in food and beverage under low ionic conditions.

Myofibrillar proteins (MPs) contributes 50% to the meat and is sparingly soluble in low ionic strength or water. Complete solubilisation of myofibrillar protein is possible when a high salt concentration is used. The investigations did in myofibrillar protein to study the solubility in low ionic solution by applying 15,000 psi HPH pressure. They concluded that HPH unfolded the MPs, and then hydrophobic and sulfhydryl groups came out to the surface, which caused the formation of myosin oligomers by forming disulphide bonds. The change in the myosin conformation prevented film formation, so the solubility of the MPs in water increased (Chen et al., 2016).

Studies were done on laccase-treated α-lactalbumin (α-LA) pre-treated with HPH to learn the effect of HPH on physicochemical, structural, and functional properties. 20, 40, 60, 80, 100, and 120 MPa for 4 cycles were applied to laccase-treated α-LA. The number of laccases treated with α-lactalbumin decreased with increased pressure; at the same time, oligomers and polymers were increased. The results of FTIR-spectroscopy showed an increase in random coil content and a decrease in α-helix with the increase in pressure of homogenisation. The degree of cross-linking, emulsifying properties, functionality, and gel-forming properties increased with HPH pre-treatment.

Studies were done to find the effect of the geometry of the disruption chamber on myofibrillar protein. They showed a decrease in myofibrillar particle size when passing through a narrow nozzle by a counterflow pattern. The cavitation produced in the narrow nozzle with a counterflow pattern employed the destruction of the secondary structure of the myofibrillar protein in water. The destruction of the secondary structure made the hydrophobic group and SH group available so that the solubility and stability of the myofibrillar protein solution increased. HPH was the method adopted for creating cavitation (Li et al., 2019). Ren et al. (2020) compared the effect of HC (using a venture tube) with ultrasonic cavitation on soy protein isolate's functional and physicochemical properties. Electrophoretic profile, pH, particle size, viscosity, surface hydrophobicity, sulfhydryl content, protein solubility, emulsifying, and foaming properties were analysed. The result showed that HC was better than ultrasound when the volume of soy protein isolate was 2 L. When 200 mL volume was treated with ultrasound, HC showed the same properties. Results indicated the usage of HC for a bulk amount of food. Studies done on lentil protein isolate (LPI) suspension showed that HPH is an effective method to change the structural, functional, and rheological properties of LPI suspension. The results of this study revealed that HPH treatment could not change the protein pattern.

Particle size is reduced up to 100 MPa pressure, and after 50 MPa, size distribution was mono-modal. When the pressure increased from 50 to 150 MPa, it resulted in the unfolding of protein detected by microstructural images. The unfolding of protein particles supports the results of water solubility, foaming, emulsifying, and particle size. LPI is a shear-thinning fluid, and the results followed the Ostwald de-Waele model. Apparent viscosity was modelled with an exponential function, and homogenisation pressure was modelled with a sigmoid function. After 51.78°C homogenisation pressure, more than 50 MPa showed improved gelation capacity (Saricaoglu, 2020). Studies were done in soybean protein isolate (SPI) emulsion gel to reveal the effect of HPH on the gelling and rheological properties showed that SPI emulsion gel strength increases with rising HPH pressure. Emulsion gel strength increases from G', 291 Pa to 528 Pa at 5 MPa and 80 MPa, respectively. The water-holding capacity of the emulsion gel improved when pressure increased from 5 to 20 MPa. Water holding capacity was 87.7% at 5 MPa and increased to 91.4% when the pressure reached 20 MPa. A further rise in pressure to 80 MPa did not affect the water-holding capacity. SPI emulsion gel's microscopic fractal dimension was from 2.96 to 2.99. The higher pressure of HPH to emulsion caused a more stable isotropic network gel structure (Bi et al., 2020).

Research is done by Saricaoglu (2020) on an edible film produced from mechanically deboned chicken meat protein after HPH treatment. 0, 25, 50, 75, 100, and 150 MPa intensities were selected to study the effect. Results revealed that particle size reduced up to 100 MPa, which provides good water vapour permeability and improved tensile strength to the film. 150 MPa did not show a smooth and homogenous surface micrograph. Pressure up to 100 MPa can be applied to the film suspensions to produce films with better barrier and mechanical properties.

12.4.5 EMULSION

Emulsifiers are used to stabilise the emulsion. HPH is an effective method to produce stable emulsions. HPH helps to give a fine texture to the emulsion. A study on the effect of HPH on oil in water emulsion, which is stabilised with muscle myofibrillar protein (MMP), showed improved properties such as an increase in emulsifying stability index (59.33–154.62 minutes), emulsifying activity index (56.93–87.68 m^2/g) and apparent viscosity. Absolute zeta potential increased from 10.67 to 37.03 mV. The droplet size of the emulsion reduced to 0.46 μm from 3.95 μm. Results showed a decreased cream index value. The solubility of the emulsion increased to 75.1% from 16.5%. The study concluded that the changes to the emulsion might result from the increased solubility and exposed hydrophobic groups (258.50 from 196.37). 40, 80, and 120 MPa pressure were applied to study the effect; 80 MPa produced the finest and most stable emulsion (Cha et al., 2019). These results give light to the natural emulsifier area. Studies showed that MMP could be used as a natural emulsifier in food and contributed a method to improve the formulation and properties of the emulsion.

Ren et al. (2020) studied the effect of HC on emulsification. Refined soybean oil was used as the oil phase. A valve with a 3 mm diameter was used to create HC. The inlet pressure was regulated by controlling the opening of the valve. A series of emulsions were prepared by changing the inlet pressure to study the effect of inlet pressure. Emulsions were created by applying 20, 50, 80, 100, 120, and 150 psi pressure. There was a decrease in droplet size when the pressure of 20 to 120 psi was applied, and the size remained the same after the 120 psi pressure. The particle size of 27 nm was obtained at 120 psi. The negative value of zeta potential and polydispersity index of less than 0.1 after 120 psi indicated the mono-dispersity of emulsion. The oil-in-water (O/W) emulsion created by HC was stable for up to 8 months. This result gives light to emulsification using HC.

The study's results on the effect of geometry and geometrical parameters on the formation and stability of mustard O/W emulsion. The droplet size of 87 nm was obtained in an orifice plate of circular shape with a single hole. This orifice device had less perimeter and higher flow area than other reactors (venturi tube and other orifice configurations). The smallest droplet size was the optimum condition to optimise the device. The smallest particle size was obtained at C_v 0.19 after 90 minutes of processing. Mustard oil in water nano emulsion made by HC was kinetically stable for up to 3 months under centrifugal and thermal stress conditions. HC was also 11 times more energy-efficient than acoustic cavitation (Carpenter, George et al., 2017). The orifice plate with a 0.0010 cm hole diameter incorporated liquid whistle HC reactor produced a submicron emulsion with a 476 nm droplet size when the optimised inlet pressure and the distance between the blade and the orifice were set accordingly (Parthasarathy et al., 2013).

12.5 SUMMARY

Among the non-thermal technologies, as mentioned in the introduction, HC works uniquely. Compared to acoustic cavitation techniques and other traditional ways, HC is an energy-efficient method. This technique offers immediate, cost-effective industrial use and higher energy efficiencies. The use of HC in large-scale processing for sterilisation, microbial cell disruption, and various other applications appears to have the potential to enhance and maintain the functional properties of food and beverages. For instance, by making subtle changes in the liquid food processing line, i.e., installing a cavitating device in the line of flow, one can achieve better results in an energy-efficient manner. The final products' quality depends on the design of the cavitators and the nature of the food material. There's still much effort needed to master the design of the cavitators that suit the need and adaptability across different food materials. With coordinated action, a broad area of study can be addressed by using this technology to produce quick and even more energy-efficient results in the food and beverage industry.

REFERENCES

Adewuyi, Y. G. (2001). Sonochemistry: Environmental science and engineering applications. *Industrial and Engineering Chemistry Research*, 40(22), 4681–4715. https://doi.org/10.1021/IE010096L

Albanese, L., Ciriminna, R., Meneguzzo, F., & Pagliaro, M. (2015). Energy efficient inactivation of saccharomyces cerevisiae via controlled hydrodynamic cavitation. *Energy Science and Engineering*, 3(3), 221–238. https://doi.org/10.1002/ESE3.62

Albanese, L., & Meneguzzo, F. (2019). Hydrodynamic Cavitation-Assisted Processing of Vegetable Beverages: Review and the Case of Beer-Brewing. In *Production and Management of Beverages*. Elsevier Inc. https://doi.org/10.1016/b978-0-12-815260-7.00007-9

Arrojo, S., & Benito, Y. (2008). A theoretical study of hydrodynamic cavitation. *Ultrasonics Sonochemistry*, 15(3), 203–211. https://doi.org/10.1016/j.ultsonch.2007.03.007

Arya, S. S., Sawant, O., Sonawane, S. K., Show, P. L., Waghamare, A., Hilares, R., & Santos, J. C. Dos. (2020). Novel, non-thermal, energy efficient, industrially scalable hydrodynamic cavitation – applications in food processing. *Food Reviews International*, 36(7), 668–691. https://doi.org/10.1080/87559129.2019.1669163

Asaithambi, N., Singha, P., Dwivedi, M., & Singh, S. K. (2019). Hydrodynamic cavitation and its application in food and beverage industry: A review. *Journal of Food Process Engineering*, 42(5), 1–14. https://doi.org/10.1111/jfpe.13144

Ashokkumar, M. (2011). The characterisation of acoustic cavitation bubbles – an overview. *Ultrasonics Sonochemistry*, 18(4), 864–872. https://doi.org/10.1016/J.ULTSONCH.2010.11.016

Azuma, Y., Kato, H., Usami, R., & Fukushima, T. (2007). Bacterial sterilisation using cavitating jet. *Journal of Fluid Science and Technology*, 2(1), 270–281. https://doi.org/10.1299/JFST.2.270

Balasundaram, B., & Harrison, S. T. L. (2006). Study of physical and biological factors involved in the disruption of *E. coli* by hydrodynamic cavitation. *Biotechnology Progress*, 22(3), 907–913.

Bi, C.-Hao, Wang, P.-Lin, Sun, D.-Yu, Yan, Z.-Ming, Liu, Y., Huang, Z.-Gang, & Gao, F. (2020). Effect of high-pressure homogenisation on gelling and rheological properties of soybean protein isolate emulsion gel. *Journal of Food Engineering*, 277, 109923.

Carpenter, J., Badve, M., Rajoriya, S., George, S., Saharan, V. K., & Pandit, A. B. (2017). Hydrodynamic cavitation: An emerging technology for the intensification of various chemical and physical processes in a chemical process industry. *Reviews in Chemical Engineering*, 33(5), 433–468. https://doi.org/10.1515/revce-2016-0032

Carpenter, J., George, S., & Saharan, V. K. (2017). Low pressure hydrodynamic cavitating device for producing highly stable oil in water emulsion: Effect of geometry and cavitation number. *Chemical Engineering and Processing – Process Intensification*, 116, 97–104. https://doi.org/10.1016/j.cep.2017.02.013

Cha, Y., Shi, X., Wu, F., Zou, H., Chang, C., Guo, Y., Yuan, M., & Yu, C. (2019). Improving the stability of oil-in-water emulsions by using mussel myofibrillar proteins and lecithin as emulsifiers and high-pressure homogenisation. *Journal of Food Engineering*, 258(April), 1–8.

Chen, X., Xu, X., Han, M., Zhou, G., Chen, C., & Li, P. (2016). Conformational changes induced by high-pressure homogenisation inhibit myosin filament formation in low ionic strength solutions. *Food Research International*, 85, 1–9.

Chen, X., Zhou, R., Xu, X., Zhou, G., & Liu, D. (2017). Structural modification by high-pressure homogenisation for improved functional properties of freeze-dried myofibrillar proteins powder. *Food Research International*, 100, 193–200.

De Bonis, M. V., & Ruocco, G. (2010). Heat and mass transfer modeling during continuous flow processing of fluid food by direct steam injection. *International Communications in Heat and Mass Transfer*, 37(3), 239–244. https://doi.org/10.1016/j.icheatmasstransfer.2009.09.008

Dhanke, P. B., & Wagh, S. M. (2020). Intensification of the degradation of acid RED-18 using hydrodynamic cavitation. *Emerging Contaminants*, 6, 20–32. https://doi.org/10.1016/J.EMCON.2019.12.001

Dong, Z., & Qin, Z. (2018). Experimental study of pathogenic microorganisms inactivated by venturi-type hydrodynamic cavitation with different throat lengths. *Journal of the Civil Engineering Forum*, 4(3). https://doi.org/10.22146/JCEF.38756

Gogate, P. R. (2011a). Cavitation in Biotechnology. In *Comprehensive Biotechnology, Second Edition* (Second Edi, Vol. 2). Elsevier B.V. https://doi.org/10.1016/B978-0-08-088504-9.00084-2

Gogate, P. R. (2011b). Hydrodynamic cavitation for food and water processing. *Food and Bioprocess Technology*, 4(6), 996–1011. https://doi.org/10.1007/s11947-010-0418-1

Gogate, P. R., & Pandit, A. B. (2001). Hydrodynamic cavitation reactors: A state of the art review. *Reviews in Chemical Engineering*, 17(1), 1–85. https://www.degruyter.com/document/doi/10.1515/REVCE.2001.17.1.1/html

Gregersen, S. B., Wiking, L., Metto, D. J., Bertelsen, K., Pedersen, B., Poulsen, K. R., Andersen, U., & Hammershøj, M. (2020). Hydrodynamic cavitation of raw milk: Effects on microbial inactivation, physical and functional properties. *International Dairy Journal*, 109, 104790.

Hernández-Hernández, H. M., Moreno-Vilet, L., & Villanueva-Rodríguez, S. J. (2019). Current status of emerging food processing technologies in Latin America: Novel non-thermal processing. *Innovative Food Science and Emerging Technologies*, 58, 102233. https://doi.org/10.1016/j.ifset.2019.102233

Joshi, R. K., & Gogate, P. R. (2012). Degradation of dichlorvos using hydrodynamic cavitation based treatment strategies. *Ultrasonics Sonochemistry*, 19(3), 532–539. https://doi.org/10.1016/j.ultsonch.2011.11.005

Jyoti, K. K., & Pandit, A. B. (2001). Water disinfection by acoustic and hydrodynamic cavitation. *Biochemical Engineering Journal*, 7(3), 201–212.

Knorr, D., & Augustin, M. A. (2021). Food processing needs, advantages and misconceptions. *Trends in Food Science and Technology*, 108, 103–110. https://doi.org/10.1016/j.tifs.2020.11.026

Kuppa, R., & Moholkar, V. S. (2010). Physical features of ultrasound-enhanced heterogeneous permanganate oxidation. *Ultrasonics Sonochemistry*, 17(1), 123–131. https://doi.org/10.1016/j.ultsonch.2009.05.011

Lee, I., & Han, J.-I. I. (2015). Simultaneous treatment (cell disruption and lipid extraction) of wet microalgae using hydrodynamic cavitation for enhancing the lipid yield. *Bioresource Technology*, 186, 246–251. https://doi.org/10.1016/j.biortech.2015.03.045

Li, Y., Chen, X., Xue, S., Li, M., Xu, X., Han, M., & Zhou, G. (2019). Effect of the disruption chamber geometry on the physicochemical and structural properties of water-soluble myofibrillar proteins prepared by high pressure homogenization (HPH). *LWT – Food Science and Technology, 105*(July 2018), 215–223. https://doi.org/10.1016/j.lwt.2019.02.036

Martynenko, A., Astatkie, T., & Satanina, V. (2015). Novel hydrothermodynamic food processing technology. *Journal of Food Engineering, 152*, 8–16. https://doi.org/10.1016/j.jfoodeng.2014.11.016

Mevada, J., Devi, S., & Pandit, A. (2019). Large scale microbial cell disruption using hydrodynamic cavitation: Energy saving options. *Biochemical Engineering Journal, 143*, 151–160. https://doi.org/10.1016/J.BEJ.2018.12.010

Mezule, L., Tsyfansky, S., Yakushevich, V., & Juhna, T. (2009). A simple technique for water disinfection with hydrodynamic cavitation: Effect on survival of Escherichia coli. *Desalination, 248*(1–3), 152–159. https://doi.org/10.1016/j.desal.2008.05.051

Milly, P. J., Toledo, R. T., Harrison, M. A., & Armstead, D. (2007). Inactivation of food spoilage microorganisms by hydrodynamic cavitation to achieve pasteurisation and sterilisation of fluid foods. *Journal of Food Science, 72*(9), 414–422. https://doi.org/10.1111/j.1750-3841.2007.00543.x

Milly, P. J., Toledo, R. T., Kerr, W. L., & Armstead, D. (2008). Hydrodynamic cavitation: Characterisation of a novel design with energy considerations for the inactivation of Saccharomyces cerevisiae in apple juice. *Journal of Food Science, 73*(6), M298–M303. https://doi.org/10.1111/j.1750-3841.2008.00827.x

Mozafari, M. R., Khosravi-Darani, K., Borazan, G. G., Cui, J., Pardakhty, A., & Yurdugul, S. (2008). Encapsulation of food ingredients using nanoliposome technology. *International Journal of Food Properties, 11*(4), 833–844. https://doi.org/10.1080/10942910701648115

Panda, D., Saharan, V. K., & Manickam, S. (2020). Controlled hydrodynamic cavitation: A review of recent advances and perspectives for greener processing. *Processes, 8*(2). https://doi.org/10.3390/pr8020220

Parthasarathy, S., Siah Ying, T., & Manickam, S. (2013). Generation and optimisation of palm oil-based oil-in-water (O/W) submicron-emulsions and encapsulation of curcumin using a liquid whistle hydrodynamic cavitation reactor (LWHCR). *Industrial and Engineering Chemistry Research, 52*(34), 11829–11837. https://doi.org/10.1021/IE4008858

Rajoriya, S., Carpenter, J., Saharan, V. K., & Pandit, A. B. (2016). Hydrodynamic cavitation: An advanced oxidation process for the degradation of bio-refractory pollutants. *Reviews in Chemical Engineering, 32*(4), 379–411. https://doi.org/10.1515/REVCE-2015-0075

Ren, X., Li, C., Yang, F., Huang, Y., Huang, C., Zhang, K., & Yan, L. (2020). Comparison of hydrodynamic and ultrasonic cavitation effects on soy protein isolate functionality. *Journal of Food Engineering, 265*(February 2019), 109697. https://doi.org/10.1016/j.jfoodeng.2019.109697

Research and Markets. (2022). *Functional Foods and Drinks – Global Market Trajectory & Analytics.* https://www.researchandmarkets.com/reports/344056/functional_foods_and_drinks_global_market#rela1-5119695

Sampath Kumar, K., & Moholkar, V. S. (2007). Conceptual design of a novel hydrodynamic cavitation reactor. *Chemical Engineering Science, 62*(10), 2698–2711.

Saricaoglu, F. T. (2020). Application of high-pressure homogenisation (HPH) to modify functional, structural and rheological properties of lentil (*Lens culinaris*) proteins. *International Journal of Biological Macromolecules, 144*, 760–769.

Save, S. S., Pandit, A. B., & Joshi, J. B. (1994). Microbial cell disruption: Role of cavitation. *The Chemical Engineering Journal and the Biochemical Engineering Journal, 55*(3), B67–B72. https://doi.org/10.1016/0923-0467(94)06062-2

Save, S. S., Pandit, A. B., & Joshi, J. B. (1997). Use of hydrodynamic cavitation for large scale microbial cell disruption. *Food and Bioproducts Processing: Transactions of the Institution of Chemical Engineers, Part C, 75*(1), 41–49. https://doi.org/10.1205/096030897531351

Sawant, S. S., Anil, A. C., Krishnamurthy, V., Gaonkar, C., Kolwalkar, J., Khandeparker, L., Desai, D., Mahulkar, A. V., Ranade, V. V., & Pandit, A. B. (2008). Effect of hydrodynamic cavitation on zooplankton: A tool for disinfection. *Biochemical Engineering Journal, 42*(3), 320–328. https://doi.org/10.1016/J.BEJ.2008.08.001

Sun, X., Chen, S., Liu, J., Zhao, S., & Yoon, J. Y. (2020). Hydrodynamic cavitation: A promising technology for industrial-scale synthesis of nanomaterials. *Frontiers in Chemistry, 8*(April), 1–7. https://doi.org/10.3389/fchem.2020.00259

Sun, X., Liu, J., Ji, L., Wang, G., Zhao, S., Yoon, J. Y., & Chen, S. (2020). A review on hydrodynamic cavitation disinfection: The current state of knowledge. *Science of the Total Environment, 737*. https://doi.org/10.1016/j.scitotenv.2020.139606

Sun, X., Xuan, X., Ji, L., Chen, S., Liu, J., Zhao, S., Park, S., Yoon, J. Y., & Om, A. S. (2021). A novel continuous hydrodynamic cavitation technology for the inactivation of pathogens in milk. *Ultrasonics Sonochemistry, 71*, 105382. https://doi.org/10.1016/J.ULTSONCH.2020.105382

Suslick, K. S., Mdleleni, M. M., & Ries, J. T. (1997). Chemistry induced by hydrodynamic cavitation. *Journal of the American Chemical Society, 119*(39), 9303–9304. https://doi.org/10.1021/ja972171i

Xue, S., Xu, X., Shan, H., Wang, H., Yang, J., & Zhou, G. (2018). Effects of high-intensity ultrasound, high-pressure processing, and high-pressure homogenization on the physicochemical and functional properties of myofibrillar proteins. *Innovative Food Science and Emerging Technologies, 45*(July 2017), 354–360.

Zhang, Z. H., Wang, L. H., Zeng, X. A., Han, Z., & Brennan, C. S. (2019). Non-thermal technologies and its current and future application in the food industry: A review. *International Journal of Food Science and Technology, 54*(1), 1–13. https://doi.org/10.1111/ijfs.13903

13 High-Pressure Processing Interventions in Marine Food Processing

B. Kamalapreetha, Anbarasan Rajan, and R. Mahendran

13.1 INTRODUCTION TO IMPORTANCE OF PROCESSING MARINE FOODS

Marine foods which include fish and shellfish (molluscs, cephalopods, crustaceans, and echinoderms) and edible sea plants are reliable sources of nutrients in the form of digestive proteins, lipid-soluble vitamins, microelements, and highly unsaturated fatty acids with lower calories for the human. Consumption of seafood varies from region to region concerning individuals, cultures, and traditions. Likewise, several factors are also influencing the consumption behavior of seafood such as various attributes of the product, characteristics of the individual (personal choice and taste) and the surrounding environment (seasonal availability, accessibility). In general, a wide range of seafood is procured and consumed by the consumer from fresh to processed food (canned, dehydrated fish, frozen, salted/smoked/pickled seafood) (Erkan & Cagiltay, 2011). According to Food and Agricultural Organization (FAO, 2021), a total of 158 million tons (89%) of fishery production are utilized for human consumption in that almost 44% are utilized in fresh or live form. Seafood undergoes certain handling and primary unit operations like washing, gut removal, filleting, shucking, etc., before entering the processing line. In this regard, assessment of quality is highly recommended to enhance the quality of seafood before and after harvest to produce a safe product and for developing value-added products.

Seafood is the most perishable commodity because of its high moisture nutritious food with loose connective tissues and neutral pH. All these factors contribute to quicker deterioration/spoilage. The reasons for the spoilage of seafood are temperature and time of processing, microbial and enzyme activity, loss in protein functionalities, and lipid oxidation, all these factors have a significant impact on the shortness of the shelf life of seafood products (Ababouch et al., 1996; Erkan, 2003). To limit spoilage and increase the keeping quality of seafood certain processing steps need to be imposed which also fosters the safety of the products. The following section will briefly explain the current processing techniques employed in marine food processing.

13.2 CURRENT PROCESSING TECHNIQUES IN MARINE FOOD PROCESSING

The linear trend in the global demand for marine foods keeps on increasing. The most commonly used method of processing involves low-temperature preservation like chilling, super-chilling, or freezing (Gómez-Guillén & Montero, 2007). In addition, alteration of the packaging, i.e., modified atmospheric packaging (MAP) (Nilsen-Nygaard et al., 2021) and heating, are also generally applied in increasing the shelf life of seafood. Moreover, curing, fermentation, salting, and smoking are the traditional methods even applicable nowadays in the production and preservation of marine foods. But all these methods have certain drawbacks which are deficient in keeping the quality and safety of the produce. All these drawbacks paved way for the novel techniques which can destroy pathogens and spoilage causing microorganisms which leads to keeping the desired quality of the food. In recent times, newly developed novels are also been widely addressed in processing seafood which aids in minimizing economic losses, sustaining freshness, provide safe and sound products

DOI: 10.1201/9781003359302-13

with reliable quality (Kykkidou et al., 2009). Some of the emerging novel techniques are irradiation, high-pressure processing (HPP), ultrasound, pulsed electric field (PEF), radio frequency, pulsed light, ohmic heating, infrared heating, and microwave processing (Arshad et al., 2021; Hassoun et al., 2020; Rathod et al., 2021).

For example, in terms of thermal processing, it prevents the spoilage of organisms and extends the shelf life of foodstuff. Recently, novel thermal techniques such as infrared, microwave, ohmic heating, and radio frequency have obtained importance in the industrial process as an alternative to conventional thermal processes (pasteurization, boiling steaming, roasting/frying). Moreover, ohmic heating (volumetric heating) is one of the novel thermal methods used for varied applications. As this method has a potential application in the field of cooking, cooling, extraction, fermentation, heating, and thawing will also aid in seafood processing (Niakousari et al., 2019). Further, ohmic heating is applied to tuna to prevent loss during thawing and reduce the time of thawing (Fattahi & Zamindar, 2017; Liu et al., 2017). Other research like textural studies done on surimi gel of Pacific whiting and Alaska pollock were studied using ohmic heating (Hoon Moon et al., 2017).

Microwave is also another thermal method where it is also applicable for cooking, extraction, thawing, and sterilization of foodstuff with negligible impact on the sensorial and nutritional properties of foods. In seafood processing, studies on microwave processing were exhibited with potential application in improving the gelling properties of fish gel (Cao et al., 2019; Yan et al., 2020). Commercialization is the major drawback as it involves higher installation costs in large-scale units.

Techniques like HPP and irradiation are promising treatments for the successful preservation of food with extended shelf life.

In non-thermal processing, the most extensively researched application on seafood is HPP, PEF, ultraviolet, cold plasma, and ionizing radiation. ultrasound (Ekonomou & Boziaris, 2021; Olatunde & Benjakul, 2018).

PEF-treated seafood have tremendous changes in oxidation and protein carbonyl due to cellular structural damage (electroporation). The use of PEF in seafood processing has prospective applications but is currently limited. Some of the studies on PEF application in the valorization of by-products are instances of calcium from fishbone (Zhou et al., 2012), protein from mussels (Zhou et al., 2017), antioxidant from fish residues (Franco et al., 2020). In addition, microbial destruction in shrimp (Shiekh & Benjakul, 2020) and textural changes in fish were studied.

Ozone is another novel, powerful disinfectant applied to seafood for preservation. For instance, Ozone in the form of ozone ice flakes on sea bass (Lu et al., 2012) and sardine (Campos et al., 2005), aqueous ozone on rainbow trout (Nerantzaki et al., 2005) and mussels (Manousaridis et al., 2005), and also as ozone pre-treatment on tilapia (Gelman et al., 2005) have proven to improve the sensory attributes and increased the shelf life by reducing the microbial growth.

Likewise, ultrasound as a sound technology was extensively researched on improving the sensory quality and shelf life of seafood (Ekonomou & Boziaris, 2021; Zhao et al., 2019), and the tenderization of jumbo squid (Hu et al., 2014). In addition, assisted in the extraction of oil (Bruno et al., 2019), gelatin from bighead carp (Huang et al., 2017), and protein from mackerel (Álvarez et al., 2018). Besides, it is also applicable to allergen reduction (Dong et al., 2020).

Similarly, cold plasma as upcoming technology in the application of perishable seafood is especially effective in microbial reduction, i.e., pathogenic and spoilage-causing organisms. This effect is due to the generation of reactive species but with shortcomings like lipid oxidation and protein disintegration (Rathod et al., 2021; Zhao et al., 2019). Further research on cold plasma along with hurdles to improve the efficiency overcome the limitation.

In the same way, a wide range of research was conducted to preserve the seafood using irradiation.

This primarily focused on extending shelf life, improving safety, delaying ripening, and controlling pests and insects (Erkan, 2014; Kontominas et al., 2021). The USA have commercially approved the use of irradiation as one of the food preservation techniques, but not yet effectively used in seafood industrial application. Studied the effect of shelf life on mackerel and retained sensory attributes of a variety of fishes.

13.3   CURRENT TREND IN HIGH-PRESSURE PROCESSING

In the current scenario, there is a higher demand for quality foods with fresher, safer, nutritive rich, and additive-free produce which in turn paved the way for the alternative technique to conventional thermal processing, i.e., non-thermal technology. This thermal processing (e.g., pasteurization) negatively impacts the quality of the foods, whereas most of the attributes like sensory, flavor, and nutrients are retained. Non-thermal techniques available to improve the quality of the foods are HPP, PEF, electron beam, and plasma. Although, HPP is highly demanding and one of the most promising commercially applied non-thermal techniques on foods to increase the shelf life and keep the quality of food products (Farkas, 2016), which was first started back in the late 1980s in Japan.

This chapter primarily summarizes the various aspects of HPP techniques applied to augment the quality of marine foods.

At room temperature, HPP kills food pathogens and lengthens the shelf life of products dispatched throughout the logistics system. These methods preserve the sensory characteristics and nutritive quality of the foods, which traditional thermal pasteurization does not. HPP has already been approved by the US Food and Drug Administration as a non-thermal pasteurization technique of preservation. Hence, HPP can replace and overcome the existing pasteurization technique in the food industry. A foster protocol with clearly defined regulations and specifications will ease the marketability of High-pressure-processed products to build up consumer trust and product quality. The extensive application of HPP has accelerated the market demand for the production of HPP equipment.

HPP technology is accepted in the packaging of dairy, fruits, meat, and seafood. More than 300 units of HPP equipment were in operation around the world by the end of 2015. Despite soaring prices and high investment barriers, the number of customized equipment manufacturers for a specific purpose has been progressively enlarging. Further in the global market, HPP has approximated $10 billion annual output value. In addition, HPP technology contributes to existing trends such as organic food, health food, or clean label to accelerate the food market.

Food is placed in flexible containers/pouches and is hermetically sealed. Further, when high pressure of 100–600 MPa is applied at atmospheric conditions, subjecting pressure-transmitting fluid (usually water) instantaneously and uniformly to all the surface of the food to achieve pasteurization (Balasubramaniam et al., 2015). For this reason, pasteurization is unaffected by the packaging form or volume of the food, where the varying volumes are processed in the same batch. Furthermore, without the use of preservatives, this processing method confirms the microbial safety of food and allows processed food to keep the natural flavors and nutritional value of the original food material.

HPP executed at atmospheric temperature and reduced energy consumption with later heating and cooling. Furthermore, the food in packed condition does not come into direct contact with the processing devices, accordingly, preventing contamination (secondary) of the food after pasteurization. Also, after processing, the pressure-transmitting fluid/medium can be reprocessed. HPP technology is an environmentally friendly processing technology because of its low energy consumption and low contamination risk (Rastogi et al., 2007).

In the United States, the first HPP experiment performed in the year 1885 reported the elimination of bacteria while high pressurization. As a result, investigators revealed that HPP treatment increased the keeping quality of milk. These studies revealed the possibility of using high pressure in the food industry. Active research is undergone on high-pressure pasteurization using pilot-scale HPP units in several government agencies, research institutions, and universities to help the food industry in developing a technical standard for food pasteurization. Before the commercialization of HPP as a new technology for pasteurization, health, safety aspects, economic value, and product forms need to be thoroughly evaluated.

Furthermore, scientific approaches and operating parameters for application feasibility needed refinement to meet the needs of food technologists in further development of the product, basic pasteurization parameters should be set up based on previous HPP research. Such pasteurization data can also be used to assess the hygienic safety of HPP products (Huang et al., 2014). Manufacturers of laboratory HPP equipment, such as Avure, Baotou, Kobelco, Stansted Fluid Power, and Toyo Koatsu, are production equipment with varying specifications and capacities ranging from 0.3 to 10 L to meet a demand for forthcoming research needs. The first and foremost commercial HPP product was developed in Japan for Jam. Following that, a variety of HPP food products were established in Europe and North American markets.

In recent years, the extensive adoption of HPP equipment has become a key driver. Around the world, a large number of machine manufacturers are also actively engaged in the research, development, and manufacture of HPP equipment. This causes better equipment manufacturing, and refined continuous production performance with long-term, stable production. Avure and Hiperbaric are the market leaders, developing HPP devices with 525 L of volume and a production capacity of 60 million tons (Balasubramaniam et al., 2008; Elamin et al., 2015).

13.4   EFFECTS OF HPP ON VARIOUS ASPECTS OF MARINE FOODS

The quality of the food is evaluated based on the various aspects of marine foods. Freshness and quality are majorly concerned with the changes in the physicochemical properties, microbial, nutritional, and sensory parameters. The changes in the different aspects of the HPP of seafoods are presented in Table 13.1. The following section will discuss the various aspects of the above-said parameters on distinct categories of seafood.

13.4.1   Effect of HPP on Physicochemical Changes

13.4.1.1   pH

pH is an essential parameter in deciding the quality of seafood. An increase in the pH value of the seafood product is a sign of microbial and enzymatic spoilages. For example, an untreated shrimp stored at 2°C for 35 days showed a linear increase in the pH value from 6.79 ± 0.10 to 8.01 ± 0.04 due to the gradual microbial spoilage. However, HPP treatments (100 MPa, 270 MPa, and 435 MPa) limited the microbial spoilage and prevented the vigorous increase (i.e., pH reached 7.79 ± 0.03 after 35 days in 100 MPa) in the pH values. Nevertheless, a slight increase in the pH values was observed immediately after treatment at all pressure levels and this could be due to the denaturation and unfolding of shrimp protein due to high pressure. However, the water activity of the samples remained unchanged and showed no significant difference with untreated samples (Kaur et al., 2013). Comparable results were obtained in squid after 300 MPa HPP treatment. The untreated squid sample's pH increased from 6.61 ± 0.10 to 7.81 ± 0.16 during the 12-day storage period. However, HPP-treated sample pH remained well below 7.0 pH even after 12 days. Nevertheless, a slight increase in the pH values was observed immediately after HPP treatment (Gou et al., 2010). Coroneo et al. (2022), also observed the pH of the treated urchins increased when compared to the fresh gonads (5.84) and further decreased while storage. The results were also similar for red mullet, where the pH change was controlled by HPP treatment till the 13th day of storage (Erkan et al., 2010).

13.4.1.2   Water Retention and Water Holding Capacity

Seafood holds up to 70% water due to the flesh's capillary action and electrostatic interactions. In this regard, the protein present (20%) in the flesh alone keeps 0.5 g of water for every gram of its weight. However, capillary force keeps the majority of water inside the intra and extracellular muscle filaments and fibers, respectively (Lakshmanan et al., 2007). During HPP treatment, the proteins present in the seafood products (i.e., fish) hydrate up to 55%. Further, the hindrance of sarcomere shortening improved myosin tail, and actin stabilities contribute to higher water retention in HPP-treated fish and other seafood products. However, waterless HPP treatment was found to

TABLE 13.1

Changes in the Different Aspects (Microbial, Sensory, and Nutritional Quality) of High-Pressure Processing of Seafood

Category	Species	Treatment Condition	Microbial Analysis	Nutritional Changes	Shelf Life	Sensory Attributes	References
			Fish				
Fish	Salmon, cod, and mackerel	Pressure (MPa) – 200 and 500 Time (min) – 2	–	Acute protein denaturation, water holding capacity is lower Higher liquid loss	–	Increase in hardness and L* value	(Christensen et al., 2017)
	Cold-smoked dolphin fish (*Coryphaena hippurus*)	Pressure (MPa) –300 Time (min) – 2	APC, H₂S producing bacteria, LAB –reached below detection limit	Water holding capacity, lipid oxidation – stable TVB-L – did not exceed 35 mg TVB-N/100 g	–	Shear strength – stable, best at 300 MPa	(Gómez-Estaca et al., 2007)
	Fresh European cat fillet	Pressure (MPa) – 200, 400, and 600 Time (min) – 1 or 5	–	–	–	Catfish became paler and looked resemble the cooked product	(Mengden et al., 2015)
	Hilsa fillet	Pressure (MPa) – 250 and 350 Time (min) – 10	–	At 350 MPa, production of TMA-N, increased lipid oxidation, decreased free fatty acid	–	Increase in hardness and L* and b* values Decrease in a*	(Kaur et al., 2016)
			Shellfish				
Molluscs	Clams	Live hard clams (*M. Lusoria*) with soy sauce	Pressure (MPa): >300 Time (min): 3	APC – 8.3 log CFU/g and PBC – 8.4 log CFU/g	pH reduced to 3.4	15 days	>300 MPa able to shuck efficiently Opaque and harder	(Lee et al., 2022)
	Oysters	Pacific oyster (*Crassostrea gigas*)	Pressure (MPa): 260, 400, 600 Time (min): 5	Significant decrease in microbial load	Increase in lipid oxidation	31 days	L*, b*, cutting strength, increased	(Cruz-Romero et al., 2008)
	Mussels	Mussels (*Perna canaliculus*)	Pressure (MPa): 500 Time (min): 3		Higher loss in water		L* Value decreased, increase in cutting strength, higher puncher strength	(Gupta et al., 2015)
			Pressure (MPa): 400 Time (min): 3	L. monocytogenes – 5 to 6 log	–	–	–	(Fletcher et al., 2008) *(Continued)*

TABLE 13.1 (Continued)
Changes in the Different Aspects (Microbial, Sensory, and Nutritional Quality) of High-Pressure Processing of Seafood

Category	Species		Treatment Condition	Microbial Analysis	Nutritional Changes	Shelf Life	Sensory Attributes	References
Cephalopods	Octopus	Octopus (Octopus vulgaris)	Pressure (MPa): 600 Time (min): 6	Reduction of Psychrotrophic bacteria days 4.7 CFU/g	Decreased the level of TMA, DMA, and BA	12 days		(Hsu et al., 2014)
	Squid	Squid (Todarodes pacificus)	Pressure (MPa): 300 Time (min): 20	TPC – 1.26 log unit	Reduced TMAO activity and DMA	12 days	Inhibit off odor	(Gou et al., 2010)
		Semi-dried squid	Pressure (MPa): 500 Time (min): 10	Morganella morganii – <1 log CFU/cm² and Klebsiella pneumonia – <1 log CFU/cm²	Reduced formation of BA, DMA, TMA	28 days		(Gou et al., 2012)
Crustaceans	Shrimp	Black Tiger Shrimp (Penaeus monodon)	Pressure (MPa): 435 Time (min): 5	Lower viable counts	A notable increase in moisture and protein content No change in TVB-N and TMA-N level	15 days	Increased whiteness value as pressure increased. Observed cooked appearance, induced hardness	(Kaur et al., 2013)
	Crab	Chinese mitten crab (Eriocheir sinensis)	Pressure (MPa): 300 Time (min): 5	–	Reduced the water loss	–	Higher meat yield with 51.33%, decreased hardness, a significant difference in aroma	(Wu et al., 2021)
		Shucked crab (Eriocheir sinensis) meat	Pressure (MPa): 300 Time (min): 20	APC – 5.71 log CFU/g Clostridium was highest on 6th day of storage	TVB-N – 24.50 mg/100 g	6 days	Sensory attributes declined progressively Texture hardness – decreased	(Ye et al., 2021)

(Continued)

TABLE 13.1 (*Continued*)

Changes in the Different Aspects (Microbial, Sensory, and Nutritional Quality) of High-Pressure Processing of Seafood

Category	Species	Treatment Condition	Microbial Analysis	Nutritional Changes	Shelf Life	Sensory Attributes	References
Echinoderms	Sea urchins	*Paracentrotus lividus*	Pressure (MPa): 350, 500 Time (min): 10	Less than 10 CFU/g after 40 days of storage	Saturated fatty acids alone changed after 40 days of storage. There is a decay of protein in pressure-treated sea urchins. 50% of free amino acids were lost after high-pressure treatment	Sea urchins were found edible even for 40 days when treated at 500 MPa	Fresh-like characteristics until 20 days were attained at 500 MPa	(Coroneo et al., 2022)
Edible Sea Plants	Microalgae	*Porphyridium cruentum*	Pressure (MPa): 100, 300, 600 Cycle: 1 or 3 Time (min): 20 Medium – propylene glycol	–	At 300 MPa, obtained extract using ethyl acetate as solvent is enriched with carotenoids, PUFA, B-phycoerythrin with 2.37 purity index	–	–	(Bueno et al., 2020)
		Spirulina platensis	Pressure: 500 MPa Time: 15 min	–	Phycocyanin yield was maximum (10.2%) with a purity of 0.909 at 280 nm and higher stability even after 14 days 83% antioxidant activity	–	–	(Seo et al., 2013)

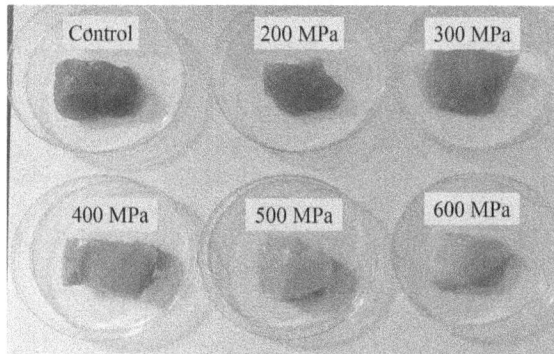

FIGURE 13.1 Appearance of tuna fish at various treatment conditions using HPP for 5 min (Tsai et al., 2022).

increase the sample moisture alone by lowering the sample water-holding capacity. The compression of muscle fibers releases the water from the sample and subsequently increases the sample's moisture (Oliveira et al., 2017). However, some seafood like sea bass, where observed a minimal surge in the water holding capacity up to 300 MPa (Chéret et al., 2005).

13.4.1.3 Color

The color of marine food and its products plays a vital role as it attracts the consumer and helps in deciding the product quality. In general, HPP-treated seafood was found to have a higher L* value than the untreated samples. When the pressure is increased (>100 MPa) linearly, the whiteness and opaque nature of the sample also increase linearly. The study on color change on the application of HPP at different pressures observed in tuna fish is presented in Figure 13.1. Similarly, changes were observed in a* (decrease) and b* (increase) values. Globin denatures (heme displacement), lipid oxidation, and muscle protein denatures are some of the reasons why color changes are observed in seafood upon HPP. Similar results were observed in seafood products such as bluefish, cod, mackerel, turbot, trout, and red mullet (Kaur et al., 2013).

13.4.1.4 Trimethylamine Nitrogen (TMA-N)

TMA-N assessment is one of the prominent ways of figuring out the quality and shelf life of fish. It is produced by TMA-oxide, a non-protein nitrogen fraction of the marine product's flesh. This off-odor (fishy) producing compound is the result of bacterial enzyme activity. Therefore, the samples treated with HPP tend to maintain their TMA values unlike the control untreated sample (Erkan et al., 2010). In the untreated samples (squid), the DMA increased continuously throughout the storage period, while the HPP-treated samples showed a minimal increase (not significant) in the DMA values. These changes can be observed even in refrigerated conditions (Gou et al., 2010).

13.4.2 Effect of HPP on Microbes

The effect of HPP processing on seafood microbes varies based on its type and nature. Among the bacteria, gram-negative bacteria are highly susceptible to HPP treatment due to their complex cell membrane structure (e.g., exceptional cases of *E. coli* O157). HPP application on microbial inactivation is widely studied for various seed foods such as cold-smoked salmon, sea bass, rainbow trout, octopus, etc. (Kaur et al., 2013). Besides microbial inactivation, HPP also preserves the sensory qualities of seafood products (e.g., oysters). HPP treatment not only induces cell damage but also denatures the enzymes linked to the microbes' cell membrane. Besides, the high pressure also alters the cell's morphology and internal organization. As a result, the microbial cell tends to lengthen and contract during pressurization and depressurization processes; further, it also induces pores in microbial cell walls. Furthermore, ribosomal damage and DNA destruction are also the major causes of microbial inactivation in HPP processing. However, a few other factors, such as microbial strain, its growth phase, and treatment temperature, affect the treatment efficiency. Apart from bacteria, HPP treatment is also effective against viruses present in seafood items. However, with bacteria, the effectiveness of treatment varies based on the virus types. For example, norovirus surrogate is susceptible to HPP treatment, and with low intense HPP treatment (275 MPa for 5 min), it is possible to inactivate these viruses. However, viruses like poliovirus are highly resistant to HPP processing, and even 600 MPa pressure treatment for a prolonged period (60 min) is not sufficient to cause the inactivation of these viruses (Murchie et al., 2005).

13.4.3 STRUCTURAL MODIFICATION

One of the important structural properties is texture in meat or seafood products. High-pressure-treated marine foods were noticed to have higher chewiness, cutting strength, hardness, and springiness (Hwang & Fan, 2015). The consequence of destruction/denaturation and aggregation of the myofibrillar protein revealed the textural modification. Higher the pressure condition higher the texture of the seafood. The high pressure applied from 250 to 500 MPa on shellfish such as lobsters, oysters, crabs, and clams causes the denatures of the protein which secures the meat to the shell. This denaturation capability facilitates the thorough separation of the flesh from the shell (He et al., 2002; Murchie et al., 2005), with superior quality and higher quantity of extracted meat from the shellfish. The protein substrates in uncooked surimi products become readily available for transglutaminase at higher pressures, with subsequent cooking creating a protein cross-linkage with higher gel strength (Ashie & Lanier, 1999). Ashie and Simpson (1996) investigated the effect of HPP on raw fish and observed a reduced toughness and flexibility in fish muscle as the consequence of the increased pressure (above 200 MPa). Depending on the applied pressure, the texture of seafood and its firmness vary. In another study (Arnaud et al., 2018), there was also an increase in the hardness of HPP-treated cod (1.4 times) and salmon (1.7 times) fillets than untreated at 150 MPa and 300 MPa, respectively. The hardness and springiness of the seafood changed due to phenomenal activity, especially the unfolding of actin as well as the sarcoplasmic protein, which creates a hydrogen link (Angsupanich & Ledward, 1998). As the consumer acceptability of marine products is dependent on textural characteristics, it is important to optimize the HPP treatment condition on various species.

13.4.4 NUTRITIONAL CHANGES

The protein content might decrease with the increase in the pressure of HPP treatment. This phenomenon might be caused due to the disruption of the myofibrillar structure resulting from leached internal fluids (Briones-Labarca et al., 2012). There is a decay of protein content in pressure-treated (350 MPa) Sea urchins, while the 500 MPa maintained the level of fresh gonads (Skipnes et al., 2008). This could be associated with the loss of free amino acids. Almost 50% of free amino acids were lost after high-pressure treatment (Coroneo et al., 2022). Sea urchin gonads were examined with 29 fatty acids, while the monounsaturated (MUFA) and polyunsaturated fatty acids (PUFA) remain unchanged except for saturated fatty acids in the high-pressure treated gonads after storing for 40 days.

Major deterioration is caused by the lipid oxidation of seafood which is a rich source of unsaturated fatty acids. Polyunsaturated fatty acid, a form of unsaturated fatty acid, gets destroyed by oxidation resulting in off flavor/off odor development and loss in sensory quality. Food with metal ions and enzymes strongly contributes to lipid oxidation. The presence of trace amounts of iron, copper, heme pigments, and hemoglobulin in fish is sensitive to lipid oxidation. On pressurization, ferric irons of heme pigments (myoglobin) get denatured, which leads to lipid oxidation (Ikeuchi, 2011).

In marine foods, HPP treatment will provoke lipid oxidation, where pressure below 300 MPa has little effect, while pressure between 300 and 400 MPa and above has a considerable impact on lipid oxidation. Treatment pressure, time, fiber type, the composition of the product, pre-treatment, and fat substances of the seafoods also contributes to the oxidation of lipids (Truong et al., 2015). A fish quality indicator, thio barbituric acid reactive substances (TBARS), is a rancidity measure. For example, in HPP treated (550 MPa) palm ruff fillets saw a 3.5-fold time surge in TBARS level than the untreated sample (Lemus-Mondaca et al., 2018). Another study (Xuan et al., 2018) on razor clam also showed a higher TBARS level on high-pressure treatment.

13.4.5   Shelf Life Extension

Seafood processing with the onset of safety and extending shelf life are the crucial targets of the seafood sector. High pressure was tremendously used as a successful treatment to inactivate pathogens. Inhibition or destruction of microorganisms to a greater level results in preservation with a higher shelf life of seafood. HPP is used as a potential pasteurization tool for producing high-quality and microbial-free produce. Examples focused on shelf life study, the shelf life of sea urchin gonad was found to maintain good organoleptic properties for 40 days when treated at 500 MPa (Coroneo et al., 2022). Smoking is the most treated method in fish processing. When smoked salmon undergoes HPP (220, 250, and 300 Mpa), it increased its shelf life to an extent of 2 more weeks than the untreated sample (Erkan et al., 2011). Karim et al. (2011) also experimented on the shelf life extension of herring and haddock to a maximum of 12 days compared to the control (6 days). Usually, the shelf life of HPP foods is like that of thermally processed food. Many research institutes are involved in perfecting the process parameters to augment the shelf life of seafoods and to commercialize the same.

13.4.6   Sensorial Quality

In most cases, HPP keeps the sensory attributes while improving the sensory attributes of seafoods. The high pressure on the sensory quality of seafood varies with the seafood product and applied treatment conditions. For example, When Sea urchins are subjected to high pressure (Coroneo et al., 2022), the sensory characteristics remain the same except for the sample treated at 350 MPa. High-pressure (350 MPa) treated sea urchins have a smooth texture but lose their firmness and texture after 20 days. This textural modification is a result of protein denaturation which is an outcome of high-pressure treatment (Sukmanov et al., 2019). Moreover, observed a color change from red-orange to dark red and also developed a fishy and musty flavor after 40 days of storage in 350 MPa-treated sea urchins. This mechanism was triggered due to enzymatic inactivation while HPP.

The flavor profile and taste of seafood are attributed to the availability of volatile and non-volatile. The development of undesirable flavor is related to lipid oxidation and the degeneration of carbohydrates and proteins (Cruz-Romero et al., 2008). Cruz-Romero et al. (2008) also studied the volatile profile of HPP-treated oysters. The main changes obtained in the HPP-treated oysters are in the Sulfur substances. The specific umami taste of squid is due to the manifestation of amino acids and nucleotides which comes under the non-volatile category. On HPP treatment (Borda, 2017) after 200, 400, and 600 MPa on squid, increased the 5′ adenosine monophosphate (most prominent flavor) and with an increase in the meaty flavor (5′-guanosine monophosphate) after 400 and 600 MPa treatment. These flavor compounds are increased because of the enzyme inactivation of HPP treatment.

13.5   COMBINATION OF HPP WITH OTHER TECHNIQUES IN THE PROCESSING OF MARINE FOODS

The advent of the combination of HPP with other technologies will provide insights into advancement in industrial application as a suitable alternative to the conventional method to overcome losses with improved quality. Pérez-Won et al. (2021) studied the combined impact of CO_2 (70%),

HPP (150 MPa, 5 min), and PEF(1–2 kV/cm, 20 s) on the physicochemical attributes of coho salmon. Pre-rigor PEF-treated salmon were found to have enhanced hardness and chewiness, whereas the combination reduces the activity of protease and lipase. A combination of ohmic heating and HPP was studied to determine the changes in thermal and structural changes of shrimp (Pérez-Won et al., 2021). This combination enhanced the peel ability of shrimp at low pressure and time. One of the research (Huang et al., 2022) was done on the combination of HPP and traditional salting (hurdle) to evaluate the microbial and physical-chemical characteristics of mackerel fillets. In this study, salting contributed to increasing the hardness and chewiness while HPP in imparting a softer texture or mackerel fillets. One more research focused on microbial destruction (*Vibrio cholerae*) in salmon and mussels using the combination of HPP (150–550 MPa for 5–13 min) with virulent bacteriophages (Ahmadi et al., 2015). This combination has effectively reduced the treatment time and pressure condition. HPP unit working above 600 MPa is costlier. In this regard, combination with other potential techniques with their benefits helps in effectively reducing the production cost of seafood processing.

13.6   LIMITATIONS AND FUTURE PROSPECTIVE

In HPP-treated foods, the sensorial properties may alter during processing. As HPPs tend to disrupt functional proteinaceous molecules like enzymes via ionic and hydrophobic-hydrophobic interactions. HPP can modify the structure and function of enzymes (Balasubramaniam et al., 2015). Still, HPP can hasten the oxidation of lipids in treated seafood upon storage. Also, while HPP treatment, the discharge of inorganic metal ions results in oxidation (Cheah & Ledward, 1997). The pressure above 200 MPa, HPP exhibits a cooked appearance. HPP can also cause the formation of formaldehyde, which causes an increase in the textural property and protein crosslinking of treated seafood (Matser et al., 2000). In addition, installation and the cost of the equipment are costly. These are some of the limitations when HPP is applied to foodstuffs.

Current technologies are quickly progressing, yet another area of interest is the use of pressure-assisted processing, such as pressure assisted thermal sterilization (PATS) because of its ability to inactivate microbial spores, as well as pressure assisted thawing (PAT) and pressure shift freezing (PSF). To meet and satisfy consumer needs with fresh, nutritive, and quality rich marine food, the evolution of diverse value-added products gets intensified. However, industrial equipment for market demand must supply new and possible sustainable solutions for implementing these new technologies.

13.7   SUMMARY

HPP has moved from lab and pilot plant to industrial application, where it is crucial and is in need to get consumer trust by proving a solid reputation as a safe and secure food processing technique. In HPP, 300–500 MPa was promising to lengthen the shelf life of frozen or refrigerated fish, lower microbial contamination, and enhance textural characteristics, but it might have some negative impacts on color and lipids with respect to seafood species and operating conditions. Future HPP in marine applications is associated with consumer desires along with its ability to reduce allergenicity, minimize salt utilization, and develop improved products with improved functionality and clean label.

The synergistic effect of HPP along with potential hurdles techniques using natural antimicrobial agents and antioxidants-rich compounds and/or other minimal processed technologies can advance product safety and quality. This will lessen the hitches associated with individual methods. The highly demanding process advances the research in developing a sustainable solution for marine processing.

REFERENCES

Ababouch, L. H., Souibri, L., Rhaliby, K., Ouahdi, O., Battal, M., & Busta, F. F. (1996). Quality changes in sardines (*Sardina pilchardus*) stored in ice and at ambient temperature. *Food Microbiology*, *13*(2), 123–132. https://doi.org/10.1006/fmic.1996.0016

Ahmadi, H., Anany, H., Walkling-Ribeiro, M., & Griffiths, M. W. (2015). Biocontrol of Shigella flexneri in ground beef and *Vibrio cholerae* in seafood with bacteriophage-assisted high hydrostatic pressure (HHP) treatment. *Food and Bioprocess Technology, 8*(5), 1160–1167. https://doi.org/10.1007/s11947-015-1471-6

Álvarez, C., Lélu, P., Lynch, S. A., & Tiwari, B. K. (2018). Optimised protein recovery from mackerel whole fish by using sequential acid/alkaline isoelectric solubilization precipitation (ISP) extraction assisted by ultrasound. *LWT, 88*, 210–216.

Angsupanich, K., & Ledward, D. A. (1998). High pressure treatment effects on cod (Gadus morhua) muscle. *Food Chemistry, 63*(1), 39–50. https://doi.org/10.1016/S0308-8146(97)00234-3

Arnaud, C., de Lamballerie, M., & Pottier, L. (2018). Effect of high pressure processing on the preservation of frozen and re-thawed sliced cod (*Gadus morhua*) and salmon (*Salmo salar*) fillets. *High Pressure Research, 38*(1), 62–79. https://doi.org/10.1080/08957959.2017.1399372

Arshad, R. N., Abdul-Malek, Z., Roobab, U., Munir, M. A., Naderipour, A., Qureshi, M. I., El-Din Bekhit, A., Liu, Z. W., & Aadil, R. M. (2021). Pulsed electric field: A potential alternative towards a sustainable food processing. *Trends in Food Science and Technology, 111*(January), 43–54. https://doi.org/10.1016/j.tifs.2021.02.041

Ashie, I. N. A., & Lanier, T. C. (1999). High pressure effects on gelation of surimi and turkey breast muscle enhanced by microbial transglutaminase. *Journal of Food Science, 64*(4), 704–708.

Ashie, I. N. A., & Simpson, B. K. (1996). Application of high hydrostatic pressure to control enzyme related fresh seafood texture deterioration. *Food Research International, 29*(5–6), 569–575.

Balasubramaniam, V. M., Farkas, D., & Turek, E. J. (2008). Preserving foods through high-pressure processing. *Food Technology, 62*(11), 32–38.

Balasubramaniam, V. M. B., Martínez-Monteagudo, S. I., & Gupta, R. (2015). Principles and application of high pressure-based technologies in the food industry. *Annual Review of Food Science and Technology, 6*(February), 435–462. https://doi.org/10.1146/annurev-food-022814-015539

Borda, D. (2017). High-Pressure Processing of Seafood. In *Trends in Fish Processing Technologies* (pp. 71–100). CRC Press. https://doi.org/10.1201/9781315120461

Briones-Labarca, V., Perez-Won, M., Zamarca, M., Aguilera-Radic, J. M., & Tabilo-Munizaga, G. (2012). Effects of high hydrostatic pressure on microstructure, texture, colour and biochemical changes of red abalone (*Haliotis rufecens*) during cold storage time. *Innovative Food Science and Emerging Technologies, 13*(January), 42–50. https://doi.org/10.1016/j.ifset.2011.09.002

Bruno, S. F., Kudre, T. G., & Bhaskar, N. (2019). Impact of pretreatment-assisted enzymatic extraction on recovery, physicochemical and rheological properties of oil from *Labeo rohita* head. *Journal of Food Process Engineering, 42*(3), 1–11. https://doi.org/10.1111/jfpe.12990

Bueno, M., Gallego, R., Chourio, A. M., Ibáñez, E., Herrero, M., & Saldaña, M. D. A. (2020). Green ultra-high pressure extraction of bioactive compounds from *Haematococcus pluvialis* and *Porphyridium cruentum* microalgae. *Innovative Food Science and Emerging Technologies, 66*, 102532. https://doi.org/10.1016/j.ifset.2020.102532

Campos, C. A., Rodríguez, Ó., Losada, V., Aubourg, S. P., & Barros-Velázquez, J. (2005). Effects of storage in ozonised slurry ice on the sensory and microbial quality of sardine (*Sardina pilchardus*). *International Journal of Food Microbiology, 103*(2), 121–130. https://doi.org/10.1016/j.ijfoodmicro.2004.11.039

Cao, H., Fan, D., Jiao, X., Huang, J., Zhao, J., Yan, B., Zhou, W., Zhang, W., Ye, W., & Zhang, H. (2019). Importance of thickness in electromagnetic properties and gel characteristics of surimi during microwave heating. *Journal of Food Engineering, 248*, 80–88. https://doi.org/10.1016/j.jfoodeng.2019.01.003

Cheah, P. B., & Ledward, D. A. (1997). Catalytic mechanism of lipid oxidation following high pressure treatment in pork fat and meat. *Journal of Food Science, 62*(6), 1135–1139.

Chéret, R., Chapleau, N., Delbarre-Ladrat, C., Verrez-Bagnis, V., & Lamballerie, M. de. (2005). Effects of high pressure on texture and microstructure of sea bass (*Dicentrarchus labrax* L.) fillets. *Journal of Food Science, 70*(8), e477–e483. https://doi.org/10.1111/j.1365-2621.2005.tb11518.x

Christensen, L. B., Hovda, M. B., & Rode, T. M. (2017). Quality changes in high pressure processed cod, salmon and mackerel during storage. *Food Control, 72*, 90–96. https://doi.org/10.1016/j.foodcont.2016.07.037

Coroneo, V., Corrias, F., Brutti, A., Addis, P., Scano, E., & Angioni, A. (2022). Effect of high-pressure processing on fresh sea urchin gonads in terms of shelf life, chemical composition, and microbiological properties. *Foods, 11*(3). https://doi.org/10.3390/foods11030260

Cruz-Romero, M., Kerry, J. P., & Kelly, A. L. (2008). Changes in the microbiological and physicochemical quality of high-pressure-treated oysters (Crassostrea gigas) during chilled storage. *Food Control, 19*(12), 1139–1147. https://doi.org/10.1016/j.foodcont.2007.12.004

Dong, X., Wang, J., & Raghavan, V. (2020). Effects of high-intensity ultrasound processing on the physiochemical and allergenic properties of shrimp. *Innovative Food Science & Emerging Technologies, 65*, 102441.

Ekonomou, S. I., & Boziaris, I. S. (2021). Non-thermal methods for ensuring the microbiological quality and safety of seafood. *Applied Science, 11*(2), 883. https://doi.org/10.3390/app11020833

Elamin, W. M., Endan, J. B., Yosuf, Y. A., Shamsudin, R., & Ahmedov, A. (2015). High pressure processing technology and equipment evolution: A review. *Journal of Engineering Science and Technology Review, 8*(5), 75–83. https://doi.org/10.25103/jestr.085.11

Erkan, Ö. N. (2003). Verderb von Fisch und Fischwaren- chemische und mikrobiologische veränderungen und Gefahren. *Rundschau Für Fleischhygiene Und Lebensmittelüberwachung, 55*(11), 254–258.

Erkan, N. (2014). Alternative seafood preservation technologies: Ionizing radiation and high pressure processing. *Journal of FisheriesSciences.Com, 8*(200), 238–251. http://acikerisim.mu.edu.tr/xmlui/handle/20.500.12809/7001

Erkan, N., & Cagiltay, F. (2011). The effect of cultural factors and consumer knowledge on fish consumption. *Food Engineering and Ingredients, 36*(2), 42–45.

Erkan, N., Üretener, G., & Alpas, H. (2010). Effect of high pressure (HP) on the quality and shelf life of red mullet (*Mullus surmelutus*). *Innovative Food Science & Emerging Technologies, 11*(2), 259–264. https://doi.org/10.1016/j.ifset.2010.01.001

Erkan, N., Üretener, G., Alpas, H., Selçuk, A., Özden, Ö., & Buzrul, S. (2011). The effect of different high pressure conditions on the quality and shelf life of cold smoked fish. *Innovative Food Science and Emerging Technologies, 12*(2), 104–110. https://doi.org/10.1016/j.ifset.2010.12.004

FAO. (2021). FAO Yearbook. Fishery and Aquaculture Statistics 2019/FAO annuaire. Statistiques des pêches et de l'aquaculture 2019/FAO anuario. Estadísticas de pesca y acuicultura 2019. In *FAO Yearbook. Fishery and Aquaculture Statistics 2019/FAO annuaire. Statistiques des pêches et de l'aquaculture 2019/FAO anuario. Estadísticas de pesca y acuicultura 2019.* https://doi.org/10.4060/cb7874t

Farkas, D. F. (2016). A short history of research and development efforts leading to the commercialization of high-pressure processing of food. *Food Engineering Series*, 19–36. https://doi.org/10.1007/978-1-4939-3234-4_2

Fattahi, S., & Zamindar, N. (2017). Effect of immersion ohmic heating on thawing rate and properties of frozen tuna fish. *Food Science and Technology International*. https://doi.org/10.1177/1082013219895884

Fletcher, G. C., Youssef, J. F., & Gupta, S. (2008). Research issues in inactivation of listeria monocytogenes associated with New Zealand greenshell mussel meat (*Perna canaliculus*) using high-pressure processing. *Journal of Aquatic Food Product Technology, 17*(2), 173–194. https://doi.org/10.1080/10498850801937208

Franco, D., Munekata, P. E. S., Agregán, R., Bermúdez, R., López-Pedrouso, M., Pateiro, M., & Lorenzo, J. M. (2020). Application of pulsed electric fields for obtaining antioxidant extracts from fish residues. *Antioxidants, 9*(2), 1–14. https://doi.org/10.3390/antiox9020090

Gelman, A., Sachs, O., Khanin, Y., Drabkin, V., & Glatman, L. (2005). Effect of ozone pretreatment on fish storage life at low temperatures. *Journal of Food Protection, 68*(4), 778–784. https://doi.org/10.4315/0362-028X-68.4.778

Gómez-Estaca, J., Gómez-Guillén, M. C., & Montero, P. (2007). High pressure effects on the quality and preservation of cold-smoked dolphinfish (*Coryphaena hippurus*) fillets. *Food Chemistry, 102*(4), 1250–1259. https://doi.org/10.1016/j.foodchem.2006.07.014

Gómez-Guillén, M. C., & Montero, M. P. (2007). Polyphenol uses in seafood conservation. *American Journal of Food Technology, 2*(7), 593–601. https://doi.org/10.3923/ajft.2007.593.601

Gou, J., Choi, G. P., & Ahn, J. (2012). Biochemical quality assessment of semi-dried squid (*Todarodes pacificus*) treated with high hydrostatic pressure. *Journal of Food Biochemistry, 36*(2), 171–178. https://doi.org/10.1111/j.1745-4514.2010.00523.x

Gou, J., Lee, H. Y., & Ahn, J. (2010). Effect of high pressure processing on the quality of squid (*Todarodes pacificus*) during refrigerated storage. *Food Chemistry, 119*(2), 471–476. https://doi.org/10.1016/j.foodchem.2009.06.042

Gupta, S., Farid, M. M., Fletcher, G. C., & Melton, L. D. (2015). Color, yield, and texture of heat and high pressure processed mussels during ice storage. *Journal of Aquatic Food Product Technology, 24*(1), 68–78. https://doi.org/10.1080/10498850.2012.758682

Hassoun, A., Ojha, S., Tiwari, B., Rustad, T., Nilsen, H., Heia, K., Cozzolino, D., El-Din Bekhit, A., Biancolillo, A., & Wold, J. P. (2020). Monitoring thermal and non-thermal treatments during processing of muscle foods: A comprehensive review of recent technological advances. *Applied Sciences (Switzerland), 10*(19). https://doi.org/10.3390/app10196802

He, H., Adams, R. M., Farkas, D. F., & Morrissey, M. T. (2002). Use of high-pressure processing for oyster shucking and shelf-life extension. *Journal of Food Science, 67*(2), 640–645. https://doi.org/10.1111/j.1365-2621.2002.tb10652.x

Hoon Moon, J., Byong Yoon, W., & Park, J. W. (2017). Assessing the textural properties of Pacific whiting and Alaska pollock surimi gels prepared with carrot under various heating rates. *Food Bioscience, 20,* 12–18. https://doi.org/10.1016/j.fbio.2017.07.008

Hsu, C. P., Huang, H. W., & Wang, C. Y. (2014). Effects of high-pressure processing on the quality of chopped raw octopus. *LWT, 56*(2), 303–308. https://doi.org/10.1016/j.lwt.2013.11.025

Hu, Y., Yu, H., Dong, K., Yang, S., Ye, X., & Chen, S. (2014). Analysis of the tenderisation of jumbo squid (*Dosidicus gigas*) meat by ultrasonic treatment using response surface methodology. *Food Chemistry, 160,* 219–225. https://doi.org/10.1016/j.foodchem.2014.01.085

Huang, C. H., Lin, C. Saint, Lee, Y. C., Ciou, J. W., Kuo, C. H., Huang, C. Y., Tseng, C. H., & Tsai, Y. H. (2022). Quality improvement in mackerel fillets caused by brine salting combined with high-pressure processing. *Biology, 11*(9), 1–12. https://doi.org/10.3390/biology11091307

Huang, H. W., Lung, H. M., Yang, B. B., & Wang, C. Y. (2014). Responses of microorganisms to high hydrostatic pressure processing. *Food Control, 40*(1), 250–259. https://doi.org/10.1016/j.foodcont.2013.12.007

Huang, T., Tu, Z.-cai, Xinchen-Shangguan, Wang, H., Zhang, L., & Sha, X.-mei. (2017). Rheological and structural properties of fish scales gelatin: Effects of conventional and ultrasound-assisted extraction. *International Journal of Food Properties, 20*(sup2), 1210–1220. https://doi.org/10.1080/10942912.2017.1295388

Hwang, C. A., & Fan, X. (2015). Processing, quality and safety of irradiated and high pressure-processed meat and seafood products. *Food Engineering Series,* 251–278. https://doi.org/10.1007/978-3-319-10677-9_11

Ikeuchi, Y. (2011). Recent advances in the application of high pressure technology to processed meat products. *Processed Meats: Improving Safety, Nutrition and Quality,* 590–616. https://doi.org/10.1533/9780857092946.3.590

Karim, N. U., Kennedy, T., Linton, M., Watson, S., Gault, N., & Patterson, M. F. (2011). Effect of high pressure processing on the quality of herring (*Clupea harengus*) and haddock (*Melanogrammus aeglefinus*) stored on ice. *Food Control, 22*(3–4), 476–484. https://doi.org/10.1016/j.foodcont.2010.09.030

Kaur, B. P., Kaushik, N., Rao, P. S., & Chauhan, O. P. (2013). Effect of high-pressure processing on physical, biochemical, and microbiological characteristics of black tiger shrimp (*Penaeus monodon*): High-pressure processing of shrimp. *Food and Bioprocess Technology, 6*(6), 1390–1400. https://doi.org/10.1007/s11947-012-0870-1

Kaur, B. P., Rao, P. S., & Nema, P. K. (2016). Effect of hydrostatic pressure and holding time on physicochemical quality and microbial inactivation kinetics of black tiger shrimp (*Penaeus monodon*). *Innovative Food Science & Emerging Technologies, 33,* 47–55.

Kontominas, M. G., Badeka, A. V., Kosma, I. S., & Nathanailides, C. I. (2021). Innovative seafood preservation technologies: Recent developments. *Animals, 11*(1), 1–40. https://doi.org/10.3390/ani11010092

Kykkidou, S., Giatrakou, V., Papavergou, A., Kontominas, M. G., & Savvaidis, I. N. (2009). Effect of thyme essential oil and packaging treatments on fresh Mediterranean swordfish fillets during storage at 4°C. *Food Chemistry, 115*(1), 169–175. https://doi.org/10.1016/j.foodchem.2008.11.083

Lakshmanan, R., Parkinson, J. A., & Piggott, J. R. (2007). High-pressure processing and water-holding capacity of fresh and cold-smoked salmon (*Salmo salar*). *LWT – Food Science and Technology, 40*(3), 544–551. https://doi.org/10.1016/j.lwt.2005.12.003

Lee, Y. C., Kung, H. F., Cheng, Q. L., Lin, C. Saint, Tseng, C. H., Chiu, K., & Tsai, Y. H. (2022). Effects of high-hydrostatic-pressure processing on the chemical and microbiological quality of raw ready-to-eat hard clam marinated in soy sauce during cold storage. *LWT, 159,* 113229. https://doi.org/10.1016/j.lwt.2022.113229

Lemus-Mondaca, R., Leiva-Portilla, D., Perez-Won, M., Tabilo-Munizaga, G., & Aubourg, S. (2018). Effects of high pressure treatment on physicochemical quality of pre- and post-rigor palm ruff (*Seriolella violacea*) fillets. *Journal of Aquatic Food Product Technology, 27*(3), 379–393. https://doi.org/10.1080/10498850.2018.1437582

Liu, L., Llave, Y., Jin, Y., Zheng, D., Fukuoka, M., & Sakai, N. (2017). Electrical conductivity and ohmic thawing of frozen tuna at high frequencies. *Journal of Food Engineering, 197,* 68–77. https://doi.org/10.1016/j.jfoodeng.2016.11.002

Lu, F., Liu, S. L., Liu, R., Ding, Y. C., & Ding, Y. T. (2012). Combined effect of ozonized water pretreatment and ozonized flake ice on maintaining quality of Japanese sea bass (*Lateolabrax japonicus*). *Journal of Aquatic Food Product Technology, 21*(2), 168–180. https://doi.org/10.1080/10498850.2011.589040

Manousaridis, G., Nerantzaki, A., Paleologos, E. K., Tsiotsias, A., Savvaidis, I. N., & Kontominas, M. G. (2005). Effect of ozone on microbial, chemical and sensory attributes of shucked mussels. *Food Microbiology, 22*(1), 1–9. https://doi.org/10.1016/j.fm.2004.06.003

Matser, A. M., Stegeman, D., Kals, J., & Bartels, P. V. (2000). Effects of high pressure on colour and texture of fish. *High Pressure Research, 19*(1–6), 109–115. https://doi.org/10.1080/08957950008202543

Mengden, R., Röhner, A., Sudhaus, N., & Klein, G. (2015). High-pressure processing of mild smoked rainbow trout fillets (*Oncorhynchus mykiss*) and fresh European catfish fillets (*Silurus glanis*). *Innovative Food Science & Emerging Technologies, 32*, 9–15.

Murchie, L. W., Cruz-Romero, M., Kerry, J. P., Linton, M., Patterson, M. F., Smiddy, M., & Kelly, A. L. (2005). High pressure processing of shellfish: A review of microbiological and other quality aspects. *Innovative Food Science and Emerging Technologies, 6*(3), 257–270. https://doi.org/10.1016/j.ifset.2005.04.001

Nerantzaki, A., Tsiotsias, A., Paleologos, E. K., Savvaidis, I. N., Bezirtzoglou, E., & Kontominas, M. G. (2005). Effects of ozonation on microbiological, chemical and sensory attributes of vacuum-packaged rainbow trout stored at 4±0.5°C. *European Food Research and Technology, 221*(5), 675–683. https://doi.org/10.1007/s00217-005-0042-x

Niakousari, M., Hedayati, S., Gahruie, H. H., Greiner, R., & Roohinejad, S. (2019). Impact of Ohmic Processing on Food Quality and Composition. In *Effect of Emerging Processing Methods on the Food Quality* (pp. 1–26). Springer.

Nilsen-Nygaard, J., Fernández, E. N., Radusin, T., Rotabakk, B. T., Sarfraz, J., Sharmin, N., Sivertsvik, M., Sone, I., & Pettersen, M. K. (2021). Current status of biobased and biodegradable food packaging materials: Impact on food quality and effect of innovative processing technologies. *Comprehensive Reviews in Food Science and Food Safety, 20*(2), 1333–1380. https://doi.org/10.1111/1541-4337.12715

Olatunde, O. O., & Benjakul, S. (2018). Nonthermal processes for shelf-life extension of seafoods: A revisit. *Comprehensive Reviews in Food Science and Food Safety, 17*(4), 892–904. https://doi.org/10.1111/1541-4337.12354

Oliveira, F. A. de, Neto, O. C., Santos, L. M. R. dos, Ferreira, E. H. R., & Rosenthal, A. (2017). Effect of high pressure on fish meat quality – A review. *Trends in Food Science & Technology, 66*, 1–19. https://doi.org/10.1016/j.tifs.2017.04.014

Pérez-Won, M., Cepero-Betancourt, Y., Reyes-Parra, J. E., Palma-Acevedo, A., Tabilo-Munizaga, G., Roco, T., Aubourg, S. P., & Lemus-Mondaca, R. (2021). Combined PEF, CO_2 and HP application to chilled coho salmon and its effects on quality attributes under different rigor conditions. *Innovative Food Science and Emerging Technologies, 74*, 102832. https://doi.org/10.1016/j.ifset.2021.102832

Rastogi, N. K., Raghavarao, K. S. M. S., Balasubramaniam, V. M., Niranjan, K., & Knorr, D. (2007). Opportunities and challenges in high pressure processing of foods. *Critical Reviews in Food Science and Nutrition, 47*(1), 69–112. https://doi.org/10.1080/10408390600626420

Rathod, N. B., Ranveer, R. C., Bhagwat, P. K., Ozogul, F., Benjakul, S., Pillai, S., & Annapure, U. S. (2021). Cold plasma for the preservation of aquatic food products: An overview. *Comprehensive Reviews in Food Science and Food Safety, 20*(5), 4407–4425. https://doi.org/10.1111/1541-4337.12815

Seo, Y. C., Choi, W. S., Park, J. H., Park, J. O., Jung, K. H., & Lee, H. Y. (2013). Stable isolation of phycocyanin from Spirulina platensis associated with high-pressure extraction process. *International Journal of Molecular Sciences, 14*(1), 1778–1787. https://doi.org/10.3390/ijms14011778

Shiekh, K. A., & Benjakul, S. (2020). Melanosis and quality changes during refrigerated storage of Pacific white shrimp treated with Chamuang (*Garcinia cowa* Roxb.) leaf extract with the aid of pulsed electric field. *Food Chemistry, 309*, 125516.

Skipnes, D., Van der Plancken, I., Van Loey, A., & Hendrickx, M. E. (2008). Kinetics of heat denaturation of proteins from farmed Atlantic cod (*Gadus morhua*). *Journal of Food Engineering, 85*(1), 51–58. https://doi.org/10.1016/j.jfoodeng.2007.06.030

Sukmanov, V., Hanjun, M., & Li, Y. (2019). Effect of high pressure processing on meat and meat products. A review. *Ukrainian Food Journal, 8*(3), 448–469. https://doi.org/10.24263/2304-974x-2019-8-3-4

Truong, B. Q., Buckow, R., Stathopoulos, C. E., & Nguyen, M. H. (2015). Advances in high-pressure processing of fish muscles. *Food Engineering Reviews, 7*(2), 109–129. https://doi.org/10.1007/s12393-014-9084-9

Tsai, Y. H., Kung, H. F., Lin, C. Saint, Hsieh, C. Y., Ou, T. Y., Chang, T. H., & Lee, Y. C. (2022). Impacts of high-pressure processing on quality and shelf-life of yellowfin tuna (*Thunnus albacares*) stored at 4°C and 15°C. *International Journal of Food Properties, 25*(1), 237–251. https://doi.org/10.1080/10942912.2022.2029483

Wu, S., Tong, Y., Zhang, C., Zhao, W., Lyu, X., Shao, Y., & Yang, R. (2021). High pressure processing pre-treatment of Chinese mitten crab (*Eriocheir sinensis*) for quality attributes assessment. *Innovative Food Science and Emerging Technologies, 73*(1800), 102793. https://doi.org/10.1016/j.ifset.2021.102793

Xuan, X. T., Cui, Y., Lin, X. D., Yu, J. F., Liao, X. J., Ling, J. G., & Shang, H. T. (2018). Impact of high hydrostatic pressure on the shelling efficacy, physicochemical properties, and microstructure of fresh razor clam (*Sinonovacula constricta*). *Journal of Food Science, 83*(2), 284–293. https://doi.org/10.1111/1750-3841.14032

Yan, B., Jiao, X., Zhu, H., Wang, Q., Huang, J., Zhao, J., Cao, H., Zhou, W., Zhang, W., Ye, W., Zhang, H., & Fan, D. (2020). Chemical interactions involved in microwave heat-induced surimi gel fortified with fish oil and its formation mechanism. *Food Hydrocolloids*, *105*, 105779. https://doi.org/10.1016/j.foodhyd.2020.105779

Ye, T., Chen, X., Chen, Z., Yao, H., Wang, Y., Lin, L., & Lu, J. (2021). Quality and microbial community of high pressure shucked crab (*Eriocheir sinensis*) meat stored at 4°C. *Journal of Food Processing and Preservation*, *45*(4), 1–11. https://doi.org/10.1111/jfpp.15330

Zhao, Y. M., de Alba, M., Sun, D. W., & Tiwari, B. (2019). Principles and recent applications of novel non-thermal processing technologies for the fish industry – A review. *Critical Reviews in Food Science and Nutrition*, *59*(5), 728–742. https://doi.org/10.1080/10408398.2018.1495613

Zhou, Y., He, Q., & Zhou, D. (2017). Optimization extraction of protein from mussel by high-Intensity pulsed electric fields. *Journal of Food Processing and Preservation*, *41*(3), e12962.

Zhou, Y., Sui, S., Huang, H., He, G., Wang, S., Yin, Y., & Ma, Z. (2012). Process optimization for extraction of fishbone calcium assisted by high intensity pulsed electric fields. *Transactions of the Chinese Society of Agricultural Engineering*, *28*(23), 265–270.

14 Applications of High-Pressure Processing in Liquid Food Processing

N. Bhanu Prakash Reddy, Yuvraj Bhosale Khasherao,
Thivya Perumal, and V. R. Sinija

14.1 INTRODUCTION

Health-conscious consumers demand better-quality food, such as nutritional value, safety, flavours, and natural freshness. Clean-label foods, which promise to be natural, fresh, and free of chemical ingredients, have gained popularity among customers over time. Beverages are one of these foods consumed more often irrespective of any conditions such as gender, age, place, time, etc. In general, juices extracted from fruits and vegetables have a very short shelf-life and must be consumed instantaneously. At times, the instantaneous consumption or transport of raw materials and processing may not be possible. The only solution is to process the juices to extend their shelf-life by killing pathogenic microorganisms and inactivating quality deteriorative enzymes, storing and transporting them wherever needed to avoid the above said problem.

Traditionally, heat has been used to inactivate and kill enzymes and microorganisms to extend the shelf-life of juice. But the thermal processing technologies were proven to adversely affect food quality regarding nutritional contents, flavours, and sensory characteristics, leading to less acceptability among consumers. Typical thermal processing (pasteurization or sterilization) is applying heat ranging between 60 and 150°C to destroy pathogenic or harmful microorganisms or enzymes, which can result in undesirable changes in the processed product. However, it was seen that the loss of volatile aroma components during such pasteurization causes a deterioration in quality and may result in a "cooked" flavour (Song et al., 2022).

For instance, tender coconut water (TCW) is naturally self-sterile as long as it is inside the nut. However, when exposed to the outside environment, there are very good chances of contamination, enhanced enzyme activity, and rapid microbial growth in storage. Ma et al. (2019) reported that with a high-temperature short time (HTST) process, coconut water achieved the lowest ratings in sensory analysis for aroma, flavour, and overall acceptability though the process yielded better colour retention. The pH of TCW (pH 5.2–6.1) also favours the activity of quality destructive enzymes such as polyphenol oxidase (PPO) and peroxidase (POD). The activity of these enzymes may accelerate the oxidation reaction in TCW, which leads to brown, pinkish, and/or yellowish colouration of TCW. At the same time, storage at both room temperature and refrigerated conditions questions the shelf-life of the TCW (Rajashri et al., 2020).

Alternative food processing technologies, i.e., non-thermal technologies, have gained extensive attention in the food industry in the last two decades. Some of the non-thermal technologies are cold plasma, high-pressure processing (HPP), hydrodynamic cavitation, pulsed electric field, pulsed light, modified atmosphere packaging, ultrasound processing, and many more; however, HPP is the most successfully commercialized non-thermal food processing technology (Huang et al., 2017).

HPP is a novel technology that possibly addresses many, if not all, of the many challenges the food industry faces today. These comprise a consistent increase in consumers' preference for premium quality and convenience, i.e., ready-to-eat (RTE) food products with natural characteristics,

DOI: 10.1201/9781003359302-14

free from external additives and preservatives, the drive towards wellness and health, and concerns about fresh food product safety. This innovation involves subjecting food to high hydrostatic pressure (HHP), normally around 100–1,000 MPa. Higher pressures (>1,200 MPa) are required to inactivate bacterial spores. HPP exposes the food commodity to elevated pressure (87,000 pounds per square inch or 600 MPa) in the presence and/or absence of heat (Gupta & Balasubramaniam, 2012). It is a promising alternative to thermal treatments for ensuring both safety and minimizing processing effects on the nutritional quality of foods. HPP has ensured uniform pressure distribution in food. Moreover, various juices and beverages, meat and vegetable products, and seafood have been commercialized. The novel functionalities of HP processed food ingredient modification led to extensive product development (Tao et al., 2014). HPP is nowadays highly acknowledged as it promises to develop a new variety of value-added products. It is a novel technique as it can either amplify the reduction of microbial load in thermal processing or use chemical preservatives that can be substituted.

The main advantages of HPP are (i) the food can be processed at an optimum temperature or below that in the presence of pressure; (ii) novel food product formulation with functional properties; and (iii) the pressure is uniformly transmitted in the food product, disregarding its shape, size, and geometry. Overall, food quality can be improved as heat damage and chemical preservatives are prevented.

However, HPP has the same effect on microbial and enzymatic inactivation as heat treatment. The high-pressure application in liquids increases temperature by 3°C/100 MPa. Additionally, if the food product consists of a considerable amount of fat, such as cream or butter, the temperature increases by 8–9°C/100 MPa (Chakraborty et al., 2014). Cell serum loss and softening of tissues lead to cell deformation or damage, which can shift the substance's pH by high-pressure compression. Certain enzymatic reactions are as well controlled as the activation energy can remain inhibited in certain reactions (Chakraborty et al., 2014). High-pressure application has a reversible effect on the proteins and enzymes at 100–400 MPa, most possibly due to association and dissociation process and conformational changes (Chakraborty et al., 2014). Microbial death during HPP occurs by cell membrane permeabilization. Nonetheless, high-pressure resistant bacterial spores require a pressure of more than 1,200 MPa for their inactivation. Thus, a high treatment and pressure combination is recommended in pasteurization and sterilization (Wu et al., 2021). HPP application also reduces browning reactions and enzymes like oxidase and POD, enhancing shelf life and stability and forming a consumer-desired product.

14.2 HIGH-PRESSURE PROCESSING IN LIQUID FOOD PROCESSING

14.2.1 PRINCIPLE OF HIGH-PRESSURE PROCESSING

The governing principles of HPP are isostatic rule which states, "the applied pressure is instantaneously and uniformly transmitted throughout the sample whether the sample is hermetically sealed in a flexible package or direct contact with the pressure medium". The effect of high pressure on microbiology and food chemistry is governed by Le Chatelier's principle, which states, "When a system at equilibrium is disturbed, the system then responds in a way that tends to minimize the disturbance" (Woldemariam & Emire, 2019). At temperatures below or above the ambient, HPP promotes the inactivation of vegetative bacteria. It has been established that HPP at refrigeration and even freezing temperatures are more successful than pressurization at ambient temperature for microbial inactivation. Similarly, combining moderate HPP treatments with moderate temperatures might result in inactivation greater than 5-log cycles in many pathogenic bacteria suspended in various laboratory media or foods.

HPP technology has also been accepted as one of the safer technologies by governmental authorities in Canada, Europe, and the USA for microbial destruction and shelf-life extension of various liquid foods. This indicated ample opportunity for HPP-treated ready-to-serve drinks and packaged coconut water, considering the adverse effects of other thermal treatments (Huang et al., 2017). Since HPP technology is a new pasteurization technology for the food industry, its health and safety aspects, economic value, and product forms should be assessed in detail before commercialized application.

14.2.2 Factors Governing the Performance of High-Pressure Processing

The major preservative effect that could be carried out in a food product is microbial inactivation which in HPP is carried out by meeting numerous intrinsic and process parameters. Major influencing parameters on HPP processing are types and number of microorganisms, their interaction with food, duration of pressure, temperature application, and composition of the food product. The pressure exerted on the food product and the exposure time of the food directly affect the microbial cell count reduction, stating that higher pressure is more effective on microbial destruction. Moreover, higher time exposure to food materials also leads to higher microbial destruction (Huang et al., 2017). This led to coining new terms such as "pressure D-value" and the same for *E. coli* ranging from 16.43 minutes to 1.21 minutes under a pressure condition of 240–400 MPa (Lavinas et al., 2008). These results showed a clear relationship between pressure value, processing time, and microbial destruction.

Along with time, process temperature is among the most important quality parameters affecting process efficiency and microbial destruction during HPP. The presence of acids in high-acid food with lower pH value add a synergistic effect on the effectiveness of HPP (Podolak et al., 2020). The effectiveness of all these parameters is mainly dependent on the bacterial strain. When exposed to any inferior environmental condition, the bacteria' basic survival instincts lead to the structural modification in the cell wall leading to resistance to particular preservation techniques. These modifications in the microbes might have led to the development of pressure resistance (Podolak et al., 2020).

14.3 HPP SYSTEMS FOR LIQUID FOOD PROCESSING

The first commercialized HPP-treated food product was a *jam* in Japan in 1993; after that, extensive use of HPP systems was adopted at the industrial level in North America and Europe (Huang et al., 2017). In the past two to three decades, the popularity of HPP systems has increased. Many machine manufacturers entered the HPP equipment design, development, and manufacturing field. In general, HPP systems are available in batch and semi-continuous modes. Besides these two types, continuous systems on a laboratory scale are under development and are not widely commercialized. Batch systems are suitable for pre-packaged solid and liquid foods. Semi-continuous and continuous ones are suitable for flowable foods (Elamin et al., 2015).

For batch systems, there are three major parts of equipment, *viz.*, vessel (where actual pressurization and food treatment happens), surrounding yoke (in most of the cases vessel is surrounded by a yoke consisting of either a wire wound steel frame or several steel plates to withstand pressure), and the hydraulics (the system of components which help to build pressure, for example, pressurizing medium, pump, etc.). In this batch system, two major geometries/orientations of the vessels can be seen. They are vertical vessels (parallel to the vertical axis) in which products can be put in and taken after the treatment, and most companies adopt this design (Fig. 14.1). The other design is a horizontal vessel type in which the unprocessed food products can push the processed foods (Fig. 14.2). In semi-continuous HPP systems, liquids are pumped into the pressure vessel using a low-pressure transport pump; then, one high-pressure pump is used to pressurize the vessel once the valves have been closed. A floating piston separates the pressure transfer medium and the product. The energy accumulated in the pressured vessel can be utilized to pressurize another vessel at ambient pressure, conserving energy and speeding up the process by connecting many pressure vessels (Balasubramaniam et al., 2016).

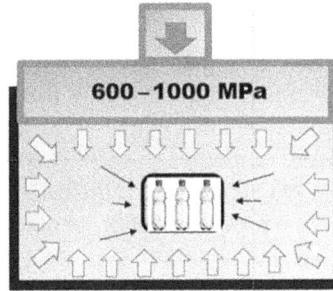

FIGURE 14.1   Schematic diagram of liquid food processing in HPP equipment.

14.3.1   GLOBAL PLAYERS IN HPP EQUIPMENT MANUFACTURING

Over the years, most companies with expertise in high-pressure systems like cold isostatic pressing and waterjet cutting (for example, Kobe steel, Japan and Flow systems, USA) have started developing HPP equipment suitable for handling high pressure, i.e., up to 1,000 MPa. After the 1990s, various manufacturers in the USA, UK, Spain, Japan, and China started developing HPP equipment. Major global players include Hyperbaric – Spain (>50% market share), Avure – USA, Multivac – Germany, and Baotou Kefa High-Pressure Technology Co., Ltd. – China). One unit of an HPP system costs approximately 4 to 12 crore INR according to the capacity, operating range, and manufacturer (Huang et al., 2017).

14.4   EFFECT OF HPP ON THE FOOD PRODUCT AND ITS ENVIRONMENT

Food safety and quality are the two utmost factors to be considered by producers and processors that play an inevitable part in deciding the acceptability of any processed food product across the consumer community. Of the above said, the quality includes composition intact after processing. The quality here includes nutritional composition, enzyme activity, microbial load or activity, and shelf-life of the processed food product. Other than these, consumer acceptance makes the ultimate decision on the adaptability of food processing. Also, some undesirable effects of HPP are discussed in this section.

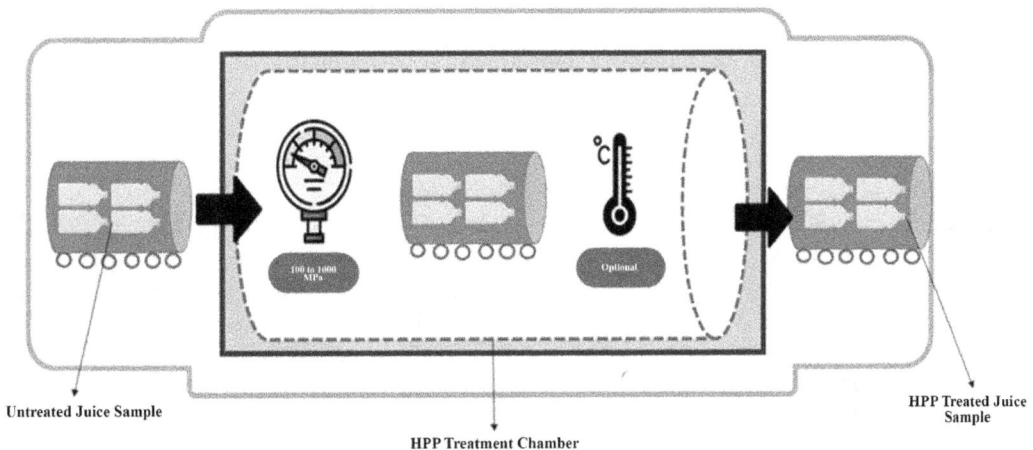

Untreated Juice Sample

HPP Treatment Chamber

HPP Treated Juice
Sample

FIGURE 14.2   Schematic diagram of batch type (horizontal vessel) liquid food processing HPP equipment.

14.4.1 NUTRITIONAL COMPOSITION

Nutritional composition is comprehensive information about important nutritional components of food items and elucidates energy values. HPP is one of the non-thermal technologies proven to be less detrimental to food quality and nutritional composition from the studies conducted so far. A study where mango slices were immersed in sorbitol solution and treated with the assistance of HPP found that HPP treatment enhanced the diffusion of solutes to fruit from the solution. The combined effect of 300 MPa pressure, 60°Brix sorbitol, and 2% calcium lactate were the most effective at conserving natural compounds of mango cubes with a good, acceptable nutritional value (Lamilla et al., 2021). Similarly, Huang et al. (2018) treated carambola juice with HPP (600 MPa for 150 seconds) and high-temperature and short-time (HTST) (110°C for 8.6 seconds) for pasteurization and found that HPP retained the colour of the juice and which was comparable to that of fresh juice.

Similarly, HPP-treated samples were comparable to fresh samples in colour, as no significant change was observed. It was also observed that the blackberry puree's anti-radical power or free radical scavenging activity increased at 600 MPa treatment by 67% compared to the unprocessed one. Also, high pressure could make phenolic compounds freely available; the purees were observed with more phenolic content than the untreated ones. Similar trends were observed in the case of anthocyanins also. But ascorbic acid contents in both the products were down by 6% at 600 MPa, which is far less destruction/loss than thermal processing, where losses were close to 23%.

In another study, Liberatore et al. (2021) compared the effects of HPP (600 MPa for 3 minutes) and thermal pasteurization (87°C for 12 minutes) on some of the quality features and volatile profiles of the cloudy apple juices from various apple cultivars. The study yielded strong supporting evidence that the colour, pH, viscosity, and total soluble solids (TSS) of HPP-treated samples were close to fresh samples, which was not the case with thermally treated samples; this proves HPP is far better technology to preserve the freshness and retain the nutritional quality of the liquid foods. Also, it revealed that HPP treatment enriched volatile components, for instance, hexanal, and helped preserve the initial volatile characteristics of all the varieties of juices, the authors reiterated that even if HPP treatment would be worth considering for stabilization of fruit juices and preservation of fresh-like characteristics despite its high processing cost than traditional thermal treatment. Nayak et al. (2017) compared the physico-chemical properties such as pH, TSS, titratable acidity (TA), viscosity, colour, antioxidant activity (AA), total phenolic content (TPC), and total flavonoid content (TFC) of the elephant apple juices processed with HPP (600 MPa for 5 minutes) and thermal processing (80°C for 60 seconds). They found no significant difference in physical characteristics such as colour, TSS, and viscosity immediately after treatment, whereas the properties deteriorated during storage. But, HPP-treated samples showed slower degradation of quality in comparison to those thermally treated juices. They also observed that the treatment enhanced the TPC values, which refers to the enhanced extractability of HPP treatment; TFC values reduced slightly during storage due to the detrimental enzyme's residual activity. The rate of AA reduction was high in the untreated sample, followed by thermal, and the slowest one was observed in HPP treatment.

Polydera et al. (2003) studied the shelf-life of orange juice processed with HPP, i.e., at 500 MPa, 35°C for 5 minutes and traditional thermal pasteurization, i.e., 80°C for 30 seconds, based on residual ascorbic acid content. It was found that HPP-treated orange juice did show greater shelf-life, i.e., > 90 days and 47 days at 5°C and 10°C conditions, respectively. At the same time, thermal treatment could extend the shelf-life to 47 and 25 days at the cost of overall nutritional quality. Thus, HPP treatment of especially liquid foods was proven to be a potentially good treatment compared to thermal treatment because the physical appeal of the treated product was so close to that of the fresh product with enhanced phytochemical properties like TPC and AA.

14.4.2 Enzyme Activity

One of the key issues in the fruits and vegetable processing and preservation industry is their perishable nature, i.e., they are prone to spoilage in lesser time. Majorly two reasons for the spoilage of fruit and vegetable products are (i) *Spoilage due to enzyme activity* and (ii) *Microbial spoilage.*

Many enzymes are involved in the quality deterioration of fruits, vegetable juices, and other unprocessed liquid foods. These include oxidoreductive enzymes like PPO – which develops brown colouration during the ripening, fermentation &/ageing process; POD – which produces off-flavours in vegetables and induces discolouration in fruit juices, lipoxygenase (LOX) – destructs essential fatty acids, vitamin A, and develops off- and odd-flavours, ascorbic acid oxidase (AAO) – destructs the ascorbic acid, i.e., vitamin C, and β-glucosidase (BGL) – degrades anthocyanins and develops off-flavour when present in excess amounts. Enzymes such as polygalacturonase (PG) and pectin methyl esterase (PME) involve in the pectin network (which surrounds the cellulose and acts as the backbone of the cell wall) breakdown and yield low viscous products having modified perceptional properties (Chakraborty et al., 2014; Lee & DeMan, 2018). Other than the abovementioned enzymes, several other enzymes like protease, lipase, thiaminase, etc., show the quality destructive nature in various other food product groups such as dairy, meat, aqua-foods, oils, etc.

Of all the enzymes mentioned, oxidoreductases are most commonly seen in liquid foods, especially fruits and vegetable juices. When the physiological barrier between the cell wall and the membranes is breached, PPO becomes more accessible (membrane-bound copper-containing enzyme) in a simple and easily accessible and reactive form. The PPO in the presence of molecular oxygen, diphenols are liberated from the vacuole and transformed into ortho-quinones. Subsequently, non-enzymatic reactions of yellow o-quinones happen in the presence of oxygen. High molecular weight polymers such as amino acids and proteins produce macromolecular complexes and aggregated melanin compounds (Riahi & Ramaswamy, 2004). Additionally, PPO can negatively affect the flavour and taste of fruit products, as phenolic compounds are responsible for these properties. POD is another enzyme responsible for most plant-based liquid foods' colour and flavour degradation. POD catalyzes single-electron oxidation reactions. Phenolic chemicals in brown goods when hydrogen peroxide is present. In addition to phenolic chemicals, they can interact with other types of hydrogen donors, such as aromatic amines, ascorbic acid, and complexes that produce brown indoles (Vernwal et al., 2006). Hydrogen peroxide as a substrate for POD is generated in the pathway of browning and anthocyanin degradation reactions catalyzed by POD.

Similarly, BGL, a glucose hydrolase, catalyzes the hydrolysis of glucoside compounds and converts them to glucose by hydrolyzing glycosidic bonds. POD also acts on anthocyanins in the presence of phenols and peroxides, yielding brown or red complexes. PME modifies the texture of liquid foods by breaking down pectin-like substances. PG catalysis hydrolysis of glycosidic bonds between galacturonic acid residues and thus depolymerizes them. The released free carboxyl groups bind to the calcium ions and neighbouring pectin molecules. These complex compounds aggregate and precipitate together, helping the phase separation in fruit juices. LOX catalyzes the oxidation of polyunsaturated fatty acids and esters in radical pathways by forming hydroperoxide intermediates (Chakraborty et al., 2014; Masood et al., 2018). However, not all enzymes are always detrimental, but they are useful in some cases. Every enzyme has different inactivation levels that vary with the type of food material and the technology used for processing (Martín-Belloso et al., 2014).

Under the high-pressure environment, the mechanism of enzyme inactivation can be hypothesized to be similar to protein denaturation (Leite Júnior et al., 2021). Enzyme inactivation by high pressure is a complex phenomenon that includes a sequence of events, such as the formation and/or disruption of numerous interactions, and modifications of the enzyme's native structure, such as folding and/or unfolding. The applied pressure can cause the native structure of the enzyme to unfold partially or completely, irreversibly, or both. As the specificity of an enzyme is correlated with the structure of its active site, this eventually changes enzyme activity (Chakraborty et al., 2014).

HPP processing of the food product aims to inhibit the activity of particular enzymes that deteriorate food quality without compromising the sensory and nutritional properties usually harmed by heat treatments. Enzymes and bacteria that cause food quality degradation are typically inactivated at pressures between 100 and 1,000 MPa. The effectiveness of high-pressure treatment depends on several factors, including the processing and operating circumstances, the food composition, and the activity of oxidative enzymes such as PPO, POD, BGL, PME, PG, etc. For instance, pressures above 400 to 450 MPa can be robust enough to inactivate PPOs (Huang et al., 2014). Fernandez et al. (2019) treated fruit and vegetable smoothie using HPP (630 MPa for 6 minutes at 20°C) and observed 83.9%, 31.4%, and 9.7% initial reduction of PME, POD, and PPO, respectively. In another study, Jayachandran et al. (2015) treated litchi-based mixed fruit beverage using HPP with the treatment conditions (pressure: 200–600 MPa; temperature: 50–70°C; time: 5–20 minutes) and found that 90% inactivation of PPO was achieved at 500 MPa/70°C/20 minutes and 600 MPa/60°C/15 minutes. Though the thermal treatment can completely inactivate the PPO and POD, it can be at the cost of the beverage's freshness, nutritional, and organoleptic properties. In another study, strawberry juice was treated with pressure (200–600 MPa), mild temperature (10–50°C), and time duration (up to 30 minutes); it has been found that close to 50% inactivation of POD was observed at 600 MPa/30°C/30 minutes and >65% inactivation was seen at 600 MPa/50°C/30 minutes, whereas 600 MPa/10°C/30 minutes could inactivate only close to 20% of initial POD (Fang et al., 2008). These studies proved that pressure combined with temperature can be more effective. However, one should be intelligent enough while selecting the extreme temperature for any food product to maintain the nutritional and sensory quality as much as possible.

Chakraborty (2015) processed pineapple puree with HPP to see the effect of pH (3, 3.5, and 4.0), pressure (100–600 MPa), temperature (20–70°C) and process time (0–30 minutes) on inactivation of PPO, PME, and bromelain (BRM). They observed that PPO was the least inactivated enzyme at all pH and HPP process conditions. In contrast, PME was observed to have been inactivated more, *viz.*, 77.3, 75.5, and 67.1% of inactivation at all three pH, i.e., 3.0, 3.5, and 4.0, respectively, with an optimized HPP condition of 600 MPa/60°C/9 to 9.5 minutes. This relation reveals that the HPP efficacy in inactivating undesirable enzyme activity depends on the surrounding media, such as pH. Similar studies were reviewed in detail by Chakraborty et al. (2014).

14.4.3 MICROBIAL LOAD AND SHELF-LIFE

In HPP, foods (containing spoilage microorganisms) are given high pressure, i.e., in the range of 100–600 MPa (sometimes even greater). Thus, the microbial organisms are inactivated by rupturing the cell wall under this uniform, higher, and unbearable pressure. Georget et al. (2015) reviewed the inactivation mechanism of HPP against microorganisms in various foods. They confirmed with enough literature evidence that the cell membrane undergoes a phase transition, changes its fluidity, and finally destroys the cell wall. Keefe et al. (2013) proved that HPP at 600 MPa for 2 minutes with water as the pressurizing medium of 4°C could inactivate *Salmonella typhimurium, L. monocytogenes,* and *E. coli* O157:H7 by at least 5-log CFU/mL in coconut water. Also, Raghubeer et al. (2020) investigated the impact of HPP (593 MPa/3 minutes/4°C) on the microbial safety of TCW. All microorganisms were killed, and HPP-treated TCW was sufficiently microbiologically safe for up to 120 days, whereas unprocessed samples at refrigerated storage could not stay 15 days.

Apart from these facts, HPP may not aggressively inactivate some spore-forming bacteria, such as *Clostridium botulinum* which produces toxic compounds called botulinum neurotoxins at even refrigerated/colder storage conditions (4–10°C); there is a chance of bacterial regrowth even after 60 days if all the *C. botulinum* is not inactivated (González-Angulo et al., 2020). Huang et al. (2018) compared carambola juice treated with HTST and HPP and found that HPP could inactivate

E. coli O157:H7 by 5.15-log reductions. Both processes brought aerobic, coliform, and psychrotrophic yeasts and moulds in the juice below the detection limit, i.e., <1.0 log CFU/mL. Also, there was no microbial regrowth over 40 days of refrigerated storage in HPP and HTST-treated samples. These works proved that HPP could achieve minimal microbial destruction/sterility, i.e., a minimum of 5-log reduction by any processing technique as per the United States Food and Drug Administration (FDA, 2004).

Nayak et al. (2017) analyzed the microbial inactivation after HPP treatment in elephant apple juice storage by periodically checking the total aerobic bacteria, yeast, and mould count for as much as 60 days. They observed that HPP treatment brought down the microbial load to <1-log CFU/mL till 60 days from initial levels of 4.24 ± 0.14 and 3.14 ± 0.08 log CFU/mL, respectively, for bacteria, yeast, and mould. Besides these, several other studies proved that HPP could be a technique capable of inactivating detrimental microorganisms with a limitation (which can be solved with hybrid treatments) of spore-forming bacteria activity after the treatment, though in smaller amounts (Table 14.1).

14.4.4 CONSUMER ACCEPTABILITY

Consumer perception of minimally processed, fresh-like, high-quality food has changed over the past few years. Consumer expectations and preferences subsequently led researchers and food processors to search for an alternative to traditional thermal processing, which induces noticeable changes in perceptional and nutritional quality. In searching for alternatives, novel thermal and non-thermal methods like HPP, PEF, and Ultrasound (US) were considered possible solutions in food processing. The use of non-thermal techniques has increased over the past decade. HPP leads from the front in the food industry across different food types, especially liquid foods and meat (Misra et al., 2017). Consumers have questions about the sensorial quality of processed food, particularly major concerns towards flavour, which is the most imperative sensory attribute in fruit- and vegetable-based products like juices. However, it is known that such loss of flavour during pasteurization can lead to deterioration and a "cooked" taste (Awuah et al., 2007). In contrast, HPP pasteurizes liquid food at comparatively low temperatures without heating the product. Thus, HPP not only addresses the downsides associated with heat treatment but also reduces energy consumption which is the case with heating and cooling (Huang et al., 2017). In a way, HPP not only preserves nutritional quality by maintaining low temperatures throughout the pasteurization but also helps consumer perception positively, particularly when it comes to fresh-like/close-to-natural food.

Song et al. (2022) reviewed the effect of HPP treatment on smoothies and juices through various sensory modalities, such as appearance, aroma and taste, mouthfeel, and overall acceptance, by taking into account a wide range of previous research related to juices and smoothies sensory studies with various sensory methods such as discriminative, descriptive, and hedonic scale methods. The thermal treatment produced a cooked-like flavour that led panellists to give a higher rating to the HPP-treated samples. Regarding mouthfeel, HPP-treated juices were superior from an astringency point of view. In contrast, the mouthfeel was close to thermally treated ones. Various studies reported similar insights regarding the acceptability of HPP-treated juice. For instance, Rao et al. (2014) reported that HPP-treated peach juice was not significantly different from untreated juice in colour and turbidity. Also, the HPP-treated one showed a better aroma than heat-treated and untreated juice after the same time, i.e., from extraction to testing. Similarly, the overall impression and quality of heat-treated juices were inferior to that of HPP-treated ones. A similar scenario can be seen in the case of the "pepper and orange juice blend" treatment (Xu et al., 2015). Thus, it can be concluded that HPP-treated food products (especially liquid foods) resemble fresh- and natural-like foods with good organoleptic quality; HPP-treated foods can address the shifting consumer demand, especially in the ready-to-drink food segment.

TABLE 14.1

Effect of High-Pressure Processing on Quality Parameters/Microbes/Enzymes of Some Liquid Foods

Sl. No.	Food Product	Processing Conditions	Quality Parameters/ Microbes/Enzymes Examined for	Outcomes	References
1	Carrot-orange juice	• Pressure – 200 to 400 MPa up to 5 minutes in different combinations • pH varied from 4 to 6	Vitamin C, carotenoids, phenolics, pH, and TSS L. innocua (ATCC 51742)	• All the physicochemical properties were similar to untreated ones. • HPP helped juice to be stable and safe after 10 days of storage. • 200 MPa/5-minute treatment could reduce L. innocua by 2-log at pH 5 and 6, whereas at pH 4, the reduction was 3.3 CFU/mL. • 300 MPa/2- minute, 400 MPa/1- minute achieved greater than 6-log reduction of L. innocua at all pH levels.	(Pokhrel et al., 2022)
2	Raw apple juice	• Static HPP: Pressure – 300, 450, and 600 MPa for 5 minutes • Pulsed HPP: 300 MPa for 5 minutes in 3 pulses	Colour, apparent viscosity, turbidity, vitamin C, and phenols Lactic acid bacteria, yeasts, moulds, and total aerobic mesophiles Polygalacturonase (PG)	• Turbidity and apparent dynamic viscosity were increased during storage, whereas vitamin C was decreased. • Phenols were degraded after the 4th week of storage and decreased antioxidant activity. • HPP treatment brought down all the microbes to a safe level. • 300 MPa*3 pulses and 600 MPa treatments could help preserve apple juice at refrigerated storage for 12 weeks. • HPP could not make a significant impact in terms of inactivation of PG (may be due to its pressure resistance till 600 MPa).	(Szczepańska et al., 2021)
3	Carrot juice	• Pressures – 200, 300, 400, and 500 MPa • Holding time – 2 minutes • Temperature – 20, 35, 50°C • Addition of Nisin 25 & 50 ppm	Shelf-life E. coli ATCC 11755™ and L. innocua ATCC 51742™	• With 25-ppm added Nicin, carrot juice was safe over 28 days without exceeding the microbial count by 2-CFU/mL. • Processing at (500 MPa/20°C/2 minutes) without nisin inactivated L. innocua and E. coli by 4- and 5-log CFU/mL, respectively. • 25-ppm Nicin added one resulted in 7-log CFU/mL reduction of both organisms.	(Pokhrel et al., 2019)
4	Pineapple puree	• Pressure – 0.1 to 600 MPa • Temperature – 30 to 70°C • Time – 1 seconds to 40 minutes	Pectin methylesterase (PME)	• PME inactivation ranged from 15% (200 MPa/30°C) to 67% (600 MPa/70°C)	(Chakraborty et al., 2019)
5	Grape juice	• Pressures – 319 to 531 MPa • Dwell time – 35 to 205 seconds	Escherichia coli O157:H7, Listeria monocytogenes, and Salmonella enterica	• E. coli O157:H7 was more resistant than S. enterica to high pressure. • >5-log reduction at highest pressure-temperature combination. • L. monocytogens were not seen for any regrowth during storage.	(Petrus et al., 2019)

(Continued)

TABLE 14.1 (Continued)

Effect of High-Pressure Processing on Quality Parameters/Microbes/Enzymes of Some Liquid Foods

Sl. No.	Food Product	Processing Conditions	Quality Parameters/ Microbes/Enzymes Examined for	Outcomes	References
6	Deionized water with microbial inoculum	• Pressure – 200 to 400 MPa till 100 seconds • Pulsed electric field – 10 to 30 kV/cm, 10 µs with pulse OFF time of 0.1 to 10 seconds	*Listeria innocua* ATCC 33090	• An 8-log reduction was seen in the microbial population after the treatment at 400 MPa for 100 seconds. • PEF (30 kV, 1 millisecond) and sonication (250 seconds) treatments resulted in only a 2-log reduction in the microbial population.	(Pyatkovskyy et al., 2018)
7	Carambola juice	• 600 MPa of pressure for 30 to 240 seconds at 10°C	*Escherichia coli* O157:H7, aerobic, psychrotrophic, yeasts, and moulds	• 600 MPa/150 seconds treatment reduced *E. coli* O157:H7 by 5.15-log reductions. • No regrowth of bacteria was seen after 40 days of refrigerated storage.	(H.-W. W. Huang et al., 2018)
8	Sugarcane juice	• Pressure – 300 to 600 MPa • Temperature – 30 to 60°C • Time – 10 to 25 minutes	Polyphenol oxidase (PPO)	• 98% inactivation of PPO was observed at 600 MPa for 25 minutes at 60°C	(Sreedevi et al., 2018)
9	Pea starch dispersion	• Pressure 0 to 600 MPa for 15 minutes at 25°C	Particle size optical microscopy, and pasting properties	• No significant difference was observed between control and processed samples (up to 400 MPa). • Viscosity increased (8 to 34 cP) when treated up to 600 MPa. • Pressures above 400 MPa reduced birefringence.	(Leite et al., 2017)
10	Elephant apple juice	• Pressure – 600 MPa for 5 minutes and refrigerated storage study	pH, TSS, TA, viscosity, colour, antioxidant capacity, TPC, TFC, and sensory analysis	• Insignificant variations in parameters like pH, TSS, and TA were observed, whereas significant differences in other parameters like antioxidant capacity, TPC, TFC, and colour were observed after 60 days of storage. • The HPP-treated sample was comparable to the untreated one immediately after the treatment.	(Nayak et al., 2017)
11	Cloudy apple juice	• Pressure – 600 MPa for 3 minutes	Sugars and acids Polyphenol oxidase (PPO) and peroxidase (POD)	• No significant difference in sugars and acids of HPP-treated and untreated apple juice was observed. • The HPP-treated sample saw a slight increase in aldehydes, alcohols, ketones, and organosulphur. • >50% residual activity was observed in the HPP-treated sample, whereas thermal treatment inactivated both enzymes completely. • Little activity in Maillard and oxidative reactions were observed after HPP treatment.	(Yi et al., 2017)

(Continued)

TABLE 14.1 (Continued)
Effect of High-Pressure Processing on Quality Parameters/Microbes/Enzymes of Some Liquid Foods

Sl. No.	Food Product	Processing Conditions	Quality Parameters/Microbes/Enzymes Examined for	Outcomes	References
12	Lentil starch dispersion	• High pressure was applied at 0.1, 400, 500, and 600 MPa for 10 minutes • Flour to water ratio as 1:4 w/w	Rheological, pasting, structural, and functional properties	• After treating at 600 MPa, resistant starch increased by 6.8%. • Starch granule size increased from 196 to 207 μm. • Complete gelatinization was seen after 500 MPa for 10-minute treatment.	(Ahmed et al., 2016)
13	Apple juice	• Pressure – 600 MPa at 75°C for 40 minutes	Spore – *Neosartorya fischeri* ascospore	• The HPP + thermal process inactivated *N. fischeri* spores by 3.3-log reduction, better than sole thermal and thermosonication treatment.	(Evelyn et al., 2016)
14	Tomato products	• Pressure – 200 to 800 MPa • Temperature – 55 to 70°C	Pectin methyl esterase (PME) and Polygalacturonase (PG)	• A combination of pressure (200–800 MPa) and temperature (55–75°C) was proven to inactivate PME and PG completely.	(Andreou et al., 2016)
15	Pepper and orange juice blend	• Blending ratio: 1:5, 1:2, 1:1, 1:0.5, and 1:0.2 (v/v) • Pressure – 550 MPa, 5 minutes • Thermal treatment – 110°C for 8.6 seconds	Total phenols, vitamin C, antioxidant capacity, and sensory acceptability; Total aerobic bacteria, yeast, and moulds	• After sensory analysis, a 1:0.5 ratio of pepper to orange juice blend was proven good. Overall acceptability was close to the untreated blend after treatment. • 77.3%, 90.8%, and 80% of TPC, vitamin C, and antioxidant capacity were retained with HPP treatment. • After HPP treatment, greater than 4-log reduction was obtained in all microorganisms and was proven safe enough.	(Xu et al., 2015)
16	Peach juice	• Pressure – 400 to 600 MPa • Treatment time – 5 to 25 minutes with 5-minute intervals	PPO and pectin methylesterase (PME)	• At 400 MPa, PPO and PME were activated by 7.3 and 2.6%, respectively. • 600 MPa for 25-minute treatment induced 81.2% inactivation of PPO and 50.4% inactivation of PME.	(Rao et al., 2014)
17	Strawberry puree	• Pressure – 100 to 690 MPa • Temperature – 24 to 90°C • Time – 5 to 15 minutes	Polyphenol oxidase (PPO) and peroxidase (POD)	• At low pressure and lower temperatures, the effect on PPO and POD activity was insignificant and vice-versa (23.2% and 96.2% inactivation of PPO and POD, respectively, at 690 MPa/90°C/15 minutes).	(Terefe et al., 2010)
18	Grape juice	• Pressure – 100 to 600 MPa • Temperature – 0 to 60°C	Polyphenol oxidase (PPO) and peroxidase (POD)	• 100 MPa/60°C and 600 MPa/60°C, the results revealed the maximum inactivation of the PPO and POD.	(Rastogi et al., 1999)

14.5 HURDLE TECHNOLOGIES, IN COMBINATION WITH HPP

As per the available literature and scientific evidence, HPP offers a wide range of applications, as discussed above. However, this technology has limitations, like the inability to inactivate highly resistant microorganisms, some enzymes, etc. To overcome these limitations, researchers tried various ways, like creating as many hurdles as possible in processing for quality retention and shelf-life extension. Some of them include pulsed pressurization, pressure-assisted thermal sterilization (PATS), and HPP along with other novel techniques such as bacteriocins, osmotic dehydration, PEF, US, Ohmic heating, and carbon dioxide (Bermúdez-Aguirre et al., 2016). The combination of HPP with the abovementioned technologies can help in the effective pasteurization of liquid foods. (Huang et al., 2017). Szczepańska et al. (2021) compared single and multi-pulse pressurization. It was observed that 3 pulses with 300 MPa had good keeping quality and no microbial regrowth over 12 weeks of storage than single pulse treatment of 600 MPa in apple juice. Similarly, Abid et al. (2014) analyzed the synergistic effect of US (25 kHz at 70% amplitude for 60 minutes at 20°C) followed by HPP (250, 350, and 450 MPa for 10 minutes at room temperature). It was seen that US-HPP 450 treatment inactivated enzymes to the maximum extent among all combinations. Also, the total plate count (TPC), yeasts, and moulds (Y&M) count was brought down to zero at the end of treatment. Besides the inactivation of enzymes and microbes, there was a significant improvement in phytochemical components and antioxidant activity. In one more study, Pyatkovskyy et al. (2018) observed greater inactivation of *Listeria innocua* (ATCC 33090) in water by exposing it to various combinations of HPP, PEF, and US. The greatest inactivation by an individual treatment was obtained with HPP, wherein exposure to 400 MPa for 100 seconds resulted in an 8-log reduction in CFU. Both PEF (30 kV, 10^{-3} s) and sonication (250 seconds) treatments gave about 2-log reduction in the Listeria population. Sequentially applied HPP-PEF and PEF-HPP treatments demonstrated mostly additive effects.

14.6 FUTURE PERSPECTIVES

The application of HPP processing on food materials extends the shelf life without decreasing the nutritional value. It has a promising future as it can target particular products in the market, including liquid and solid foods. However, HPP processed foods have higher production cost than thermally processed foods. This problem can be tackled by incorporating HPP technology with organic foods, directly providing consumers with natural, high-quality, and adequate nutritional value. Therefore, this would enhance the acceptability of high prices and improve the growth of both types of foods. HPP has relatively improved nutrient preservation; however, more research is needed to completely grasp its effect under diverse operating situations. Data collection is also required to establish the role of HP in toxicity, allergenicity, loss of digestibility, and food's eating quality (Woldemariam & Emire, 2019).

Moreover, clear specification of processing/storage conditions is necessary, such as maintenance of proper storage and transportation conditions under refrigeration, to reduce potential attacks by microbes such as *Clostridium spores*. Since only plastic packaging can be used for HPP processed foods due to its minimum compressibility of at least 15%, this imposes the establishment of pertinent laws and regulations, facilitating better consumer acceptability, improved food quality, and development. HPP of foods makes them free of chemical and chemical additives and contain simple additives. This labelling system increases consumer trust and maintains food safety and quality. Maintaining the original quality and reducing the use of chemical preservatives and additives could be an effective strategy to combat Global ageing. HPP can also be used as hurdle technology with other processes, such as reducing the duration of wine fermentation by enhancing wine ageing. Henceforth, unlike thermal techniques, HPP meets the consumer demand for high-quality, organic, safe, and wholesome food.

14.7   SUMMARY

The processing of food materials is an important concern to attain food security. However, the quantity of processed food expels a large amount of CO_2, increasing the preservation process's carbon footprint. HPP plays an important role in preserving food, keeping the food material's quality in terms of its nutrients, and decreasing the emission of CO_2 throughout the process. Over the past few decades, HPP has been adopted to process different food categories like smoothies, purees, and fruit juices to improve safety and shelf-life while reducing the undesirable effects of thermal processing, especially sensory and nutritional qualities. HPP preserves nutritional quality in many ways and, at the same time, inactivates microbial load to the safest levels to consume, along with considerable inactivation of enzymes. Besides these advantages, HPP offers a greater shelf-life of juices with sensory properties retention without adding chemical preservatives. It can be foreseen that HPP can be used along with other non-thermal technologies as a hurdle technology to inactivate enzymes effectively, which was a bit of concern when juices or liquid foods were processed with HPP alone.

REFERENCES

Abid, M., Jabbar, S., Hu, B., Hashim, M. M., Wu, T., Wu, Z., Khan, M. A., & Zeng, X. (2014). Synergistic impact of sonication and high hydrostatic pressure on microbial and enzymatic inactivation of apple juice. *LWT – Food Science and Technology, 59*(1), 70–76. https://doi.org/10.1016/J.LWT.2014.04.039

Ahmed, J., Thomas, L., Taher, A., & Joseph, A. (2016). Impact of high pressure treatment on functional, rheological, pasting, and structural properties of lentil starch dispersions. *Carbohydrate Polymers, 152*, 639–647. https://doi.org/10.1016/j.carbpol.2016.07.008

Andreou, V., Dimopoulos, G., Katsaros, G., & Taoukis, P. (2016). Comparison of the application of high pressure and pulsed electric fields technologies on the selective inactivation of endogenous enzymes in tomato products. *Innovative Food Science and Emerging Technologies, 38*, 349–355. https://doi.org/10.1016/j.ifset.2016.07.026

Awuah, G. B. B., Ramaswamy, H. S. S., & Economides, A. (2007). Thermal processing and quality: Principles and overview. *Chemical Engineering and Processing: Process Intensification, 46*(6), 584–602. https://doi.org/10.1016/j.cep.2006.08.004

Balasubramaniam, V. M., Barbosa-Cánovas, G. V, & Lelieveld, H. L. M. (2016). High-Pressure Processing Equipment for the Food Industry. In V. Balasubramaniam, G. Barbosa-Cánovas, & H. Lelieveld (Eds.), *High Pressure Processing of Food* (Food Engineering Series), pp. 39–65). https://doi.org/10.1007/978-1-4939-3234-4_3

Bermúdez-Aguirre, D., Corradini, M. G., Candoğan, K., & Barbosa-Cánovas, G. V. (2016). High Pressure Processing in Combination with High Temperature and Other Preservation Factors. In V. Balasubramaniam, G. Barbosa-Cánovas, & H. Lelieveld (Eds.), *High Pressure Processing of Food* (Food Engin, pp. 193–215). Springer, New York, NY. https://doi.org/10.1007/978-1-4939-3234-4_11

Chakraborty, S. (2015). *High Pressure Processing of Pineapple (Ananas comosus L.) Puree : Effect on Quality Attributes and Shelf-Life* (Issue August). Indian Institute of Technology Kharagpur.

Chakraborty, S., Kaushik, N., Rao, P. S., & Mishra, H. N. (2014). High-pressure inactivation of enzymes: A review on its recent applications on fruit purees and juices. *Comprehensive Reviews in Food Science and Food Safety, 13*(4), 578–596. https://doi.org/10.1111/1541-4337.12071

Chakraborty, S., Rao, P. S., & Mishra, H. N. (2019). Modeling the inactivation of pectin methylesterase in pineapple puree during combined high-pressure and temperature treatments. *Innovative Food Science & Emerging Technologies, 52*(April 2018), 271–281. https://doi.org/10.1016/j.ifset.2019.01.008

Elamin, W. M., Endan, J. B., Yosuf, Y. A., Shamsudin, R., & Ahmedov, A. (2015). High pressure processing technology and equipment evolution: A review. *Journal of Engineering Science and Technology Review, 8*(5), 75–83. https://doi.org/10.25103/jestr.085.11

Evelyn, Kim, H. J., & Silva, F. V. M. (2016). Modeling the inactivation of Neosartorya fischeri ascospores in apple juice by high pressure, power ultrasound and thermal processing. *Food Control, 59*, 530–537. https://doi.org/10.1016/j.foodcont.2015.06.033

Fang, L., Jiang, B., & Zhang, T. (2008). Effect of combined high pressure and thermal treatment on kiwifruit peroxidase. *Food Chemistry, 109*(4), 802–807. https://doi.org/10.1016/j.foodchem.2008.01.017

FDA (2004). Guidance for Industry: Juice Hazard Analysis Critical Control Point Hazards and Controls Guidance, First Edition. *Center for Food Safety and Applied Nutrition.* https://www.regulations.gov/document/FDA-2002-D-0298-0004

Fernandez, M. V, Denoya, G. I., Jagus, R. J., Vaudagna, S. R., & Agüero, M. V. (2019). Microbiological, antioxidant and physicochemical stability of a fruit and vegetable smoothie treated by high pressure processing and stored at room temperature. *LWT – Food Science and Technology, 105*, 206–210. https://doi.org/10.1016/j.lwt.2019.02.030

Georget, E., Sevenich, R., Reineke, K., Mathys, A., Heinz, V., Callanan, M., Rauh, C., & Knorr, D. (2015). Inactivation of microorganisms by high isostatic pressure processing in complex matrices: A review. *Innovative Food Science and Emerging Technologies, 27*, 1–14. https://doi.org/10.1016/j.ifset.2014.10.015

González-Angulo, M., Clauwers, C., Harastani, R., Tonello, C., Jaime, I., Rovira, J., & Michiels, C. W. (2020). Evaluation of factors influencing the growth of non-toxigenic *Clostridium botulinum* type E and Clostridium sp. in high-pressure processed and conditioned tender coconut water from Thailand. *Food Research International, 134*(April), 109278. https://doi.org/10.1016/j.foodres.2020.109278

Gupta, R., & Balasubramaniam, V. M. (2012). High-pressure processing of fluid foods. *Novel Thermal and Non-Thermal Technologies for Fluid Foods*, 109–133. https://doi.org/10.1016/B978-0-12-381470-8.00005-0

Huang, H. W., Chen, B. Y., & Wang, C. Y. (2018). Comparison of high pressure and high temperature short time processing on quality of carambola juice during cold storage. *Journal of Food Science and Technology, 55*(5), 1716–1725. https://doi.org/10.1007/s13197-018-3084-3

Huang, W., Ji, H., Liu, S., Zhang, C., Chen, Y., Guo, M., & Hao, J. (2014). Inactivation effects and kinetics of polyphenol oxidase from *Litopenaeus vannamei* by ultra-high pressure and heat. *Innovative Food Science & Emerging Technologies, 26*, 108–115. https://doi.org/10.1016/J.IFSET.2014.10.005

Huang, H. W., Wu, S. J., Lu, J. K., Shyu, Y. T., & Wang, C. Y. (2017). Current status and future trends of high-pressure processing in food industry. *Food Control, 72*(12), 1–8. https://doi.org/10.1016/j.foodcont.2016.07.019

Jayachandran, L. E., Chakraborty, S., & Rao, P. S. (2015). Effect of high pressure processing on physicochemical properties and bioactive compounds in litchi based mixed fruit beverage. *Innovative Food Science and Emerging Technologies, 28*, 1–9. https://doi.org/10.1016/j.ifset.2015.01.002

Keefe, S. O., Williams, R., & Lukas, A. R. (2013). *Use of High Pressure Processing to Reduce Foodborne Pathogens in Coconut Water* [Virginia Tech]. https://vtechworks.lib.vt.edu/handle/10919/24760

Lamilla, C. P., Vaudagna, S. R., Alzamora, S. M., Mozgovoj, M., & Rodriguez, A. (2021). Effect of the high-pressure assisted-infusion processing on nutritional and antioxidant properties of mango cubes. *Innovative Food Science & Emerging Technologies, 71*(January), 102725. https://doi.org/10.1016/j.ifset.2021.102725

Lavinas, F. C., Miguel, M. A. L., Lopes, M. L. M., & Valente Mesquita, V. L. (2008). Effect of high hydrostatic pressure on cashew apple (*Anacardium occidentale* L.) juice preservation. *Journal of Food Science, 73*(6), M273–M277.

Lee, C. Y., & DeMan, J. M. (2018). Enzymes. In J. M. deMan Finley, J. W. Lee, W. J. Hurst, & C. Yong (Eds.), *Principles of Food Chemistry* (4th ed., pp. 397–433). Springer International Publishing AG. https://doi.org/10.1007/978-3-319-63607-8_10

Leite, T. S., de Jesus, A. L. T., Schmiele, M., Tribst, A. A. L., & Cristianini, M. (2017). High pressure processing (HPP) of pea starch: Effect on the gelatinization properties. *LWT – Food Science and Technology, 76*, 361–369. https://doi.org/10.1016/j.lwt.2016.07.036

Leite Júnior, B. R. C., Kubo, M. T. K., Augusto, P. E. D., & Tribst, A. A. L. (2021). High-Pressure Homogenization on Food Enzymes. In *Innovative Food Processing Technologies* (pp. 293–314). Elsevier. https://doi.org/10.1016/B978-0-08-100596-5.22999-5

Liberatore, C. M., Cirlini, M., Ganino, T., Rinaldi, M., Tomaselli, S., & Chiancone, B. (2021). Effects of thermal and high-pressure processing on quality features and the volatile profiles of cloudy juices obtained from Golden Delicious, Pinova, and Red Delicious Apple cultivars. *Foods, 10*(12), 3046. https://doi.org/10.3390/foods10123046

Ma, Y., Xu, L., Wang, S., Xu, Z., Liao, X., & Cheng, Y. (2019). Comparison of the quality attributes of coconut waters by high-pressure processing and high-temperature short time during the refrigerated storage. *Food Science and Nutrition, 7*(4), 1512–1519. https://doi.org/10.1002/fsn3.997

Martín-Belloso, O., Marsellés-Fontanet, Á. R., Elez-Martínez, P., Martin-Belloso, O., Robert Marsell's-Fontanet, A., Elez-Martinez, P., Martín-Belloso, O., Marsellés-Fontanet, Á. R., & Elez-Martínez, P. (2014). Enzymatic Inactivation by Pulsed Electric Fields. In *Emerging Technologies for Food Processing* (pp. 155–168). Elsevier. https://doi.org/10.1016/B978-0-12-411479-1.00009-7

Masood, H., Diao, Y., Cullen, P. J., Lee, N. A., & Trujillo, F. J. (2018). A comparative study on the performance of three treatment chamber designs for radio frequency electric field processing. *Computers and Chemical Engineering*, *108*, 206–216. https://doi.org/10.1016/j.compchemeng.2017.09.009

Misra, N. N. N., Koubaa, M., Roohinejad, S., Juliano, P., Alpas, H., Inácio, R. S., Saraiva, J. A., & Barba, F. J. (2017). Landmarks in the historical development of twenty first century food processing technologies. *Food Research International*, *97*(May), 318–339. https://doi.org/10.1016/j.foodres.2017.05.001

Nayak, P. K., Rayaguru, K., & Radha Krishnan, K. (2017). Quality comparison of elephant apple juices after high-pressure processing and thermal treatment. *Journal of the Science of Food and Agriculture*, *97*(5), 1404–1411. https://doi.org/10.1002/jsfa.7878

Petrus, R., Churey, J., & Worobo, R. (2019). Searching for high pressure processing parameters for *Escherichia coli* O157:H7, *Salmonella enterica* and *Listeria monocytogenes* reduction in Concord grape juice. *British Food Journal*, *122*(1), 170–180. https://doi.org/10.1108/BFJ-06-2019-0446

Podolak, R., Whitman, D., & Black, D. G. (2020). Factors affecting microbial inactivation during high pressure processing in juices and beverages: A review. *Journal of Food Protection*, *83*(9), 1561–1575.

Pokhrel, P. R., Boulet, C., Yildiz, S., Sablani, S., Tang, J., & Barbosa-Cánovas, G. V. (2022). Effect of high hydrostatic pressure on microbial inactivation and quality changes in carrot-orange juice blends at varying pH. *LWT*, *159*, 113219. https://doi.org/10.1016/j.lwt.2022.113219

Pokhrel, P. R., Toniazzo, T., Boulet, C., Oner, M. E., Sablani, S. S., Tang, J., & Barbosa-Cánovas, G. V. (2019). Inactivation of *Listeria innocua* and *Escherichia coli* in carrot juice by combining high pressure processing, nisin, and mild thermal treatments. *Innovative Food Science & Emerging Technologies*, *54*, 93–102. https://doi.org/10.1016/j.ifset.2019.03.007

Polydera, A. C., Stoforos, N. G., & Taoukis, P. S. (2003). Comparative shelf life study and vitamin C loss kinetics in pasteurized and high pressure processed reconstituted orange juice. *Journal of Food Engineering*, *60*(1), 21–29. https://doi.org/10.1016/S0260-8774(03)00006-2

Pyatkovskyy, T. I., Shynkaryk, M. V, Mohamed, H. M., Yousef, A. E., & Sastry, S. K. (2018). Effects of combined high pressure (HPP), pulsed electric field (PEF) and sonication treatments on inactivation of *Listeria innocua*. *Journal of Food Engineering*, *233*, 49–56. https://doi.org/10.1016/j.jfoodeng.2018.04.002

Raghubeer, E. V, Phan, B. N., Onuoha, E., Diggins, S., Aguilar, V., Swanson, S., & Lee, A. (2020). The use of high-pressure processing (HPP) to improve the safety and quality of raw coconut (*Cocos nucifera* L) water. *International Journal of Food Microbiology*, *331*(February), 108697. https://doi.org/10.1016/j.ijfoodmicro.2020.108697

Rajashri, K., Rastogi, N. K., & Negi, P. S. (2020). Non-thermal processing of tender coconut water – A review. *Food Reviews International*, *38*(sup1), 1–22. https://doi.org/10.1080/87559129.2020.1847142

Rao, L., Guo, X., Pang, X., Tan, X., Liao, X., & Wu, J. (2014). Enzyme activity and nutritional quality of peach (*Prunus persica*) juice: Effect of high hydrostatic pressure. *International Journal of Food Properties*, *17*(6), 1406–1417. https://doi.org/10.1080/10942912.2012.716474

Rastogi, N. K., Eshtiaghi, M. N., & Knorr, D. (1999). Effects of combined high pressure and heat treatment on the reduction of peroxidase and polyphenoloxidase activity in red grapes. *Food Biotechnology*, *13*(2), 195–208. https://doi.org/10.1080/08905439909549971

Riahi, E., & Ramaswamy, H. S. (2004). High pressure inactivation kinetics of amylase in apple juice. *Journal of Food Engineering*, *64*(2), 151–160. https://doi.org/10.1016/j.jfoodeng.2003.09.025

Song, Q., Li, R., Song, X., Clausen, M. P., Orlien, V., & Giacalone, D. (2022). The effect of high-pressure processing on sensory quality and consumer acceptability of fruit juices and smoothies: A review. *Food Research International*, *157*, 111250. https://doi.org/10.1016/j.foodres.2022.111250

Sreedevi, P., Jayachandran, L. E., & Rao, P. S. (2018). Kinetic modeling of high-pressure induced inactivation of polyphenol oxidase in sugarcane juice (*Saccharum officinarum*). *Journal of the Science of Food and Agriculture*, *99*(5), 2365–2374. https://doi.org/10.1002/jsfa.9443

Szczepańska, J., Pinto, C. A., Skąpska, S., Saraiva, J. A., & Marszałek, K. (2021). Effect of static and multi-pulsed high pressure processing on the rheological properties, microbial and physicochemical quality, and antioxidant potential of apple juice during refrigerated storage. *LWT*, *150*, 112038. https://doi.org/10.1016/j.lwt.2021.112038

Tao, Y., Sun, D., Hogan, E., & Kelly, A. L. (2014). High-Pressure Processing of Foods. In *Emerging Technologies for Food Processing* (pp. 3–24). Elsevier. https://doi.org/10.1016/B978-0-12-411479-1.00001-2

Terefe, N. S., Yang, Y. H., Knoerzer, K., Buckow, R., & Versteeg, C. (2010). High pressure and thermal inactivation kinetics of polyphenol oxidase and peroxidase in strawberry puree. *Innovative Food Science & Emerging Technologies*, *11*(1), 52–60. https://doi.org/10.1016/j.ifset.2009.08.009

Vernwal, S., Yadav, R., & Yadav, K. (2006). Purification of a peroxidase from *Solanum melongena* fruit juice. *Indian Journal of Biochemistry and Biophysics*, *43*(4), 239–243.

Woldemariam, H. W., & Emire, S. A. (2019). High pressure processing of foods for microbial and mycotoxins control: Current trends and future prospects. *Cogent Food & Agriculture, 5*(1), 1622184. https://doi.org /10.1080/23311932.2019.1622184

Wu, W., Xiao, G., Yu, Y., Xu, Y., Wu, J., Peng, J., & Li, L. (2021). Effects of high pressure and thermal processing on quality properties and volatile compounds of pineapple fruit juice. *Food Control, 130*(June). https://doi.org/10.1016/j.foodcont.2021.108293

Xu, Z., Lin, T., Wang, Y., & Liao, X. (2015). Quality assurance in pepper and orange juice blend treated by high pressure processing and high temperature short time. *Innovative Food Science & Emerging Technologies, 31*, 28–36. https://doi.org/10.1016/j.ifset.2015.08.001

Yi, J., Kebede, B. T., Hai Dang, D. N., Buvé, C., Grauwet, T., Van Loey, A., Hu, X., & Hendrickx, M. (2017). Quality change during high pressure processing and thermal processing of cloudy apple juice. *LWT, 75*, 85–92. https://doi.org/10.1016/j.lwt.2016.08.041

15 Sub and Supercritical Food Processing

K. Krishna Priya, L. Kavitha, K. Najma, and D. V. Chidanand

15.1 INTRODUCTION

Engineers have recently faced difficulties with the creation of new products with specific features or the design of environmentally friendly procedures. The industrial processes used to create various products were often run at atmospheric pressure. About 20 years ago, research on supercritical extraction technology began. Hundreds of supercritical extraction facilities have been built across the world to function at pressures as high as 2000 bar (Knez et al., 2018). Traditional applications of supercritical fluids and their liquefied counterparts include extraction, fractionation, and utilizing neat SC-CO$_2$ or with the required modifiers. Supercritical fluid derived extracts or products were made possible by the development of columnar and chromatographic procedures in the middle of the 1980s, which expanded the use of a critical fluids processing platform beyond SFE (Majid et al., 2019). The research of these more new advances was prompted by the diversity of many natural compounds, matrices and the urge to focus specific target elements to be utilized in food and other commercial uses (King & Srinivas, 2009b).

In power generation units, liquefied gases like liquid air, liquid nitrogen (LN$_2$), or liquefied natural gas (LNG) can be used as cost-effective energy vectors as well as emission-free "fuels" for transportation. As an example, energy from renewable sources can be used to "cool" air or nitrogen until they turn into liquids (Mahdi et al., 2018). When these liquids are subsequently injected into a heated environment, they quickly re-gas and significantly expand in volume. This can either be utilized emerging ultra-low emission combustion systems to optimize the compression stroke and lower emissions, as in the case of the Cryopower split-cycle engine, or it can drive a turbine or piston engine even without combustion (Gopal et al., 2020).

Traditional extraction techniques used to extract these compounds have a number of disadvantages; they take a long time, require a lot of work, have less selectivity, and/or have low extraction yields. Additionally, these conventional methods use a lot of hazardous solvents (Meireles, 2008). Currently, researchers are looking at extraction techniques that can get over the problems mentioned above; among these, supercritical fluid extraction (SFE) and subcritical water extraction (SWE) are among the more promising procedures (Herrero et al., 2006). These extraction methods require less time and less harmful organic solvents while offering higher selectivity. SFE is used to extract materials from solids and liquids at specific temperatures and pressures (20-60°C and 50–250 bar, respectively) above their critical points. At this location, supercritical fluids disperse like gases and liquefy the analytes. By altering the conditions of the dissolving capacity over the critical threshold, it is possible to change its power. Figure 15.1 illustrates the triple point and critical point, two important places. Triple points, where three states coexist simultaneously, are located at boiling points, when two states (liquid and gas) simultaneously exist. The area is known as the supercritical point above this point. Supercritical fluids can dissolve compounds that are partially or totally insoluble in separate gases or liquids because they have both the physical and chemical characteristics of both liquid and gas states. Additionally, supercritical fluids' improved diffusivity and lower viscosity in the case of viscous, dense liquids boost the high mass transfer rate of analytes into fluids, which improves the effectiveness and speed of chemical extraction from food samples (Majid et al., 2019).

DOI: 10.1201/9781003359302-15

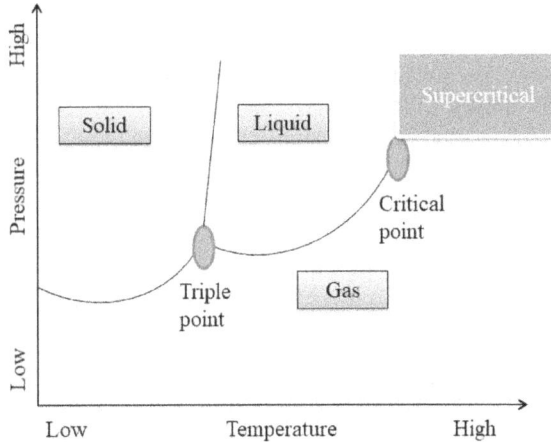

FIGURE 15.1 Phase diagram of carbon dioxide with reference to temperature and pressure variation.

As a result, the objective of this review is to present an updated advance of two eco-friendly safe techniques, SFE and SWE, as well as their use in the food industry.

15.2 NATURE AND CHARACTERISTICS OF SUB AND SUPERCRITICAL FLUIDS

15.2.1 Supercritical Fluid Extraction

Common supercritical fluids include water (critical conditions = 374.096°C and 217.55 atm), carbon dioxide (critical conditions = 31.1°C and 72.8 atm), methane (critical conditions = −82.6°C and 45.4 atm), ethane (critical conditions = 32.3°C and 48.1 atm), propane (critical conditions = 96.8°C and 41.9 atm), ethylene, propylene, methanol, ethanol, acetone, etc (Majid et al., 2019). Carbon dioxide is one of these that is frequently employed as an extracting solvent. Carbon dioxide is cheap, harmless, and has been verified as safe by governing bodies like the FDA and EFSA. Since carbon dioxide is a gas at ambient temperature, it must be removed from the mixture via decompression after the extraction process (Majid et al., 2019). SFE-produced extract remained residue-free. Furthermore, carbon dioxide may be recycled on an industrial basis. SFE has a number of advantages over other extraction methods because of their high diffusivity and low viscosity.

With improved transport efficiency and easier diffusion within solids and in highly dense liquids, supercritical fluids can increase yields. By modifying the adjustable density at different temperature and pressure circumstances, the solubility is increased (Varzakas & Tzia, 2014).

15.2.1.1 Principles and Instrumentation

A fluid becomes a supercritical fluid when it is pushed to pressure and temperature over its critical point. The fluid's unique properties in these conditions lie in between those of a gas and that of a liquid. Although the density and viscosity of a supercritical fluid are similar to those of a liquid and a gas, respectively, their diffusivity lies between those two states. Therefore, the concept of a fluid's supercritical state entails either a state where liquid and gas cannot be characterized from each other or a state where the fluid is compressible (acting like a gas) and yet had a density equivalent to that of a liquid and, consequently, a related solvating power (Herrero et al., 2006).

15.2.1.2 Properties

Supercritical fluids (SCFs) are well suited for bioseparation procedures due to their beneficial characteristics. The following are some of the main benefits: a) SCFs offer greater mass transfer properties than liquid solvents, such as low surface tension and high diffusion, which make it easier to penetrate porous solid matrices and increase solvation power, while low viscosities promote better flow. b) Altering pressure and temperature above critical limits has an impact on SCFs' physicochemical characteristics, including density, viscosity, diffusivity, and dielectric constant, and improves their extractability (Yildiz-Ozturk & Yesil-Celiktas, 2019). SCFs' variable solvation power may produce a remarkable increase in selectivity. c) SC-CO$_2$ is a prime option for the extraction of thermally labile and oxidizing chemicals due to its relatively low critical temperature and pressure. d) SC-CO$_2$ disintegrates non-polar or hardly polar substances. However, the addition of co-solvents can take advantage of the process' selectivity and polar compounds' solubility to boost yields, shorten processing times, and provide a softer processing environment (Knez, 2016). e) SC-CO$_2$ has a strong solvation power for low molecular weight molecules, however, this power declines as molecular weight increases. By increasing pressure, it is feasible to segregate high molecular weight and less volatile molecules. f) Because SC-CO$_2$ and sub- and supercritical water products are stable, non-toxic, and lack carcinogens, they have positive effects on health. In fact, it has received approval from the European Food Safety Authority and the US Food and Drug Administration as a "generally regarded as safe" procedure (Knez et al., 2014). g) In contrast to conventional extraction techniques, depressurization allows for the total removal of solvents from the extract phase while also maintaining better compound stability due to the absence of organic solvent. At low working temperatures, SC-CO$_2$ extraction enables the separation and fractionation of individual components from total extracts (Yildiz-Ozturk & Yesil-Celiktas, 2019).

The substantial initial investment and ongoing maintenance expenses associated with the high pressures used in supercritical fluid processing are some of its downsides. As a result, the costs are somewhat higher than for those produced using more traditional methods. The non-polar character of SC-CO$_2$ is another drawback; in some cases, adding 10% to 20% of a co-solvent may not be enough to extract the target molecules, which are highly polar. In order to enhance the solubility of the compounds of interest, it seems sense to concentrate on various methods to reduce their polarity. Chemical in situ derivatization of a group of chemicals has increased the method' selectivity for this goal (Yildiz-Ozturk & Yesil-Celiktas, 2019).

15.2.2 SUBCRITICAL WATER EXTRACTION

15.2.2.1 Principles and Instrumentation

Subcritical water extraction (SWE), or extraction with hot water under pressure, has recently come to be recognized as a viable alternative to conventional extraction techniques. SWE is an environmental safe method which can increase the output of extraction from solid samples (Herrero et al., 2006). Subcritical water is defined as liquid water that is below the critical point (Tc = 374.15°C, Pc = 22.1 MPa) of both temperature and pressure. To maintain water in the liquid state at a specific temperature, the subcritical water pressure needs to be greater than the vapour pressure. The physic-chemical characteristics of subcritical water alter dramatically as temperature rises (Cheng et al., 2021). In order to keep the water in the liquid condition, SWE uses hot water (between 100°C and 374°C, the latter being the water critical temperature) under high pressure (often between 10 and 60 bar) (Herrero et al., 2006). The variation of the dielectric constant with temperature is the most crucial aspect to take into account in this kind of extraction method. With a dielectric constant close to 80, water is a strongly polar solvent when it is at normal temperature. However, by using the proper pressure and heating the water to a temperature of 250°C, this number can be greatly reduced to levels near to 27. This dielectric constant is comparable to ethanol in terms of value (King & Srinivas, 2009b).

The instrumentation basically consists of an extraction cell enclosed in an oven is installed and extraction will hell, a reservoir of water connected to a high-pressure pump to inject the solvent into the system, and a restrictor or valve to maintain pressure. The vial at the bottom of the extraction mechanism is where the extracts are collected. The system can also be fitted with a coolant

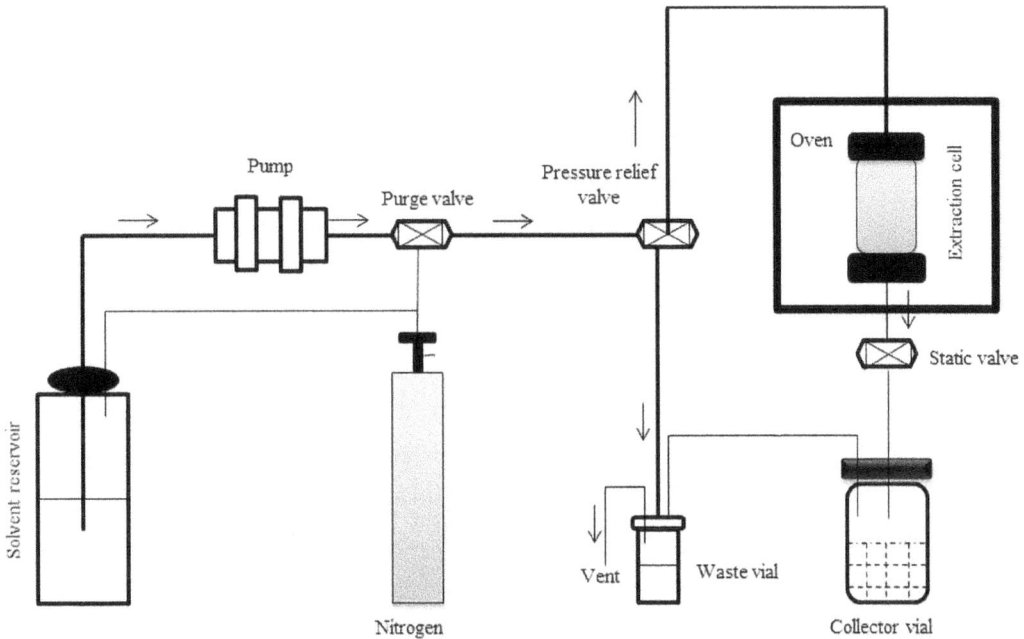

FIGURE 15.2   Process flow of subcritical water extraction system.

mechanism to allow for the quick cooling of the extracted product (Cheng et al., 2021). The equipment design of the subcritical fluid extraction is provided in Figure 15.2.

Applications for subcritical water could be seen in areas: (1) reversed-phase liquid chromatography with subcritical water as the only mobile phase-subcritical water chromatography; (2) environmental sample extraction-such as the identification of organic pollutants in solid wastes, soils, sediments, and airborne particles; (3) hydrolysis, degradation, polymerization, and synthesis reactions-using subcritical water as both a solvent and a reactant; and (4) bioremediation as washing polluted sewages and soils, decomposing pollutants (pesticides, polycyclic aromatic hydrocarbons, and polychlorinated biphenyls) and explosives; and (5) extraction of active compounds from medicinal and seasoning herbs, vegetables, fruits, and other plant-related matrices (Cheng et al., 2021).

Although experiments on the online coupling of a SWE system to an HPLC apparatus via solid phase trapping have been published, this technology has primarily been used as a batch process.

15.2.2.2   Properties
Subcritical water is liquid at pressures greater than the normal boiling point of water, which is 100°C (212°F). It is also referred to as "pressurized hot water" or "subcritical water." Water is kept liquid at the subcritical point by applying pressure. Therefore, at saturation vapour pressure, liquid water and vapour are in equilibrium (Kari, 2018). There are some property changes:

- High solvating power
- Low viscosity and surface tension
- High diffusivity
- Low polarity
- High self-ionization of water
- Light acidification (acid hydrolysis)
- The duration of extraction is reduced and the solvent volume too which give more concentrated extract

15.2.2.3 Additional Benefits Associated with Its Use Are

- No toxicity, "green" solvent
- No residual organic solvent in the final product/extract
- Non-flammable, non-explosive: less expensive installation
- Modulation of properties by adjusting the temperature of assay – high temperature allows chemical modification of the material (e.g., hydrolysis)

15.3 DIFFERENT PROCESS INVOLVED IN SCF

15.3.1 OLEOCHEMICAL PROCESSING

Oleochemical processing including such hydrogenation or fat splitting are the traditional operations that are frequently carried out by using sub or supercritical techniques. The most popular Industrial processing methods for fat splitting are cold emery or Twitchell processes (Barnebey & Brown, 1948). These processes use the pressure and temperature that are higher than the boiling temperature of water under desired pressure, but below the critical point of water, the hydrolysis conversion of triglycerides into fatty acids takes place (King & Srinivas, 2009a).

The Twitchell process was mainly performed at milder conditions by utilizing a catalyst (i.e., Twitchell reagent) to accelerate the hydrolysis. This Twitchell reagent mainly comprises of oleic acids, concentrated sulphuric acid, and hydrocarbons. It involves splitting or hydrolyzing of fats by boiling it with steam by adding roughly 1% sulphuric acid and certain quantity of water under atmospheric pressure in a closed chamber. In addition, a small quantity of Twitchell reagent is utilized and performs as a catalyst to segregate the fat into glycerine and fatty acids (Hartman, 1951).

The cold emery process was operated at higher temperature conditions and it doesn't require catalyst as that of the Twitchell process. The cold emery process was carried out at working temperature ranging from 250°C to 330°C and pressure of about 5–6 MPa, respectively (Barnebey & Brown, 1948). This method of processing is highly efficient with a triglyceride conversion rate of more than 98%. Water, particularly in subcritical and supercritical forms, is exceedingly efficient, and the hydrolysis happens very quickly at temperatures ranging between 330°C and 340°C, producing a conversion rate of 97% or more (Holliday et al., 1997; J. King et al., 1999). Because of the usage of higher temperatures, this process makes triglycerides to experience unfavourable dehydration, interesterification, and oxidation which causes fatty acids to deteriorate (Murty et al., 2002; Rooney & Weatherley, 2001).

15.3.2 COUPLED PROCESSING

SFE is not the only technique employed in recent years to separate certain constituents from natural matrices. Supercritical fluids are being used more frequently in fractionation and reaction modes, which may be coupled with SFE to provide a specific end product or fraction (King, 2000). Figure 15.3 illustrates the options of coupled processing for critical fluids. Therefore, one may acquire several

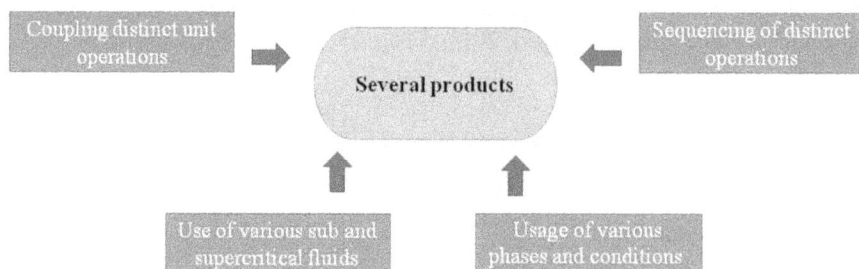

FIGURE 15.3 Choices of coupled processing for critical fluids.

TABLE 15.1

Options for Coupling Processes for Producing Essential Oils with Pressured Fluids

Processes Involved

(a) SFE (SC-CO$_2$)

(b) SFF (SC-CO$_2$) – reduction of pressure

(c) SFF (SC-CO$_2$) – column deterpenation

(d) SFC (SC-CO$_2$/co-solvent)

(e) SFF – (subcritical H$_2$O deterpenation)

(f) SFM – (aqueous extract/SC-CO$_2$)

Combination of Processes

(a) + (b)

(a) + (b) + (c)

(a) + (d)

(a) + (e)

(a) + (e) + (f)

Source: J. W. King and Srinivas (2009a).

products and improve the extraction or reaction procedures by the combination of various unit operations and sequencing them using a combination of multiple fluids hold on to operational densities by the utilization of distinct pressures and temperatures (King & Srinivas, 2009a).

For example, utilized six-unit operations for the processing of essential oil of citrus obtained by using pressurized liquids (Table 15.1). For instance, combining unit operations (a) and (b) from the table listed might result in a more precise composition of extraction. Unit process (a), when combined with a succession of lowering pressures (unit process (b)), has the potential to magnify the SFF effect. A more particular end product will be produced from the initial citrus oil by merging unit operations (a) and (b), which use SC-CO$_2$ and subcritical H$_2$O, respectively, to determine the extraction from the unit procedure (a). To achieve a specific concentrated or enriched end product, unit operations (f) and (e) might be added (King, 2000; King & Srinivas, 2009a).

15.3.3 INTEGRATION OF MULTIPLE UNITS

Supercritical fluid extraction or pressurized liquid extraction (PLE), chemical reactions, formation of particles, and improvement of these methods can all be used as single project to enhance the products' final quality, or to improve operations efficacy or to extract different products from single processing line (Viganó et al., 2015). With the knowledge gained over the past several years, it is now conceivable to study and envision uses for sub-and supercritical fluid operations that go beyond conventional extractions, such as coupling these techniques with other unit operations or processing methods. This chapter deals with the SCF coupling with other processing methods such as adsorption and membrane separation (Sarrade et al., 2002, 2023).

Adsorption is the mass transfer phenomenon in which the solids contain liquid or gas molecules on their surface. It is feasible to fractionate a solution since the adsorption mechanism is frequently selective. Adsorbents are typically solid materials that include porous particles. The most often utilized adsorbents include silicon, activated carbon, molecular sieves, polyacrylamide cryogels, and activated alumina (Soto et al., 2011).

Ambrogi et al. (2003) fractionated the carotenoids in powdered *Capsicum annuum* peppers by combining SFE with CO$_2$ and segregation using silica gel adsorbent. The research was carried out in a pilot plant with the operating range of 50 MPa and 373 K. To assess the concentration of carotenoid during the liquid phase and to regulate the various stages of the process, a photometer was utilized in the line. The combination of adsorption and extraction processes results in targeted adsorption. The yellow colour pigments (carotenoids) were retained in the supercritical extraction stage was retrieved after segregation under low pressure whereas esterified carotenoids (red colour) were adsorbed.

Membrane separation techniques and their uses have become more widespread in a number of industrial sectors, including the treatment of wastewater and chemicals, minerals, metals, and food (Mulder & Mulder, 1996). In order to achieve membrane separation, components are often excluded due to their distinct forms, molecular weights, and contact with the surface of the membrane. A driving force provides the flowability of solute or solvent when the membrane is positioned in between the phases. This force could be brought on by temperature, pressure, or concentration gradients (Lyman, 1983; Mulder & Mulder, 1996).

Researchers have lately been interested in combining membrane filtration techniques for sub- and supercritical processes because of the benefits of both technologies. Applications often improve the manufacturing process by making it simpler, using less energy, producing fewer waste fluids, and maybe scaling it up. Another anticipated benefit is a strong level of control over the entire extraction-separation procedure(Viganó et al., 2015).

For instance, a reverse osmosis membrane composed of cellulose acetate was put to the examination by Spricigo et al. (2001) for the extraction of essential oil from nutmeg with SC-CO$_2$. The membrane retained 96.4% of the oil on average and was unaffected by transmembrane pressure, temperature, or content of oil in the feed.

15.4   PROCESS ENHANCEMENT OF CRITICAL FLUIDS

The SFE was applied in food processing discussed in the preceding sections often utilizes heat with other parameters to improve the process effectiveness. Technologies that are regarded as non-thermal, notably pulsed electric field (PEF), ultrasound, high pressure, and microwave-assisted extraction (MAE), are considered potentially applicable for improving extraction from a wide range of compounds present in food processing.

15.4.1   HIGH-PRESSURE TREATMENT

High-pressure processing (HPP) is considered a non-thermal method of processing, which preserves food quality and organoleptic properties while prolonging the shelf life of products that have undergone treatment (Torres-Ossandón et al., 2018). HPP employs water as a medium to transfer pressures ranging from 0 to 800 MPa. By applying HPP, cells become more permeable, allowing metabolites to diffuse from the internal to the exterior of the cells. One of this technique's key benefits is the virtually immediate transfer of isostatic pressure to the final product, which produces very homogenous results regardless of product shape, size, or food content (Patras, Brunton, Da Pieve, & Butler, 2009). Higher amounts of bioactive compounds are retained after high-pressure treatment as compared to conventional thermal treatment (Patras, Brunton, Da Pieve, Butler, et al., 2009). There is no much literature available on SFE coupled with HPP, whereas for extracting bio-compounds, HPP could employ as a pretreatment before going for SFE. Combining HPP and SFE technology has demonstrated that the Allium species more effectively recovered sulphur biologically active compounds than conventional extraction techniques (Poojary et al., 2017). Torres-Ossandón et al. (2018) reported that the HPP pretreatment at 400 MPa for 3 minutes assisted with SFE (SC-CO$_2$) resulted in greater extraction of phenolic compounds from cape gooseberry.

15.4.2   PULSED ELECTRIC FIELD

A non-thermal technique called PEF involves applying high voltage electric pulses (up to 80 kV/ cm) for a short period of time to a substance that is placed between the two electrodes (Deeth et al., 2007). PEF involves modifying the cell membrane structure in order to achieve improved extraction. A temporary or permanent permeability of the cell membranes results from an electric collapse that happens when the critical electrostatic potential of the cell membrane relies on the strength of external field strength is surpassed. This occurs as a consequence of pore generation on the cell membrane's weak spots, which transforms the food substrate into a highly porous substance. This significantly increases mass transfer in various operations, which enhances yield and speeds up the demise of microorganisms (Nowosad et al., 2021).

In recent decades, an increasing number of studies have concentrated on the utilization of PEF to enhance transfer of mass, increase cell membrane permeability, and shorten extraction periods. In order to extract components sensitive to heat, including such carotenoids, anthocyanins, and chlorophyll, this approach should be taken into account in the pharmaceutical and food industry sectors (Knorr et al., 2011; Puértolas et al., 2012). However, as a coupled strategy to extract bioactive chemicals from residues, SFE or PLE supported by PEF has not yet been investigated. According to certain studies, PEF may be used effectively as a pretreatment before or in conjunction with traditional extraction methods from grape wastes (Boussetta et al., 2009; Corrales et al., 2008). PEF may well be applied to enhance supercritical operations from food products by lowering the extraction time. Pataro et al. (2010) reported that the inactivation of *Saccharomyces cerevisiae* was improved when the PEF was applied with the high-pressure CO_2 treatment. Corrales et al. (2008) reported that PEF showed greater efficacy than high pressure and ultrasound-assisted extraction for extraction of anthocyanin compounds from grapes. Hence, PEF might be utilized as sample pretreatment to enhance the SFE and PLE. PEF is confined to extractions from aqueous medium, which is unusual in SFE. As a result, this method could be more appropriate for improving SFE or subcritical operations which use water as a co-solvent.

15.4.3 MICROWAVES AND ULTRASOUND

The application of ultrasonic frequency in the food industries has grown over the past years because it offers a potentially effective means of improving mass transfer operations. Ultrasonic energy may be employed as an auxiliary method in SCF processes, for instance, without compromising or even enhancing the primary qualities and attributes of the end products. Ultrasound was majorly been used in the extraction processes (Riera et al., 2004).

Ultrasound employs the sound waves of high frequency which is greater than 20 kHz. These ultrasound waves induce the formation of cavitation bubbles in the food product, but when the energy-absorbing capacity of this created bubbles is eventually exceeded, causing them to collapse and produces microjets (Roseiro et al., 2013). Additionally, this acoustic cavitation allows for the penetration of food surfaces, which facilitates enhanced mass transfer and makes it simpler to produce the solvent into the cells. Furthermore, ultrasound can accelerate the intracellular material to release, increasing extraction rates and yields (Wijngaard et al., 2012). Reátegui et al. (2014) reported that the extraction of antioxidants was improved by using SC-CO_2 assisted by ultrasound from blackberry bagasse. The enhanced phenolic compounds and antioxidants were achieved at 200 W, 15 MPa, and 60°C and also improved extraction rate (6.84–9.87%). Santos et al. (2015) observed that the capsaicinoids were extracted from Malagueta pepper by using SFE with ultrasound. The ultrasound was used for 60 minutes at 360 W, the extraction yield rose by up to 77% (Santos et al., 2015).

Similarly, MAE methods became popular for extracting smaller molecules or organic compounds from food matrices (Kaufmann & Christen, 2002). Electromagnetic radiation produces microwave (MW) energy with frequencies ranging from 0.3 to 300 GHz. The major mechanism involved during MW contact with the food material is ionic polarization and dipolar rotation. This mechanism causes polar molecules such as water to release heat which leads to temperature rose thus, resulting in improved extraction efficiency (Teo et al., 2013).

As a result, unlike traditional heating techniques via conduction, natural bioactive chemicals may be extracted effectively using less time, energy, and solvent volume when employing microwaves since they heat the entire sample uniformly (Joana Gil-Chávez et al., 2013; Teo et al., 2013).

The synergistic interaction of mass and heat transfer processes happening in the same direction may result in acceleration and improved extraction rates (Mason et al., 2011). Due to these properties, MAE has become one of the most sophisticated methods for extracting phytochemicals from diverse substrates. Research findings show that the MAE has been used as a preliminary or later process with SFE for extracting polar molecules like phenolic compounds (Dejoye et al., 2011). Furthermore, the MAE was also coupled with pressurized liquids for extracting the inorganic compounds in food matrices (Matusiewicz & Ślachciński, 2014).

Dejoye et al. (2011) discovered that the MW as a pretreatment before SFE (SC-CO$_2$) helps in improved extraction of lipids from the *Chlorella vulgaris* (microalgae). The extraction rate of lipids without MW pretreatment was obtained only 1.81% whereas, with MW pretreatment results in 4.73%. The lipids extracted by combining two technologies were showed enhanced fatty acid contents. Da Porto et al. (2016) extracted oil from *Moringa oleifera* seeds by combining two methods, MW as a preliminary step for 30 seconds at 100 W followed by SFE (SC-CO$_2$). The MW pretreated samples found to have maximum extraction yield of about 35.28%.

The research discussed in this section suggests that using pre-or post-treatment or coupling microwave energy can enhance sub-and supercritical operations. However, the restriction of employing supercritical CO$_2$ is the necessity to handle samples with low water content since moist or other polar liquids often reduce the performance of the process. The residual liquid might be removed in a subsequent stage, or co-solvents could be employed in conjunction with supercritical fluids to solve the issue.

15.5 APPLICATIONS OF SUB AND SUPERCRITICAL FLUIDS IN VARIOUS FOOD PROCESSING SECTORS

15.5.1 DAIRY PRODUCTS

Around 36% increase in dairy consumption is expected all over the world by the year 2024. In supercritical carbon dioxide (CO$_2$) technology, the combination of pressure and carbon dioxide is supplied for microbial inactivation without nutrient loss in food (Amaral et al., 2017). Supercritical carbon dioxide technology is applied in dairy industries for the processing of milk products in a non-thermal way. Alkaline phosphatase, endogenous enzyme present in the milk is inactivated at higher rate during supercritical carbon dioxide treatment. The promising rate of alkaline phosphatase inactivation supercritical CO$_2$ was found to be 98.2% at 8 MPa, 70°C for 30 minutes reaction time (Amaral et al., 2017). Action mechanism of supercritical CO$_2$ on microorganisms in dairy products is given in Figure 15.4.

In order to develop less fat cheese, Yee et al. (2008) studied the extraction of fat from cheese by SFE, which varies based on the cheese variety and age. About 55.07% fat was reduced in young (9–10 months old cheese) parmesan cheese when compared with young cheddar cheese of 53.56%. Major amount of lipids (triglycerides, cholesterol) gets removed from cheese using the supercritical CO$_2$ as non-polar solvent. The key flavouring substances extracted from milk are aldehydes, ketones, and lactones by supercritical CO$_2$ process (Gandhi et al., 2018).

15.5.2 FRUITS AND VEGETABLES

Some of the widely used solvents for supercritical techniques are carbon dioxide, water, and ethanol. Large amount of food wastes generated from fruit processing industries are treated by supercritical extraction for the isolation of fatty acids and phenolic compounds (Viganó et al., 2015). Few bioactive compounds like gallic acid, catechin, anthocyanin, coumarins, lignan, polyphenols, and quercetin were extracted through supercritical CO$_2$ process from rice, berries, blueberry, sweet grass, five-flavour berry, cocoa, and grapes, respectively (Knez et al., 2014).

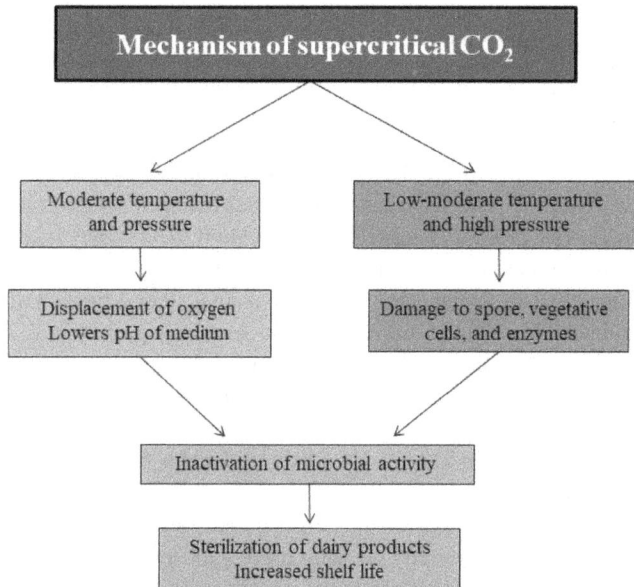

FIGURE 15.4 Mechanism of supercritical CO_2 in dairy products.

Devani et al. (2019) extracted oleoresin from the onion waste using supercritical CO_2 technique through response surface methodology. Oleoresin finds its use in the food industry as a flavouring compound, especially in dairy food products. The optimum conditions for achieving quality yield of oleoresin from onion waste are at a temperature of 80°C, 0.53 mm particle size, and pressure of 400 bar for 60-minute time duration. Another example of SFE in vegetables is lycopene from tomato waste with process maintained at 80°C, using CO_2 as a solvent of 4 g CO_2/min flow at 40 MPa pressure (Zhou et al., 2021). Fabian et al. (2019) researched the use of solvent mixture (ethanol & water) to extract bioactive antioxidant compounds from the beetroot residues (stem and leaves). The resulted bioactive compounds can be used in food and pharmaceutical industries via green extraction process.

Another study dealt with the extraction of flavonoids from peels of brown onion through SFE at 40°C and 10 MPa. The antioxidant activity test unveiled the presence of high antioxidant content in the supercritical fluid extracts (Zhou et al., 2021). In a study carried out by Shi et al. (2013), the carotenoid substances were extracted from the pumpkin flesh by the application of SFE technique with maximum yield at 50°C. de Aguiar et al. (2018) used supercritical carbon dioxide for extracting oleoresin from the capsicum with no traces of solvent residues. The capsaicinoids present in capsicum are highly soluble in carbon dioxide in supercritical extraction of oleoresin. Thus, high yield of bioactive compound is achieved in short period at 40°C and 15 MPa with minimal production cost.

15.5.3 Cereals and Pulses

Spices are the most majorly used ingredient in the food industry for distinct purposes. The presence of bioactive substances provides antioxidant, antimicrobial, and anti-inflammatory activities to the spices. These bioactive compounds can be used as functional food or even used in the pharmaceutical industries for the production of therapeutic drugs (Nagavekar & Singhal, 2019).

Using the SFE technique, the oleoresins, volatile compounds, and curcuminoids are extracted from the two varieties of spices namely *Curcum amada* and *Curcuma longa* using carbon dioxide as solvent. These bioactive compounds are only responsible for many functional properties. Similarly, omega 3 and 6 fatty acids rich oil was extracted from the flax seeds with the help of CO_2 by supercritical extraction (Gandhi et al., 2018).

15.5.4 FISH AND MEAT

The hydrolysis using the subcritical water has been employed in the preparation of antimicrobial and antioxidant substances from the tuna fish skin and isolated collagen (Ahmed & Chun, 2018). Only water is used in the hydrolysis process at subcritical condition instead of any harmful substances. The hydrolysis product showed higher maximum at 280°C when studied at varied temperature and pressure (Ahmed & Chun, 2018). The subcritical state water acts as an acid or a base catalyst and helps in the breakdown of higher molecular weight proteins into low molecular peptides. Thus, the subcritical water hydrolysis gained more interest in the food, pharmaceutical, and cosmetic industries.

Collagen protein is isolated from the mackerel skin and bone using subcritical water hydrolysis. The skin and bone of mackerel are the large amount of by-products generated during the fish processing stage. In cosmetic industries, the lower molecular (<3 kDa) weighted collagens are responsible for proper penetration in skin cells for effective results. The subcritical water hydrolysis method is used especially for the production of low molecular weight collagen (Asaduzzaman et al., 2019). This collagen is pepsin solubilized one, with more yield from skin (8.10%) when compared with fish bone (1.75%). The molecular weight of the peptides obtained from the above collagen was less than 1650 Da.

Fang et al. (2018) studied the extraction of oil from tuna liver using subcritical dimethyl ether as solvent. This tuna liver (high moisture substance) oil is an excellent source of polyunsaturated fatty acids and vitamins A and D. The two main advantages of using dimethyl ether in oil extraction are free from freeze drying and the very low operation pressure (Fang et al., 2018). Since, more amount of cost is incurred in energy consumption stage. Even the extraction percentage was also seemed to be high in subcritical dimethyl ether (98.57%) when compared with wet reduction (56.76%), supercritical CO_2 (98.45%), and enzymatic extractions (85.25%). With the subcritical oil extraction process, the by-products of fish processing industries can be effectively valourized.

Similarly, omega-3 polyunsaturated fatty acid rich oil was extracted from the by-products (head, shrimp, tail) of northern shrimp variety (Zhou et al., 2021). Carotenoids rich oil was extracted from the pink shrimp at 333.15 K, 300 bar with 13.3 g/min flow rate of CO_2 for successful SFE. All these essential compounds (omega-3 fatty acids, carotene, lycopene, astaxanthin, alpha tocopherol, oleic acid) rich oil in food intake is helpful in overcoming various diseases like obesity, cardiovascular diseases, Alzheimer, liver dysfunctions, etc.

15.6   FUTURE PERSPECTIVES AND CHALLENGES

The non-thermal inactivation of microbes and enzymes present in dairy products using sub and supercritical CO_2 is affected by the parameters like pH, time, operating temperature, operating pressure, kind of food, and the state of CO_2, i.e., whether the CO_2 is subcritical, supercritical, gaseous or high-pressure liquid (Amaral et al., 2017).

Further upscale in the yield can be achieved by integrating the sub and supercritical extraction processes within or with other existing techniques. For example, oleuropeine rich extract was collected from the olive leaves by the coupling of subcritical and supercritical techniques (Viganó et al., 2015). In this method, CO_2 is combined with 5% ethanol and used as extraction solvent. This effect resulted in highly concentrated bioactive compound upon removal of other unwanted solvents.

15.7   SUMMARY

The sub and SFE techniques were found to be the most promising methods for extracting valuable bioactive compounds as well as oils from the food wastes. Thereby providing a green technique as an alternative for the harmful chemical extraction procedures. It is also economically manageable when compared with the conventional extraction techniques. In most of the cases, sub and SFE attains the maximum and quality yield of extracts. This chapter provides the detailed insights into the properties and applications of subcritical and SFE in various sectors. This field of research could bring extensive application in the food sector in future.

REFERENCES

Ahmed, R., & Chun, B. (2018, March). The journal of supercritical fluids subcritical water hydrolysis for the production of bioactive peptides from tuna skin collagen. *The Journal of Supercritical Fluids*, 0–1. https://doi.org/10.1016/j.supflu.2018.03.006

Amaral, G. V, Keven, E., Cavalcanti, R. N., Cappato, L. P., Alvarenga, O., Esmerino, E. A., Guimaraes, J. T., Silva, C., Raices, R. S. L., Ana, A. S. S., Meireles, M. A. A., & Cruz, A. G. (2017). Dairy processing using supercritical carbon dioxide technology: Theoretical fundamentals, quality and safety aspects. *Trends in Food Science & Technology, 64*, 94–101. https://doi.org/10.1016/j.tifs.2017.04.004

Ambrogi, A., Cardarelli, D. A., & Eggers, R. (2003). Separation of natural colorants using a combined high pressure extraction-adsorption process. *Latin American Applied Research, 33*(3), 323–326.

Asaduzzaman, A. K. M., Getachew, A. T., Cho, Y., Park, J., Haq, M., & Chun, B. (2019). Characterization of pepsin-solubilised collagen recovered from mackerel (*Scomber japonicus*) bone and skin using subcritical water hydrolysis. *International Journal of Biological Macromolecules.* https://doi.org/10.1016/j.ijbiomac.2019.10.104

Barnebey, H. L., & Brown, A. C. (1948). Continuous fat splitting plants using the Colgate-Emery process. *Journal of the American Oil Chemists' Society, 25*(3), 95–99.

Boussetta, N., Lebovka, N., Vorobiev, E., Adenier, H., Bedel-Cloutour, C., & Lanoiselle, J.-L. (2009). Electrically assisted extraction of soluble matter from chardonnay grape skins for polyphenol recovery. *Journal of Agricultural and Food Chemistry, 57*(4), 1491–1497.

Cheng, Y., Xue, F., Yu, S., Du, S., & Yang, Y. (2021). Subcritical water extraction of natural products. *Molecules, 26*(13), 4004.

Corrales, M., Toepfl, S., Butz, P., Knorr, D., & Tauscher, B. (2008). Extraction of anthocyanins from grape by-products assisted by ultrasonics, high hydrostatic pressure or pulsed electric fields: A comparison. *Innovative Food Science & Emerging Technologies, 9*(1), 85–91.

Da Porto, C., Decorti, D., & Natolino, A. (2016). Microwave pretreatment of *Moringa oleifera* seed: Effect on oil obtained by pilot-scale supercritical carbon dioxide extraction and Soxhlet apparatus. *The Journal of Supercritical Fluids, 107*, 38–43.

de Aguiar, A. C., Osorio-Tobón, J. F., Silva, L. P. S., Barbero, G. F., & Martínez, J. (2018). Economic analysis of oleoresin production from malagueta peppers (*Capsicum frutescens*) by supercritical fluid extraction. *Journal of Supercritical Fluids, 133*, 86–93. https://doi.org/10.1016/j.supflu.2017.09.031

Deeth, H. C., Datta, N., Ross, A. I. V, & Dam, X. T. (2007). *Pulsed Electric Field Technology: Effect on Milk and Fruit Juices' Advances in Thermal and Non-Thermal Food Preservation.* Blackwell Publishing (pp. 241–269). Seite.

Dejoye, C., Vian, M. A., Lumia, G., Bouscarle, C., Charton, F., & Chemat, F. (2011). Combined extraction processes of lipid from Chlorella vulgaris microalgae: microwave prior to supercritical carbon dioxide extraction. *International Journal of Molecular Sciences, 12*(12), 9332–9341.

Devani, B. M., Jani, B. L., Balani, P. C., & Akbari, S. H. (2019).Optimization of supercritical CO_2 extraction process for oleoresin from rotten onion waste *Food and Bioproducts Processing.* 199 (January), 287–295. https://doi.org/10.1016/j.fbp.2019.11.014

Fabian, H., Lasta, B., Lentz, L., Mezzomo, N., Regina, S., & Ferreira, S. (2019). Biocatalysis and agricultural biotechnology supercritical CO_2 to recover extracts enriched in antioxidant compounds from beetroot aerial parts. *Biocatalysis and Agricultural Biotechnology, 19*(March), 101169. https://doi.org/10.1016/j.bcab.2019.101169

Fang, Y., Liu, S., Hu, W., Zhang, J., Ding, Y., & Liu, J. (2018). Extraction of oil from high-moisture tuna livers by subcritical dimethyl ether: A comparison with different extraction methods. *European Journal of Lipid Science and Technology. 121*(1), 1800087. https://doi.org/10.1002/ejlt.201800087

Gandhi, K., Arora, S., & Kumar, A. (2018). Industrial applications of supercritical fluid extraction: A review Industrial applications of supercritical fluid extraction : A review. *Energy, 77*(1), 235–243. https://doi.org/10.1016/j.energy.2014.07.044

Gopal, J. M., Tretola, G., Morgan, R., De Sercey, G., Atkins, A., & Vogiatzaki, K. (2020). Understanding sub and supercritical cryogenic fluid dynamics in conditions relevant to novel ultralow emission engines. *Energies, 13*(12), 3038.

Hartman, L. (1951). Kinetics of the Twitchell hydrolysis. *Nature, 167*(4240), 199.

Herrero, M., Cifuentes, A., & Ibañez, E. (2006). Sub-and supercritical fluid extraction of functional ingredients from different natural sources: Plants, food-by-products, algae and microalgae: A review. *Food Chemistry, 98*(1), 136–148.

Holliday, R. L., King, J. W., & List, G. R. (1997). Hydrolysis of vegetable oils in sub-and supercritical water. *Industrial & Engineering Chemistry Research, 36*(3), 932–935.

Joana Gil-Chávez, G., Villa, J. A., Fernando Ayala-Zavala, J., Basilio Heredia, J., Sepulveda, D., Yahia, E. M., & González-Aguilar, G. A. (2013). Technologies for extraction and production of bioactive compounds to be used as nutraceuticals and food ingredients: An overview. *Comprehensive Reviews in Food Science and Food Safety, 12*(1), 5–23.

Kari, H. (2018). Subcritical water. *Celabor, 50265.* https://www.celabor.be/site/FCK_STOCK/Media/2016-08%20Eau%20subcritique%20EN.pdf

Kaufmann, B., & Christen, P. (2002). Recent extraction techniques for natural products: microwave-assisted extraction and pressurised solvent extraction. *Phytochemical Analysis: An International Journal of Plant Chemical and Biochemical Techniques, 13*(2), 105–113.

King, J. W. (2000). Advances in critical fluid technology for food processing. *Food Science & Technology Today : The Information Journal of the Institute of Food Science & Technology (UK), 14*, 186–191.

King, J., Holliday, R., & List, G. (1999). Hydrolysis of soybean oil. in a subcritical water flow reactor. *Green Chemistry, 1*(6), 261–264.

King, J. W., & Srinivas, K. (2009a). Multiple unit processing using sub- and supercritical fluids. *Journal of Supercritical Fluids, 47*(3), 598–610. https://doi.org/10.1016/j.supflu.2008.08.010

King, J. W., & Srinivas, K. (2009b). Multiple unit processing using sub-and supercritical fluids. *The Journal of Supercritical Fluids, 47*(3), 598–610.

Knez, Ž. (2016). Food Processing Using Supercritical Fluids. In *Emerging and Traditional Technologies for Safe, Healthy and Quality Food* (pp. 413–442). Springer.

Knez, M., Cör, D., Verboten, M. T., & Knez, Ž. (2018, May). Application of supercritical and subcritical fluids in food processing. *Food Quality and Safety, 2*(2), 59–67. https://doi.org/10.1093/fqsafe/fyy008

Knez, Ž., Markočič, E., Leitgeb, M., Primožič, M., Hrnčič, M. K., & Škerget, M. (2014). Industrial applications of supercritical fluids: A review. *Energy, 77*, 235–243.

Knorr, D., Froehling, A., Jaeger, H., Reineke, K., Schlueter, O., & Schoessler, K. (2011). Emerging technologies in food processing. *Annual Review of Food Science and Technology, 2*, 203–235.

Lyman, D. J. (1983). Membranes. In *Replacement of Renal Function by Dialysis* (pp. 97–105). Springer.

Mahdi, S., Zahra, P., Ali, S., & Mohammad, E. (2018). Application of supercritical fluids in cholesterol extraction from foodstuffs: A review. *Journal of Food Science and Technology.* https://doi.org/10.1007/s13197-018-3205-z

Majid, A., Phull, A. R., Khaskheli, A. H., Abbasi, S., Sirohi, M. H., Ahmed, I., Ujjan, S. H., Jokhio, I. A., & Ahmed, W. (2019). Applications and opportunities of supercritical fluid extraction in food processing technologies: A review. *International Journal of Advanced and Applied Sciences, 6*, 99–103.

Mason, T. J., Chemat, F., & Vinatoru, M. (2011). The extraction of natural products using ultrasound or microwaves. *Current Organic Chemistry, 15*(2), 237–247.

Matusiewicz, H., & Ślachciński, M. (2014). Development of a one-step microwave-assisted subcritical water extraction for simultaneous determination of inorganic elements (Ba, Ca, Cu, Fe, Mg, Mn, Na, Pb, Sr, Zn) in reference materials by microwave induced plasma spectrometry. *Microchemical Journal, 115*, 6–10.

Meireles, M. A. A. (2008). *Extracting Bioactive Compounds for Food Products: Theory and Applications.* CRC press.

Mulder, M., & Mulder, J. (1996). *Basic Principles of Membrane Technology.* Springer Science & Business Media.

Murty, V. R., Bhat, J., & Muniswaran, P. K. A. (2002). Hydrolysis of oils by using immobilized lipase enzyme: A review. *Biotechnology and Bioprocess Engineering, 7*(2), 57–66.

Nagavekar, N., & Singhal, R. S. (2019). Supercritical fluid extraction of Curcuma longa and Curcuma amada oleoresin: Optimization of extraction conditions, extract profiling, and comparison of bioactivities. *Industrial Crops & Products, 134*(November 2018), 134–145. https://doi.org/10.1016/j.indcrop.2019.03.061

Nowosad, K., Sujka, M., Pankiewicz, U., & Kowalski, R. (2021). The application of PEF technology in food processing and human nutrition. *Journal of Food Science and Technology, 58*(2), 397–411.

Pataro, G., Ferrentino, G., Ricciardi, C., & Ferrari, G. (2010). Pulsed electric fields assisted microbial inactivation of *S. cerevisiae* cells by high pressure carbon dioxide. *The Journal of Supercritical Fluids, 54*(1), 120–128.

Patras, A., Brunton, N. P., Da Pieve, S., & Butler, F. (2009). Impact of high pressure processing on total antioxidant activity, phenolic, ascorbic acid, anthocyanin content and colour of strawberry and blackberry purées. *Innovative Food Science & Emerging Technologies, 10*(3), 308–313.

Patras, A., Brunton, N., Da Pieve, S., Butler, F., & Downey, G. (2009). Effect of thermal and high pressure processing on antioxidant activity and instrumental colour of tomato and carrot purées. *Innovative Food Science & Emerging Technologies, 10*(1), 16–22.

Poojary, M. M., Putnik, P., Kovačević, D. B., Barba, F. J., Lorenzo, J. M., Dias, D. A., & Shpigelman, A. (2017). Stability and extraction of bioactive sulfur compounds from Allium genus processed by traditional and innovative technologies. *Journal of Food Composition and Analysis, 61*, 28–39.

Puértolas, E., Luengo, E., Álvarez, I., & Raso, J. (2012). Improving mass transfer to soften tissues by pulsed electric fields: Fundamentals and applications. *Annual Review of Food Science and Technology, 3*, 263–282.

Reátegui, J. L. P., da Fonseca Machado, A. P., Barbero, G. F., Rezende, C. A., & Martínez, J. (2014). Extraction of antioxidant compounds from blackberry (*Rubus* sp.) bagasse using supercritical CO_2 assisted by ultrasound. *The Journal of Supercritical Fluids, 94*, 223–233.

Riera, E., Golas, Y., Blanco, A., Gallego, J. A., Blasco, M., & Mulet, A. (2004). Mass transfer enhancement in supercritical fluids extraction by means of power ultrasound. *Ultrasonics Sonochemistry, 11*(3–4), 241–244.

Rooney, D., & Weatherley, L. R. (2001). The effect of reaction conditions upon lipase catalysed hydrolysis of high oleate sunflower oil in a stirred liquid–liquid reactor. *Process Biochemistry, 36*(10), 947–953.

Roseiro, L. B., Duarte, L. C., Oliveira, D. L., Roque, R., Bernardo-Gil, M. G., Martins, A. I., Sepúlveda, C., Almeida, J., Meireles, M., & Gírio, F. M. (2013). Supercritical, ultrasound and conventional extracts from carob (Ceratonia siliqua L.) biomass: Effect on the phenolic profile and antiproliferative activity. *Industrial Crops and Products, 47*, 132–138.

Santos, P., Aguiar, A. C., Barbero, G. F., Rezende, C. A., & Martínez, J. (2015). Supercritical carbon dioxide extraction of capsaicinoids from malagueta pepper (*Capsicum frutescens* L.) assisted by ultrasound. *Ultrasonics Sonochemistry, 22*, 78–88.

Sarrade, S., Guizard, C., & Rios, G. M. (2002). Membrane technology and supercritical fluids: chemical engineering for coupled processes. *Desalination, 144*(1–3), 137–142.

Sarrade, S., Guizard, C., & Rios, G. M. (2003). New applications of supercritical fluids and supercritical fluids processes in separation. *Separation and Purification Technology, 32*(1–3), 57–63.

Shi, X., Wu, H., Shi, J., Xue, S. J., Wang, D., Wang, W., Cheng, A., Gong, Z., Chen, X., & Wang, C. (2013). Effect of modifier on the composition and antioxidant activity of carotenoid extracts from pumpkin (*Cucurbita maxima*) by supercritical CO_2. *LWT, 51*(2), 433–440. https://doi.org/10.1016/j.lwt.2012.11.003

Soto, M. L., Moure, A., Domínguez, H., & Parajó, J. C. (2011). Recovery, concentration and purification of phenolic compounds by adsorption: A review. *Journal of Food Engineering, 105*(1), 1–27.

Spricigo, C. B., Bolzan, A., Machado, R. A. F., Carlson, L. H. C., & Petrus, J. C. C. (2001). Separation of nutmeg essential oil and dense CO_2 with a cellulose acetate reverse osmosis membrane. *Journal of Membrane Science, 188*(2), 173–179.

Teo, C. C., Chong, W. P. K., & Ho, Y. S. (2013). Development and application of microwave-assisted extraction technique in biological sample preparation for small molecule analysis. *Metabolomics, 9*(5), 1109–1128.

Torres-Ossandón, M. J., Vega-Gálvez, A., López, J., Stucken, K., Romero, J., & Di Scala, K. (2018). Effects of high hydrostatic pressure processing and supercritical fluid extraction on bioactive compounds and antioxidant capacity of Cape gooseberry pulp (*Physalis peruviana* L.). *The Journal of Supercritical Fluids, 138*, 215–220.

Varzakas, T., & Tzia, C. (2014). *Food Engineering Handbook: Food Process Engineering*. CRC Press.

Viganó, J., Machado, A. P. D. F., & Martínez, J. (2015). Sub- and supercritical fluid technology applied to food waste processing. *Journal of Supercritical Fluids, 96*, 272–286. https://doi.org/10.1016/j.supflu.2014.09.026

Wijngaard, H., Hossain, M. B., Rai, D. K., & Brunton, N. (2012). Techniques to extract bioactive compounds from food by-products of plant origin. *Food Research International, 46*(2), 505–513.

Yee, J. L., Walker, J., Khalil, H., & Jiménez-Flores, R. (2008). Effect of variety and maturation of cheese on supercritical fluid extraction efficiency. *Journal of Agricultural and Food Chemistry*, *56*(13), 5153–5157. https://doi.org/10.1021/jf703673n

Yildiz-Ozturk, E., & Yesil-Celiktas, O. (2019). Supercritical Fluid Technology in Bioseperation. *Comprehensive Biotechnology*, *2*(3), 713–724. https://doi.org/10.1016/B978-0-444-64046-8.00122-1

Zhou, J., Gull, B., Wang, M., Gull, P., Lorenzo, J. M., & Barba, F. J. (2021). The application of supercritical fluids technology to recover healthy valuable compounds from marine and agricultural food processing by-products: A review. *Processes*, *9*(2), 357. https://doi.org/10.3390/pr9020357

16 Recent Advances in Membrane Processing

*Addanki Mounika, L. Kavitha, S. Akalya,
and D. V. Chidanand*

16.1 INTRODUCTION

The membrane processing technology is a separation technique that separates the molecules with the help of semi-permeable membranes; these membranes can transport the components under hydrostatic pressure, making the separation effective. These membranes have two sides: the feed and the other one is permeated, where the solution passes through the feed stream with all the components and gets the permeate with selectable features. The waste materials left behind can be reused to get the maximum permeate, and these waste fluids are called retentate. Because of the partial pressure on the permeate side, liquid components that diffuse through the membrane evaporate (Conidi et al., 2020b). The semi-permeable membrane has a more significant advantage in the food industry after the water treatment plants. Membranes of wide varieties are used to separate various components; most repeatedly using membranes include microfiltration (MF), ultrafiltration (UF), nanofiltration (NF), reverse osmosis (RO), dialysis (DY), and electrodialysis (ED).

All these membranes are classified based on driving forces, such as pressure-driven, diffusional, and electrical processes. Under these, MF, UF, NF, and RO are pressure-driven, pervaporation (PV), DY, membrane absorption, and membrane extraction are diffusional, whereas ED, electro static liquid membrane, are electrically driven process. The primary membranes in descending order of their pore sizes are MF, UF, NF, and RO (Nissar et al., 2018). There are five different types of membranes with a more significant number of physical structures: microporous, homogeneous, homogeneous ion exchange, asymmetric integral, and composite. Generally, microporous is often used in the food and beverage industries to remove colloidal matter and suspended solids. There are two types of homogenous membranes. One contains dense films that separate the molecules with similar dimensions, solubilities, and diffusivities in the matrix.

Examples are gas permeation and PV; another type is an ion-exchange membrane used in ED. Another type of membrane is an asymmetric membrane that contains high permeation and selectivity with a thin dense layer or active layer that acts as a surface filter retaining the rejected material at the membrane surface and not inside the pores, which eventually avoids the blocking. All these membranes are substitutes for conventional membranes and can be used in sugar, dairy, fruit juice, and beverages. The material from which membranes are made is classified into four generations for various purposes, explained in Table 16.1.

According to (Maria Norbeta de Pinho and Miguel Minhalma) the external pressure, power, and driving forces help in membrane processing, and different membranes have different pore sizes for the separation of other application purposes. MF works on a sieving mechanism; it contains pore sizes ranging from 0.1 μm to 10 μm and can separate based on the pressure of 0.1–1 bar. Similarly, UF also works on the sieving mechanism, and the interactions between membrane, solvent, and solutes have a pore size of 0.02–0.05 μm at the pressure of 0.5–0.8 bar. NF helps to separate the salt and organic solute at 10–40 bar pressure and has a pore size range from 0.1 to 1 nm. RO can separate the organic microsolutes at 20–100 bar pressure; these have a pore size of approximately 0.0001 μm and can be used for diffusion. Another method for separation is DY; it works based on

TABLE 16.1

Membrane Generations and Their Types

S. No	Name of Generation	Type of Membrane	Applications
1	First generation	Cellulose derived	Desalination of seawater
2	Second generation	Synthetic polymers (polysulfone)	Removal of organic and natural gas enrichment
3	Third generation	Mineral membranes	Demineralization of dairy products (cheese) (Kotsanopoulos & Arvanitoyannis, 2015)
4	Fourth generation	Hybrid membranes (different pore size membranes, ED + MF, etc.)	Recovery of oil seed protein (Olaru et al., 2021)

ED, electrodialysis, *MF*, microfiltration.
Source: Adopted and modified from Nissar et al. (2018), Kotsanopoulos and Arvanitoyannis (2015), and Olaru et al. (2021).

the concentration gradient and helps to separate solutes at a micro level ranging from 0.1 to 10 μm (Galanakis, 2018).

Similarly, ED's advanced DY version helps desalter ionic solutions based on the electric potential gradient; it has a pore size of 10–100 Å (angstrom). Gas permeation is one of the dual separation processes that work on the driving force of pressure and concentration gradient, helps separate gas mixtures, and has a pore size of 3–4 Å. PV is the final separation process; it helps separate small organic solute concentrations; it works on the principle of a concentration gradient and has a pore size of 4 Å. These PV membranes are mostly non-porous and used for separating liquid mixtures; the separation characteristic is based on affinity for membrane materials; molecules with greater affinity are adsorbed and permeate through the membrane, whereas the membrane retains molecules with lower affinity (Jyoti et al., 2015). Among all the filtrations mentioned above, MF and UF are under hydrodynamic sieving control, and the rest comes under sorption diffusion. Transferring aromatics from an aqueous solution to a membrane by sorption (Jyoti et al., 2015). The two most common membrane types are dead-end and cross-flow. The fluid to be filtered is forced through the membrane pores in the dead-end mode, typically by applying pressure to the feed side. In a cross-flow configuration, the feed flows parallel to the membrane surface and penetrates the membrane due to a pressure differential. The cross-flow reduces the formation of the filter cake, keeping it at a low level (Charcosset, 2021).

16.1.1 Membrane Contactors

Membrane contactors do not control the permeate flow rate, but capillary forces prevent the phases from mixing. The principle is that the driving force is the pressure difference between the feed and permeate side. Membranes used are hydrophobic porous membranes with small pore sizes where capillary forces help the phases prevent from mixing. Membrane distillation (MD) and osmotic distillation (OD) are viable alternatives to thermal evaporation and RO in the concentration of liquid foods within membrane contactors. The membrane separates two liquid phases with different solute concentrations in the OD process: a dilute solution on one side and a hypertonic salt solution on the other (Conidi et al., 2018). The essential factor is the suitable porous membrane and pore size with various surface tension. These contractors were also used in the separation tow immiscible phases or liquids. The principle involved in this is the membrane is filled with water by one of the two immiscible phases, which then makes contact with the other phase at the pore mouth on the opposite side of the membrane. A higher transmembrane pressure from the non-wetting phase side keeps the wetting stage within the membrane pores, preventing it from dispersing into the other phase (Giorno, Drioli, et al., 2015a).

16.1.2 INTEGRATED MEMBRANE PROCESS

It is an alternative process involving various stages of membrane filtration or a combination of multiple membranes, such as microfiltration followed by NF. The combination of the membrane process gives additional benefits such as lower economic cost and clarified and higher concentrated juice with increased antioxidant polyphenol content. In a study, the combination of MF and OD was done on blood orange juices and pomegranate juices; prickly juice was first crushed and processed through MF and then clarified liquid was passed through NF, and finally, the pre-concentrated fluid was passed through the OD to get the concentrated juice at 50–60° Brix. The resulting sample shows the inhibitory value IC_{50} value is 44.36 ± 0.85 and 214.65 ± 2.03 µg/mL, which works actively against pancreatic lipase and α-amylase inhibitory activity, respectively. The clarified and concentrated juice was achieved at the pressure of 4.5 bar and temperature of $20 \pm 5°C$; the permeate fluxes are about 140 L/m²h for MF and NF the pressure 34 ± 2 bar, and at the temperature of $20 \pm 5°C$, the feed flowrate is 7.8 ± 0.1 m³/h and permeate flux range is 0.7–2.2 L/m²h (Morelli et al., 2022).

In another study, proteins and carbohydrates were recovered from whey water using integrated membrane process filters, ultra, and NF. Whey water was initially passed through the UF membrane at a pressure of 2 bar via dead-end filtration flow, and then the UF permeate was passed through the NF at a pressure of 20 bar. After the integrated process, the protein and carbs recovered were 91.48% and 97.2%, respectively. UF 50 and NF 200 membranes are utilized for the maximum component recovery from whey water. Water flow is reduced with time due to fouling and cake layer development (Lakra & Basu, 2021).

16.2 MEMBRANE PROCESSING

Membrane methods have been proposed to handle numerous food industry concerns, including food processing, fruit juices, and drinks. It also helps reduce the food processing industries' organic and chemical load with the help of MF, UF, and NF (Castro-Muñoz, Barragán-Huerta et al., 2018).

16.2.1 PRINCIPLE AND MECHANISM

The MF process uses the physical separation principle, containing an average pore size of 0.1 µm and 10 µm. Particles larger than 10 µm will pass through the filter also helps to separate particles and dissolved components, bacteria, and suspended solids represented in Figure 16.1(a) by a sieving mechanism. The pressure gradient is the driving force for the mass transport across the membrane pressure difference between the feed and permeates independent of the diffusion of the particles (Giorno, Piacentini, et al., 2015).

UF is the process of considering the two pressure differences between the feed and permeates side of the membrane as the driving force; under the influence of static pressure, the solvent or solute in the liquid passes via the membrane from the tiny pores from higher pressure to low pressure or higher concentration to the lower concentration larger molecules will trap inside the membrane pores represented in the Figure 16.1(b), and this causes cake layer due to more accumulation of the larger molecules this means when the sea water or wine with the larger molecules or solids don't pass through the membrane and form as a cake layer. Water, inorganic salts, and smaller substances ions will pass through the membrane to achieve higher purification, concentration, and separation of the selected liquids. The main steps in UF are the solutes absorption on the surface of the membranes and in the pores, retention in the pores, and removal of pores on the surface of the membrane (Li et al., 2018).

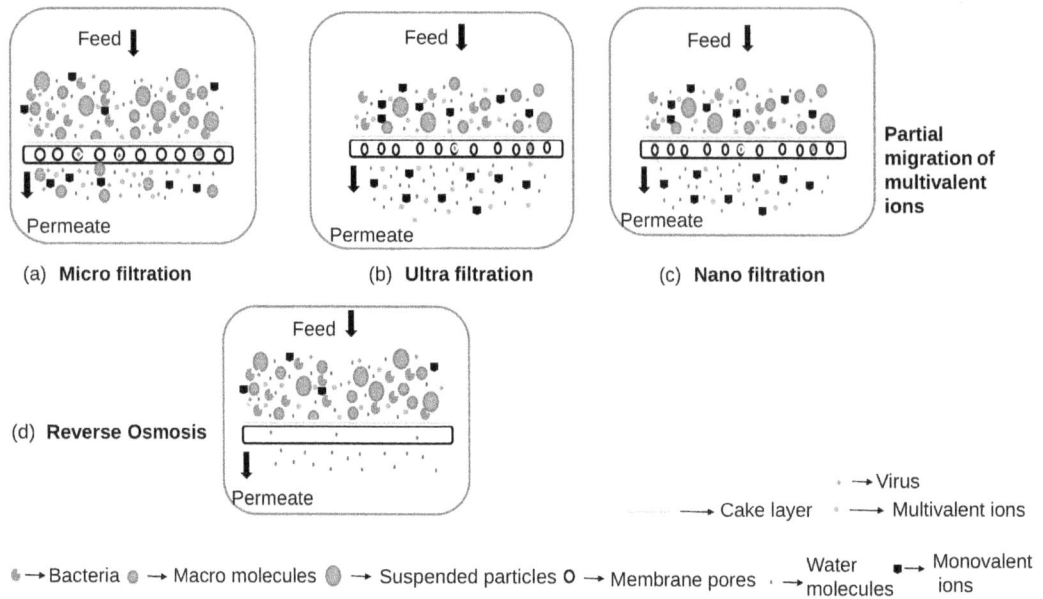

FIGURE 16.1 Principle for MF, UF, NF, and RO. (Adapted and modified from Woo, 2018.)

NF is one of the pressure-driven asymmetric membranes with a dense skin layer, and it's similar to RO; earlier, it was called loose RO; it only passes the monovalent ions represented in Figure 16.1(c) and rejects multivalent ions by suppressing the osmotic pressure to get lower specific energy consumption. Components are permeated in this solvent and have low molecular weight; the remaining components are retained. The only difference between the RO and NF is the transportation mode in the membrane (Bruggen, 2018; Giorno, Drioli, et al., 2015b).

RO is the opposite of osmosis, in which water separation from high concentration to low concentration under high pressure (hydrostatic) is more than osmosis. Membranes used in this are non-porous. The operation of RO consists of three main stages: (i) stage one, (ii) stage two, and (iii) semi batch RO. In step one, the feed solution is separated in the membrane into a concentrated brine solution and desalinated permeate solution. It is because, under minimum pressure, the solution passes through the membrane once; the lowest applied pressure is the exit brine osmotic pressure, determined by the water recovery rate. In a one-stage RO process, the strong driving force at the module's entry causes a high degree of thermodynamic irreversibility and significant specific energy consumption. In the stage – two methods of reusing concentration brine solution from stage one as the feed solution, the change in the pressure applied in this stage eventually reduces the energy consumption and permeate collected from both stages. In the semi-batch RO, the effluent is first mixed with the feed water, which is then circulated to the module as in a stream; as the rejected salts collect in the closed circuit, the brine concentration in the course rises over time. The principle of RO is represented in Figure 16.1(d) (Lin & Elimelech, 2017).

PV involves elective and partial evaporation of a two or multicomponent liquid stream via a membrane. The word PV is generally referred to as "per" (across) and "vaporation" (vapor formation). The feed is in direct contact with the upstream side of the membrane, while the opposite side removes a solute-enriched vapor phase that permeates, represented in Figure 16.2(a). In the desalination example, the heated salt solution comes in contact with the upstream side of the membrane. At the same time, a vacuum or sweep gas is applied to the downstream side of the membrane, and water is continually extracted from the permeate side in the form of vapor (Xie et al., 2018). PV's physical mechanism is based on non-specific interactions, such as Vanderwall's, involving the solution-diffusion and molecular sieving mechanisms. In contrast,

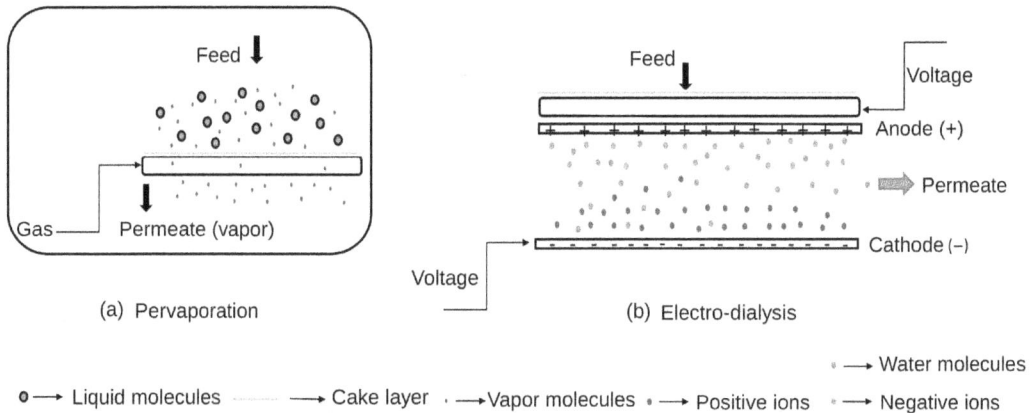

FIGURE 16.2 Diagrammatic representation of the principle of PV and ED. (Adapted and modified from Zhu et al., 2020 and Islam et al., 2018.)

the chemical agent is based on reversible chemical reactions involving chemical bonds forming where electron sharing occurs between permeate molecules and membrane materials (Song et al., 2019).

The ED principle is based on separating ions from an aqueous solution using a membrane and electricity. Figure 16.2(b) depicts a stabilized anion and cation membrane system between negative and positive electrodes. As ED membranes are impermeable to the ions, they accumulate on both sides of the system depending on their load during the passage of NaCl solution, desalination Na^+ moves toward the negative electrode, and Cl^- ions move toward the positive electrode. At the same time, the ionic charge in the remaining position is reduced (Kotsanopoulos & Arvanitoyannis, 2015).

Sieving mechanism: It is mentioned as if the size of the particle is greater than the membrane pore size, it deposits on the surface of the membrane, and as the particle is smaller in size than the membrane pore size, it will pass through the permeate side, but if the particle is the same size as the membrane size, they will be stuck in the membrane pores and needs a lot of pressure to pass through this results in the cake layer formation and causes fouling of the membrane. This sieving mechanism happens in MF, NF, and UF filtration methods (Ullah et al., 2013).

Donnan exclusion mechanism is also called a non-sieving mechanism based on the charge and electrostatic interactions. It occurs during the ions' separation into a material with higher charge density. NF and RO membranes contain the Donnan exclusion mechanism. Their membranes are two types: negative and positive charged membranes. This mechanism varies based on the type of the membranes; on the membranes with negative control, the rejection and allowing of negative and positive ions will occur and vice versa. This transportation is caused due to diffusion, convection, and electromigration (Suhalim et al., 2022).

16.2.2 MEMBRANE FOULING

Membrane fouling is a severe problem during filtration; it happens due to the accumulation of suspended solids and macromolecules on the surface of the membrane or inside the membrane pores. In other words, molecules more significant than the membrane's pore size get stuck in the pore and cause blockage of the pores, resulting in the surface layer formation called cake formation. It leads to an increase in the economic, maintenance, and clean-up costs. Decreased cake layer formation increases the membranes' efficiency, and the membranes' shelf life increases the permeate, reducing the rejection rate. This fouling can be indicated by viewing the permeate flow and quantity; the less permeate, the more fouling. It can be considered in the initial stage of the filtration process, as there will be many cake formations. Maintaining the process conditions is essential to decrease the cake layer formation. It helps slow permeate flow with less cake formation, high efficiency, and a low rejection rate. As a result, the decrease in permeate leads to a fouling index, denoted as FI, which can be computed using the following formula:

$$FI = \left(1 - \frac{L_{p1}}{L_{p0}}\right) \times 100,$$

whereas L_{p1} is the permeate after filtration and L_{p0} is the permeate before filtration (Lu et al., 2021).

In the case of membrane fouling in the proteins, there are four main effects: (a) membrane hydrophobicity, (b) membrane surface charge, (c) protein hydrophobicity, and (d) protein surface charge. These physicochemical parameters, surface charge, and hydrophobicity influence the aqueous system's adsorption. Protein absorption on the surface of the membrane is due to proteins being hydrophobic at isoelectric points. The adsorption to the membrane increases their hydrophobic interactions and causes fouling in the system. Also, aggregations of the proteins in the solution to their isoelectric point cause reduced electrostatic repulsions, causing protein fouling on the surface of the membranes. During membrane filtration, the decrease in the flux is due to the increase of polarization, fouling by adsorption, and material deposition. Another explanation is that the deposition of smaller particles on the surface creates more significant fouling on the membrane surface than bigger particles (Wen-qiong et al., 2019).

The relevant approach or methodology for measuring membrane fouling characteristics was Brunauer-Emmett-Teller (BET), which assists in estimating the microporous and mesoporous material's porous porosity and surface area. These two characteristics enable the identification of potential membrane fouling spots during the procedure. It should be emphasized that the BET technique assesses all accessible holes for nitrogen adsorption. Hence, a sample may have a high porosity despite poor permeability due to multiple linked dead-end pores. This method helps measure the pore's availability to nitrogen adsorption, which helps define the membrane's permeability (Virtanen et al., 2020).

Membrane fouling monitoring can be detected by using the laser technique in which the laser should contact the cake surface layer. It worked on the principle of the angle at which reflected laser light strikes a detector, and the deviation of the laser beam from its original path can be used to study variations in the surface of a cake layer. The main point is that the laser should not be composed of photosensitive material (Rudolph et al., 2019).

16.2.3 EQUIPMENT

Membrane modules are used to prevent the fouling of membranes during the processing; there are four different types of membrane modules: (a) plate and frame; (b) Tubular module; (c) spiral wound module; and (d) hollow – fiber module (Bera et al., 2022).

(a) Plate and frame module: Membrane structures consisting of liquid and product spacers join together in a flat position called a frame and are considered plate and frame. This help prevents the sticking of the membranes and provides free flow of the feed or product. This frame consists of packaging density of 148–492 (m^2/m^3) along with good cleaning performance and high manufacturing cost due to moderate prevention of fouling the schematic representation given in Figure 16.3(a) (Ezugbe & Rathilal, 2020).

Castro-Muñoz et al. mentioned that the apple juice was clarified by using a plate and frame in the RO filtration mode at 25° Brix achieved. Evaporation was used to bring the Brix concentration up to 72° Brix. Nevertheless, by employing the plate and frame modules, the total enrichment factors and mass transfer rate were raised to 100–1000 kg m^{-2}, 5–500 kg m^{-2}, respectively, with low fouling

(a)

(b)

(c)

(d)

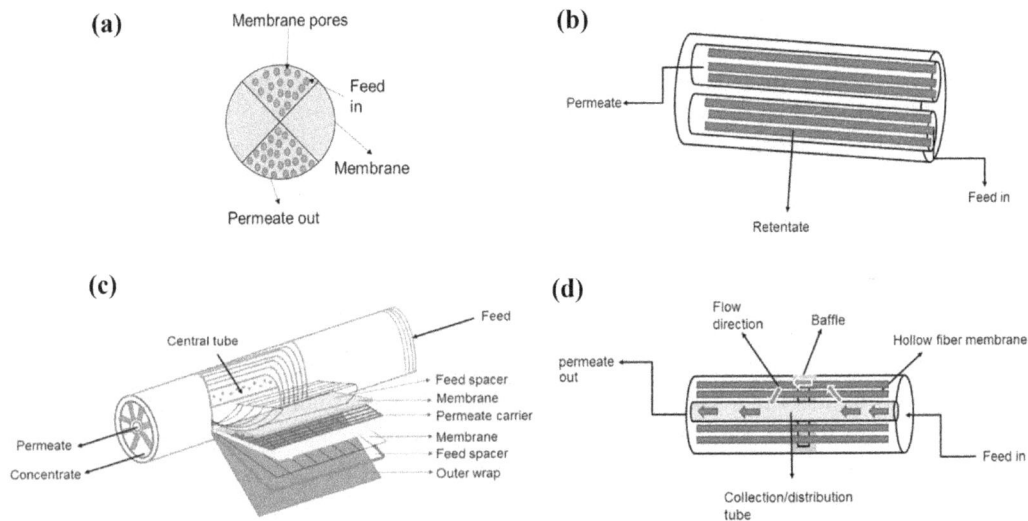

FIGURE 16.3 Diagrammatic representation of plate and frame (a). (Adapted and modified from Harlacher & Wessling, 2015.) Diagrammatic representation of tubular modules (b). (Adapted and modified from Xue et al., 2021.) Diagrammatic representation of spiral wound module (c). (Adapted and modified from Hasani et al., 2019.) Diagrammatic representation of hollow fiber (d). (Adapted and modified from Bazhenov et al., 2018.)

and no flux fluctuations during the operation. The temperature and pressure were 20°C and 2 bar, respectively (Castro-Muñoz et al., 2021).

(b) **Tubular module:** In this module, semi-permeable membrane is embedded in a tubular shell consisting of perforated stainless steel, and these are the most usable membranes in the food processing industries because of their high flexibility and easy to clean, high potential for fouling along with the packaging density of 6–12 (m^2/m^3), the schematic representation given in the Figure 16.3(b). The disadvantages are high energy consumption, weak thermal resistance, and high cost. The main advantage of this module is that treating high viscous or high solid content samples can be quickly done (Guiga & Lameloise, 2019).

Urošević et al. mentioned clarifying sugarcane juice using tubular ceramic membranes embedded in MF with a pore size of 0.1 μm. The initial flux and pressure of the membrane were 55.6 L/m^{2}h, 100 kP. Increasing the membrane pressure increased the fluctuations to 87.9 L/m^{2}h at 200 kPa, and further pressure increments showed no variation in the flux (Urošević et al., 2017).

(c) **Spiral wound module** concentrates food products; milk and displacement help remove concentrates (Kieferle & Kulozik, 2021). The overview of the spiral module consists of four sides, of which three are closed and packed within the membrane layers fast. The fourth side was attached to the permeate tube, and the whole module was surrounded by a plastic mesh of the central permeate tube, which serves as a feed channel; the schematic representation is in Figure 16.3(c). These modules will have a packaging density of 150–380 (m^2/m^3), along with poor cleaning quality, prone to high fouling, and can bear the cost at a moderate level. These have difficulty separating or filtering the highly viscous products, difficulty cleaning, dead fouling, and higher pressure drops. It has the advantage of lower energy consumption (Shahir, 2021; Guiga & Lameloise, 2019).

In a study, ethanol production using spiral module membranes in the mode of PV was done from red grape wine. In this process, two stages of separation were done; in the first stage, aroma substance and ethanol 30 volume percentage were separated. Further, it was concentrated in the second stage and achieved 50% volume and was considered grape spirit at the temperature of 45°C, and higher retentate was observed. Still, the efficiency was decreased, which affected the product's quality; it is advisable to maintain the proper temperatures. The initial flow rate and pressure were 12 m³/h and 0.37 inlet pressure, respectively. Outlet and pressure drop were 0.23 and 0.14, respectively. The spiral wound thickness was 30 μm, and the pore size of 50 nm, along with an area of 13 m². The activated layer thickness was 2 μm (Sun et al., 2020).

(d) Hollow fiber module: These are mechanically self-supporting membrane modules with good flexibility; they consist of two ways: one is outside, and another is inside, in which feed can be fed. They have asymmetric shapes and are mainly used for RO applications. These can be arranged as shells and tubes to appear mounted in the schematic representation in Figure 16.3(d) to create a high surface area and packaging density of 150–1500 (m²–m³). They have a very high potential for fouling and are challenging to clean because of their structure and arrangement. Still, they can be easily replaceable due to their low cost and frequent availability with a thickness of greater than 40 nm to withstand the high-pressure supporting layer with a thickness of 20 μm is necessary. The significant drawbacks of this hollow fiber membrane are frequent fouling and weak thermal resistance, and the advantages are low energy consumption, good mechanical properties support, and greater compactness (Balster, 2014).

In a study, skim milk protein fractionation was done using MF embedded with hollow fiber membranes. It showed that greater flux values were achieved in a low feed flow rate of 20 m³ h⁻¹ compared to spiral wound membrane, and feed was given in the cross-flow rates. Results showed that the whey protein transmission percentage was 50% for hollow and 46% for spiral membranes, and the mass flow was 86 and 64 g m⁻² h⁻¹. Similarly, the whey protein mass flow per module was calculated and showed higher for the spiral followed by hollow membranes, which were 1050 and 80 g h⁻¹, respectively (Schopf et al., 2021).

16.2.4 FACTORS AFFECTING MEMBRANE PROCESSING

16.2.4.1 Liquid Entry Pressure (LEP)

Different parameters affect the membrane usability during the MD by using seawater as a source. One factor is LEP; certain membrane types, such as hydrophobic membranes, won't allow water inside the pores, but they will allow the water into the pores at a specific pressure. This process is called LEP and shows the hydrophobicity of that membrane. Membranes will only enable the vapor out of the membrane pores. The water will affect the membrane's pores, damaging the membrane. By controlling the incoming pressure, we can kind the LEP value using the Laplace-Young equation (My et al., 2018):

$$LEP = \Delta P = P_f - P_d = \frac{-2B\gamma_1\cos\theta}{r_{max}}$$

P_f and P_d are feed and permeate hydraulic pressure
B= Geometric factor of membrane pore
γ_1 = liquid surface tension
θ = contact angle
r_{max} = largest pore radius (My et al., 2018)

16.2.4.2 Membrane Pore Size

In a study of the separation of anthocyanin from grape marc extract, this extract was obtained from ultrasound-assisted extraction with the help of three different pore-size membrane filters, 0.2 μm, 0.8 μm, and 0.14 μm. The filtration process was done. For this filtration, cross-flow filtration was chosen, and observed the fouling mechanism, and there is no significant difference in permeate flux in all three different pore size membranes. Mora et al. observed that complete pore blocking occurred in the 0.8 μm membrane, whereas incomplete pore blocking was observed in the other two pore-sized pore blocks. The separation took about 180 minutes, and the results supported the three stages of fouling concentration polarization, solute deposition, and consolidation of deposits. Based on the three stages, it was stated that 0.14 μm membrane has a longer concentration polarization time. Also, based on the total phenols rejection, this particular size membrane has shown the lowest sacrifice among others. So 0.14 μm membrane pore size is the best for anthocyanin separation. Other pore sizes exhibited the highest phenols and sugar rejections. It stated that pore size would impact the filtration process and varies for different membranes and foods (Mora et al., 2019).

16.2.4.3 Membrane Thickness

It is essential to know the thickness of the membrane in filtration or processing. Filtration time and membrane process will affect because of the higher viscosity of the membrane. Consistency can be measured using micrometers during the thickness measurement; five locus points need to consider, which are top, bottom, middle edges, left, and right, with an accuracy of 0.01 mm (Utoro et al., 2019). Flow rates influence the thickness of the boundary layer at the membrane surface and, as a result, the resistance to mass transfer. The concentrations of brine and feed affect the vapor pressure gradient across the membrane, which is proportional to the magnitude of the driving force (Cassano, 2016). To comprehend the fouling layer thickness, the inner diameter of the membrane is combined with the fouling layer thickness (Wang et al., 2017).

16.2.4.4 Contact Angle

In membrane, processing to get the characteristics of the membranes, such as hydrophobicity and hydrophilicity, can be measured using the contact angle. A change in contact angle indicates fouling formation. Less contact angle denotes higher affinity, and a higher contact angle implies lesser association. The contact angle changes affect the membrane surface's charge and the liquid macromolecules. The contact angle for various polymer ultra, nano, and micro filter membranes in droplet volume was studied using a goniometer. An increase in droplet volume did not impact the contact angle (Kertész et al., 2014). Pore size and porosity will also affect the contact angle; in one study using Pluronic F127, increased the hydrophilicity of the cellulose acetate membrane in this, the contact angle reduced from 55° to 35.6° by increasing this angle pure water flux was changed from 3.24 to 93.14 $Lm^{-2}h^{-1}$ (Chan & Ng, 2018).

16.2.5 APPLICATIONS AND ADVANTAGES OF MEMBRANE PROCESSING IN THE FOOD INDUSTRY

Membrane technology has been used in the food business in foreign and local markets since the 1960s; among the advanced characteristics are high separation precision, increased selectivity, room temperature operation, and no high temperature or chemical damage. In this application sector, the key benefits of membrane technology are economic energy, decreased cost, exclusive resource use, and reduced pollution.

MF, UF, RO, and NF are often utilized in the food business. Some of the applications include the removal of microorganisms from food, purification or concentration of water, fruits, and vegetable juices, clarification of soy, wine, and beer, concentration or separation-purification of proteins, polypeptides, fats, sugars, and purification, discoloration of oil, and wastewater treatment (Nissar et al., 2018).

16.2.5.1 Clarification of Juices

In one study, sugarcane juice was clarified using an UF process. Before the clarification, the liquid was subjected to carbonation using calcium hydroxide, and then adding sodium bicarbonate into the carbonated juice after discarding the cake layer deposition. It was then subjected to the UF membrane for clarification, which resulted in avoiding sucrose destruction and improved yield. This process helped remove polysaccharides from the sugarcane juice that inhibited the colorant formations, resulting in decreased sucrose crystal color; also, calcium cations were removed successfully.

The UF treatment resulted in decreased viscosity and turbidity, quicker crystal development, larger crystal size, and more color reduction. When compared to commercial procedures, the UF technique has the potential to be used in the manufacturing of refined sugar at reduced energy, operational, and capital cost; it also offers environmental benefits and no compromise in product quality. Finally, a UF system will provide an alternative to the sulfidation process, enhancing sugar product quality while reducing sulfur dioxide emissions and creating a hostile working environment for mill personnel (Vu et al., 2020).

In a similar study, cross-flow UF using a hollow fiber membrane clarified the Mexican Xoconostle fruit. 81.65% of the initial juice was recovered as clarified juice with a more significant proportion of antioxidant activity and a high phenolic content using membrane filtering. Removing phenolic components on the surface of the membrane increases the juice's color properties and turbidity (Castro-Muñoz, Fíla et al., 2018).

16.2.5.2   Purification of Juices

Pressurized liquid extraction (PLE) and membrane separation help recover concentrated monomeric anthocyanin grape marc. The extracts were obtained using PLE and then subjected to 0.20 μm dead-end filtration of MF. The permeate was then forwarded to the cross-flow filtration of the nano-filtration method. In MF, the removal of larger particles was achieved. The final product is rich in antioxidant capacity with a concentration of 52%, along with antioxidants total phenolic and antho-cyanin are also performed in higher concentrations of 71% and 78.2%, respectively, which can be used in food, cosmetics, and other applications (Tamires Vitor Pereira et al., 2020).

In another study, the purification and concentration of goji leaf extract were achieved with the help of an aqueous extraction process, and the section was subjected to the UF membrane in the cross-flow filtration mode. At the end of the filtration process, there are suspended solids compared to the extract, and there is a 0.1 decrease in the pH value, and total phenols before and after filtration are 1870.7 ± 4.5 to 1530.2 ± 2.8 mg/GAE/L about 80% of phenols were achieved (Carmela Conidi et al., 2020a).

In another study from the peel of beetroot, the extraction of juice and using ethanol-water and only ethanol was then subjected to the RO membrane process. In this study, two RO membranes were chosen one is RO99 and X20. The juice was extracted using water at a temperature of 22°C for 60 minutes; in the case of ethanol-water solvent, 15% v/v ethanol must be added for the extrac-tion process. These extracts were subjected to the RO membrane at the pressure of 40 bars, the steam temperature was 30°C, and samples were collected for every 500 mL of permeate. Different analytical measurements measured the concentrations of betaxanthin, betacyanin, total phenolic content, and antioxidant. The attention of betaxanthin before and after the membrane processing of RO99 is 75.46 ± 0.45 mg/L (water solvent) and 87.27 ± 0.93 mg/L (ethanol-water solvent), 292.47 ± 1.93 mg/L, and 337.26 ± 4.31 mg/L in water and ethanol-water extracts, respectively. For X20, the concentration of betaxanthins for water extract was 84.44 ± 0.45 mg/L. In contrast, aqueous ethanol extract contained 94.71 ± 0.19 mg/L; after the membrane process, the concentrations were 253.33 ± 1.52 mg/L for water extract 366.78 ± 0.00 mg/L for a water-ethanol quote. It shows that the RO99 has demonstrated the highest efficiency for the concentration process (Zin et al., 2020).

16.2.5.3   Removal of Bacteria from Food

Milk contains bacterial spores that impair dairy products' quality and shelf life. Using MF, spores of *Bacillus licheniformis* and *Geobacillus sp.* were removed from skim milk using membranes with pore diameters of 1.4 and 1.2 m, respectively. Skim milk was injected with 106 cfu/mL spores before being micro-filtered through ceramic membranes at six °C, 4.1 m/s cross-flow velocity, and 69 to 74 kPa transmembrane pressure. The spores were shorter after filtering, measuring 1.37 to 1.59 m in length and 0.64 to 0.81 m in breadth. *Licheniformis* reduction was enhanced by lowering the pore size from 2.17 logs for 1.4-m pore size to 4.57 logs for 1.2-m pore size. The smallest quan-tity of *B. Licheniformis* was passed through the two-pore diameters, whereas *Geobacillus* sp. was completely removed (Griep et al., 2018).

16.2.6   ADVANTAGES OF MEMBRANE PROCESSING IN THE FOOD INDUSTRY

There are several advantages of membrane processing over non-thermal or conventional techniques that work under high hydrostatic pressure and homogenization. Greater output under mild operation conditions such as flow rate, pressure, low operational cost, feasibility, process reliability, and relative insensitivity. The temperature of all the membranes can work at room temperature and get higher efficiency; yield also maintains the functional properties of the food.

Each membrane filtration has its advantage over other filtration types (Vieira et al., 2020). Membrane technology can be applied in many areas besides the food industry, such as cosmetics and pharmaceuticals. It contains only some moving parts, does not need many human resources, is easily operated, and is easy to clean; at the end of the process, the product can be recovered unchanged. It can be used with three or more filters for better output without adding additives; the system is easily scalable and can be integrated with other treatment processes. Permeate quality depends on the type of the membrane, its properties, and pore size; permeate quality is frequently independent of the feed stream (Gadlula et al., 2019).

Microfiltration has the advantage of consuming lower energy for the higher separation process, high productivity, absence of phase transition, no need for changing the temperature and pH of the solution or liquid, and separation of compounds such as carbohydrates and proteins without any use of solvents (Vieira et al., 2020). A study used microfiltration membranes to process goat whey orange juice. In this, both the pasteurization and microfiltration process were done. It was observed that microfiltration processed juice has higher acceptance when filtered at the temperature of 30°C and showed higher Smith's Salience Index values (SSI value) for orange color (0.533) and fermented aroma (0.361), which are higher than the pasteurized technique (0.438) and (0.295), respectively. This process showed that microfiltration helps maintain the product's color better than pasteurization (Vieira et al., 2020).

16.2.6.1   Ultrafiltration Advantages

It has higher turbidity and accuracy of filtration, UF has a higher turbidity removal rate up to 0.1 nephelometric turbidity unit (NTU) compared to conventional methods, little changes in the quality index during water purification, effective removal of particulate matter up to 99.9% along with effective removal of microorganisms such as Giardia, Cryptosporidium up to 4.4 log removal and some bacteriophages. No chemical agents are added after the UF process; in the case of water filtration after coagulation and sediments, on these, water can be directly carried to the UF without conventional filtration. The surface area is only one-fifth of the original and traditional process (Li et al., 2018).

A comprehensive pore-negative charge UF membrane is the first study using broad pore-charged tangential flow membranes with 100 kDa; the milk protein concentration was separated. It provides a fourfold higher permeate flux than the unmodified membranes with 30 kDa. When negatively charged proteins were placed on the surface of an UF membrane, they were rejected by electrostatic repulsion rather than size-based sieving (Arunkumar & Etzel, 2018).

16.2.6.2   Nanofiltration Advantages

NF requires higher energy than the other membrane filtrations, but this energy is usually low compared to the other conventional techniques. This NF has the advantages like energy saving, cost-effectiveness, compact equipment, higher permeate flux, sustainability, and retaining most of the natural color. It won't allow any suspended and colloidal particles into the permeate side. Also, the reduction of microorganisms can be seen compared to UF and MF. Among the other advantages, it helps remove heavy metals, and turbidity is higher than MF because of the smaller pore size. The operation can be done at low pressures, with retention of multivalent anions and higher salts and removal of organic materials, primarily used to desalinate seawater and brackish water. Thanks to NF membranes, the water permeates after the water filtering, is highly soft, and can hold more excellent water feed (Shah et al., 2021).

16.2.6.3   Reverse Osmosis Advantages

RO has been used in drinking water purification effectively because RO membranes don't contain any pores by applying pressure at the feed side. Only water molecules will pass through, which was discussed in detail in 2.1. The main advantage of using RO is the microorganisms can be removed up to 90%, and water recovery can be up to 80% in the integrated membrane mode. It decreases labor and cost and helps control fouling with regular cleaning compared to the UF and MF processes. With simple controls, the system achieves maximum recovery batch after batch. Permeate water quality can be improved by total/partial second-pass treatment and controlled by stopping the process. When the feed water quality changes, batch RO is a flexible system, and cleaning is simple and easy to implement in this type of system (Patel et al., 2021).

16.2.6.4   Electro Dialysis Advantages

It aids in removing salts at higher rates and produces a more concentrated brine solution that is less prone to scaling. It boasts a higher water recovery rate compared to RO and enjoys a longer membrane life. It is also easy to operate at higher temperatures and requires no pre- or post-treatment like RO. Additionally, it assists in the separation of monovalent ions from multivalent ions. The desalination using ED is 4%. It won't produce noise or pollution, has low maintenance costs, is easy to transport, and is feasible and environmentally friendly. Effective cleaning of the membrane helps maintain the properties of the membranes. Solution with low ions gives higher permeability (Al-Amshawee et al., 2020).

16.2.6.5   Pervaporation Advantages

The most crucial advantage of PV is the separation of the azeotropes, which reduces costs by lowering the heat or removing higher amounts of heat during the process. Easy operation, no need for labor work, meager maintenance cost up to 0.75–1.25 US cent/dm^3 compared to ED of produced ethanol. PV, along with distillation, helps reduce the cost and save energy, and gives much more efficiency; it also helps retain the aroma of particular liquids and less waste. The process's flexibility in particle load and changing product and feed concentrations is a significant advantage over other separation techniques (Kaminski & Marszalek, 2006).

16.3.   RECENT ADVANCES IN MEMBRANE PROCESSING IN FOOD APPLICATIONS

16.3.1   RECENT ADVANCES IN MEMBRANE PROCESSING OF FOODS

Due to the increased energy efficiency and avoidance of additives and chemicals, which benefit public health and the environment, membrane processing technologies are regarded as green technology (Akin et al., 2012).

Membranes are crucial in the current dairy processing sectors because they help to clarify the milk, improve the concentrations of certain constituents, and separate the essential elements from the dairy or milk wastes. Evaporation, centrifugation, demineralization of whey, bactofugation, and other crucial steps in the milk production process in the dairy industry may all be replaced with membrane processes practically and cost-effectively (Pouliot, 2008). Caseins obtained by MF concentration are used to make cheese because they are more effective than casein from skim milk. Additionally, a superior whey product with exceptional qualities may be made from the permeation product, and serum proteinic purification as a feed was employed (Lawrence et al., 2008).

To concentrate whey protein solutions, Wang et al. used forward osmosis. In this osmosis type, a semi-permeable membrane separates concentrated and diluted feed solutions because the osmotic pressure causes the liquid to pass from the feed solution to the concentrated solution. When processing viscous solutions with a high total solid, forward osmosis needs less pressure than RO (Gadlula et al., 2019).

The separation process does not employ heat treatment or chemicals; MF and UF are particularly effective at retaining juice freshness, flavor, and nutritional content while producing natural, high-quality, and additive-free goods (Carmela Conidi et al., 2017). These procedures divide the juice into a fibrous pulp and a cleared fraction entirely stable and devoid of spoilage microbes. Proteins and pectin having higher molecular weight are recovered by applying these membrane systems. At the same time, substances such as sucrose, salts, and flavor compounds are transferred through the membrane due to their lower molecular weight (Qaid et al., 2017). According to (Nourbakhsh et al. the velocity of the feed climbed when the clarification process was done with PVDF membrane in the red plum juice; the total resistance dropped around 45% due to a change in the resistance of the cake (Nourbakhsh et al., 2014).

Additionally, the temperature in the juice rose from 20°C to 30°C, reducing fouling resistance. Whereas the feed velocity increased in the case of watermelon juice, resulting in a surge of cake resistance. Because of the hydrophilic nature, the membrane made of Mixed cellulose ester showed lower cake resistance compared to the PVDF membrane.

Applying UF-NF membrane technology (Tsui & Cheryan et al. studied methods of recovering xanthophylls (lutein and zeaxanthin) from corn using ethanol extraction. The UF stage was intended to segregate ethanol-soluble proteins, while the NF stage was designed to restore carotenoid content. The higher protein concentration as UF retentate was seen in membranes with a molecular weight of 1–2 kDa. More excellent stability and fluxes were observed in the 300 Da membrane at the last stage of NF (Tsui & Cheryan, 2007).

16.3.2 RECENT ADVANCES IN MEMBRANE PROCESSING IN THE RECOVERY OF PROTEINS AND BIO-ACTIVES FROM THE WASTE STREAM

Protein can be obtained from various resources with generous health benefits. When it comes to the use of membrane processing, milk proteins have gotten the most significant interest. Alternative sources of protein that have gone through membrane processing include rapeseed, egg yolk, rice bran, and soy. It has been demonstrated that bioactive peptides, often produced by hydrolyzing natural proteins, have various positive health effects, notably anti-cholesterol, antioxidant, immuno-modulating, and antihypertensive properties (Kotsanopoulos & Arvanitoyannis, 2015).

A cheap and excellent source of protein for culinary consumption is rice bran. The protein recovered by the hydrolyzation of rice bran was investigated (Hamada, 2000). Two distinct molecular weight cut-off (MWCO) UF membranes, 1 kDa and 3 kDa, were utilized to retrieve the party hydrolyzed protein from the bran. The membranes with size 3 kDa generated yield loss since they could pass through a significant proportion of hydrolysates. However, despite having a greater membrane surface area and longer working times, the 1 kDa membrane proved ineffective for removing protein hydrolysates. As a result, membranes having 2 kDa were suggested for even more effective protein hydrolysate purification (Hamada, 2000).

For a massive purification process, rapeseed proteins such as napkin, cruciferin, and lipid-transfer proteins were investigated (Bérot et al., 2005). Despite being underutilized, rapeseed proteins must be studied in large amounts to realizing their potential benefits as functional components fully. It is due to their restricted use in livestock feed – protein purification methods combined with the help of NF membranes and chromatographic techniques. When desalting and removing phytoconstituents from proteins with an NF membrane system having about 1 kDa, MWCO removes 98.6% of the salt and 88.5% of the colorant. After the procedure, recoveries of 40 and 18% for napkin and crucerin proteins were achieved (Bérot et al., 2005).

Several UF membrane technologies with different clarifying pre-treatment techniques were investigated by Gao et al. before UF. Researchers evaluated two distinct MWCO membranes having a size of about 10 and 30 kDa, using spiral wound modules after being pretreated with centrifugation. The content of protein isolates in the discharge of bag filtration rose by 38.1 to 84.4 g/100 g. It is equivalent to 15 to 52 g of protein per 100 g of pea retrieved using a 10 kDa UF method. Because of its low molecular weight, the membrane size of 30 kDa did not collect albumin protein. Additionally, pea whey proteins showed more excellent functional qualities than traditional product (Gao et al., 2001).

It has been proven that membrane-based techniques may successfully recover and extract bioactive compounds from various food products, including phenolics, flavonoids, and aromatic substances (Castro-Muñoz et al., 2020). Multiple forms of phytochemicals are abundant in grape seeds. Nawaz et al. investigated the usage of UF membranes with various MWCOs for extracting phytochemicals in grape seed during processing. After extraction, membranes with pore sizes of 0.22 and 0.45 m were interpreted. As it retained bigger solutes and molecular-size particles, a membrane with a 0.22 m pore size produced improved performance. High recovery rates and accuracy with a reduced extraction time were delivered via membrane concentration. The process could recover a total of 11.4% of the weight of the seeds (Nawaz et al., 2006).

Biologically active compounds in pomegranate juice were purified using NF and UF membranes with MWCO varying between 1 kDa and 4 kDa. Compared with other selected membranes, 2 kDa showed a reduced index of fouling, more significant permeate fluxes, and superior efficacy in separating sugars from polyphenol substances (Carmela Conidi et al., 2020b). In another study by de Moraes et al., the extraction obtained from graviola leaves (containing 50 or 70% ethanol (v/v)) was permeated to UF, NF, and a combination of both procedures. To determine its selectivity for each stage of concentration, the fractions obtained from these processes were assessed for their total phenolics and antioxidant activities. The UF and NF combined procedures achieved retention coefficients greater than 75% regarding the entire phenolic content. This sequential approach also produced higher NF fluxes compared to direct NF processing. For extracts containing 50% and 70% ethanol, the fluxes rose by roughly seven and three times, respectively (Moraes et al., 2018). Various recovery components from different food products are mentioned in Table 16.2.

TABLE 16.2

Various Membrane Filtration and Modules Are Used in Multiple Areas of Food Industries to Recover Components

S. No	Types of Filtration	Membrane Module	Food Product	Recovered Component	Operating Parameters	References
1	Microfiltration (diafiltration)	Spiral wound module	Skim milk	Micellar casein concentration	Pressure: 34.5 kPa Diafiltration level: 150% Serum protein removed: 81.45%	(Marella et al., 2021)
2	Microfiltration (diafiltration)	Tubular ceramic module (cross-flow)	Cashew apple	Carotenoid extract	Pressure: 3 bar Flux: 100 Lh^{-1}m^{-2} Concentration: 0.2–0.3 gkg^{-1}	(Servent et al., 2020)
3	Ultrafiltration	Hollow module	Broccoli juice	Sulforaphane, malic acid, and citric acid	Permeate flux: 141 kg/m^{2}H Concentration: 64.7° Brix Pressure: 1.5 bar	(Yilmaz & Bagci, 2018)

(Continued)

TABLE 16.2 (*Continued*)
Various Membrane Filtration and Modules Are Used in Multiple Areas of Food Industries to Recover Components

S. No	Types of Filtration	Membrane Module	Food Product	Recovered Component	Operating Parameters	References
4	Ultrafiltration (two-stage)	Tubular module membrane	Pineapple	Bromelain purification	Stage 1 Pressure: 2 bar Velocity: 0.30 m/s Rejection: 76.9% (1.74 purity) Stage 2 Pressure: 1.5 bar Velocity: 0.24 m/s Purity: 1.4 fold Enzyme recovery: 96.1%	(Nor et al., 2017)
5	Nanofiltration	Flat sheet membrane	Silage juice	Lactic acid	Rejection: 71% Pressure: 32 bar Yield: 93%	(Cabrera-González et al., 2022)
6	Nanofiltration	Plate and frame	Red grape pomace extract	Phenolic compounds	Permeate flux: 50.58 L/m^2 Pressure: 20 bar Total phenolic content: 92%	(Arboleda Mejia et al., 2020)
7	Reverse osmosis	Spiral wound	Whey water recovery	Biofilm communities	Filamentous yeasts *Magnusiomyces spicifer* and *Saprochaete clavata, Sporopachydermia lactativora* and Gram-negative bacteria: *Pseudomonas sp., Raoultella sp., Escherichia sp.*, and *Enterobacter sp.*	(Vitzilaiou et al., 2019)
8	Reverse osmosis		Beetroot peel and flesh	Concentrate betalains	Pressure: 40 bar; Flow rate: 400 Lh^{-1} Temperature: 27°C. Total betalains compound: 4.16 and 5.52 for peel and flesh, respectively	(Zin et al., 2022)
9	Pervaporation	Spiral wound membrane	Wine	Reduction of alcohol content	Total flux: 0.073 kg/h/m^2 Final alcohol degree: 10.2–10.5 volume percentage Recovered compounds: hexanal, 2-phenyl ethanol, isoamyl alcohol	(Castro-Muñoz, 2019)

(Continued)

TABLE 16.2 (*Continued*)
Various Membrane Filtration and Modules Are Used in Multiple Areas of Food Industries to Recover Components

S. No	Types of Filtration	Membrane Module	Food Product	Recovered Component	Operating Parameters	References
10	Pervaporation	Plate and frame	Oyster boiling juice	Volatile organic compounds	Temperature – 30 and 40°C Pressure: 17450 Pa Aroma compounds: benzaldehyde and 1-octen-3-ol	(Soares et al., 2020)
11	Electrodialysis	Hollow fiber	Vinasses from grape marc or wine lees	Recovery of tartaric acid	Tartaric acid produced: 9.9 g/L and potassium hydroxide: 5.6 g/L Purity: 69.7 ± 1.3	(Vecino et al., 2020)
12	Electrodialysis	Hollow fiber	Lactose free milk (nanofiltration retentate)	Demineralization	Level of demineralization: 90% Time: 120 minutes Voltage: 5–15 volts Demineralized minerals: K^+, Na^+, Mg^+, Ca^+ Electrical conductivity: 5.52, 4.69, and 3.71, respectively	(Rasmussen et al., 2020)

Note: K^+ (potassium), Na^+ (sodium), Mg^+ (magnesium), Ca^+ (calcium).

16.4 CONCLUSION

Membrane processing techniques mentioned above can be operated via pressure and ion based. Change in the pore size, membrane module, and material of the filtration product varies; also, they will affect the product quality, quantity, retentate, and permeation. It provides a lot of potential changes compared to the conventional and most environmentally friendly, not only in the food industry. Other sectors, such as pharma, cosmetics, and oil, will benefit due to their easy operational and low-cost conditions and excellent product quality. Especially in the field of protein which is very sensitive to heat, it can be achieved at maximum yield by using the techniques mentioned earlier either in a single or integrated manner.

REFERENCES

Akin, O., Temelli, F., & Köseoğlu, S. (2012). Membrane applications in functional foods and nutraceuticals. *Critical Reviews in Food Science and Nutrition, 52*(4), 347–371. https://doi.org/10.1080/10408398.2010.500240

Al-Amshawee, S., Yunus, M. Y. B. M., Azoddein, A. A. M., Hassell, D. G., Dakhil, I. H., & Hasan, H. A. (2020). Electrodialysis desalination for water and wastewater: A review. *Chemical Engineering Journal, 380*. https://doi.org/10.1016/j.cej.2019.122231

Arboleda Mejia, J. A., Ricci, A., Figueiredo, A. S., Versari, A., Cassano, A., Parpinello, G. P., & de Pinho, M. N. (2020). Recovery of phenolic compounds from red grape pomace extract through nanofiltration membranes. *Foods, 9*(11). https://doi.org/10.3390/foods9111649

Arunkumar, A., & Etzel, M. R. (2018). Milk protein concentration using negatively charged ultrafiltration membranes. *Foods, 7*(9). https://doi.org/10.3390/foods7090134

Balster, J. (2014). Hollow fiber membrane module. *Encyclopedia of Membranes*, 1–2. https://doi. org/10.1007/978-3-642-40872-4_1583-2

Bazhenov, S. D., Bildyukevich, A. V., & Volkov, A. V. (2018). Gas-liquid hollow fiber membrane contactors for different applications. *Fibers*, 6(4). https://doi.org/10.3390/fib6040076

Bera, S. P., Godhaniya, M., & Kothari, C. (2022). Emerging and advanced membrane technology for wastewater treatment: A review. *Journal of Basic Microbiology*, 62(3–4), 245–259. https://doi.org/10.1002/ jobm.202100259

Bérot, S., Compoint, J. P., Larré, C., Malabat, C., & Guéguen, J. (2005). Large scale purification of rapeseed proteins (*Brassica napus* L.). *Journal of Chromatography B: Analytical Technologies in the Biomedical and Life Sciences*, 818(1 Spec. Iss.), 35–42. https://doi.org/10.1016/j.jchromb.2004.08.001

Bruggen, B. Van der. (2018). Microfiltration, ultrafiltration, nanofiltration, reverse osmosis, and forward osmosis. *Fundamental Modeling of Membrane Systems – Membrane and Process Performance*, 25–70. https://doi.org/10.1016/B978-0-12-813483-2.00002-2

Cabrera-González, M., Ahmed, A., Maamo, K., Salem, M., Jordan, C., & Harasek, M. (2022). Evaluation of nanofiltration membranes for pure lactic acid permeability. *Membranes*, 12(3). https://doi.org/10.3390/ membranes12030302

Cassano, A. (2016). Integrated membrane processes in the food industry. *Integrated Membrane Systems and Processes*, 35–60. https://doi.org/10.1002/9781118739167.ch3

Castro-Muñoz, R. (2019). Pervaporation-based membrane processes for the production of non-alcoholic beverages. *Journal of Food Science and Technology*, 56(5), 2333–2344. https://doi.org/10.1007/ s13197-019-03751-4

Castro-Muñoz, R., Ahmad, M. Z., & Cassano, A. (2021). Pervaporation-aided processes for the selective separation of Aromas, Fragrances and Essential (AFE) solutes from agro-food products and wastes. *Food Reviews International*. https://doi.org/10.1080/87559129.2021.1934008

Castro-Muñoz, R., Barragán-Huerta, B. E., Fíla, V., Denis, P. C., & Ruby-Figueroa, R. (2018). Current role of membrane technology: From the treatment of agro-industrial by-products up to the valorization of valuable compounds. *Waste and Biomass Valorization*, 9(4), 513–529. https://doi.org/10.1007/ s12649-017-0003-1

Castro-Muñoz, R., Boczkaj, G., Gontarek, E., Cassano, A., & Fíla, V. (2020). Membrane technologies assisting plant-based and agro-food by-products processing: A comprehensive review. *Trends in Food Science and Technology*, 95, 219–232. https://doi.org/10.1016/j.tifs.2019.12.003

Castro-Muñoz, R., Fíla, V., Barragán-Huerta, B. E., Yáñez-Fernández, J., Piña-Rosas, J. A., & Arboleda-Mejía, J. (2018). Processing of Xoconostle fruit (*Opuntia joconostle*) juice for improving its commercialization using membrane filtration. *Journal of Food Processing and Preservation*, 42(1). https://doi. org/10.1111/jfpp.13394

Chan, M., & Ng, S. (2018). Effect of membrane properties on contact angle. *AIP Conference Proceedings*, 2016. https://doi.org/10.1063/1.5055437

Charcosset, C. (2021). Classical and recent applications of membrane processes in the food industry. *Food Engineering Reviews*, 13(2), 322–343. https://doi.org/10.1007/s12393-020-09262-9

Conidi, C., Cassano, A., Caiazzo, F., & Drioli, E. (2017). Separation and purification of phenolic compounds from pomegranate juice by ultrafiltration and nanofiltration membranes. *Journal of Food Engineering*, 195, 1–13. https://doi.org/10.1016/j.jfoodeng.2016.09.017

Conidi, C., Drioli, E., & Cassano, A. (2018). Membrane-based agro-food production processes for polyphenol separation, purification and concentration. *Current Opinion in Food Science*, 23, 149–164. https://doi. org/10.1016/j.cofs.2017.10.009

Conidi, C., Drioli, E., & Cassano, A. (2020a). Biologically active compounds from goji (*Lycium barbarum* L.) leaves aqueous extracts: Purification and concentration by membrane processes. *Biomolecules*, 10(6), 1–18. https://doi.org/10.3390/biom10060935

Conidi, C., Drioli, E., & Cassano, A. (2020b). Perspective of membrane technology in pomegranate juice processing: A review. *Foods*, 9(7). https://doi.org/10.3390/foods9070889

Ezugbe, E. O., & Rathilal, S. (2020). Membrane technologies in wastewater treatment: A review. *Membranes*, 10(5), 89. https://doi.org/10.3390/membranes10050089

Gadlula, S., Ndlovu, L. N., Ndebele, N. R., & Ncube, L. K. (2019). Membrane technology in tannery wastewater management: A review. *Zimbabwe Journal of Science and Technology*, 14(1), 57–72.

Galanakis, C. M. (2018). *Separation of Functional Molecules in Food by Membrane Technology* (pp. 1–415). https://doi.org/10.1016/C2017-0-02808-2

Gao, L., Nguyen, K. D., & Utioh, A. C. (2001). Pilot scale recovery of proteins from a pea whey discharge by ultrafiltration. *LWT*, 34(3), 149–158. https://doi.org/10.1006/fstl.2000.0743

Giorno, L., Drioli, E., & Strathmann, H. (2015a). The principle of membrane contactors. *Encyclopedia of Membranes*, 1–6. https://doi.org/10.1007/978-3-642-40872-4_2230-1

Giorno, L., Drioli, E., & Strathmann, H. (2015b). The principle of nanofiltration (NF). *Encyclopedia of Membranes*, 1–5. https://doi.org/10.1007/978-3-642-40872-4_2234-1

Giorno, L., Piacentini, E., & Bazzarelli, F. (2015). The principle of microfiltration. *Encyclopedia of Membranes*, 1–4. https://doi.org/10.1007/978-3-642-40872-4_2233-1

Griep, E. R., Cheng, Y., & Moraru, C. I. (2018). Efficient removal of spores from skim milk using cold microfiltration: Spore size and surface property considerations. *Journal of Dairy Science*, *101*(11), 9703–9713. https://doi.org/10.3168/jds.2018-14888

Guiga, W., & Lameloise, M. L. (2019). Membrane separation in food processing. *Green Food Processing Techniques: Preservation, Transformation and Extraction*, 245–287. https://doi.org/10.1016/B978-0-12-815353-6.00009-4

Hamada, J. S. (2000). Ultrafiltration of partially hydrolyzed rice bran protein to recover value-added products. *JAOCS, Journal of the American Oil Chemists' Society*, *77*(7), 779–784. https://doi.org/10.1007/s11746-000-0124-3

Harlacher, T., & Wessling, M. (2015). Gas- gas separation by membranes. *Progress in Filtration and Separation*, 557–584. https://doi.org/10.1016/B978-0-12-384746-1.00013-6

Hasani, S. M. F., Sowayan, A. S., & Shakaib, M. (2019). The effect of spacer orientations on temperature polarization in a direct contact membrane distillation process using 3-d CFD modeling. *Arabian Journal for Science and Engineering*, *44*(12), 10269–10284. https://doi.org/10.1007/s13369-019-04089-x

Islam, M. S., Sultana, A., Saadat, A. H. M., Islam, M. S., Shammi, M., & Uddin, M. K. (2018). Desalination technologies for developing countries: A review. *Journal of Scientific Research*, *10*(1), 77–97. https://doi.org/10.3329/jsr.v10i1.33179

Jyoti, G., Keshav, A., & Anandkumar, J. (2015). Review on pervaporation: Theory, membrane performance, and application to intensification of esterification reaction. *Journal of Engineering (United Kingdom)*, *2015*. https://doi.org/10.1155/2015/927068

Kaminski, W., & Marszalek, J. (2006). Pervaporation for drying and dewatering. *Drying Technology*, *24*(7), 835–847. https://doi.org/10.1080/07373930600733994

Kertész, S., De Freitas, T. B., & Hodúr, C. (2014). Characterization of polymer membranes by contact angle goniometer. *Analecta Technica Szegedinensia*, *8*(2), 18–22. https://doi.org/10.14232/analecta.2014.2.18-22

Kieferle, I., & Kulozik, U. (2021). Model representation of flow patterns in the displacement of non-Newtonian products from spiral-wound membranes. *Journal of Membrane Science*, *624*. https://doi.org/10.1016/j.memsci.2020.118983

Kotsanopoulos, K. V., & Arvanitoyannis, I. S. (2015). Membrane Processing technology in the food industry: Food processing, wastewater treatment, and effects on physical, microbiological, organoleptic, and nutritional properties of foods. *Critical Reviews in Food Science and Nutrition*, *55*(9), 1147–1175. https://doi.org/10.1080/10408398.2012.685992

Lakra, R., & Basu, S. (2021). Recovery of protein and carbohydrate from whey water using integrated membrane processes. *Abstracts of International Conferences & Meetings*, *1*(6), 7–7. https://doi.org/10.5281/zenodo.5586351

Lawrence, N. D., Kentish, S. E., O'Connor, A. J., Barber, A. R., & Stevens, G. W. (2008). Microfiltration of skim milk using polymeric membranes for casein concentrate manufacture. *Separation and Purification Technology*, *60*(3), 237–244. https://doi.org/10.1016/j.seppur.2007.08.016

Li, X., Jiang, L., & Li, H. (2018). Application of ultrafiltration technology in water treatment. *IOP Conference Series: Earth and Environmental Science*, *186*(3). https://doi.org/10.1088/1755-1315/186/3/012009

Lin, S., & Elimelech, M. (2017). Kinetics and energetics trade-off in reverse osmosis desalination with different configurations. *Desalination*, *401*, 42–52. https://doi.org/10.1016/j.desal.2016.09.008

Lu, C., Bao, Y., & Huang, J. Y. (2021). Fouling in membrane filtration for juice processing. *Current Opinion in Food Science*, *42*, 76–85. https://doi.org/10.1016/j.cofs.2021.05.004

Marella, C., Sunkesula, V., Hammam, A. R. A., Kommineni, A., & Metzger, L. E. (2021). Optimization of spiral-wound microfiltration process parameters for the production of micellar casein concentrate. *Membranes*, *11*(9). https://doi.org/10.3390/membranes11090656

Mora, F., Pérez, K., Quezada, C., Herrera, C., Cassano, A., & Ruby-Figueroa, R. (2019). Impact of membrane pore size on the clarification performance of grape marc extract by microfiltration. *Membranes*, *9*(11). https://doi.org/10.3390/membranes9110146

Moraes, I. V. M. d., Rabelo, R. S., Pereira, J. A. de L., Hubinger, M. D., & Schmidt, F. L. (2018). Concentration of hydroalcoholic extracts of graviola (*Annona muricata* L.) pruning waste by ultra and nanofiltration: Recovery of bioactive compounds and prediction of energy consumption. *Journal of Cleaner Production*, *174*, 1412–1421. https://doi.org/10.1016/j.jclepro.2017.11.062

Morelli, R., Conidi, C., Tundis, R., Loizzo, M. R., D'Avella, M., Timpone, R., & Cassano, A. (2022). Production of high-quality red fruit juices by athermal membrane processes. *Molecules*, *27*(21). https://doi.org/10.3390/molecules27217435

My, N. T. T., Nhi, V. T. Y., & Thanh, B. X. (2018). Factors affecting membrane distillation process for seawater desalination. *Journal of Applied Membrane Science & Technology*, *22*(1).

Nawaz, H., Shi, J., Mittal, G. S., & Kakuda, Y. (2006). Extraction of polyphenols from grape seeds and concentration by ultrafiltration. *Separation and Purification Technology*, *48*(2), 176–181. https://doi.org/10.1016/j.seppur.2005.07.006

Nissar, N., Hameed, O., & Nazir, F. (2018). Application of membrane technology in food processing industries: A review. *International Journal of Advance Research in Science and Engineering*, *07*(04), 2106–2114.

Nor, M. Z. M., Ramchandran, L., Duke, M., & Vasiljevic, T. (2017). Integrated ultrafiltration process for the recovery of bromelain from pineapple waste mixture. *Journal of Food Process Engineering*, *40*(3). https://doi.org/10.1111/jfpe.12492

Nourbakhsh, H., Alemi, A., Emam-Djomeh, Z., & Mirsaeedghazi, H. (2014). Effect of processing parameters on fouling resistances during microfiltration of red plum and watermelon juices: A comparative study. *Journal of Food Science and Technology*, *51*(1), 168–172. https://doi.org/10.1007/s13197-011-0472-3

Olaru, M., Bordianu, I., & Simionescu, B. C. (2021). Polymers in membrane science. Systems Membranes-complex roadmaps towards functional devices and coupled processes SYSMEM, Editura PRINTECH Bucharest, ROMANIA, 1, 191-240. https://www.researchgate.net/profile/Gheorghe-Nechifor/publication/314257931_9_BCS-PPI/data/58be3c5ba6fdcc2d14eb58c3/9-BCS-PPI.pdf

Patel, D., Mudgal, A., Patel, V., & Patel, J. (2021). Water desalination and wastewater reuse using integrated reverse osmosis and forward osmosis system. *IOP Conference Series: Materials Science and Engineering*, *1146*(1), 012029. https://doi.org/10.1088/1757-899x/1146/1/012029

Pouliot, Y. (2008). Membrane processes in dairy technology-From a simple idea to worldwide panacea. *International Dairy Journal*, *18*(7), 735–740. https://doi.org/10.1016/j.idairyj.2008.03.005

Qaid, S., Zait, M., EL Kacemi, K., EL Midaoui, A., EL Hajji, H., & Taky, M. (2017). Ultrafiltration for clarification of Valencia orange juice: Comparison of two flat sheet membranes on quality of juice production. *Journal of Materials and Environmental Science*, *8*(4), 1186–1194.

Rasmussen, P. K., Suwal, S., van den Berg, F. W. J., Yazdi, S. R., & Ahrné, L. (2020). Valorization of side-streams from lactose-free milk production by electrodialysis. *Innovative Food Science and Emerging Technologies*, *62*. https://doi.org/10.1016/j.ifset.2020.102337

Rudolph, G., Virtanen, T., Ferrando, M., Güell, C., Lipnizki, F., & Kallioinen, M. (2019). A review of in situ real-time monitoring techniques for membrane fouling in the biotechnology, biorefinery and food sectors. *Journal of Membrane Science*, *588*. https://doi.org/10.1016/j.memsci.2019.117221

Schopf, R., Schmidt, F., Linner, J., & Kulozik, U. (2021). Comparative assessment of tubular ceramic, spiral wound, and hollow fiber membrane microfiltration module systems for milk protein fractionation. *Foods*, *10*(4). https://doi.org/10.3390/foods10040692

Servent, A., Abreu, F. A. P., Dhuique-Mayer, C., Belleville, M. P., & Dornier, M. (2020). Concentration and purification by crossflow microfiltration with diafiltration of carotenoids from a by-product of cashew apple juice processing. *Innovative Food Science and Emerging Technologies*, *66*. https://doi.org/10.1016/j.ifset.2020.102519

Shah, M., Rodriguez-Couto, S., & Biswas, J. K. (2021). *Development in Wastewater Treatment Research and Processes: Removal of Emerging Contaminants from Wastewater through Bio-Nanotechnology* (pp. 1–698). https://doi.org/10.1016/B978-0-323-85583-9.00037-5

Shahir, S. (2021). Alternative thermal processing technique for liquid foods-membrane processing. *International Journal of Plant & Soil Science*, 68–79. https://doi.org/10.9734/ijpss/2021/v33i2430752

Soares, L. S., Vieira, A. C. F., Fidler, F., Fritz, A. R. M., & Di Luccio, M. (2020). Pervaporation as an alternative for adding value to residues of oyster (*Crassostrea gigas*) processing. *Separation and Purification Technology*, *232*. https://doi.org/10.1016/j.seppur.2019.115968

Song, Y., Pan, F., Li, Y., Quan, K., & Jiang, Z. (2019). Mass transport mechanisms within pervaporation membranes. *Frontiers of Chemical Science and Engineering*, *13*(3), 458–474. https://doi.org/10.1007/s11705-018-1780-1

Suhalim, N. S., Kasim, N., Mahmoudi, E., Shamsudin, I. J., Mohammad, A. W., Zuki, F. M., & Jamari, N. L. A. (2022). Rejection mechanism of ionic solute removal by nanofiltration membranes: An overview. *Nanomaterials*, *12*(3). https://doi.org/10.3390/nano12030437

Sun, X., Dang, G., Ding, X., Shen, C., Liu, G., Zuo, C., Chen, X., Xing, W., & Jin, W. (2020). Production of alcohol-free wine and grape spirit by pervaporation membrane technology. *Food and Bioproducts Processing*, *123*, 262–273. https://doi.org/10.1016/j.fbp.2020.07.006

Tamires Vitor Pereira, D., Vollet Marson, G., Fernández Barbero, G., Gadioli Tarone, A., Baú Betim Cazarin, C., Dupas Hubinger, M., & Martínez, J. (2020). Concentration of bioactive compounds from grape marc using pressurized liquid extraction followed by integrated membrane processes. *Separation and Purification Technology, 250.* https://doi.org/10.1016/j.seppur.2020.117206

Tsui, E. M., & Cheryan, M. (2007). Membrane processing of xanthophylls in ethanol extracts of corn. *Journal of Food Engineering, 83*(4), 590–595. https://doi.org/10.1016/j.jfoodeng.2007.03.041

Ullah, A., Starov, V. M., Naeem, M., Holdich, R. G., & Semenov, S. (2013). Filtration of suspensions using slit pore membranes. *Separation and Purification Technology, 103*, 180–186. https://doi.org/10.1016/j.seppur.2012.10.038

Urošević, T., Povrenović, D., Vukosavljević, P., Urošević, I., & Stevanović, S. (2017). Recent developments in microfiltration and ultrafiltration of fruit juices. *Food and Bioproducts Processing, 106*, 147–161. https://doi.org/10.1016/j.fbp.2017.09.009

Utoro, P. A. R., Sukoyo, A., Sandra, S., Izza, N., Dewi, S. R., & Wibisono, Y. (2019). High-throughput microfiltration membranes with natural biofouling reducer agent for food processing. *Processes, 7*(1). https://doi.org/10.3390/pr7010001

Vecino, X., Reig, M., Gibert, O., Valderrama, C., & Cortina, J. L. (2020). Integration of monopolar and bipolar electrodialysis processes for tartaric acid recovery from residues of the winery industry. *ACS Sustainable Chemistry and Engineering, 8*(35), 13387–13399. https://doi.org/10.1021/acssuschemeng.0c04166

Vieira, A. H., Balthazar, C. F., Guimaraes, J. T., Rocha, R. S., Pagani, M. M., Esmerino, E. A., Silva, M. C., Raices, R. S. L., Tonon, R. V., Cabral, L. M. C., Walter, E. H. M., Freitas, M. Q., & Cruz, A. G. (2020). Advantages of microfiltration processing of goat whey orange juice beverage. *Food Research International, 132.* https://doi.org/10.1016/j.foodres.2020.109060

Virtanen, T., Rudolph, G., Lopatina, A., Al-Rudainy, B., Schagerlöf, H., Puro, L., Kallioinen, M., & Lipnizki, F. (2020). Analysis of membrane fouling by Brunauer-Emmet-Teller nitrogen adsorption/desorption technique. *Scientific Reports, 10*(1). https://doi.org/10.1038/s41598-020-59994-1

Vitzilaiou, E., Stoica, I. M., & Knøchel, S. (2019). Microbial biofilm communities on reverse osmosis membranes in whey water processing before and after cleaning. *Journal of Membrane Science, 587.* https://doi.org/10.1016/j.memsci.2019.117174

Vu, T., LeBlanc, J., & Chou, C. C. (2020). Clarification of sugarcane juice by ultrafiltration membrane: Toward the direct production of refined cane sugar. *Journal of Food Engineering, 264.* https://doi.org/10.1016/j.jfoodeng.2019.07.029

Wang, Y. N., Wang, R., Li, W., & Tang, C. Y. (2017). Whey recovery using forward osmosis – evaluating the factors limiting the flux performance. *Journal of Membrane Science, 533*, 179–189. https://doi.org/10.1016/J.MEMSCI.2017.03.047

Wen-Qiong, W., Yun-Chao, W., Xiao-Feng, Z., Rui-Xia, G., & Mao-Lin, L. (2019). Whey protein membrane processing methods and membrane fouling mechanism analysis. *Food Chemistry, 289*, 468–481. https://doi.org/10.1016/j.foodchem.2019.03.086

Woo, C. H. (2018). Research trend of membranes for water treatment by analysis of patents and papers' publications. *Desalination and Water Treatment.* https://doi.org/10.5772/intechopen.76694

Xie, Z., Li, N., Wang, Q., & Bolto, B. (2018). Desalination by pervaporation. *Emerging Technologies for Sustainable Desalination Handbook*, 205–226. https://doi.org/10.1016/B978-0-12-815818-0.00006-0

Xue, Y. L., Zhang, R., Cao, B., & Li, P. (2021). Tubular membranes and modules. *Hollow Fiber Membranes*, 431–448. https://doi.org/10.1016/B978-0-12-821876-1.00011-1

Yilmaz, E., & Bagci, P. O. (2018). Production of phytotherapeutics from broccoli juice by integrated membrane processes. *Food Chemistry, 242*, 264–271. https://doi.org/10.1016/j.foodchem.2017.09.056

Zhu, H., Liu, G., & Jin, W. (2020). Recent progress in separation membranes and their fermentation coupled processes for biobutanol recovery. *Energy and Fuels, 34*(10), 11962–11975. https://doi.org/10.1021/acs.energyfuels.0c02680

Zin, M. M., Alsobh, A., Nath, A., Csighy, A., & Bánvölgyi, S. (2022). Concentrations of beetroot (*Beta vulgaris* L.) peel and flesh extracts by reverse osmosis membrane. *Applied Sciences (Switzerland), 12*(13). https://doi.org/10.3390/app12136360

Zin, M. M., Márki, E., & Bánvölgyi, S. (2020). Evaluation of reverse osmosis membranes in concentration of beetroot peel extract. *Periodica Polytechnica Chemical Engineering, 64*(3), 340–348. https://doi.org/10.3311/PPch.15040

17 Low-Temperature Drying Technologies

Pramila Murugesan, J. A. Moses,
and C. Anandharamakrishnan

17.1 INTRODUCTION

Drying is one of the traditional methods for the preservation of food by eliminating moisture (dewatering) from unprocessed or processed food products and it helps to maintain food quality and stability and inhibits contamination and spoilage during storage as well as other decomposition reactions including hydrolysis, non-enzymatic browning, and lipid oxidation. Dewatering is the process of eliminating water from a product without changing its phase. According to Vega-Mercado et al. (2001), dewatering includes removing water from the product without causing any phase changes, therefore, the liquid water is removed using mechanical or chemical pressures (Vega-Mercado et al., 2001). Conduction, convection, and radiation-based drying techniques have been adapted for decades. The most popular method of food dehydration is convective drying with hot air due to its low cost and simplicity. However, the high-temperature drying routes could lead to changes in product quality by altering the nutritional content, color, flavor, defects in textures, and decomposition of thermolabile compounds. Hence, the selection of effective drying routes for the removal of free moisture from the product has become significant. Low-temperature drying techniques are considered a potential drying technique for food products. Low-temperature drying is the process of removing water at temperatures lower than those seen in a typical room, such as below 20°C.

To date, different low-temperature drying methods have been adapted for drying food products such as vacuum, osmotic, freeze, spray freeze, conductive, electrohydrodynamic, low temperature-low humidity, low temperature-high velocity fluidized bed, and low-temperature desiccant-based drying. Drying at a low-temperature alone consumes more energy and exhibits a low dehydration rate when dried for a prolonged period. In addition, low-temperature drying takes longer than conventional drying because the mass transfer rate also slows down when the air temperature is reduced (Jariyawaranugoon, 2015). The combination of the low-temperature drying process with other conventional methods is a promising alternative promoting the drying rate, reducing energy utilization without altering the quality of the dried sample. For instance, Szadzińska et al. (2019) demonstrated that the hybridization of convective ultrasound drying with microwave drying could improve the product quality of red beetroot while reducing overall drying time and energy usage. This chapter discusses the general characteristics (concepts, influencing factors) of the low-temperature drying process and its combination with other conventional drying techniques as they relate to the drying and dehydration of solid product material.

17.2 NEED AND IMPORTANCE OF LOW-TEMPERATURE DRYING

Conventional drying techniques such as solar and hot air drying could affect the product quality by altering the nutritional content, color, flavor, and texture. For instance, the high-temperature drying process leads to changes in the color and aromatic profile of the product material. Moreover, increased hot air temperatures promoted starch gelatinization, which lowered relative crystallinity.

The food market is seeing an increase in demand for high-quality goods that retain fresh product sensory and nutritional qualities. Low-temperature drying can be a viable solution in this regard. The low-temperature drying techniques have the following benefits as

- Preservation of food quality (nutritional)
- Reduction in microbial count
- Limited scope for oxidation processes
- Minimal impact on product's sensory attributes
- Reduces heat-sensitive components' susceptibility to destruction

17.3 TYPES OF LOW-TEMPERATURE DRYING

17.3.1 VACUUM DRYING

17.3.1.1 Concept of the Vacuum Drying Process

Vacuum drying is the unit operation, removing the moisture content from wet products in a reduced pressure, low-temperature, and oxygen-deficient environment, which reduces the amount of heat required for quick drying and lowers the oxidation potential. There are two main steps involved in the vacuum drying process: (1) the process of heat transfer: transferring heat from the source to the material to encourage moisture evaporation. In the meantime, vaporization has little effect on the material temperature. (2) Process of mass transfer: as the liquid water inside the material migrates to the drying chamber, the water vapor migrates there as well and is expelled from the drying chamber by the vacuum system (Egas-Astudillo et al., 2010). The pressure is reduced around the surface of the product to be dried by a vacuum pump. This lowers the water boiling within that product, increasing the evaporation rate of water at a constant temperature. Thus promoting the drying rate of the products. Besides, the mass transfer resistance at the product surface is diminished due to an increase in the hydraulic conductivity of the material in vacuum conditions. Moreover, the porosity of products has increased during the vacuum process, which results in higher rehydration and less shrinkage (Porciuncula et al., 2016).

17.3.1.2 Components and Process Considerations

Generally, cast iron or stainless steel was used to construct the vacuum dryer. The basic components involved in a vacuum dryer are trays, shelves, condenser, vacuum pump, and water trap (Figure 17.1). To enhance the surface area for heat conduction, the oven is divided into hollow trays. The wet material is placed on heated shelves, the vacuum pump helps eliminate the water from the product material, and the eliminated moisture has been condensed using the condenser. Typically, heat is provided by running hot water or steam through hollow shelves. Generally, off-shelf pumps are used for the vacuum drying process. A roots blower, a positive displacement air pump, may be utilized in addition to the primary vacuum pump to boost pumping efficiency and speed in lower pressure levels. The roots blower is located on the vacuum pump inlet side and employs positive displacement of air to raise the air pressure at the vacuum pump inlet, which improves the effectiveness of the vacuum pump. The water trap tank is adopted into a vacuum drying system as a means of shielding the vacuum pumps from destruction. This allows the air-water mixture in the vacuum lines to pass through the water trap tank inlet while the water trap tank is valved into the vacuum path. While permitting the gases to pass through the exhaust port and on to the vacuum pump, liquid water is deposited in the tank. The boiling point of water and pressure maintained in the vacuum drying process are 25–30°C and 0.0296–0.059 atm, respectively. The vacuum drying process necessitates a residence time of 20–100 hours. As a result, the majority of the food's nutritional and sensory properties are preserved.

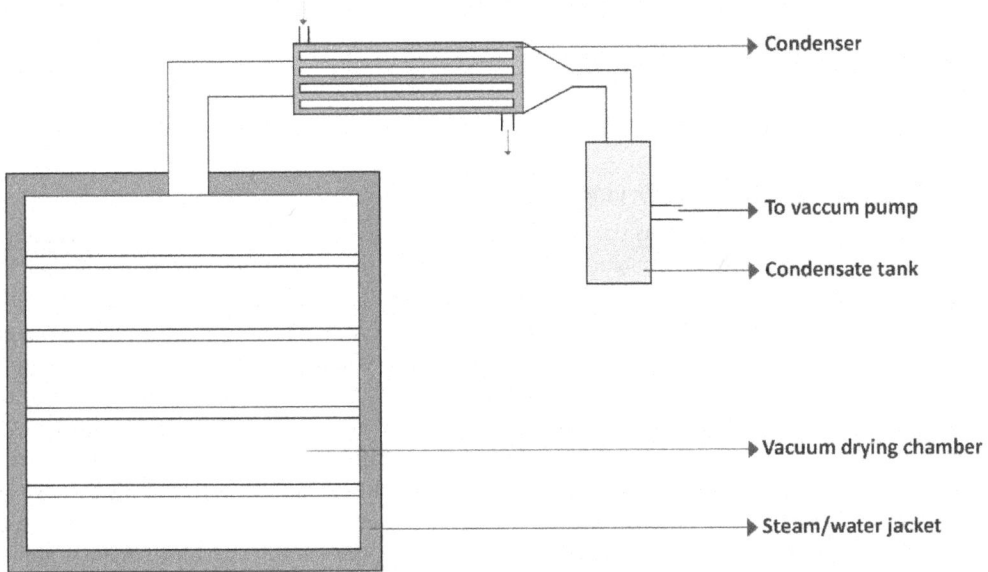

FIGURE 17.1   Schematic illustration of vacuum drying setup.

17.3.1.3   Parameters Influencing the Vacuum Drying Process

The heating plate temperature, pressure, water content in the product, and exposed product thickness are the main parameters that influence the drying kinetics. Higher heating plate temperatures resulted in faster drying times or a lower dewatering rate. The rapid decrease in moisture content at higher temperatures is mainly attributed to the increase in temperature deviation between the heating element and samples (Devahastin et al., 2004). Jaya and Das observe that the rate of drying is higher at 75°C as compared with 65°C. In another study, the water removal rates for drying honey at 40, 45, and 50°C were 0.14%, 0.16%, and 0.28%, respectively. Hence, to prevent the quality of the honey from deteriorating throughout the dehydration process, dewatering honey at 45°C was strongly advised (Halim et al., 2021). No significant changes in moisture content and drying rate at different pressures ranging from 10 kPa to 20 kPa as reported by Soliman et al. (2017). Due to the evaporation of the remaining free water content, the dehydration rate increases with an increase in the water content of the product. Lower water content required more energy to evaporate the bound water, which caused the dewatering rate to slow down (Vaxelaire & Puiggali, 2002). The increase in the thickness of the product to be dried could increase the drying rate and reduce color alterations (Das, 2013).

17.3.1.4   Assisted Vacuum Drying Techniques

The rate and efficiency of vacuum drying of wet products can be enhanced by combining them with other methods. Microwave-vacuum drying is the combination of microwave drying and vacuum drying. Water evaporation is caused by the action of microwave energy absorbed by a product in a high-frequency electric field on the water in the product as well as the reduced water boiling temperature in the low-pressure environment. It is a low-temperature process, with less drying time, more reduction in the microbial count, and reduces the unwanted changes in the chemical structure of dried products (Dak & Pareek, 2014). Microwave vacuum drying systems can be classified as static and rotary systems. The product thickness is the main key parameter in a static system. The temperature would be uneven if the product had numerous layers made up of different components because microwave radiation would take a while to penetrate each layer. Concerns with the static system include uneven drying of the product, inadequate rehydration, and color changes. The rotary system-based microwave drying provides uniform distribution of temperature and avoids electric arc damage (Kumar & Shrivastava, 2017).

17.3.2 OSMOTIC DRYING

17.3.2.1 Concept of Osmotic Drying

Osmotic dehydration also known as the "dewatering impregnation soaking process" is a technique that involves reducing the water content of the product to be dried by immersing it in a hypertonic solution of sugar, salt, or both with high osmotic pressure. The hypertonic solution drives water withdrawal from the product into the osmo-active solution because it has a larger osmotic pressure but less water activity. Water and some natural solutes of product passed into the osmotic solution, while a certain amount of the osmotic solute penetrates the food. The process is accelerated by the concentration difference between the interstitial fluid and osmotic solutions (Agnelli et al., 2005). The osmotic dehydration process could alter the nutritional value of dried product materials, improve the overall quality of finished products, and demonstrate the potential for energy savings (Shi & Maguer, 2002). Furthermore, it reduces the water activity of the dried sample, which reduces microbial proliferation (Rastogi et al., 2014). Papaya, grapes, citrus fruits, bananas, apples, mangoes, and other fruits and vegetables have all been subjected to osmotic dehydration (Ahmed et al., 2016).

The osmotic dehydration process exhibits three kinds of counterflow mass transfer phenomena. (1) The flow of water from the product into the osmotic liquid, initially more quickly and then more slowly; (2) the instantaneous movement of water-soluble solutes from the hypertonic solution into the product. This enables the addition of the selected quantity of an active ingredient, a preservative, a nutritional interest, a solute, or an enhancement of the product's sensory quality; and (3) the migration of solutes (minerals, volatiles, reducing sugar, organic acids, and vitamins) (Ciurzyńska et al., 2016).

17.3.2.2 Process Considerations

The substances used for the osmotic dehydrations are sodium chloride, glucose, corn syrup, sorbitol, and sugar (glucose, fructose, and sucrose) solutions. During osmotic dehydration, the low molecular weight of the solute migrates into the osmotic solution. The mass transfer process during osmotic dehydration is conceptually represented in Figure 17.2. The first layer of cells in a cellular material that is submerged in an osmotic solution starts to lose water as a result of the concentration gradient between the cells and the hypertonic solution. The first layer cells begin to contract as a result of

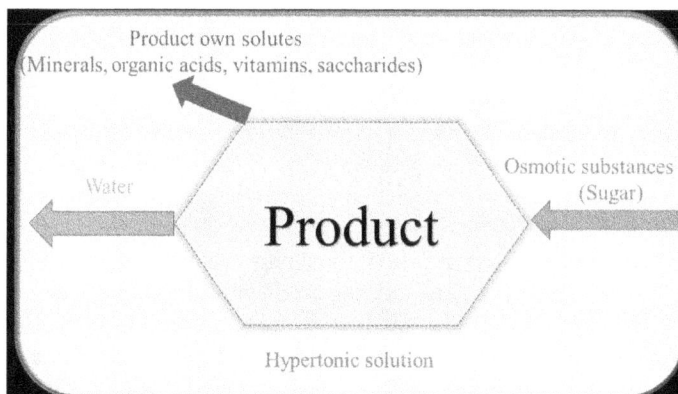

FIGURE 17.2   Osmotic dehydration principles and relative mass flow of water and solute.

the water being pumped to them by the second layer cells. After prolonged solid-liquid contact, the product loses its water and the mass transfer mechanism tends to equilibrate. Nearly 10–70% of the water in the product can be eliminated during the dehydration process. Osmotic dehydration is an energy-efficient route, and it does not change the quality (color, texture, and flavor) of food products. All volatile compounds in the frozen osmotic-treated kiwi fruits decreased considerably after 30 days of being frozen osmotic-treated kiwi fruit reported by Talens et al. (2003). Moreover, osmotic drying causes changes in the cell organization and structure (Talens et al., 2003).

17.3.2.3 Factors Influencing the Osmotic Drying Process

The kinetics of mass transfer in osmotic drying is affected by several factors, including pretreatment, product-related (size and shape), osmotic solution, and other environmental-related factors (temperature and time). The permeability of the cell membrane has a significant impact on the rate of dehydration. More membrane permeability will cause osmotic dehydration to occur more quickly. Pretreatment conditions before the osmotic dehydration process influence the product's inherent integrity, which improve the mass transfer process. Different pretreatment techniques have been adapted, including blanching, peeling, coating, ultrasound treatment, freezing, applying high-intensity pulses, an electric field, ohmic heating, and vacuum impregnation (Allali et al., 2009). The product size and shape have an impact on the kinetics of mass transfer. Due to the variations in surface area to volume ratio and diffusion length between the solutes and product, the water loss and solid gain decrease with increased product size (Rastogi et al., 2002). Larger solids will dehydrate more slowly because there is less surface area available for solid exchange and a longer diffusion path, which causes the solids to enter the sample more slowly. Moreover, the porosity of the product could significantly affect the ratio of rehydration, rate of mass transfer, and shrinkage occurrence. Falade and Igbeka (2007) reported that vegetable and fruit slices in sucrose syrup release water at a rate mostly dependent on their size. The structure of the cell membrane and chemical composition of the product also affect the mass transfer kinetics (Falade et al., 2007).

The osmotic process is also impacted by the physicochemical characteristics of the solutes used. As variations in the efficiency of dehydration are mostly caused by solubility in water, the ionic state, and variations in the solute's molecular weight. The solute penetration rate is proportional to solution concentration and inversely proportional to solute size. Due to the rapid speed of molecule penetration in low molecular mass saccharides, sugar absorption is favored, leading to the major outcome of this process being solid enrichment rather than dehydration. Kowalska et al. (2008) reported that low molecular weight substances showed higher diffusion and solid gain as compared with high molecular weight substances. The concentration of osmotic substances strongly affects the dehydration kinetics, solid gain, and equilibrium water content. The rise in osmotic solution concentration could increase the mass transfer rate resistance in the solution at the sample surface. In addition, the osmotic solution becomes more viscous as the concentration increases, which hinders the penetration of the solute. Giraldo et al. (2003) noted that the concentration of sucrose had a significant impact on the mass transport characteristics of fruit tissue in osmotic drying. The liquid phase effective diffusion coefficient of fruit increased with a drop in the sugar concentration. The utilization of viscous sugar solutions with highly concentrated sugar causes significant issues such as floating food pieces, which prevent food material from making contact with the osmotic solution and lower mass transfer rates. The weight loss and dehydration compounds of the process are promoted by increasing the molar mass of the solute, while equilibrium and drying rates rise with rising osmotic pressure.

Moreover, the mixer of solute solution exhibits more diminished water activity of the product than a single solute. Furthermore, the acidification of solutions accelerates water removal by changing the physicochemical properties of products. The rate of dehydration and diffusion of solute are significantly affected by environmental conditions. Treatment time and temperature are the main factors that influence water loss and solute gain. Due to the elevations in diffusion rate, the water removal rate increases with increasing temperatures, while temperature has less impact on solid gain.

Temperature plays a crucial role during the drying process. The mass exchange process is accelerated when the temperature rises resulting in a decrease in the viscosity of the solution. The rise in the osmotic pressure gradient was seen along with a rise in temperature, which boosted dehydration. Higher temperatures are likely to promote water loss by swelling and plasticizing cell membranes. Besides, the dehydration time could be greatly reduced at a temperature in the range of 20°C to 50°C and a further rise in the temperature leads to the loss of cell membrane permeability. Torreggiani (1993) suggested that the osmotic process is more effective at room temperature. Phisut (2012) demonstrated that the viscosity of the osmotic solution is reduced as the temperature rises and decreases in the exterior resistance to the mass transfer rate at the surface of the product. This allows for high rates of solute diffusion into the product as well as the outflow of water from the product. Time affects both the water content loss and the increase in soluble salt content. The longer drying time could increase the mass transfer rate and provide highly enhanced moisture loss and solid gain (Phisut, 2012).

The agitation of high viscous hypertonic solution leads to an improved dehydration rate due to the minimization of mass transfer resistance in the liquid side and continuous driving force. Higher water loss may have indirectly affected the solute concentration inside the solid particle, causing the agitation-induced drop in the solid gain rate during the prolonged osmosis time (Mavroudis et al., 1998). The mass ratio between the hypertonic solution and product displays the influence on the dehydration process. A larger mass ratio can be employed to prevent the medium from being significantly diluted by the water absorption and the solute loss from the sample, resulting in a decrease in the osmotic driving force during osmotic dehydration. With a rise in the osmotic solution and product mass ratio to a certain level, both solid gain and water loss increased and then declined later (Yadav & Singh, 2014).

17.3.2.4 Assisted Osmotic Drying Techniques

Osmotic dehydration decreases the water activity of the product thereby extending its shelf life. Since microbial growth is partially hindered, this approach is frequently used in conjunction with other methods. The main benefits are higher product quality and cost savings. Moreover, osmotically pretreated frozen samples showed noticeably enhanced firmness for a longer storage period. The osmotic pretreatment resulted in samples that were thawed with the most ideal sensory qualities, including appealing appearance, acceptable texture, and pleasant taste (Agnelli et al., 2005). Before osmotic dehydration, freezing the banana helped shorten the soaking period by at least 1 h as compared to the untreated sample. Banana slices that had been air dried were of higher quality in terms of color, texture, and shrinkage after being frozen, osmotically dehydrated, and then dried at 70°C (Jariyawaranugoon, 2015).

A developing method called osmotic dehydration in a microwave setting offers significant promise for improving product quality, limiting the solids gain, and increasing the rate of moisture loss from the product. The microwaves create the heat in the product to be dried by friction owing to the polarization of ionic salts and the rotation of dipolar molecules, which attempt to align themselves with the microwave field. In the context of microwave-assisted osmotic dehydration, there can be rapid and differential heat generation within the product because of microwave absorption. This results in a pressure buildup within the product thereby accelerating moisture loss (Azarpazhooh & Ramaswamy, 2012).

The use of osmotic drying as a pretreatment step in the convective air drying process allows for a reduction in the energy required for heating and product moisture evaporation. For instance, when the pulsed fluid bed or convective vibrating fluid bed drying techniques are used in conjunction with the osmotic dehydration process, an energy savings of around 2150 kJ per kg of fresh berries can be attained (Kudra, 2009). In another study, the osmotic dehydration kinetics and the impact of sucrose migration on the thermal air-drying of vegetable slices were studied by Garcia et al. (2007), who suggested that the water diffusion coefficient increases with increasing drying temperature.

Vacuum osmotic drying promotes the rate of mass transfer. The rate of mass transfer is improved as the pores are loaded with a hypertonic solution and the gas contained in them expands and escapes with a decrease in pressure. The ultrasound could also influence the mass transfer rate during the osmotic drying process. The continuous compression and release of ultrasonic waves develop microchannels in the porous products which may accelerate the rate of mass transfer. Additionally, there is the development of bubbles in the liquid via cavitation that have the potential to explode, collapse, and generate localized pressure. As a result, degassing is accelerated and completed, resulting in more diffusion in the osmotic process. Moreover, osmotic dehydration under vacuum conditions protects against enzymatic and osmotic stress (Torreggiani, 1993).

17.3.3 Freeze Drying

17.3.3.1 Concept of Freeze Drying

Freeze drying is an effective route for preserving a wide variety of heat-sensitive products since it prevents shrinkage. Moreover, it also produces high porosity products with unaltered flavor, aroma, taste, and improved rehydration properties (Krokida et al., 1998). The freeze-drying process consists of three steps such as freezing (the solution is cooled until ice nucleation begins), primary drying (removal of frozen water under defined temperature and pressure), and secondary drying (removal of moisture from the solute phase by the desorption process) (Abla & Mehanna, 2022). Freeze-dried material possesses more bioactive substances and antioxidant properties (Chumroenphat et al., 2021).

17.3.3.2 Components and Process Considerations

Typically, freeze dryers have four different components: a drying chamber, condenser, heat source, and vacuum pump, as shown in Figure 17.3. The frozen samples are put inside the drying chamber. The chamber is vacuum-tight and the inner shelves have temperature control. Additionally, a sufficient vacuum pump is needed to eliminate the non-condensable gases from the drying platform and attain the desired vacuum. The latent heat of sublimation is provided by the heating source. After the frozen product has already reached a high vacuum, the heating source releases the latent heat of sublimation, and the temperature of heating source is varied between −30°C and 150°C (243.15 K and 423.15 K (Garcia-Amezquita et al., 2015). Lastly, the condenser is used to collect the vapor discharged from water when ice is sublimated together with the finished product. It must have enough cooling power and surface area to completely freeze all the steam produced throughout the sublimation process. Water vapor is converted into ice crystals as it comes into contact with the condenser surface, which releases energy; the crystals are then eliminated from the system. In commercial freeze dryers, condensers typically operate at a temperature of roughly 208.15 K. Freezing and drying are the two phases occurring during the freeze-drying process, as illustrated in Table 17.1.

17.3.3.3 Parameters Influencing the Drying Process

Different types of operational variables could influence freeze-drying efficiency. The reduction in shelf temperature at constant pressure could reduce the product temperature and sublimation rate (Assegehegn et al., 2020). The samples that were freeze-dried at ambient temperature (shelf temperature) resulted in a larger mean pore size as compared with the highest temperature (40°C). The alteration in chamber pressure could significantly influence the sublimation process. The ice sublimation is large at low pressure. On the other hand, ice sublimation could not occur at low chamber pressure unless the product temperature is too high (Assegehegn et al., 2020). The porosity of the dried product increases with decreasing pressure and diminishes when the pressure is increased (Garcia-Amezquita et al., 2015) (Figure 17.3b). At low applying pressure the bulk density is reduced (Krokida et al., 1998).

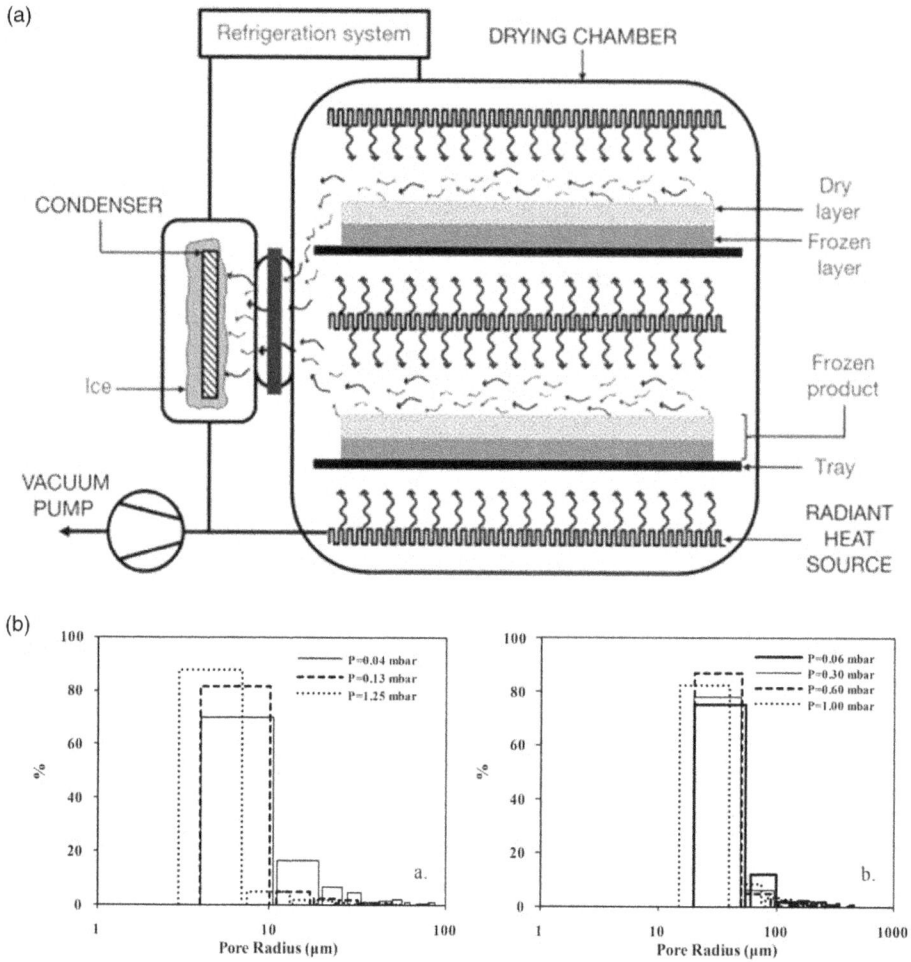

FIGURE 17.3   (a) basic components of typical freeze drier system (Garcia-Amezquita et al., 2015); (b) effect of applied pressure on pore size distribution of freeze-dried products (rice and potato) (Oikonomopoulou et al., 2011).

TABLE 17.1
Phases of the Freeze-Drying Process

Phase	Stages	Functions
Freezing stage	Cooling	Liquid-to-solid transition
		• To completely freeze any mobile water that was initially present in the food.
		• Notably, the temperature of the food should be below the glass transition temperature for moisture level.
Drying	Primary drying	Solid-to-gas phase transition
		Nearly 90% of free water and some bound water could be removed from the food by the sublimation process.
	Secondary drying	Liquid-to-gas phase transition
		The unfrozen bound water could be removed from the product via the desorption process.

17.3.3.4 Assisted Freeze-Drying Techniques

Conventional freeze drying requires a long processing time because of conductive or radiant heat applied in this process which increases the overall process cost. The combination of freeze drying with different drying technologies (infrared, ultrasonic, and microwave) has been investigated. The use of infrared and freeze dryers together has shown the effectiveness of drying the products quickly while maintaining high quality. The infrared radiation permeates the wet porous product surface layer, causing molecular vibration and supplying heat energy for a faster drying rate during freeze drying (Chakraborty et al., 2011). Utilizing the power ultrasound before freeze drying can also enhance drying by improving mass and heat transfer owing to the introduction of mechanical energy into the medium. The ultrasonic waves cause the compression and decompression of solid materials as well as the formation of microchannels that can facilitate the passage of water from an inner region to a solid surface. Notably, the use of ultrasonic waves does not create any significant rise in temperature (Cárcel et al., 2012). Microwave-assisted freeze drying could reduce the production cost, enhance the quality of final products, and reduce the energy cost. Microwave-assisted freeze drying can be applied in two distinct forms. The first one is freeze drying and assisted microwave/vacuum microwave drying in two sequentially distinct stages consisting of freeze drying and microwave or vacuum microwave drying stages. The second form is freeze drying with concurrently assisted microwave application. The heat source of this process is the microwave field which provides the necessary sublimation heat for freeze drying, which heats the product volumetrically under vacuum conditions (Bhambhani et al., 2021; Duan et al., 2010).

17.3.4 SPRAY FREEZE-DRYING TECHNIQUE

17.3.4.1 Concepts of Drying

The three primary processes in a conventional spray freeze-drying process are the production of droplets from the dispersed bulk liquid, the solidification of droplets with direct contact with the liquid, and solidified droplet sublimation at low pressure and temperature (Ishwarya et al., 2015). Atomization is the initial step in the spray freeze-drying process that breaks down the bulk liquid into small droplets. Two-fluid, three-fluid, four-fluid, and ultrasonic nozzles have all been employed for the atomization process (Kowalska et al., 2008). Surface tension, sample viscosity, sample flow rate, and atomization energy are variables that affect atomization. After atomization, freezing is done at temperatures below zero with the use of liquid nitrogen. The freezing rate of atomized particles is mainly influenced by the mode of exposure, the quantity of cryogen, and the cryogen type. Droplet temperatures are first lowered below the freezing point. Rapid crystal development is followed by nucleation and then gradual freezing occurs with a decrease in cryogen heat transmission. After being frozen, particles are dried using the freeze-drying technique. The spray freeze-drying offers low processing times and enhances the keeping of highly volatile flavonoid compounds in the products. Spray freeze-drying results in highly stable spherical-shaped porous particles with improved quality, as shown in Figure 17.4(a). The generated particle size ranges from 10 μM to 30 μM.

17.3.4.2 Components and Process Considerations

The main components in the spray freeze dryer are an atomizer, liquid nitrogen, and a freeze dryer. The basic process of spray freeze drying comprises the feeding solution being pumped into a spraying unit, where they are subsequently atomized with a nozzle in a chamber filled with liquid nitrogen. Then, the frozen products were dried and sublimed at lower temperatures. The spray freeze-drying approach can be classified into three different routes (spray freezing into vapor over liquid, spray freezing drying into liquid, and spray freezing vapor into liquid) (Table 17.2). The steps involved in spray freeze drying are illustrated in Figure 17.4(b).

(a)

Magnification at 1000x   Magnification at 3000x   Magnification at 5000x

(b)

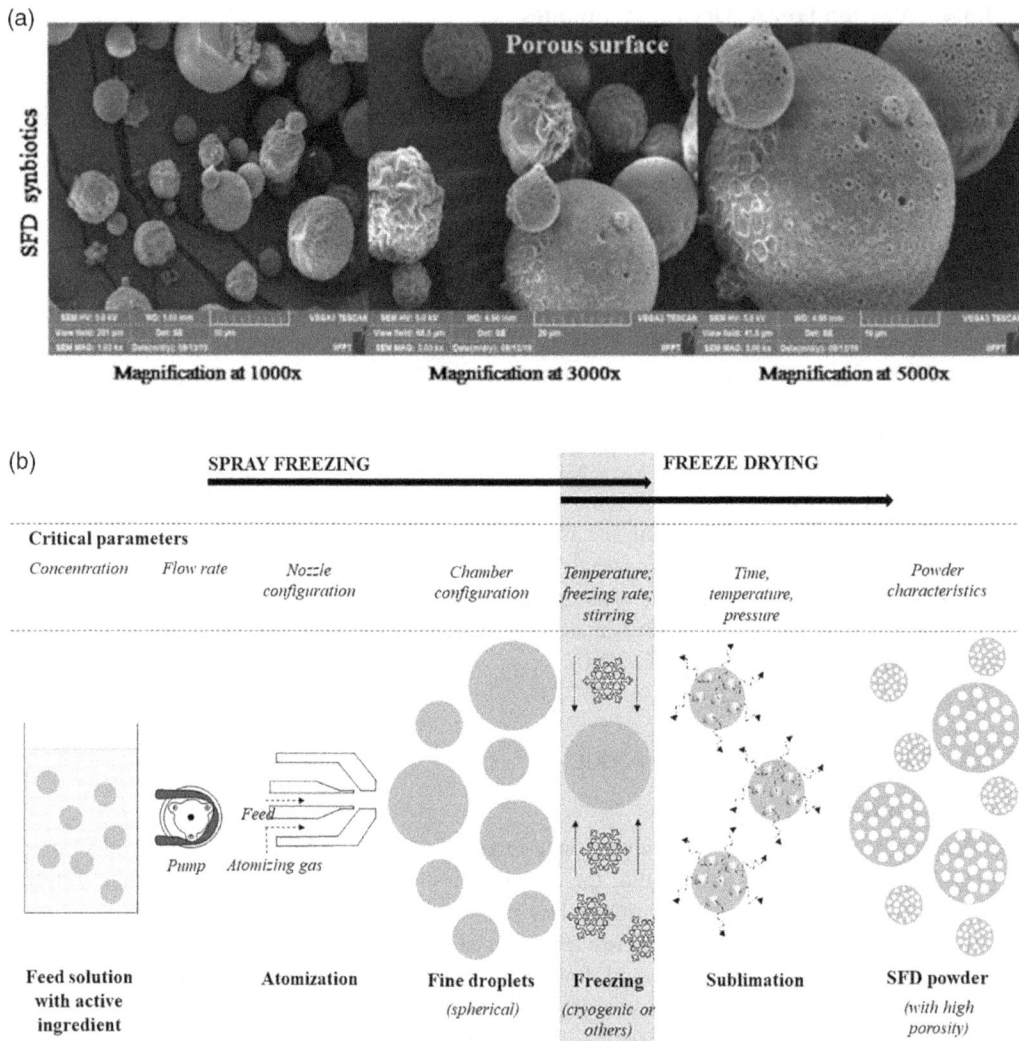

FIGURE 17.4 (a) morphological structure of spray freeze-dried powder (Yoha et al., 2020b); (b) steps involved in a spray freeze-drying process (Vishali et al., 2019).

TABLE 17.2
Classification of Spray Freezing Process

Techniques	Description
Spray freezing into vapor over liquids	**Spray freeze-drying technique a** nozzle is used to atomize the solution that is located just above a boiling cryogenic liquid at a close distance. Droplets in the vapor gap start to harden before becoming frozen as they come into contact with the liquid.
Spray freeze drying into liquid	The liquid leaving the atomizer collides sharply with the cryogen.
Spray freezing vapor into liquid	A process that involves atomizing liquid using spray nozzles and coming into contact with the desiccated gas directly.

17.3.4.3   Parameters Influencing the Drying Process

Nozzle geometry, atomization conditions, and feed characteristics are the main factors that influence the spray freezing drying route. The size of the particle is significantly affected by changing the atomization conditions. Surface tension, feed viscosity, feed flow rate, and atomization energy are the main atomization conditions. The droplet size is reduced with increased atomization pressure. On the other hand, improving the feed flow rate results in larger droplets. While increasing feed viscosity results in smaller droplets. Larger viscous forces are overcome by the atomization energy delivered to the nozzle which leads to the generation of small droplets. The surface tension of the feed material could play a significant role during atomization. High-surface tension liquid is challenging to atomize because the atomizer must overcome the surface tension of the feed material to increase the surface area of the droplets. In these circumstances, the addition of an emulsifier followed by a homogenization process could induce the atomization process. The mass ratio of the atomization gas and the flow rate of the liquid also influence the droplet size. The droplet size is decreased with increasing the mass ratio of gas to liquid.

17.3.5   Conductive Hydro Drying

17.3.5.1   Concept of Conductive Hydro Drying

Conductive hydro-drying, also known as Refractance Window (RW) drying, employs hot water heated to 90–95°C as the heat transfer medium and is mostly used to turn purees, liquid, or semi-solid products into flakes or powders. The procedure involves placing the moist material to be dried in a semitransparent mylar sheet floating over hot water. A transparent mylar (polyethylene terephthalate) sheet transports heat from flowing hot water (between 94°C and 98°C) to produce thermal energy. The water from the product (solid, liquid, or semiliquid) to be dried is evaporated and deposited on the thin plastic film surface. The hot water transfers its thermal energy via different proportions of conduction and radiation mechanisms. Through this method, infrared radiation may reach the liquid slurry directly. The "window" that permits the quick passage of infrared radiation closes when the substance loses moisture. In addition, heat is transferred through the film, which helps in moisture evaporation from the product. The conducted and infrared heat promotes the drying rate at atmospheric pressure (Haron et al., 2017).

Typically, the product temperature during the procedure ranges from 70°C to 80°C. The RW dryer often takes less time to dry products than traditional hot air-drying techniques, and the finished goods typically have a consistent thickness and a smooth flaky texture. RW-based drying technology has been shown more energy efficient than other conventional drying methods and freeze drying along with high retention of heat-sensitive quality criteria (vitamins, antioxidants, and color) (Abonyi et al., 2001; Hernández-Santos et al., 2016). Several fruits and vegetables have been reported to be dried using this approach. Besides, the production of a variety of other products, including food ingredients, nutraceutical ingredients, dietary, seafood, herbal extracts, and egg mixtures, has also been done successfully using this drying method. Moreover, Abony et al. (2002) reported that RW dried strawberry purees contained fewer esters and alcohols and higher heat-induced ketones and aldehydes (Abonyi et al., 2002).

The RW dryers maintain the temperature throughout the operation by continuously circulating the warmed water. The conductive hydro-drying process exhibits many advantages including simplicity, improved thermal efficiency, retention of all bioactive compounds, low operating cost, low energy consumption, and low cross-contamination. RW is an extremely effective drying technique, as evidenced by its evaporation capability of up to 10 kg m^{-2} h^{-1} (Zotarelli et al., 2015).

17.3.5.2   Components and Process Considerations

The components of RW dryers are a water heater, pump, adjustable blade, exhaust fan, mylar conveyor belt, and adjustable blade, as shown in Figure 17.5(a). The puree was evenly spread during the

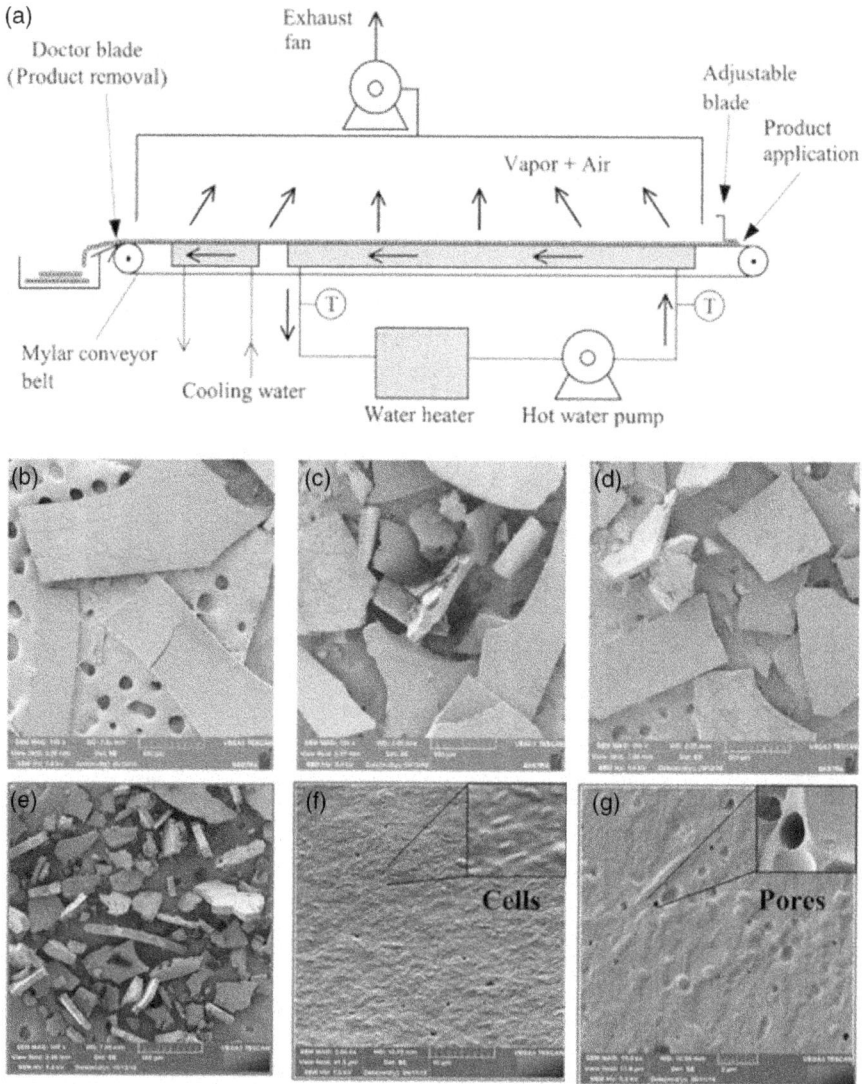

FIGURE 17.5 (a) Schematic representation of Refractance Window drying system (Baeghbali et al., 2016); Scanning electron microscopic images of conductive hydro-dried egg powder at different temperatures (b) 40°C, (c) 45°C and (d) 50°C; (e–g) synbiotic powder (Murugesan et al., 2019; Yoha et al., 2020a).

drying process on the top surface of a plastic conveyor belt, with hot water flowing over the bottom surface. To prevent bubble formation and keep the hot water in consistent contact with the conveyor belt, the water temperature was kept slightly below the boiling point. This allowed thermal energy to be transferred from the water directly into the product. Suction air was blown to remove any moisture that emerged from the product during drying.

17.3.5.3 Parameters Influencing the Drying Process

The thickness of the product to be dried influences the RW drying time. With an increase in sample thickness, the drying time lengthens considerably as the drying time is directly related to the weight of the sample (Shende & Datta, 2020). RW drying is more effective in drying pulp of thickness 3 mm (Zotarelli et al., 2015). Moreover, the ability of wall materials to absorb water is influenced by their molecular weight, that in turn directly influences the sample's moisture content (Hernández-Carrión et al., 2021). The temperature of drying significantly varies the product properties. The moisture content, drying time, contact angle, and water activity of the dried product decrease with the increased temperature of drying (Hernández-Carrión et al., 2021). Furthermore, the decrease in pressure and the increase in temperature could reduce the quantity of evaporated water (Morison et al., 2006). Besides, the water temperature was the factor that had a direct impact on the three critical components of RW drying: heat emission, heat propagation, and heat absorption by the drying source, medium, and product, respectively (Hernández-Carrión et al., 2021). The retention of ascorbic acid improves with an increase in drying temperature, which may be owing to a reduction in the sample's exposure time to heat and a consequent decrease in oxidation (Abonyi et al., 2002). As the drying temperature rises, cellular components break down more quickly and bound phenolic compounds are released, increasing the total phenolic content of mango leather (Shende & Datta, 2020). Rising conductive hydro-drying temperature, intense moisture loss, and quick heating rates accelerate crack development. Notably, if the sample has a high moisture content, the water vapor pressure inside the product will cause the formation of fractures on the product surface (Figure 17.5b–d) (Preethi et al., 2020). Moreover, conductive hydro-dried synbiotic powders showed evidence of the creation of porous structures (Figure 17.5e–g). Internal moisture on the bottom layer causes capillary action owing to continuous conduction during RW drying at lower temperatures, resulting in a porous structure with less surface shrinkage (Yoha et al., 2020a).

17.3.5.4   Assisted Conductive Hydro Drying Techniques

The combination of a conductive hydro-drying process with other drying methods could enhance the rate of drying, the quality of the product, and moisture-diffusive efficiency. Infrared radiation-assisted RW drying is a quick and easy way to dry food products with good quality. A rapid outflow of water vapor due to the rapid evaporation of water is caused by quick energy absorption by the water molecules, which appear in the product. The RW drying assisted with the infrared radiation route could reduce the drying process time by roughly 60% to achieve the equilibrium moisture content of 10% (Puente-Díaz et al., 2020). Numerous studies have demonstrated that employing ultrasound in conjunction with other drying techniques can speed up the drying process. Due to the synergistic, vibration, and heating effects, the ultrasound waves accelerated the drying process (Kowalski & Mierzwa, 2015). These effects included a temperature rise, turbulence in the air surrounding the drying region that boosted heat and mass transfer, and an increase in drying speed. An energy-efficient procedure with good quality retention capabilities can be achieved using conductive hydro-drying in conjunction with ultrasound and infrared technology as reported by Baeghbali et al. (2018).

17.3.6   Electrohydrodynamic Drying

17.3.6.1   Concepts of Electrohydrodynamic Drying

Electrohydrodynamic drying is now attracting great interest in drying heat-sensitive plant-based food products. The electric field and ionic wind are two important driving factors that act on the surface of the product to be dried during electrohydrodynamic drying and control the drying kinetics. Ionic wind with an aerodynamic flow is used in this drying process to remove moisture from a product (Moses et al., 2014; Ni et al., 2019). The ionic wind explains a mixing of the stream of airflow with ion flow. The applied electric field affects the induced ion flow, which in turn affects the ion velocity (Anukiruthika et al., 2021; Martynenko et al., 2019). The large voltage differential between an emitter electrode and a grounded collecting electrode produces ionic wind. The air is ionized by the high voltage, which causes ionic movement and airflow toward the collection electrode. It improves the rate of drying and the ability to retain nutritional and sensory qualities. Additionally, this innovative drying method results in minimal energy use, which lowers the carbon

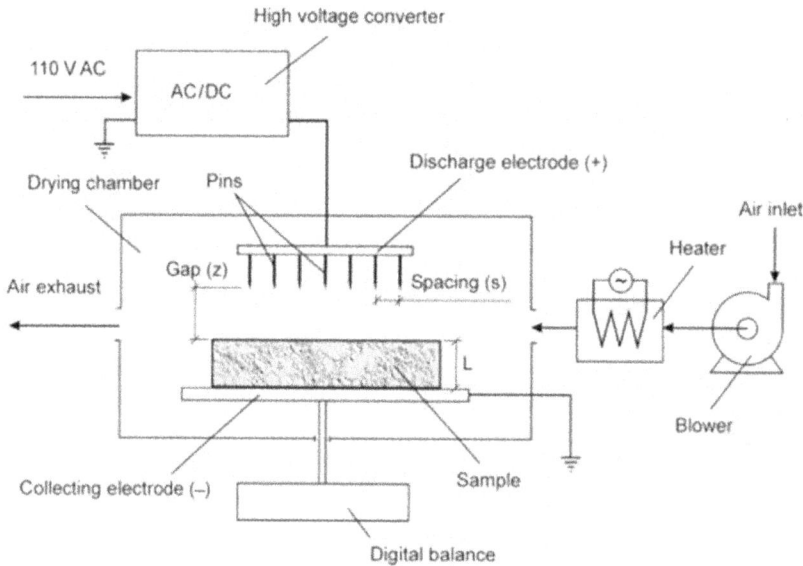

FIGURE 17.6 Experimental setup of an electrohydrodynamic drying system.

impact of processed goods (Bajgai et al., 2006a; Ni et al., 2019). Different configuration routes have been adapted, such as wire to mesh and wire to the plate.

17.3.6.2 Components and Process Considerations

Typically, the electrohydrodynamic drying system comprises two different electrodes such as ground electrodes and emitting electrodes also known as corona electrodes (Figure 17.6). The emitter electrode can be a wired or a point electrode. The food item to be dried was placed in front of the vertically moveable ground electrode, which projected the fixed horizontal electrode. A high-voltage source is connected to the emitting electrode. A transformer is connected to the voltage regulator to achieve the desired voltage values. Typically, the electrode gap varies from 1 cm to 10 cm in the electrohydrodynamic drying of products depending on the applied voltage, which is a crucial parameter to achieve adequate drying effects. The top emitter electrode can be moved with respect to the fixed ground electrode to change the electrode gap. The evaporation of water from the product to be dried is caused by this non-uniform electric field between the emitter and ground electrode.

17.3.6.3 Parameters Influencing the Drying Process

The type of electrode material has a direct impact on the efficiency of electrohydrodynamic drying systems. Generally, high-conductive materials such as copper, tungsten, and stainless steel are commonly adapted in the electrohydrodynamic drying process. Because the electric field that energizes positive and negative ion movement is frequently created with highly conducting materials (Ramaswamy & Ramachandran, 2017). Another factor that affects the efficiency of ionic wind and the related phenomena is the electrode geometry, including its thickness, sharpness, and shape. The point with the conical end electrode displays faster drying rates as compared with the hemispherical and flat electrodes (Zheng et al., 2011). Moreover, the long curvature and sharp needle-type electrodes provide a higher evaporation rate of water (Bajgai et al., 2006b; Dalvi-Isfahan et al., 2016). Another design element that is closely connected to the electrostatic and aerodynamic flow conditions of an electrohydrodynamic system is the configuration of electrodes. Wire to mesh type or wire to a plate or perforated emitting electrode with single or multiple needles with a solid collector electrode is different electrode configurations used commonly. Chen et al. (1994) discovered that drying potato slabs without the use of heat is possible using a single point-to-plate corona electrode method (Chen et al., 1994). Defraeye and

Martynenko concluded that for consistent drying of food components, the wire-to-mesh arrangement was beneficial over the wire-to-plate method (Chen et al., 1994). The distance between two needles or wires in setups with several needles or wires has a significant role in regulating the drying rate. The corona wind that is being produced would collapse before it reached the sample surface, if the tips of the electrodes are too close, causing considerable energy losses that would slow the drying rates. However, larger spaces between the needles or wire may also cause problems because the corona wind effects might not reach the sample surface (Dalvi-Isfahan et al., 2016).

Increasing the applied voltage at the emitter could reduce the drying time and increase the specific energy consumption (Martynenko & Zheng, 2016). The ionic wind velocity rises with applied voltage and shortens the time needed for the wet sample to acclimate to the surrounding air (Zamani et al., 2019). Generally, a voltage increase causes the current to rise, which boosts the ionic wind and speeds up the drying process. The drying time increases noticeably as the distance between the electrodes increases. The reason is that air speed in the product domain, charge density, and electric field is suppressed when increasing the distance between the emitter and the product. Because of this impact, there are fewer ion currents near the sample, which leads to slower airflow around the sample and decreased convective transfer rates (Defraeye & Martynenko, 2018; Ramaswamy & Ramachandran, 2017). Conversely, regardless of the amount of dried product utilized, increasing the electrode spacing resulted in a considerable decrease in the specific energy consumption (Dinani et al., 2014). The airflow and air temperature are the two significant factors that influence the electrohydrodynamic drying efficiency (Defraeye & Martynenko, 2018). The applied voltage causes a quadratic rise in airflow. The effect of the applied voltage is considerably minimized with an increase in air temperature (Martynenko et al., 2019).

17.3.6.4 Assisted Electrohydrodynamic Drying Techniques

Electrohydrodynamic drying has been used with other drying techniques such as oven drying, freeze drying, heat pump drying, and spray drying to enhance drying times, product quality, and energy usage. Electrohydrodynamic conjugation with hot air drying provides the potential for lower energy consumption and a faster drying rate. Moreover, products dried using the needle-type electrohydrodynamic drying and hot air drying in combination had a greater rehydration ratio than each method used independently (Polat & Izli, 2022). Electrohydrodynamic drying aided spray drying is another well-known electrohydrodynamic conjugation. The most advantageous feature of this combination technology is its ability to manage end-product properties such as uniform particle size and morphology with enhanced product yield (Bhushani, 2014; Lastow et al., 2007).

In another study, electrohydrodynamic drying and the freeze-drying combination can reduce drying durations by roughly 15%, reduce energy consumption, the better quality of dried products, and lower processing costs than the freeze-drying method alone (Hu et al., 2013). In another case, the impact of electrohydrodynamic drying as a pretreatment for freeze drying and sunlight drying has been reported. For instance, the drying characteristics were enhanced by electrohydrodynamic pretreatment of Chilean sea cucumber for 30 minutes with 30 kV followed by freeze drying (Tamarit-Pino et al., 2004). Electrohydrodynamic as a pretreatment was used for coating bananas before they were dried convectively. Electrohydrodynamic drying induces the breakdown of the cell wall, which enhances the water removal rate and decreases the shrinkage of the dried product. Thus, it reduces the drying time and reduces the rate of degradation of functional elements (Nadery Dehsheikh & Taghian Dinani, 2020).

17.3.7 Low-Temperature, Low-Relative Humidity Drying

17.3.7.1 Concepts of Low-Temperature, Low-Relative Humidity Drying

Drying at high temperatures causes physical and chemical transformations that can change the color of the product after it has dried. Similar to hot air grain drying, low-temperature, low-relative humidity drying does not use heat to reduce the relative humidity of the drying air. In low-temperature, low-humidity drying, the relative humidity of air is reduced by circulation through desiccant substances (montmorillonite clay or silica) which eliminate the moisture present in the air. The average relative humidity and temperature for a solar dryer with a low temperature and low relative humidity drying system are 42.58% and 29.25°C, respectively (Wisaiprom et al., 2018).

17.3.7.2 Components and Process Considerations

The low air temperature and relative humidity drying system consist of the microcontroller unit, heating coil, relative humidity and temperature sensors, air-conditioner, and humidity control unit (Figure 17.7) (Ondier et al., 2010a). The samples were taken from cold storage and allowed to acclimate to room temperature for 24 hours in sealed plastic bags before drying processes. This process brought the product into thermal equilibrium with the ambient temperature, eliminating any temporary impacts of heat transfer on the drying rates and inhibiting condensation on the samples when it was positioned in the drying platform. The food products from each cultivar were put on individual trays on one side of the dryer after the drying air conditions were settled. Through holes at the bottom of each tray holding the product, the drying air was routed back to the air conditioning unit at a fixed temperature and relative humidity. To track the temperature and relative humidity of the drying air during the drying process, temperature, and relative humidity sensors were installed in the drying chamber.

17.3.7.3 Effect of Process Parameters

Drying air conditions (relative humidity and velocity) could influence the drying efficiency. Relative humidity has a significant impact on drying time. Comparatively, at low temperatures (<34°C), drying air dehumidification can greatly speed up the drying process. Moreover, there is no considerable effect of relative humidity on the color and product yield (Ondier et al., 2010b). Under the drying conditions of low humidity, low temperature, and maximum air velocity, the ultimate equilibrium moisture content is superior to the value at a limited velocity of air. This is because low humidity and high air velocity allow moisture on a product surface to evaporate quickly, causing a case hardening that obstructs the transmission of heat to the product from hot air as well as moisture transfer from the product to the air. At low temperatures, the moisture content of the sample has a low diffusion coefficient. The resulting moisture balance is, therefore, high (Namkanisorn & Murathathunyaluk, 2020). An ultrasonic-assisted low-temperature drying process could increase the rate of mass transfer and reduce the time for drying. However, it should be noted that the ultrasonic pretreatment

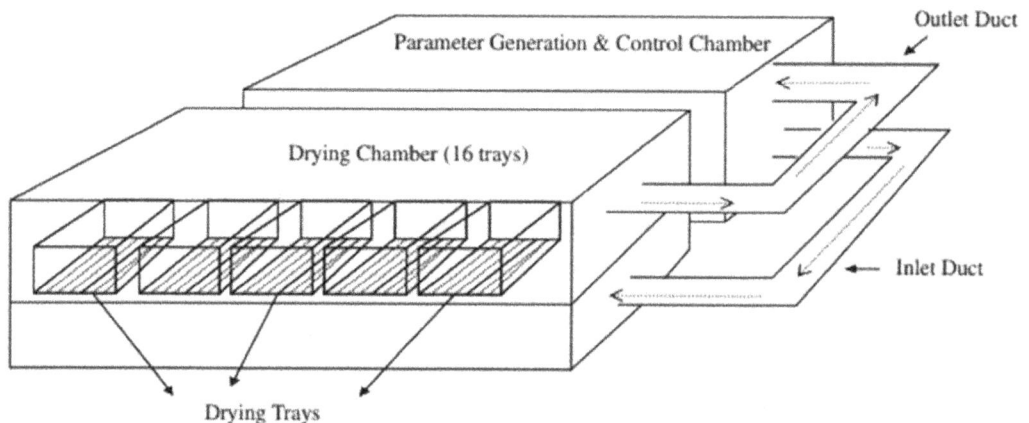

FIGURE 17.7   Schematic illustration of thin layer low temperature and low humidity drying unit (Ondier et al., 2010a).

technique could harm biological elements because of the mechanical stress brought on by the acoustic waves (Santacatalina et al., 2014).

17.3.8 LOW TEMPERATURE – HIGH VELOCITY FLUIDIZED BED DRYING

17.3.8.1 Concept of Low Temperature – High Velocity Fluidized Bed Drying

Fluidized beds are employed in many industrial operations such as coating, chilling, and granulation. It can be used on both heat-sensitive and non-heat-sensitive materials. The phenomenon of fluidization or two-phase flow occurs when liquid or gas interacts with a solid product. The solid products remain undisturbed at low flow rates and are referred to as a fixed bed. When the flow rate reaches a point where the particles begin to move, they become suspended in the flow because an outside force acting on the particles is counterbalanced by their weight. The flow rate is known as the minimum fluidized velocity, and the regime is known as the incipiently fluidized bed. The benefits of fluidized bed drying include uniform reduction of moisture at a short drying time, an improved rate of drying, and rapid heat and mass transfer through fluid, mixing, and solid particles. Fluidized bed dryers maintain a constant temperature during the drying process and extend the time at a constant drying rate (Haron et al., 2017).

17.3.8.2 Components and Process Considerations

A blower set, a heater, and a drying column are the main components of the fluidized bed dryer (Figure 17.8) (Rongchai et al., 2022). A drying pad with a load cell for weight measurement was installed and managed by a computer and microcontroller. A temperature controller was used to regulate the drying temperature. With the use of a three-phase inverter, the airflow was altered. An acrylic pipe was used to create a drying chamber. When a fluid flow is delivered at the bottom of

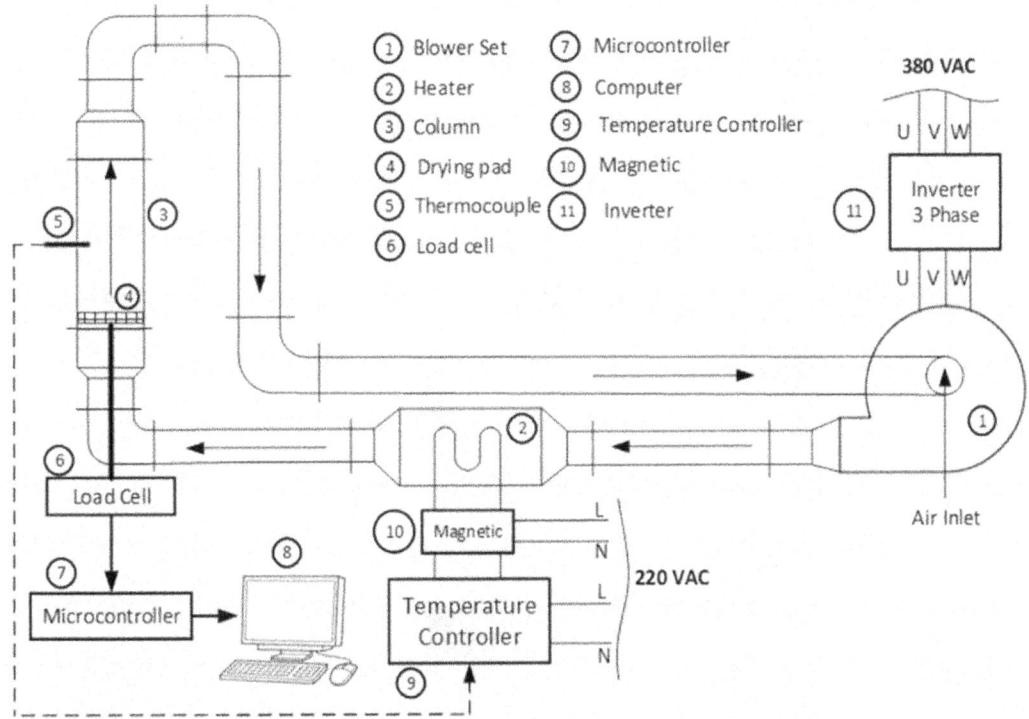

FIGURE 17.8   Schematic of the fluidized bed drying experimental setup (U, V, W indicates the three phase power supply) (Rongchai et al., 2022).

the fluidizing bed column, it disturbs and suspends the solid grains inside, causing fluidization to occur. The product materials are continuously in contact with air and allow the rapid extraction of moisture.

17.3.8.3 Parameters Influencing the Drying Process

Sample size, drying temperature, and gas flow rate are the factors that commonly influence drying kinetics. A sample size between 50 μm and 5000 μm is commonly used. When the size of the sample exceeds its range, it requires extra forces such as mechanical vibration to fluidize. If the airflow temperature is raised steadily, weight loss will first show a dramatic decline, then a progressive decline until the trial is complete (Dejahang, 2015). The samples become more equally dispersed and independently suspended as the flow rate increases. The solid samples will eventually escape the column if the flow is increased further; this flow speed is known as the terminal velocity. It is possible to reduce drying periods by increasing the drying air flow speed (Rongchai et al., 2022). Moreover, lower bed height and higher frequency are beneficial for drying as reported by Zhao et al. (2014).

17.3.9 Low Temperature – Desiccant-Based Drying

17.3.9.1 Concepts of Low Temperature-Desiccant-Based Drying

Materials that absorb or adsorb moisture have been used to dry food products. In this procedure, the air is dried more effectively by removing moisture using a desiccant. Solar energy is then used to renew the desiccant. Desiccants are hygroscopic compounds that have the potential to absorb moisture. Desiccant-based drying system deals with the latent and sensible components of vapor pressure. Desiccant-based drying uses low temperatures and quick drying to preserve the color and texture of vegetables. It can also assist in offering low-cost drying with less energy usage. Desiccant may occasionally be made from a component of the commodity itself, particularly grain. Some of the over-dried grain is taken out of the desiccant bin and mixed with wet grain to create a combination with an average moisture content that is close to the desired level (Mora et al., 2019; Nagaya et al., 2006).

17.3.9.2 Components and Process Considerations

The desiccant-based drying system consists of (i) desiccant wheels that are used to remove moisture from the air, (ii) desiccant wheel regeneration heaters, (iii) heaters that are used to heat air, and (iv) drying bin/structure (Figure 17.9). The dehumidifier has a silica gel containing desiccant rotor. The rotor comprises three different zones. The operational zone is where silica gel removes air moisture and supplies dry air to the rotor side; a recovery zone is where moist air is heated and expelled, and a region of heat collection where heat is released by silica gel into the outside. Additionally, there are thermometers placed in the drying chamber for temperature visual validation. The product was

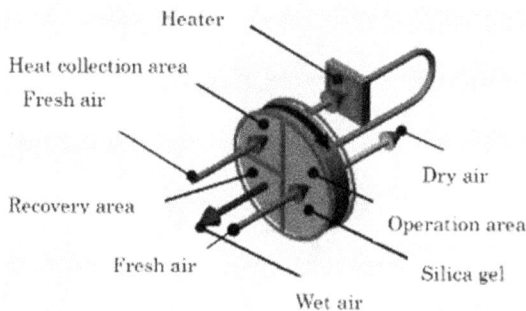

FIGURE 17.9 Nine setups of desiccant drying rotor (Nagaya et al., 2006).

placed in the drying chamber (acrylic sheet) and enabling the dry air to rise from the chamber's base to pass through the product and exit the outlet. The chamber temperature was fixed at below 49°C (Nagaya et al., 2006).

17.3.9.3 Parameters Influencing the Drying Process

Airflow and temperature control are used to quicken the drying process and the chamber's drying conditions should be consistent throughout drying (Nagaya et al., 2006). High drying temperatures have a significant negative impact on product quality, for example, by affecting vitamin C, which is present in many vegetables and fruits and is thought to be a measure of nutritional loss, to deteriorate. Grain may be over-dried in a desiccant system during the summer if solar energy is employed for drying (Dai et al., 2002).

17.4 SUMMARY

As compared with the other conventional drying routes, the low-temperature drying process shows superiority in terms of the dried product quality and, in certain cases, operational cost and energy requirements. This approach is best for foods with heat-sensitive components. Different low-temperature drying techniques such as vacuum, osmotic, freeze, spray freeze, conductive, electro-hydrodynamic, low temperature-low humidity, low-temperature-high velocity fluidized bed, and low-temperature desiccant-based drying have been discussed. Overall, these remain topics that require extensive studies in aspects of energy, time, environmental sustainability, and cost calculations in comparison with conventional processes. In other words, from the perspective of being employed as sustainable food processing approaches, low-temperature drying techniques must be viewed and better analyzed.

REFERENCES

Abla, K. K., & Mehanna, M. M. (2022). Freeze-drying: A flourishing strategy to fabricate stable pharmaceutical and biological products. *International Journal of Pharmaceutics*, *628*, 122233. https://doi.org/10.1016/j.ijpharm.2022.122233

Abonyi, B. I., Feng, H., Tang, J., Edwards, C. G., Chew, B. P., Mattinson, D. S., & Fellman, J. K. (2002). Quality retention in strawberry and carrot purees dried with Refractance Window™ system. *Journal of Food Science*, *67*(3), 1051–1056.

Agnelli, M. E., Marani, C. M., & Mascheroni, R. H. (2005). Modelling of heat and mass transfer during (osmo) dehydrofreezing of fruits. *Journal of Food Engineering*, *69*(4), 415–424. https://doi.org/10.1016/j.jfoodeng.2004.08.034

Ahmed, I., Qazi, I. M., & Jamal, S. (2016). Developments in osmotic dehydration technique for the preservation of fruits and vegetables. *Innovative Food Science & Emerging Technologies*, *34*, 29–43. https://doi.org/10.1016/j.ifset.2016.01.003

Allali, H., Marchal, L., & Vorobiev, E. (2009). Effect of blanching by ohmic heating on the osmotic dehydration behavior of apple cubes. *Drying Technology*, *27*(6), 739–746. https://doi.org/10.1080/07373930902827965

Anukiruthika, T., Moses, J. A., & Anandharamakrishnan, C. (2021). Electrohydrodynamic drying of foods: Principle, applications, and prospects. *Journal of Food Engineering*, *295*, 110449.

Assegehegn, G., Brito-de la Fuente, E., Franco, J. M., & Gallegos, C. (2020). Freeze-drying: A relevant unit operation in the manufacture of foods, nutritional products, and pharmaceuticals. *Advances in Food and Nutrition Research*, *93*, 1–58. https://doi.org/10.1016/bs.afnr.2020.04.001

Azarpazhooh, E., & Ramaswamy, H. S. (2012). Evaluation of factors influencing microwave osmotic dehydration of apples under continuous flow medium spray (MWODS) conditions. *Food and Bioprocess Technology*, *5*(4), 1265–1277. https://doi.org/10.1007/s11947-010-0446-x

Baeghbali, V., Niakousari, M., & Farahnaky, A. (2016). Refractance Window drying of pomegranate juice: Quality retention and energy efficiency. *LWT-Food Science and Technology*, *66*, 34–40.

Baeghbali, V., Niakousari, M., Ngadi, M. O., & Hadi, M. (2018). Combined ultrasound and infrared assisted conductive hydro-drying of apple slices. *Drying Technology*, *37*(14), 1–13. https://doi.org/10.1080/07373937.2018.1539745

Bajgai, T. R., Raghavan, G. S. V., Hashinaga, F., & Ngadi, M. O. (2006a). Electrohydrodynamic drying – A concise overview. *Drying Technology, 24*(7), 905–910. https://doi.org/10.1080/07373930600734091

Bhambhani, A., Stanbro, J., Roth, D., Sullivan, E., Jones, M., Evans, R., & Blue, J. (2021). Evaluation of microwave vacuum drying as an alternative to freeze-drying of biologics and vaccines: The power of simple modeling to identify a mechanism for faster drying times achieved with microwave. *AAPS PharmSciTech, 22*(1), 52. https://doi.org/10.1208/s12249-020-01912-9

Bhushani, J. A. (2014). Electrospinning and electrospraying techniques : Potential food based applications. *Trends in Food Science & Technology, 38*(1), 21–33. https://doi.org/10.1016/j.tifs.2014.03.004

Martynenko, A., & Kudra, T. (2019). Electrohydrodynamic drying. In *Advanced drying technologies for food*, CRC Press, 1, 1–30. https://doi.org/10.1201/9780367262037

Cárcel, J. A., García-Pérez, J. V., Benedito, J., & Mulet, A. (2012). Food process innovation through new technologies: Use of ultrasound. *Journal of Food Engineering, 110*(2), 200–207.

Chakraborty, R., Bera, M., Mukhopadhyay, P., & Bhattacharya, P. (2011). Prediction of optimal conditions of infrared assisted freeze-drying of aloe vera (*Aloe barbadensis*) using response surface methodology. *Separation and Purification Technology, 80*(2), 375–384. https://doi.org/10.1016/j.seppur.2011.05.023

Chen, Y., Barthakur, N. N., & Arnold, N. P. (1994). Electrohydrodynamic (EHD) drying of potato slabs. *Journal of Food Engineering, 23*(1), 107–119. https://doi.org/10.1016/0260-8774(94)90126-0

Chumroenphat, T., Somboonwatthanakul, I., & Saensouk, S. (2021). Changes in curcuminoids and chemical components of turmeric (*Curcuma longa* L.) under freeze-drying and low-temperature drying methods. *Food Chemistry, 339*(September 2020), 128121. https://doi.org/10.1016/j.foodchem.2020.128121

Ciurzyńska, A., Kowalska, H., Czajkowska, K., & Lenart, A. (2016). Osmotic dehydration in production of sustainable and healthy food. *Trends in Food Science & Technology, 50*, 186–192. https://doi.org/10.1016/j.tifs.2016.01.017

Dai, Y., Wang, R. Z., & Xu, Y. (2002). Study of a solar powered solid adsorption–desiccant cooling system used for grain storage. *Renewable Energy, 25*, 417–430. https://doi.org/10.1016/S0960-1481(01)00076-3

Dak, M., & Pareek, N. K. (2014). Effective moisture diffusivity of pomegranate arils under going microwave-vacuum drying. *Journal of Food Engineering, 122*, 117–121.

Dalvi-Isfahan, M., Hamdami, N., Le-Bail, A., & Xanthakis, E. (2016). The principles of high voltage electric field and its application in food processing: A review. *Food Research International, 89*, 48–62. https://doi.org/10.1016/j.foodres.2016.09.002

Das, S. J., and H. (2013). A vacuum drying model for mango pulp. *Drying Technology, 7*(January 2003), 1215–1234. https://doi.org/10.1081/DRT-120023177

Defraeye, T., & Martynenko, A. (2018). Electrohydrodynamic drying of food: New insights from conjugate modeling. *Journal of Cleaner Production, 198*, 269–284.

Dejahang, T. (2015). Low Temperature Fluidized Bed Coal Drying : Experiment, Analysis and Simulation. ERA eucation & research archieve (University of Alberta Library), https://doi.org/10.7939/R3K931D1N

Devahastin, S., Suvarnakuta, P., Soponronnarit, S., & Mujumdar, A. S. (2004). A comparative study of low-pressure superheated steam and vacuum drying of a heat-sensitive material. *Drying Technology, 22*(8), 1845–1867. https://doi.org/10.1081/DRT-200032818

Dinani, S. T., Hamdami, N., Shahedi, M., & Havet, M. (2014). Mathematical modeling of hot air/electrohydrodynamic (EHD) drying kinetics of mushroom slices. *Energy Conversion and Management, 86*, 70–80.

Duan, X., Zhang, M., Mujumdar, S., & Wang, R. (2010). Trends in microwave-assisted freeze drying of foods. *Drying Technology, 28*(4), 444–453. https://doi.org/10.1080/07373931003609666

Egas-Astudillo, L. A., Silva, A., Uscanga, M., Martínez-Navarrete, N., & Camacho, M. M. (2010). Impact of shelf temperature on freeze-drying process and porosity development. *World Science and Technology, 12*(April), 843–850. https://doi.org/10.4995/ids2018.2018.7481

Falade, K. O., Igbeka, J. C., & Ayanwuyi, F. A. (2007). Kinetics of mass transfer, and colour changes during osmotic dehydration of watermelon. *Journal of Food Engineering, 80*(3), 979–985.

Garcia, C. C., Mauro, M. A., & Kimura, M. (2007). Kinetics of osmotic dehydration and air-drying of pumpkins (*Cucurbita moschata*). *Journal of Food Engineering, 82*(3), 284–291. https://doi.org/10.1016/j.jfoodeng.2007.02.004

Garcia-Amezquita, L. E., Welti-Chanes, J., Vergara-Balderas, F. T., & Bermúdez-Aguirre, D. (2015). Freeze-drying: The basic process. In *Encyclopedia of Food and Health* (pp. 104–109). https://doi.org/10.1016/B978-0-12-384947-2.00328-7

Giraldo, G., Talens, P., Fito, P., & Chiralt, A. (2003). Influence of sucrose solution concentration on kinetics and yield during osmotic dehydration of mango. *Journal of Food Engineering, 58*(1), 33–43.

Halim, N. F., Basrawi, M. F., Azman, N. A. M., Sulaiman, S. A., & Razak, S. B. A. (2021). The effect of temperature on low temperature vacuum drying with induced nucleate boiling for stingless bees honey. *IOP Conference Series: Materials Science and Engineering*, *1078*(1), 012018. https://doi.org/10.1088/1757-899x/1078/1/012018

Haron, N. S., Zakaria, J. H., & Mohideen Batcha, M. F. (2017). Recent advances in fluidized bed drying. *IOP Conference Series: Materials Science and Engineering*, *243*(1), 012038. https://doi.org/10.1088/1757-899X/243/1/012038

Hernández-Carrión, M., Moyano-Molano, M., Ricaurte, L., Clavijo-Romero, A., & Quintanilla-Carvajal, M. X. (2021). The effect of process variables on the physical properties and microstructure of HOPO nano-emulsion flakes obtained by refractance window. *Scientific Reports*, *11*(1), 1–14. https://doi.org/10.1038/s41598-021-88381-7

Hernández-Santos, B., Martínez-Sánchez, C. E., Torruco-Uco, J. G., Rodríguez-Miranda, J., Ruiz-López, I. I., Vajando-Anaya, E. S., Carmona-García, R., & Herman-Lara, E. (2016). Evaluation of physical and chemical properties of carrots dried by Refractance Window drying. *Drying Technology*, *34*(12), 1414–1422. https://doi.org/10.1080/07373937.2015.1118705

Hu, Y., Huang, Q., & Bai, Y. (2013). Combined electrohydrodynamic (EHD) and vacuum freeze drying of shrimp. *Journal of Physics: Conference Series*, *418*(1), 12143. https://doi.org/10.1088/1742-6596/418/1/012143

Ishwarya, S. P., Anandharamakrishnan, C., & Stapley, A. G. F. (2015). Spray-freeze-drying: A novel process for the drying of foods and bioproducts. *Trends in Food Science and Technology*, *41*(2), 161–181. https://doi.org/10.1016/j.tifs.2014.10.008

Jariyawaranugoon, U. (2015). Effect of Freezing on Quality of Osmotically Dehydrated Banana Slices. *Advance Journal of Food Science and Technology*, *35*(3), 22–24.

Kowalska, H., Lenart, A., & Leszczyk, D. (2008). The effect of blanching and freezing on osmotic dehydration of pumpkin. *Journal of Food Engineering*, *86*(1), 30–38. https://doi.org/10.1016/j.jfoodeng.2007.09.006

Kowalski, S. J., & Mierzwa, D. (2015). US-assisted convective drying of biological materials. *Drying Technology*, *33*(13), 1601–1613. https://doi.org/10.1080/07373937.2015.1026985

Krokida, M. K., Karathanos, V. T., & Maroulis, Z. B. (1998). Effect of freeze-drying conditions on shrinkage and porosity of dehydrated agricultural products. *Journal of Food Engineering*, *35*(4), 369–380. https://doi.org/10.1016/S0260-8774(98)00031-4

Kudra, T. (2009). Energy Aspects in Food Dehydration. In *Advances in Food Dehydration (CRC Press, NW, U.S.A)* (pp. 423–445). ISBN: 9780429147296. https://doi.org/10.1201/9781420052534

Kumar, V., & Shrivastava, S. L. (2017). Vacuum-assisted microwave drying characteristics of green bell pepper. *International Journal of Food Studies*, *6*(1), 67–81. https://doi.org/10.7455/ijfs/6.1.2017.a7

Lastow, O., Andersson, J., Nilsson, A., & Balachandran, W. (2007). Low-voltage electrohydrodynamic (EHD) spray drying of respirable particles. *Pharmaceutical Development and Technology*, *12*(2), 175–181.

Martynenko, A., & Zheng, W. (2016). Electrohydrodynamic drying of apple slices: Energy and quality aspects. *Journal of Food Engineering*, *168*, 215–222.

Mavroudis, N. E., Gekas, V., & Sjöholm, I. (1998). Osmotic dehydration of apples – effects of agitation and raw material characteristics. *Journal of Food Engineering*, *35*(2), 191–209.

Mora, A., Sufi, U., Roach, J. I., Thompson, J. F., & Donis-Gonzalez, I. R. (2019). Evaluation of a small-scale desiccant-based drying system to control corn dryness during storage. *AIMS Agriculture and Food*, *4*, 136–148.

Morison, K. R., Worth, Q. A. G., & O'dea, N. P. (2006). Minimum wetting and distribution rates in falling film evaporators. *Food and Bioproducts Processing*, *84*(4), 302–310. https://doi.org/10.1205/fbp06031

Moses, J. A., Norton, T., Alagusundaram, K., & Tiwari, B. K. (2014). Novel drying techniques for the food industry. *Food Engineering Reviews*, *6*(3), 43–55.

Murugesan, P., Moses, J. A., & Anandharamakrishnan, C. (2019). Photocatalytic disinfection efficiency of 2D structure graphitic carbon nitride-based nanocomposites: A review. *Journal of Materials Science*, *54*(19), 12206–12235. https://doi.org/10.1007/s10853-019-03695-2

Nadery Dehsheikh, F., & Taghian Dinani, S. (2020). Influence of coating pretreatment with carboxymethyl cellulose in an electrohydrodynamic system on convective drying of banana slices. *Journal of Food Process Engineering*, *43*(2), e13308.

Nagaya, K., Li, Y., Jin, Z., Fukumuro, M., Ando, Y., & Akaishi, A. (2006). Low-temperature desiccant-based food drying system with airflow and temperature control. *Journal of Food Engineering*, *75*(1), 71–77.

Namkanisorn, A., & Murathathunyaluk, S. (2020). Sustainable drying of galangal through combination of low relative humidity, temperature and air velocity. *Energy Reports*, *6*, 748–753.

Ni, J., Ding, C., Zhang, Y., Song, Z., Hu, X., & Hao, T. (2019). Electrohydrodynamic drying of Chinese wolfberry in a multiple needle-to-plate electrode system. *Foods*, *8*(5), 152. https://doi.org/10.3390/foods8050152

Oikonomopoulou, V. P., Krokida, M. K., & Karathanos, V. T. (2011). The influence of freeze drying conditions on microstructural changes of food products. *Procedia Food Science*, *1*, 647–654. https://doi.org/10.1016/j.profoo.2011.09.097

Ondier, G. O., Siebenmorgen, T. J., & Mauromoustakos, A. (2010b). Low-temperature, low-relative humidity drying of rough rice. *Journal of Food Engineering*, *100*(3), 545–550. https://doi.org/10.1016/j.jfoodeng.2010.05.004

Phisut, N. (2012). Factors affecting mass transfer during osmotic dehydration of fruits. *International Food Research Journal*, *19*(1), 7–18.

Polat, A., & Izli, N. (2022). Determination of drying kinetics and quality parameters for drying apricot cubes with electrohydrodynamic, hot air and combined electrohydrodynamic-hot air drying methods. *Drying Technology*, *40*(3), 527–542. https://doi.org/10.1080/07373937.2020.1812633

Porciuncula, B. D. A., Segura, L. A., & Laurindo, J. B. (2016). Processes for controlling the structure and texture of dehydrated banana. *Drying Technology*, *34*(2), 167–176. https://doi.org/10.1080/07373937.2015.1014911

Preethi, R., Shweta, D., Moses, J. A., & Anandharamakrishnan, C. (2020). Conductive hydro drying as an alternative method for egg white powder production. *Drying Technology*, *39*(3), 1–13. https://doi.org/10.1080/07373937.2020.1788073

Puente-Díaz, L., Spolmann, O., Nocetti, D., Zura-Bravo, L., & Lemus-Mondaca, R. (2020). Effects of infrared-assisted Refractance Window™ drying on the drying kinetics, microstructure, and color of physalis fruit purée. *Foods (Basel, Switzerland)*, *9*(3). https://doi.org/10.3390/foods9030343

Ramaswamy, R., & Ramachandran, R. P. (2017). Electric field analysis of different compact electrodes for pulsed electric field applications in liquid food. *2016 IEEE International Power Modulator and High Voltage Conference, IPMHVC 2016*, 160–165. https://doi.org/10.1109/IPMHVC.2016.8012788

Rastogi, N. K., Raghavarao, K. S. M. S., & Niranjan, K. (2014). Recent Developments in Osmotic Dehydration. In *Emerging Technologies for Food Processing* (2nd ed., pp. 181–212). Elsevier Ltd. https://doi.org/10.1016/B978-0-12-411479-1.00011-5

Rastogi, N. K., Raghavarao, K. S. M. S., Niranjan, K., & Knorr, D. (2002). Recent developments in osmotic dehydration: Methods to enhance mass transfer. *Trends in Food Science & Technology*, *13*(2), 48–59. https://doi.org/10.1016/S0924-2244(02)00032-8

Rongchai, K., Somboon, T., & Charmongkolpradit, S. (2022). Fluidized bed drying behaviour of moringa leaves and the influence of temperature on the calcium content. *Case Studies in Thermal Engineering*, *40*, 102564. https://doi.org/10.1016/j.csite.2022.102564

Santacatalina, J. V., Rodríguez, O., Simal, S., Cárcel, J. A., Mulet, A., & García-Pérez, J. V. (2014). Ultrasonically enhanced low-temperature drying of apple: Influence on drying kinetics and antioxidant potential. *Journal of Food Engineering*, *138*, 35–44. https://doi.org/10.1016/j.jfoodeng.2014.04.003

Shende, D., & Datta, A. K. (2020). Optimization study for Refractance Window drying process of Langra variety mango. *Journal of Food Science and Technology*, *57*(2), 683–692. https://doi.org/10.1007/s13197-019-04101-0

Shi, J., & Maguer, M. Le. (2002). Osmotic dehydration of foods: Mass transfer and modeling aspects. *Food Reviews International*, *18*(4), 305–335. https://doi.org/10.1081/FRI-120016208

Soliman, J. L., Lim, N. R. E. G., Mercado, J. P. D., Lagurin, M. R. C., Felix, C. B., Ubando, A. T., & Culaba, A. B. (2017). Effects of temperature and pressure on the vacuum drying characteristics of red seaweed kappaphycus alvarezii. *HNICEM 2017 - 9th International Conference on Humanoid, Nanotechnology, Information Technology, Communication and Control, Environment and Management, 2018, January*, 1–5. https://doi.org/10.1109/HNICEM.2017.8269458

Szadzińska, J., Mierzwa, D., Pawłowski, A., Musielak, G., Pashminehazar, R., & Kharaghani, A. (2019). Ultrasound- and microwave-assisted intermittent drying of red beetroot. *Drying Technology*. *38*(1–2), 93–107. https://doi.org/10.1080/07373937.2019.1624565

Talens, P., Escriche, I., Martínez-Navarrete, N., & Chiralt, A. (2003). Influence of osmotic dehydration and freezing on the volatile profile of kiwi fruit. *Food Research International*, *36*(6), 635–642.

Tamarit-Pino, Y., Batías-Montes, J. M., Segura-Ponce, L. A., & Díaz-Álvarez, R. E. (2004). Effect of electrohydrodynamic pretreatment on drying rate and rehydration properties of Chilean sea cucumber (*Athyonidium chilensis*). *Foo Bioprod. Processes*, *123*, 264–295.

Torreggiani, D. (1993). Osmotic dehydration in fruit and vegetable processing. *Food Research International*, *26*(1), 59–68. https://doi.org/10.1016/0963-9969(93)90106-S

Vaxelaire, J., & Puiggali, J. R. (2002). Analysis of the drying of residual sludge: From the experiment to the simulation of a belt dryer. *Drying Technology, 20*(4–5), 989–1008. https://doi.org/10.1081/DRT-120003773

Vega-Mercado, H., Góngora-Nieto, M. M., & Barbosa-Cánovas, G. V. (2001). Advances in dehydration of foods. *Journal of Food Engineering, 49*(4), 271–289.

Vishali, D. A., Monisha, J., Sivakamasundari, S. K., Moses, J. A., & Anandharamakrishnan, C. (2019). Spray freeze drying: Emerging applications in drug delivery. *Journal of Controlled Release, 300*, 93–101.

Wisaiprom, N., Kasayapanand, N., Pratinthong, N., Songprakorp, R., Thepa, S., & Deeto, S. (2018). The study of shrimp drying by greenhouse drying combined with low humidity air. *International Journal of Smart Grid and Clean Energy, 7*(4), 303–313. https://doi.org/10.12720/sgce.7.4.303-313

Yadav, A. K., & Singh, S. V. (2014). Osmotic dehydration of fruits and vegetables: A review. *Journal of Food Science and Technology, 51*(9), 1654–1673.

Yoha, K. S., Moses, J. A., & Anandharamakrishnan, C. (2020a). Conductive hydro drying through Refractance Window drying – An alternative technique for drying of Lactobacillus plantarum (NCIM 2083). *Drying Technology, 38*(5–6), 610–620. https://doi.org/10.1080/07373937.2019.1624972

Yoha, K. S., Moses, J. A., & Anandharamakrishnan, C. (2020b). Effect of encapsulation methods on the physicochemical properties and the stability of Lactobacillus plantarum (NCIM 2083) in synbiotic powders and in-vitro digestion conditions. *Journal of Food Engineering, 283*(October), 110033. https://doi.org/10.1016/j.jfoodeng.2020.110033

Zamani, E., Mehrabani-Zeinabad, A., & Sadeghi, M. (2019). Sensitivity analysis pertaining to effective parameters on electrohydrodynamic drying of porous shrinkable materials. *Drying Technology, 37*(14), 1821–1832. https://doi.org/10.1080/07373937.2018.1541902

Zhao, P., Zhao, Y., Luo, Z., Chen, Z., Duan, C., & Song, S. (2014). Effect of operating conditions on drying of Chinese lignite in a vibration fluidized bed. *Fuel Processing Technology, 128*, 257–264. https://doi.org/10.1016/j.fuproc.2014.07.014

Zheng, D.-J., Liu, H.-J., Cheng, Y.-Q., & Li, L.-T. (2011). Electrode configuration and polarity effects on water evaporation enhancement by electric field. *International Journal of Food Engineering, 7*(2). https://doi.org/doi:10.2202/1556-3758.1794

Zotarelli, M. F., Carciofi, B. A. M., & Laurindo, J. B. (2015). Effect of process variables on the drying rate of mango pulp by Refractance Window. *Food Research International, 69*, 410–417. https://doi.org/10.1016/j.foodres.2015.01.013

18 Emerging Trends in Biological Control Agents for Food Safety Applications

Arunkumar Elumalai and S. Shanmugasundaram

18.1 INTRODUCTION

Dietician strongly recommends the consumption of fresh vegetables and fruits for human as they contain essential nutrients such as vitamins, minerals, and dietary fiber at optimal level. Studies have shown that their consumption can help improve the health in human. They can be consumed either fresh or minimally processed or boiled in water (Nawawi et al., 2023). They are typically eaten as raw, with little processing, pasteurized, or cooked in the microwave or by boiling in water. Thermal treatments may decrease the nutritional value and sensory qualities of the raw materials while increasing the shelf-life and safety of these goods. Though fresh food items with little processing are good for human health, they have a short shelf life due to microbial decomposition. Studies have shown that compared to the raw food materials, minimally processed food products are highly vulnerable for decomposition (Tewari et al., 2017). High availability of the nutrients at the cut surface, metabolic activity at tissue level, confinement of the food items within the packages, failure in the employment of the thermal treatment to the food items for the destruction of the microbial population and cross-contamination during cutting and peeling strongly supports the microbial growth on fruits and vegetables (Ali et al., 2017). Although minimally processed vegetables and fruits only have a relatively brief storage life of 4 to 10 days, raw product is believed to have a storage period of few weeks or months. Their shelf-life is influenced by a number of variables, including the quality of the vegetables and fruits, production methods, and the number and interactions of microbial groups (Wilson et al., 2019).

Mesophilic bacteria are very common in conventional and organic fruits and vegetables. It has been revealed that both conventional and organic fruits and vegetables possess between 6 and 7 log10 CFU/g of these bacteria (Szczech et al., 2018). Furthermore, at the retail level, these figures could increase even further. Disinfection techniques have been found to increase product quality, shelf life, and safety by lowering the microbial loads in fresh fruits and vegetables. However, using chemicals as antimicrobial agents for raw food materials is inadequate to prevent microbial deterioration or to guarantee the safety of the final product. Fresh fruits and vegetables frequently go through chlorine-based disinfection procedures to improve their shelf-life and safety profiles (Feliziani et al., 2016). But the usage of chlorine disinfection leads to the generation of carcinogenic chlorinated derivatives, the release of harmful vapors, and the raise in chlorine-resistant bacteria. Moreover, chemical disinfectants that replace chlorine have shown to be ineffective in eliminating or killing microbes on fresh fruits and vegetables as well as having potential for toxicity and negative impacts on the sensory qualities of the food products (Asadollahi et al., 2017). Consumer concern over the usage of synthetic disinfectants on fresh produce has inspired the research into alternate strategies for minimizing the deterioration of fresh and minimally processed fruits and vegetables and enhancing product safety. One of the promising options for preventing the growth of harmful and pathogenic microorganisms is the utilization of generally recognized as safe (GRAS) microbes and their natural metabolites (Ibrahim et al., 2021). The consumer also considers this

DOI: 10.1201/9781003359302-18

approach as a natural way to preserve food. This book chapter discussed the usage of biocontrol agents and their action mode against harmful or spoilage microorganisms typically found in fresh fruits and vegetables.

18.2   WHAT EXACTLY IS BIOCONTROL?

The practice of using one or more microbes to attenuate or regulate the growth of other microbes is termed as biocontrol. Biocontrol agents are classified into two types: Usage of live microbes including phages and the usage of byproducts derived from the microbes, like bacteriocins. In a broad sense, biocontrol can be defined as utilization of lactic acid bacteria (LAB), phages, toxins such as bacteriocins, endolysins, and "protective cultures" to control the microbial deterioration of food (Stiling & Cornelissen, 2005). Biocontrol is purely based on the capacity of factor to cause interference in microbial growth. Non-specific suppression, attenuation, or killing of one microbe by other in a similar ecosystem is generally termed as microbial interference. Biocontrol, unlike other microorganisms control strategies, has the advantage of being obtained from natural and sustainable sources. In the existing economy, biocontrol materials are usually less harsh and commonly accepted by the general population. These agents are safe and non-toxic, natural, and capable of withstanding physiological state while maintaining a narrow spectrum of cytotoxic activity. While bacteriophages are an emerging biocontrol strategy, some biocontrol methodologies are already extensively used (Adetunji et al., 2019). The bacteriocin nisin was first commercially launched around 60 years before and has since been employed in a variety of beverages and food products including milk, pasteurized liquid eggs, and canned vegetables (Lopresti et al., 2021). Endolysins are the proteins obtained directly from bacteriophages used as biocontrol agents to destroy bacteria exogenously. Usually, endolysins are used in combination with holins, a phage protein that punctured the peptidoglycan layer of Gram-positive bacteria to grant access into the cytosol of the cell (Roach et al., 2015). Endolysins have been used as biocontrol agents to control the growth of Clostridium perfringens, Clostridium fallax, and Listeria monocytogenes. Proteins from the bacteriophage have previously been used as biocontrol agents for food systems in the past, but the advantages and benefits of using the entire bacteriophage have recently gained attention (Xu, 2021).

18.3   HISTORICAL PERSPECTIVE ON THE USE OF BIOLOGICAL CONTROL AGENTS IN FOOD SAFETY

The use of biological control agents (BCAs) in food safety has a long and rich history dating back thousands of years. In ancient China, for example, farmers used parasitic wasps to control the spread of caterpillars in crops such as mulberry trees (Benelli et al., 2016). Similarly, in ancient Egypt, farmers used cats to control the spread of rodents in granaries and other food storage facilities (Edoh Ognakossan et al., 2016). In the early 1900s, researchers began to investigate the use of bacteria and other microorganisms as BCAs in food systems (Tomberlin et al., 2017). One of the earliest examples of this was the utilization of lactic acid microbes to ferment dairy foods such as yogurt and cheese, which helped to attenuate the growth of harmful microbes and raise the safety and shelf life of these food products. In the mid-20th century, the use of chemical pesticides became widespread in agriculture, and BCAs fell out of favor. However, concerns about the environmental and health impacts of chemical pesticides led to renewed interest in BCAs in the 1970s and 1980s (Mesnage et al., 2021). Since then, an increasing number of research on the application of biocontrol agents in food safety has been conducted. This has led to the emerge of different types of biocontrol agents, including bacteria, fungi, viruses, and parasitic wasps, as well as a better understanding of their mechanisms of action and effectiveness in different food systems. Today, biocontrol agents are used in a wide range of food systems, from fresh fruits and vegetables to processed foods such as dairy products and meat. They are particularly valuable in organic and sustainable agriculture systems, where the use of chemical pesticides is limited or prohibited (Siroli et al., 2015). Overall, the historical perspective on the utilization of biocontrol agents in food safety underscores the importance of sustainable and environmentally friendly approaches to food production. By harnessing the power of nature to control pests and improve food safety, we can create a healthier and more sustainable food system for all.

18.4 TYPES OF BIOLOGICAL CONTROL AGENTS

18.4.1 BACTERIOPHAGES AS BIOLOGICAL CONTROL AGENTS

Bacteriophages, also known as phages, are a type of virus that exhibit a narrow host range, exclusively infecting and lysing bacterial cells. These viruses are ubiquitous in nature and possess considerable potential for utilization as biocontrol agents for food safety. In terms of taxonomy, phages constitute the largest viral group and rely on bacterial and archaebacterial species as their hosts. They exhibit a size range between 20 and 200 nm and are estimated to be the most abundant microbes on Earth, with a count of 10^{31} phages in the biosphere (Ackermann & Prangishvili, 2012). Based on their morphology, phages can be divided into tailed, polyhedral (quasiicosahedral or icosahedral), filamentous, and pleomorphic categories. Tailed phages attach to bacterial cell walls through tail fibers that recognize specific molecular markers on the surface of host cells. Most tailed phages are stable in a pH range of 5–9, and their activity is lost upon heating at 60°C for 30 minutes. The replication cycle of a phage can either be active (virulent) or passive (lysogenic) (Ackermann & Prangishvili, 2012).

Phages are categorized as either lytic (virulent) or lysogenic (temperate) based on their replication cycle. Lytic phages attack bacterial cells, leading to the attenuation of host metabolism and the redirection of cellular machinery toward the synthesis of phage progeny. The lytic cycle culminates in the lysis of the host bacterium and the liberation of multiple phage fragments, which subsequently infect other cells. The entire replication cycle typically occurs within 1–2 hours, and the phage progeny population released varies depending on the phage varieties (Kohm & Hertel, 2021). Some phages can integrate their nucleic acid into the host genome or persist as episomal elements, resulting in a permanent association with the host cell and all its descendants as a prophage. In contrast to lytic phages, temperate phages infect bacterial cells and enter a dormant state during lysogeny, during which they neither produce virions nor lyse bacteria. The host cells that harbor a prophage are referred to as lysogenic (Gill & Abedon, 2003).

18.4.1.1 Mechanism of Action

The use of phages as bactericidal agent is based on their ability to infect bacterial cells and replicate within them. Upon attachment to specific receptors on the bacterial surface, phages inject their genomic DNA into the host cell, which subsequently provides the necessary materials for the multiplication of phage genome and virion assembly. Late in the lytic cycle, phage-encoded holins and lysins are expressed, causing lysis of the bacterial cell and release of viral progeny. Virulent phages, which rapidly replicate and kill infected cells, are promising candidates to control the bacterial growth (Nobrega et al., 2018; Weinbauer, 2004). In contrast, temperate phages do not immediately produce phages but instead integrate their DNA into the host chromosome or form an extrachromosomal plasmid-like replicon, becoming a prophage within the bacterial lysogen. The prophage replicates with the host genome for multiple generations, resulting in a stable relationship between the phage and the host cell (Feiner et al., 2015).

Phage proteins such as lysins can also be used as antimicrobials. Lysins are hydrolytic enzymes which break down the cell wall peptidoglycan in bacteria and facilitate the release of phage. Currently, recombinant lysins have been prepared and applied successfully against Gram-positive bacteria but need to be combined with peptide sequences for the penetration of outer region to destroy the Gram-negative bacteria (Fischetti, 2008). Lysins showed specificity for some peptidoglycan types which results in the antimicrobial action of specific Gram-positive genera or species.

Numerous lysins with various cleavage sites can be merged to increase the efficiency, and these lysins can be modified through various methods to enhance bactericidal activity (Gerstmans et al., 2016). Chimeric lysins against Listeria spp., Staphylococcus spp. And *S. pneumonia* has exhibited improved bactericidal action compared to parental proteins. Additionally, lysins can act against bacteria growth in biofilms, which are generally more resistant to traditional antibiotics. Lysostaphin, for example, was effective in eradicating biofilms produced by *Staphylococcus aureus* (Wu et al., 2003).

18.4.1.2   Biocontrol Approaches of Phages

In recent years, phages have gained attention as a potential alternative to chemical and physical methods for controlling pathogenic bacteria in food, making them an environmentally friendly and sustainable approach to food safety. Due to their specificity in targeting and leaving no residues, phages are considered safe for human consumption and have been granted GRAS status by the US Food and Drug Administration. Phages can be applied to various food products, such as raw and processed meats, dairy products, fruits and vegetables, and ready-to-eat foods, in different forms, including liquid, powder, or incorporated into packaging materials (Moye et al., 2018). Once the phages come into contact with the target bacteria, they attach to the bacterial surface and inject their genetic material, leading to the death of the bacterial cell and reducing the risk of foodborne illness.

Phages have shown effectiveness against a variety of pathogenic bacteria, including Listeria monocytogenes, Salmonella spp., *Escherichia coli*, and *S. aureus*. Furthermore, phages can be customized to target specific bacterial strains, which is particularly important for controlling antibiotic-resistant bacteria, a growing concern in the food industry (García et al., 2008). However, the usage of phages as biocontrol representative for food safety is not without limitations. One of the challenges is ensuring that the phages remain stable and active throughout the food processing and storage conditions. Another challenge is the potential for the development of phage-resistant bacteria, which can reduce the effectiveness of the treatment over time (Vikram et al., 2020). Ongoing research aims to address these limitations and develop more effective phage-based control strategies for food safety.

18.4.2   Bacteria-Based Biological Control Agents

Bacteria-based BCAs, also known as probiotics, are becoming increasingly popular as a natural and sustainable approach to controlling pathogenic bacteria in food products. The process of fermentation involves the growth of microbes, either naturally or through addition, and is known for producing numerous beneficial products. Through fermentation, bacteria can reduce food spoilage and make it free from pathogenic microorganisms and metabolites. This ecologically friendly approach has become increasingly popular. The primary microorganisms used in fermentation are LAB, which produce organic acids that have antimicrobial properties and add unique flavors and textures to food products. The LAB category typically consists of Gram-positive, non-motile, non-spore-forming microorganisms that can ferment carbohydrates. This category includes Lactobacillus, Leuconostoc, Lactococcus, Pediococcus, Streptococcus, Oenococcus, Enterococcus, Carnobacterium, Vagococcus, Aerococcus, Tetragenococcus, and Weisella genera of microorganisms. The main product obtained after fermentation is lactic acid (Trias et al., 2008). Fermentation has been used for centuries to preserve various food products, and today, approximately 60 percent of industrialized food items undergo fermentation to ensure consistency, quality, and safety. The food industry, in its pursuit of excellence, leans toward utilizing the indigenous microbiota in a meticulously controlled environment to imbue its products with an optimal texture and flavor. These microorganisms are employed in the manufacture of milk, meat, and vegetable items, and the selection of bio-preservatives is contingent upon their classification as GRAS, with absolutely no pathogenic or toxic effects on food. To this end, biological agents used in food production are categorized as either starter cultures or protective cultures. The former is a collective of microorganisms that initiate the fermentation process and generate specific compounds that endow fermented products with their characteristic texture and flavoring (Lucera et al., 2012). The food industry has a paramount interest in ensuring the highest standards of food safety, which entails the incorporation of protective cultures to effectively monitor and curb the antimicrobial activity of pathogenic microorganisms. To this end, protective cultures are employed to efficiently curtail the survival and growth of these harmful microbes in food products. Nonetheless, it is widely acknowledged that a judicious combination of both starter and protective cultures is the optimal approach for the food industry, as it guarantees not only superior food quality but also reinforces the food's inherent safety measures.

18.4.2.1   LAB and Its Action as Biocontrol Agent

Bacteria, as biocontrol agents for food safety, exert their mechanism of action through several different pathways. One of the main mechanisms of action is through competition for nutrients. Probiotic bacteria can consume the same nutrients as pathogenic bacteria, which can reduce nutrient availability for the disease causing bacteria, inhibiting their growth and proliferation. Another mode of action of probiotics is through the production of antimicrobial compounds. Probiotic bacteria can produce a range of antimicrobial substances such as bacteriocins, hydrogen peroxide, and organic acids which can attenuate the growth of pathogenic bacteria.

LAB is known to produce bacteriocins that can reduce microbial growth in food systems. These bacteriocins are cationic molecules that can withstand high temperatures and consist of up to sixty amino acid residues and hydrophobic groups. The bactericidal activity of LAB bacteriocins is attributed to their interaction with negatively charged phosphate groups present on the target cell membranes, which leads to binding, pore formation, and cell death (Mokoena, 2017). Subsequently, the cellular wall is digested by autolysin activation, causing lethal damage. However, bacteriocins are prone to inactivation by proteases present in the gut. Nonetheless, LAB bacteriocins are viewed as superior bio-preservatives due to their non-toxic, non-immunogenic, and thermo-resistant characteristics and broad spectrum of bactericidal activity (García-Bayona et al., 2017). Nisin is the most common bacteriocin produced by LAB and has many applications in the food industry. The Food and Drug Administration (FDA) has approved the use of nisin-like bacteriocins produced by GRAS microorganisms such as *Lactococcus lactis* in various processed dairy products, canned foods, and cheese. Nisin has potential against Gram-positive spoilage microorganisms and foodborne pathogens like *L. monocytogenes* (Wiedemann et al., 2001).

Despite the potential benefits of using probiotics as biocontrol agents for food safety, there are some limitations to their use. One of the challenges is ensuring that the probiotics remain stable and active throughout the food processing and storage conditions. Another challenge is the potential for the development of antibiotic resistance among probiotic bacteria, which can reduce the effectiveness of the treatment over time (Jankovic et al., 2010). In conclusion, bacteria-based BCAs, or probiotics, offer a natural and sustainable approach to controlling pathogenic bacteria in food products. They are safe for human consumption, have a long history of usage in the food processing industry, and can be used in a variety of food applications. While there are still challenges to overcome, the use of probiotics as biocontrol agents for food safety is a rapidly evolving area of research and has the potential to have a significant impact on the food industry and public health.

18.4.3   Fungi-Based Biological Control Agents

Fungi-based BCAs are a promising alternative to synthetic pesticides in controlling pests and pathogens in crops. These agents are living organisms that can be used to control the growth and proliferation of pathogenic fungi in food products without causing harm to the environment or human health (Nabi et al., 2017). Fungi-based agents are versatile and can be applied in different forms, such as spores, mycelia, and enzymes, depending on the type of application and the target pathogen. Numerous studies have been conducted to investigate the potential of fungi as biocontrol agents for managing post-harvest diseases. Some examples of these applications are outlined below. One study discovered that applying specific strains of *Pichia anomala* was a secure and efficient method of controlling *Diplodia* post-harvest rot in guava fruit, which is caused by *Lasiodiplodia theobromae* (Pat.) (Mohamed et al., 2009). Another study showed that *Trichoderma* species exhibited superior biocontrol potential for managing the post-harvest crown rot complex in bananas caused by several fungal pathogens, including *Fusarium verticillioides, Lasiodiplodia theobromae* and *Colletotrichum musae* (Alvindia & Natsuaki, 2008; Sangeetha et al., 2009). The investigation for effective biological control methods has mostly been ongoing over the past 50 years, and there has been a lot of interest in using antagonistic microbes to prevent post-harvest illnesses. In addition, other fungi-based agents that have been studied for use in food products include *Aspergillus, Penicillium*, and *Beauveria* species (Devi et al., 2020). One of the advantages of using fungi-based BCAs in food products is that they are safe for human consumption. Fungi-based BCAs can be used in a range of food products, such as fruits, vegetables, grains, and dairy products, to control the growth and proliferation of pathogenic fungi. They are also environmentally friendly, as they do not leave harmful residues in the environment or cause harm to non-target organisms.

18.4.3.1 Mechanism of Action of Fungi-Based Biocontrol Agents

Fungi-based biocontrol agents have a range of mechanisms of action in controlling the growth and proliferation of pathogenic fungi in fruits, vegetables, and crops. One of the most common mechanisms is the production of enzymes, such as glucanases and chitinases, that can degrade the cell walls of pathogenic fungi. This mechanism can inhibit the growth and proliferation of pathogenic fungi, reducing the risk of foodborne illness (Cao et al., 2009). Another mechanism of action of fungi-based biocontrol agents is the production of secondary metabolites, such as trichokonins and trichorzianines, that can inhibit the growth of pathogenic fungi. These metabolites can disrupt the cell membranes and interfere with the cellular processes of the pathogenic fungi, leading to their death or inhibition of growth (Al-Ani, 2019). Mycoparasitic fungi, which can infect and kill other fungi, including pathogenic fungi, can also be used as biocontrol agents in fruits and vegetables. These fungi penetrate the cell walls of the pathogenic fungi and release toxic metabolites, leading to their death or inhibition of growth (Mukherjee et al., 2022). Aspergillus and Penicillium species produce a range of secondary metabolites, such as aflatoxins and patulin, that can inhibit the growth of pathogenic fungi (Izzo et al., 2022). Overall, the mechanism of action of fungi-based biocontrol agents in fruits and vegetables involves a range of enzymatic, metabolic, and parasitic mechanisms, which can inhibit the growth and proliferation of pathogenic fungi, leading to improved food safety and quality.

18.4.4 PARASITOIDS AS BIOLOGICAL CONTROL AGENTS

Parasitoids are a type of BCA that are used to control pests in agricultural and horticultural crops. These BCAs are insects that lay their eggs on or inside the bodies of other insects, which eventually kill the host insect. This parasitoid attacks the larvae of several pests, such as corn borers, tomato fruit worms, and cabbage worms. The wasp lays its eggs inside the host larvae, and the larvae eventually die. The braconid wasp is a natural enemy of several pests and is widely used in biological control programs. Parasitoids can be classified into two main groups: endoparasitoids, which lay their eggs inside the host's body, and ectoparasitoids, which lay their eggs on the host's body (Wang et al., 2019). Parasitoid used in biological control is the *Trichogramma* wasp. *Trichogramma* species are tiny wasps that act as endoparasites of lepidopteran eggs. Trichogramma's small size is by far their greatest advantage as BCAs. Egg-parasitoids are virtually invisible to the human eye because they are only 0.3 mm long as adults. The best use for Trichogramma spp. is the prevention of infestation in packed goods (Sumer et al., 2009). Trichogramma species deposit their eggs inside lepidopteran eggs, leading to the death of the moth embryo before hatching and thereby stops the harmful larval stage. The parasitoid larva feeds on the contents of the moth egg, undergoes pupation, and emerges as an adult wasp within 7 to 14 days. Upon emerging, the adult parasitoids mate, and in her adult lifespan of 3 to 14 days, a single female wasp is capable of parasitizing up to 50 eggs. *Trichogramma* spp. usually forage while moving on a substrate. The eggs are frequently released with the wasps as a card with an attached parasitized egg which can be preserved at 8--12°C for up to seven days (Brunner et al., 2001).

Parasitoids are also effective in controlling pests in stored food products. The Indian meal moth is a common pest in stored grains, nuts, and cereals. The parasitoid wasp, *Habrobracon hebetor*, attacks the larvae of the Indian meal moth and other stored-product pests. The wasp lays its eggs inside the host larvae, and the larvae eventually die. The use of parasitoids in stored-product pest management can reduce the need for chemical insecticides and improve the safety and quality of stored food products (Johnson et al., 2000). The effectiveness of parasitoids as BCAs can be influenced by several factors, such as the timing and frequency of release, the host's population density, and the environmental conditions. Parasitoids are most effective when released at the early stages of pest infestation when the pest population is low. Parasitoids are also effective in reducing the spread of pests in crops and can prevent the transformation of pest resistance to chemical insecticides. In conclusion, parasitoids are a promising alternative to chemical insecticides in controlling pests in agricultural and horticultural crops. These BCAs are specific in their target pest, environmentally friendly, and have a long lifespan (Bellone et al., 2023). The use of parasitoids in biological control programs can improve food safety and quality, reduce the need for chemical insecticides, and promote sustainable agriculture. However, the efficacy of parasitoids as BCAs can be influenced by several factors, and their use requires careful monitoring and management.

18.5 CHALLENGES ASSOCIATED WITH THE USE OF BIOLOGICAL CONTROL AGENTS

The use of biological pesticides in food systems offers many benefits, including improved food safety and reduced reliance on chemical pesticides. However, there are also several challenges associated with the use of these agents. One of the main challenges is the limited effectiveness of some BCAs. The effectiveness of BCAs can be affected by several factors, such as environmental conditions, host population density, and the timing and frequency of application (Sharma, 2023). In some cases, BCAs may not be able to attain complete control of the specific pest population, leading to economic losses for farmers. Another challenge associated with the use of BCAs is the potential for unintended effects on non-target species. BCAs are often specific to their target pest, but there is always a risk that they may also harm beneficial insects or other non-target organisms. This can disrupt the balance of ecosystems and have unintended consequences for the environment. The regulatory framework for BCAs can also be a challenge (Andersen & Winding, 2004). The use of BCAs is often subject to complex regulations and may require lengthy approval processes. This can be a barrier to the widespread adoption of BCAs in food systems, especially for small-scale farmers who may not have the resources to navigate these regulatory hurdles.

Another challenge is the limited availability of BCAs. Many BCAs are produced on a small scale and may be expensive or difficult to obtain. This can limit their use in food systems, especially in developing countries where resources may be limited (Bale et al., 2008). However, there is a need for more research to better understand the interactions between BCAs and their target pests. This includes research on the mechanisms of action of different BCAs, as well as the factors that influence their effectiveness in different environments. More research is also needed to identify new BCAs and to optimize the use of existing agents.

18.6 SUMMARY

In conclusion, the use of BCAs in food systems offers many benefits, but there are also several challenges that must be addressed. These include the limited effectiveness of some BCAs, the potential for unintended effects on non-target species, regulatory hurdles, limited availability, and the need for more research. Addressing these challenges will require collaboration between researchers, regulators, farmers, and other stakeholders, and a commitment to promoting sustainable and environmentally friendly food systems.

REFERENCES

Ackermann, H. W., & Prangishvili, D. (2012). Prokaryote viruses studied by electron microscopy. *Archives of Virology*, *157*(10), 1843–1849. https://doi.org/10.1007/S00705-012-1383-Y

Adetunji, C. O., Kumar, D., Raina, M., Arogundade, O., & Sarin, N. B. (2019). Endophytic microorganisms as biological control agents for plant pathogens: A panacea for sustainable agriculture. *Plant Biotic Interactions: State of the Art*, 1–20. https://doi.org/10.1007/978-3-030-26657-8_1

Al-Ani, L. K. T. (2019). Bioactive secondary metabolites of *Trichoderma* spp. for efficient management of phytopathogens. *Secondary Metabolites of Plant Growth Promoting Rhizomicroorganisms: Discovery and Applications*, 125–143. https://doi.org/10.1007/978-981-13-5862-3_7

Ali, A., Yeoh, W. K., Forney, C., Siddiqui, M. W., & Wasim, M. (2017). Advances in postharvest technologies to extend the storage life of minimally processed fruits and vegetables. *Taylor & Francis*, *58*(15), 2632–2649. https://doi.org/10.1080/10408398.2017.1339180

Alvindia, D. G., & Natsuaki, K. T. (2008). Evaluation of fungal epiphytes isolated from banana fruit surfaces for biocontrol of banana crown rot disease. *Crop Protection*, *27*(8), 1200–1207. https://doi.org/10.1016/J.CROPRO.2008.02.007

Andersen, K. S., & Winding, A. (2004). Non-target effects of bacterial biological control agents on soil Protozoa. *Biology and Fertility of Soils*, *40*(4), 230–236. https://doi.org/10.1007/S00374-004-0774-Y

Asadollahi, M., Bastani, D., & Musavi, S. A. (2017). Enhancement of surface properties and performance of reverse osmosis membranes after surface modification: A review. *Desalination*, *420*, 330–383. https://doi.org/10.1016/J.DESAL.2017.05.027

Bale, J. S., Van Lenteren, J. C., & Bigler, F. (2008). Biological control and sustainable food production. *Philosophical Transactions of the Royal Society B: Biological Sciences*, *363*(1492), 761–776. https://doi.org/10.1098/RSTB.2007.2182

Bellone, D., Gardarin, A., Valantin-Morison, M., Kergunteuil, A., & Pashalidou, F. G. (2023). How agricultural techniques mediating bottom-up and top-down regulation foster crop protection against pests. A review. *Agronomy for Sustainable Development*, 43(1). https://doi.org/10.1007/s13593-023-00870-3

Benelli, G., Jeffries, C. L., & Walker, T. (2016). Biological control of mosquito vectors: Past, present, and future. *Insects*, *7*(4), 1–18. https://doi.org/10.3390/insects7040052

Brunner, J. F., Dunley, J. E., Doerr, M. D., & Beers, E. H. (2001). Effect of pesticides on *Colpoclypeus florus* (Hymenoptera: Eulophidae) and *Trichogramma platneri* (Hymenoptera: Trichogrammatidae), parasitoids of leafrollers in Washington. *Journal of Economic Entomology*, 94(5), 1075–1084. https://academic.oup.com/jee/article-abstract/94/5/1075/2217465

Cao, R., Liu, X., Gao, K., Mendgen, K., Kang, Z., Gao, J., Dai, Y., & Wang, X. (2009). Mycoparasitism of endophytic fungi isolated from reed on soilborne phytopathogenic fungi and production of cell wall-degrading enzymes in vitro. *Current Microbiology*, *59*(6), 584–592. https://doi.org/10.1007/S00284-009-9477-9

Devi, R., Kaur, T., Guleria, G., Rana, K. L., Kour, D., Yadav, N., Yadav, A. N., & Saxena, A. K. (2020). Fungal secondary metabolites and their biotechnological applications for human health. *New and Future Developments in Microbial Biotechnology and Bioengineering*, 147–161. https://doi.org/10.1016/B978-0-12-820528-0.00010-7

Edoh Ognakossan, K., Affognon, H. D., Mutungi, C. M., Sila, D. N., Midingoyi, S. K. G., & Owino, W. O. (2016). On-farm maize storage systems and rodent postharvest losses in six maize growing agro-ecological zones of Kenya. *Food Security*, *8*(6), 1169–1189. https://doi.org/10.1007/s12571-016-0618-2

Feiner, R., Argov, T., Rabinovich, L., Sigal, N., Borovok, I., & Herskovits, A. A. (2015). A new perspective on lysogeny: Prophages as active regulatory switches of bacteria. *Nature Reviews Microbiology*, 13(10), 641–650. https://doi.org/10.1038/nrmicro3527

Feliziani, E., Lichter, A., Smilanick, J. L., & Ippolito, A. (2016). Disinfecting agents for controlling fruit and vegetable diseases after harvest. *Postharvest Biology and Technology*, *122*, 53–69. https://doi.org/10.1016/J.POSTHARVBIO.2016.04.016

Fischetti, V. A. (2008). Bacteriophage lysins as effective antibacterials. *Current Opinion in Microbiology*, *11*(5), 393–400. https://doi.org/10.1016/J.MIB.2008.09.012

García, P., Martínez, B., Obeso, J. M., & Rodríguez, A. (2008). Bacteriophages and their application in food safety. *Letters in Applied Microbiology*, *47*(6), 479–485. https://doi.org/10.1111/J.1472-765X.2008.02458.X

García-Bayona, L., Guo, M. S., & Laub, M. T. (2017). Contact-dependent killing by *Caulobacter crescentus* via cell surface-associated, glycine zipper proteins. *ELife*, *6*. https://doi.org/10.7554/ELIFE.24869

Gerstmans, H., Rodríguez-Rubio, L., Lavigne, R., & Briers, Y. (2016). From endolysins to Artilysin®s: Novel enzyme-based approaches to kill drug-resistant bacteria. *Biochemical Society Transactions*, *44*(1), 123–128. https://doi.org/10.1042/BST20150192

Gill, J., & Abedon, S. T. (2003). Bacteriophage ecology and plants. *APSnet Feature Articles*. https://www.apsnet.org/edcenter/apsnetfeatures/Documents/2003/BacteriophageEcology.pdf

Ibrahim, S. A., Ayivi, R. D., Zimmerman, T., Siddiqui, S. A., Altemimi, A. B., Fidan, H., Esatbeyoglu, T., & Bakhshayesh, R. V. (2021). Lactic acid bacteria as antimicrobial agents: Food safety and microbial food spoilage prevention. *Foods*, *10*(12), 3131. https://doi.org/10.3390/FOODS10123131

Izzo, L., Mikušová, P., Lombardi, S., Sulyok, M., & Ritieni, A. (2022). Analysis of mycotoxin and secondary metabolites in commercial and traditional Slovak cheese samples. *Toxins*, *14*(2), 134. https://doi.org/10.3390/toxins14020134

Jankovic, I., Sybesma, W., Phothirath, P., Ananta, E., & Mercenier, A. (2010). Application of probiotics in food products—challenges and new approaches. *Current Opinion in Biotechnology*, *21*(2), 175–181. https://doi.org/10.1016/J.COPBIO.2010.03.009

Johnson, J. A., Valero, K. A., Hannel, M. M., & Gill, R. F. (2000). Seasonal occurrence of postharvest dried fruit insects and their parasitoids in a culled fig warehouse. *Journal of Economic Entomology*, *93*(4), 1380–1390. https://doi.org/10.1603/0022-0493-93.4.1380

Kohm, K., & Hertel, R. (2021). The life cycle of SPβ and related phages. *Archives of Virology*, *166*(8), 2119–2130. https://doi.org/10.1007/s00705-021-05116-9

Lopresti, F., Botta, L., La Carrubba, V., Di Pasquale, L., Settanni, L., & Gaglio, R. (2021). Combining carvacrol and nisin in biodegradable films for antibacterial packaging applications. *International Journal of Biological Macromolecules*, *193*, 117–126. https://doi.org/10.1016/J.IJBIOMAC.2021.10.118

Lucera, A., Costa, C., Conte, A., & Del Nobile, M. A. (2012). Food applications of natural antimicrobial compounds. *Frontiers in Microbiology*, *3*(August), 1–13. https://doi.org/10.3389/fmicb.2012.00287

Mesnage, R., Székács, A., & Zaller, J. G. (2021). Herbicides: Brief history, agricultural use, and potential alternatives for weed control. *Herbicides: Chemistry, Efficacy, Toxicology, and Environmental Impacts*, 1–20. https://doi.org/10.1016/B978-0-12-823674-1.00002-X

Mohamed, H., & Saad, A. (2009). The biocontrol of postharvest disease (*Botryodiplodia theobromae*) of guava (*Psidium guajava* L.) by the application of yeast strains. *Postharvest Biology and Technology*, *53*(3), 123–130. https://doi.org/10.1016/j.postharvbio.2009.04.001

Mokoena, M. P. (2017). Lactic acid bacteria and their bacteriocins: Classification, biosynthesis and applications against uropathogens: A mini-review. *Molecules*, *22*(8), 1255. https://doi.org/10.3390/MOLECULES22081255

Moye, Z. D., Woolston, J., & Sulakvelidze, A. (2018). Bacteriophage applications for food production and processing. *Viruses*, *10*(4), 205. https://doi.org/10.3390/V10040205

Mukherjee, P. K., Mendoza-Mendoza, A., Zeilinger, S., & Horwitz, B. A. (2022). Mycoparasitism as a mechanism of Trichoderma-mediated suppression of plant diseases. *Fungal Biology Reviews*, *39*, 15–33. https://doi.org/10.1016/J.FBR.2021.11.004

Nabi, S. U., Raja, W. H., Kumawat, K. L., Mir, J. I., Sharma, O. C., Singh, D. B., & Sheikh, M. A. (2017). Post harvest diseases of temperate fruits and their management strategies-a review. *International Journal of Pure & Applied Bioscience*, *5*(3), 885–898. https://doi.org/10.18782/2320-7051.2981

Nawawi, N. I. M., Ijod, G., Senevirathna, S. S. J., Aadil, R. M., Yusof, N. L., Yusoff, M. M., Adzahan, N. M., & Azman, E. M. (2023, February). Comparison of high pressure and thermal pasteurization on the quality parameters of strawberry products: a review. *Food Science and Biotechnology*. https://doi.org/10.1007/s10068-023-01276-3

Nobrega, F. L., Vlot, M., de Jonge, P. A., Dreesens, L. L., Beaumont, H. J. E., Lavigne, R., Dutilh, B. E., & Brouns, S. J. J. (2018). Targeting mechanisms of tailed bacteriophages. *Nature Reviews Microbiology*, *16*(12), 760–773. https://doi.org/10.1038/s41579-018-0070-8

Roach, D. R., & Donovan, D. M. (2015). Antimicrobial bacteriophage-derived proteins and therapeutic applications. *Bacteriophage*, *5*(3), e1062590. https://doi.org/10.1080/21597081.2015.1062590

Sangeetha, G., Usharani, S., & Muthukumar, A. (2009). Biocontrol with Trichoderma species for the management of postharvest crown rot of banana. *Phytopathologia Mediterranea*, *48*(2), 214–225. https://www.jstor.org/stable/26463346

Sharma, P. (2023). Biocontrol strategies—retrospect and prospects. *Indian Phytopathology, 76*(1), 47–59. https://doi.org/10.1007/S42360-023-00601-4

Siroli, L., Patrignani, F., Serrazanetti, D. I., Gardini, F., & Lanciotti, R. (2015). Innovative strategies based on the use of bio-control agents to improve the safety, shelf-life and quality of minimally processed fruits and vegetables. *Trends in Food Science & Technology, 46*(2), 302–310. https://doi.org/10.1016/J.TIFS.2015.04.014

Stiling, P., & Cornelissen, T. (2005). What makes a successful biocontrol agent? A meta-analysis of biological control agent performance. *Biological Control, 34*(3), 236–246. https://doi.org/10.1016/J.BIOCONTROL.2005.02.017

Sumer, F., Tuncbilek, A. S., Oztemiz, S., Pintureau, B., Rugman-Jones, P., & Stouthamer, R. (2009). A molecular key to the common species of Trichogramma of the Mediterranean region. *BioControl, 54*(5), 617–624. https://doi.org/10.1007/S10526-009-9219-8

Szczech, M., Kowalska, B., Smolińska, U., Maciorowski, R., Oskiera, M., & Michalska, A. (2018). Microbial quality of organic and conventional vegetables from Polish farms. *International Journal of Food Microbiology, 286*, 155–161. https://doi.org/10.1016/j.ijfoodmicro.2018.08.018

Tewari, S., Sehrawat, R., Nema, P. K., & Kaur, B. P. (2017). Preservation effect of high pressure processing on ascorbic acid of fruits and vegetables: A review. *Journal of Food Biochemistry, 41*(1). https://doi.org/10.1111/jfbc.12319

Tomberlin, J. K., Crippen, T. L., Tarone, A. M., Chaudhury, M. F. B., Singh, B., Cammack, J. A., & Meisel, R. P. (2017). A review of bacterial interactions with blow flies (Diptera: Calliphoridae) of medical, veterinary, and forensic importance. *Annals of the Entomological Society of America, 110*(1), 19–36. https://doi.org/10.1093/aesa/saw086

Trias, R., Bañeras, L., Montesinos, E., & Badosa, E. (2008). Lactic acid bacteria from fresh fruit and vegetables as biocontrol agents of phytopathogenic bacteria and fungi. *International Microbiology, 11*(4), 231–236. https://doi.org/10.2436/20.1501.01.66

Vikram, A., Woolston, J., & Sulakvelidze, A. (2021). Phage biocontrol applications in food production and processing. *Current Issues in Molecular Biology, 40*(1), 267–302. https://doi.org/10.21775/CIMB.040.267

Wang, Z.-zhi, Liu, Y.-quan, Shi, M., Huang, J.-hua, & Chen, X.-xin. (2019). Parasitoid wasps as effective biological control agents. *Journal of Integrative Agriculture, 18*(4), 705–715. https://doi.org/10.1016/S2095-3119(18)62078-7

Weinbauer, M. G. (2004). Ecology of prokaryotic viruses. *FEMS Microbiology Reviews, 28*(2), 127–181. https://doi.org/10.1016/j.femsre.2003.08.001

Wiedemann, I., Breukink, E., Van Kraaij, C., Kuipers, O. P., Bierbaum, G., De Kruijff, B., & Sahl, H. G. (2001). Specific binding of nisin to the peptidoglycan precursor lipid II combines pore formation and inhibition of cell wall biosynthesis for potent antibiotic activity. *Journal of Biological Chemistry, 276*(3), 1772–1779. https://doi.org/10.1074/jbc.M006770200

Wilson, M. D., Stanley, R. A., Eyles, A., & Ross, T. (2019). Innovative processes and technologies for modified atmosphere packaging of fresh and fresh-cut fruits and vegetables. *Critical Reviews in Food Science and Nutrition, 59*(3), 411–422. https://doi.org/10.1080/10408398.2017.1375892

Wu, J. A., Kusuma, C., Mond, J. J., & Kokai-Kun, J. F. (2003). Lysostaphin disrupts *Staphylococcus aureus* and *Staphylococcus epidermidis* biofilms on artificial surfaces. *Antimicrobial Agents and Chemotherapy, 47*(11), 3407–3414. https://doi.org/10.1128/AAC.47.11.3407-3414.2003

Xu, Y. (2021). Phage and phage lysins: New era of bio-preservatives and food safety agents. *Journal of Food Science, 86*(8), 3349–3373.

19 Novel Disinfectant Technologies Applications for the Food Industries

V. Monica, Anbarasan Rajan, and R. Mahendran

19.1 INTRODUCTION

Cleaning and disinfection are the two major processing steps that can help in producing contaminant-free foods by removing impurities and microbes. These terminologies get differentiated from each other based on their mode and target of action. Cleaning removes impurities present on the top surface of food and reduces the microbial load to a certain extent, while disinfection disrupts the infection mechanism of the microorganisms and pathogens that reside in the food (Joshi et al., 2013). The efficacy of the disinfectants depends upon the nature of the impurity (organic and inorganic), target microorganisms, characteristics of the disinfectant, the temperature of the solution, nature of the surface or inmate object, and specific requirements of heat in the case of physical disinfection methods (Animal and Plant Health Inspection Service, 2020). The characteristics of an ideal disinfectant must have the following characteristics as shown in Figure 19.1 (Centre for Disease Control and Prevention, 2016).

Physical disinfection methods like heat, ultraviolet (UV), electron beam, and chemical disinfectants like alcohols, sodium bicarbonate, quaternary ammonium compounds, peracetic acid, etc., destroy pathogenic microscopic organisms (Meireles et al., 2016). Broadly speaking, chemical disinfectants have both positive and negative sides. For example, alcohols are fast in action but are non-sporicidal and flammable in nature. On the other hand, iodophors have intermediate action on bacteria but take a longer time to kill fungi. Whereas, the quaternary ammonium compounds were responsible for several outbreaks and water hardness during water disinfection. Further, these disinfectants are less stable and categorized under low-level disinfectants that can inactivate bacteria, viruses, and fungi but not spores (Rutala & Weber, 2019). Further, chemical disinfectants like biocide are not capable of removing biofilm formed on the equipment surface (Alvarez-Ordóñez et al., 2019). Therefore, employing novel technologies in place of existing disinfectant technologies would be efficacious and non-toxic.

Presently, physical methods including, cold plasma (CP), ionizing radiation, high power intensity ultrasound, pulsed light (PL), high pressure, and chemical methods comprising ozone, chlorine dioxide, electrolyzed water, and high-pressure carbon dioxide are the emerging disinfectant technologies with all the above-mentioned ideal characteristics and appreciable biocidal efficacy. Further, these technologies do not cause any biochemical and physiological deterioration in foods, particularly, fruits and vegetables (Deng et al., 2020). Indeed, these novel disinfectant technologies increase the shelf life of fresh produce by removing the soil, dirt, and microbes from the food contact surfaces.

This chapter lists out applications and exigencies of various novel disinfectant technologies in food industries including their role in disinfecting food processing equipment surfaces.

19.2 CHLORINE DIOXIDE

Chlorine dioxide (ClO_2) is a yellowish to reddish gas, readily soluble in water even at high concentrations at room temperature. ClO_2 has been conceded for disinfection, sterilization, fumigation,

DOI: 10.1201/9781003359302-19

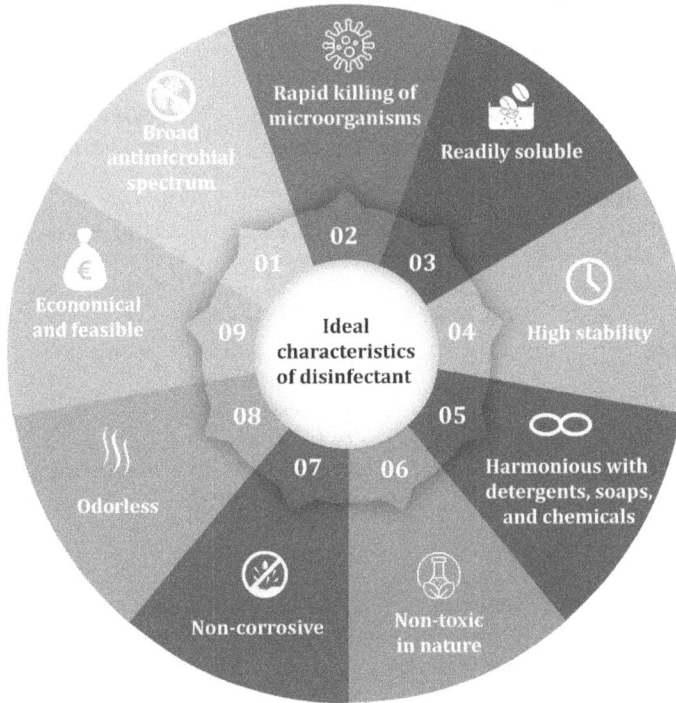

FIGURE 19.1 Characteristics of an ideal disinfectant.

and preservation in the food processing sector due to its high oxidation capacity. The oxidizing power of ClO_2 is 2.5 times better than chlorine hydroxide (HOCl) due to its structural characteristics. Further, it exists in aqueous and gaseous forms which facilitate its application in fruits and vegetables, edible flowers, dairy products, condiments, and aquatic and poultry products. However, ClO_2 gas is widely used due to its high penetrating ability than aqueous ClO_2 (Joshi et al., 2013).

When applied on contaminated surfaces, ClO_2 gas oxidizes the proteins present in the cell surface of the target microorganisms by diffusing through the cell membrane. As a result, the conformational structure of proteins gets altered and induces cell metabolic disorder. Similarly, lipids are oxidized by ClO_2 gas. On top of all these, ClO_2 penetration damages the trans-membrane ion gradient and respiratory pathway of the microorganisms which ultimately causes death (Han et al., 2018). The applications of chlorine dioxide disinfection in food industries are summarized in Table 19.1.

When aqueous ClO_2 was assessed for its disinfection efficiency along with chlorine (Cl_2) in fresh-cut iceberg lettuce, ClO_2 showed to reduce disinfectants' need by ten times that of chlorine (Cl_2) to perform the same task. Further, just 5 mg/L of ClO_2 inactivated psychotropic microorganisms (>3 CFU/mL in 2 min) and *Escherichia coli* (>5 log in 3 minutes). Besides, ClO_2 is also more stable than chlorine and hydrogen peroxide (but less stable than peracetic acid in disinfection action). However, degradation of ClO_2 to ClO_2^- is the only concern (~77% decay of ClO_2 was reported in studies). Therefore, ClO_2 must be used with sufficient water for a safe disinfectant process in controlled conditions (Van Haute et al., 2017). Similar to aqueous ClO_2 disinfectant, gaseous ClO_2 was applied to spices to inactivate *Salmonella* present in the black peppercorns and cumin seeds. The gas composition, RH, and exposure time had a positive linear effect on the bactericidal mechanism. A high ClO_2 gas concentration of 15 mg/L with 80% RH for 300 minutes brought more than a 5-log reduction of *Salmonella* in both spices. Under high RH, moisture accumulated on the bacterial surface of the spices, and this made bacteria sensitive to ClO_2 gas. ClO_2 gas disinfection efficiency increased as it got oxidized by surface moisture (Wei et al., 2021). ClO_2 disinfection is

TABLE 19.1

Novel Disinfection Application in Food Industries

Disinfectant Technology	Product	Target Microorganism	Disinfectant Condition	Effect on the Pathogens	References
Chlorine Dioxide Disinfection					
Gaseous ClO_2	Spices	*Salmonella*	15 mg/L for 300 minutes	>5 log reduction	(Wei et al., 2021)
	Baby carrots	*Salmonella* *Escherichia coli*	6 g for 24.6 minutes 6 g for 31.3 minutes	3 log reductions	(Guan et al., 2021)
Aqueous ClO_2	Beef	*E. coli*	15 ppm, sprayed or for 4 seconds every 15 minutes for 36 cycles	3 log reductions	(Kocharunchitt et al., 2020)
Gaseous ClO_2 + Heat	Cabbage and radish seeds	*E. coli O157:H7*	3000 ppm for 90 minutes at 60°C and 85% RH	0.5 log CFU/g	(Yeom et al., 2021)
Ozone Disinfection					
Gaseous ozone	Sugar cane juice	Total plate count Yeast and mold	6.5 L/min for 20 minutes	3.72 log reduction 2.43 log reduction	(Panigrahi et al., 2020)
Aqueous ozone	Fresh cut onions	Total plate count	5 ppm for 8 minutes	6.396 log CFU/g	(Aslam et al., 2022)
Aqueous ozone + sodium metasilicate	Cabbage	*E. Coli* Bacteria Yeast and mold	2 ppm of ozone with 0.4% of SM for 2 minutes	$3.05 \log_{10}$ CFU g^{-1} $3.03 \log_{10}$ CFU g^{-1} $<1 \log_{10}$ CFU g^{-1}	(Nie et al., 2020)
Electrolyzed Water Disinfection					
SAEW	Cabbage	*Pectobacterium carotovorum*	22.17 ppm of free available chlorine for 180 seconds	$5.94 \pm 0.07 \log_{10}$ CFU/g	(Song et al., 2021)
SAEW	Wheat grains Clear flour Bran flour	Total plate count bacteria	70 ppm for 90 seconds	0.65 log CFU/g 0.91 log CFU/g 0.93 log CFU/g	(Chen et al., 2020)
Ultraviolet Disinfection					
UV	Orange peel Orange juice	Bacterial spores	Intensity: 18.0 kJ/cm^2 Time: 3 minutes Intensity: 278.8 kJ/cm^2 Time: 45 minutes	2 log units 2 log units	(Colás-Medà et al., 2021)
UV + UHPH	Whole milk and skim milk	*Bacillus subtilis*	91.2 J/mL + 200 MPa at 80°C	5 log CFU/mL	(Martinez-Garcia et al., 2019)
UV + OZONE + IR	Black pepper Onion flakes	*E. coli*	UV and IR Wavelength: 254 nm Intensity: 100 W Time: 2.5 minutes Ozone Flow rate: 300 mg/h Time: 10 minutes	4.20 log reduction 2.69 log reduction	(El Darra et al., 2021)

(Continued)

TABLE 19.1 (*Continued*)
Novel Disinfection Application in Food Industries

Disinfectant Technology	Product	Target Microorganism	Disinfectant Condition	Effect on the Pathogens	References
Pulsed Light Disinfection					
PL	Chicken	*Campylobacter jejuni*	Energy: 9.68 ± 0.15 J/cm^2 Voltage: 2.28 kV Distance: 3 cm	4.5 ± 0.1 log CFU/g	(Baptista et al., 2022)
	Fermented salami	*E. coli 0157:H7* *Salmonella typhimurium*	Energy: 3 J/cm^2 Voltage: 3 kV	2.29 log CFU/g 2.25 log CFU/g	(Rajkovic et al., 2017)
Plasma Activated Water Disinfection					
PAW (hollow fiber-based cold microplasma jet)	Apples	Aerobic plate count Yeast and mold Coliforms	Type: Immersion, 5 minutes	$1.05 \log_{10}$ CFU/g 0.64 and 1.04 $\log_{10}$ CFU/g $0.86 \log_{10}$ CFU/g	(Liu et al., 2020)
PAW (surface barrier discharge)	Baby spinach leaves	Total bacterial count	Type: Immersion, 2 minutes	1 log CFU/g	(Vaka et al., 2019)
PAW (atmospheric pressure plasma jet)	Mung bean sprouts	Total bacterial count Total yeast and mold	Type: Immersion, 30 minutes	$2.32 \log_{10}$ CFU/g $2.84 \log_{10}$ CFU/g	(Xiang et al., 2019)
Ultrasound Disinfection					
US	Chicken breast meat	*Staphylococcus aureus* *Salmonella* spp.	40 kHz, intensity 9.6 W/cm^{-2} for 50 minutes	1.26 log CFU/mL 1.83 log CFU/mL	(Piñon et al., 2020)
US + ClO$_2$	Alfalfa seeds sprouts Mung beans sprouts	*Salmonella enteritidis* *E. coli* *S. enteritidis* *E. coli*	26 kHz, 200 W + 3 ppm for 5 minutes	1.94 ± 0.42 log CFU/g 2.06 ± 0.23 log CFU/g 2.62 ± 0.02 log CFU/g 2.08 ± 0.2 log CFU/g	(Millan-Sango et al., 2017)

claimed to be a non-toxic residue-free process in food processing industries (Feliziani et al., 2016). In addition, ClO$_2$ can also act as an alternate sanitizing agent for phosphine and methyl bromide in inactivating fungi and insects in grain storage (Han et al., 2018) (Figure 19.2).

In the meat processing industry, the application of ClO$_2$ on beef meat caused a consistent antimicrobial effect on the fat surface than on the lean surface. The reason behind this remains unknown (Kocharunchitt et al., 2020). The above results show that ClO$_2$ has an effective antimicrobial intervention in inactivating *E. coli* in meat. Apart from all these, ClO$_2$ also has the potential to produce a synergistic effect along with other disinfectants. Following are some examples of hurdle technologies of ClO$_2$ disinfection in food industries.

Apart from microbial inactivation, ClO$_2$ also increases the keeping quality of fruits and vegetables by maintaining their biochemical characteristics during storage. The combination of aqueous ClO$_2$ and passive modified atmospheric packaging (21% O$_2$, 0.03% CO$_2$, and 79% N$_2$) reduced the respiration rate and maintained the firmness, pH, and total soluble solids of sweet cherries.

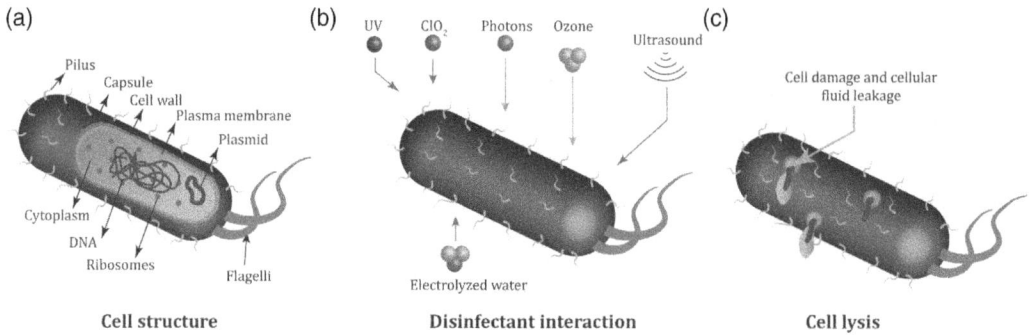

FIGURE 19.2 Overview on disinfectant interaction with bacterial cell (a) cell structure, (b) disinfectant interaction, and (c) cell lysis.

However, adverse effects were seen on respiration rate, weight loss, pH, anthocyanin profile, and sensory scores with the ClO_2 concentration exceeding 25 mg/L (Colgecen & Aday, 2015). Fresh quail eggs (surface) being carriers of *Salmonella* are responsible for foodborne illness. A combination of X-ray irradiation followed by ClO_2 disinfection inactivated *Salmonella typhimurium* with no negative impact on the color and thickness of the eggshell. The synergistic effect of X-ray and ClO_2 damaged the genetic material responsible for protein synthesis and improved the keeping quality of eggs by destroying the biofilms of *Salmonella*. Therefore, ClO_2 is an effective disinfectant in egg production, processing, and storage areas to ensure the product's safety and quality (Park et al., 2018).

In the food packaging industry, gaseous ClO_2 is used as an antimicrobial agent, even though it has strong oxidation power. In fruit and vegetable packaging, ClO_2 is used to prevent tissue softening by inhibiting the enzymes such as polyphenol oxidase and peroxidase. In the MAP, gaseous ClO_2 disinfected packages are used to store meat, poultry, and fish, where ClO_2 acts as a causative agent against microorganism growth. However, ClO_2 may also alter the properties and integrity of packaging material and might lower the shelf life of food products (Singh et al., 2021).

Though FDA permits ClO_2 as a disinfecting agent in fruits, vegetables, grain, meat, poultry processing, and packaging industries, it has the following disadvantages (Joshi et al., 2013):

    a. *Toxicity and explosiveness*: The explosion of ClO_2 occurs above partial pressure of 0.1 bar and it is also toxic to humans
    b. *Not suitable for low-moisture foods*: Aqueous ClO_2 is not recommended for low-moisture foods
    c. ClO_2 has a short life and requires an on-site production process
    d. ClO_2-treated products require water treatment after the disinfection process

19.3 OZONE (O_3)

Ozone (O_3) is a triatomic oxygen molecule and is regarded as generally recognized as safe (GRAS) by FDA because of its residue-free nature (Joshi et al., 2013). Normally, Ozone becomes colorless at higher concentrations of air and blue with a pungent odor in the presence of cold water and alkali conditions. The stability of Ozone is less and so the half-life of Ozone is about 20 to 50 minutes, however, it varies with temperature. O_3 can be continuously generated in presence of air and it cannot be stored for a long time due to its instability. Corona discharge, UV radiation, and electrolysis are common Ozone generation methods. Corona discharge and UV are preferred methods to generate ozone because of their cost-effectiveness (Boopathy et al., 2022).

Both gaseous and aqueous phases of ozone are used in disinfecting foodstuffs like fruits, vegetables, meat, fish, spices, and others. Ozone dissolves ten times higher in water than in oxygen and aqueous ozone dissolves faster than gaseous ozone. Disinfecting properties of ozone have broad spectra antimicrobial intervention like killing gram-negative and gram-positive bacteria and their spores. The mechanism of ozone disinfection involves an attack on the cell membrane followed by cytoplasm, spore coats, and virus capsids. Damage to the cell membrane insists intracellular content leakage from the contaminating microorganism and leading to death. The mechanism also involves the oxidation of the sulfhydryl group, proteins, amino acids, and peptides into small peptides, and polyunsaturated fatty acids into acid peroxides (Brodowska et al., 2018).

Ozone acts as an alternative to thermal and chemical disinfectant technologies in food processing industries. Ozone disinfection application in food industries is summarized in Table 19.1. Ozone concentration, pH, temperature, bubble size, gas flow rate, and contact time influence its disinfection efficiency. An increase in the ozone concentration and exposure time increases microbial inactivation. Further, it also improves the total soluble solids, polyphenol content, and flavonoid content, and inactivates polyphenols oxidase and peroxidase enzymes in a liquid food product such as sugarcane juice (Panigrahi et al., 2020). Sarooei et al. (2019) stated that it might be due to an unknown lag phase of microorganisms or high humidity during ozonation. Comparative evaluation of sanitizing efficiency of chlorine and ozone treatment on freshly cut onions showed better microbial inactivation for ozone treatment than in chlorine-treated samples. Conversely, ozonated refrigerated onions have an increased microbial count than chlorine-treated refrigerated onions due to the higher disinfecting efficiency of aqueous ozone on mesophiles than psychotropics (Aslam et al., 2022).

In the meat processing industry, ozone treatment was applied for 8 hours, targeting total mesophiles, Enterobacteriaceae, yeast, and mold present in the turkey breast meat. By the end of the 4th and 6th hour, maximum inactivation of yeast, mold, and Enterobacteriaceae was achieved. Extending the treatment for another 2 hours inactivated mesophiles in meat. During the last 2 hours, ozone couldn't reach deeper into the breast meat to cause the further inactivation of yeast and mold and Enterobacteriaceae. Therefore, optimizing the ozone treatment condition is essential to utilize O_3 as a decontaminating agent in meat processing industries (Ayranci et al., 2020).

Hurdle technology plays an important role in increasing the disinfection potential of ozone. For example, L-lactate increased the decontamination power of ozone while disinfecting chicken drumsticks (Megahed et al., 2020). The combination of aqueous ozone with GRAS sodium metasilicate (SM) preservative enhanced the shelf life of freshly cut cabbage by reducing the population of spoiling microbes. Ozone treatment alone caused a 1.78 log reduction of E. coli and a 1.67 log reduction of bacteria. But the combination treatment brought more than a 2-fold reduction in microbial load. This is because SM acts as a complementary additive to ozone in E. coli and bacterial inactivation. As ozone works effectively with SM, it may be used as a processing intervention to control spoiling microorganisms in vegetables (Nie et al., 2020).

Though ozone is an alternative to thermal and chemical disinfectants, it possesses the following disadvantages (Boopathy et al., 2022):

a. *Instability of ozone*: Ozone either decomposes or oxidizes into oxygen in the form of free radicals at a temperature above 50°C
b. *Safety limit of ozone*: Exceeding the safe limit (>0.10 ppm) causes breathing difficulties, nasal irritation, asthma attacks, and headaches
c. *Exceeded ozonation*: Beyond the threshold limit of ozone, it causes discoloration in pigment-rich fruit and vegetables. It also affects the stability of fat-rich foods such as dairy products

19.4 ELECTROLYZED WATER

Electrolyzed water (EW) is a novel disinfection technology with a broad antimicrobial spectrum. EW is composed of regular water without any harmful chemicals but contains NaCl. The principle of EW production consists of electrolysis of NaCl solution involving the movement of ions to the opposite charge electrode. Oxidation occurs at the anode electrode with the production of oxygen gas, hypochlorite ions (OCl^-), hypochlorous acid (HOCl), and hydrochloric acid. Meanwhile, reduction in the cathode forms hydrogen and NaOH (Joshi et al., 2013).

The bactericidal mechanism of EW disinfection is brought by the by-products formed during EW production. Hypochlorite formed disrupts the cell membrane and damages the proteins in the plasma membrane. Hypochlorous acid in EW damages the cell wall, inhibits enzymatic activity, and attacks the protein structure of the microbe. This gradually affects the metabolism and leads to the death of the microorganism. Active chlorine (Cl_2^-) in EW also follows a similar mechanism to inactivate microbes (Deng et al., 2020).

Intrinsic factors like pH, ORP, exposure time, treatment temperature, free available chlorine content (ACC), and type of EW affect EW disinfection efficiency. For example, during SAEW disinfection of celery and cilantro, an increase in ACC reduced the total aerobic bacteria, yeast, and mold below the detectable limit. Effective inactivation of causative microorganisms was achieved by longer exposure to the SAEW, while the temperature did not affect the disinfection efficiency at any point (Zhang et al., 2016).

In fruit processing industries, norovirus remains a spoilage microorganism and causes foodborne illness. Washing strawberries with neutral electrolyzed water (NEW) before and after storage have resulted in the reduction of non-enveloped RNA noroviruses. The efficiency of NEW in virus inactivation is evaluated by a surrogate, MS2 bacteriophage. Inactivation of the surrogate was observed during 30 days of strawberry storage. NEW resulted in ~1 log reduction of MS2 bacteriophage in strawberries which is similar to inactivation brought by 50 ppm of chlorine wash (Huang et al., 2019).

In grain processing, slightly acidic electrolyzed water (SAEW) is owing to stabilize final flour products by altering the rheology of straight and clear flour, total polyphenols, and enzymatic activities of bran and clear flour. On top, the microbial profile comprising the total plate count of bacteria, yeast, and mold got significantly decreased in wheat grains and milled flours. SAEW is used along with distilled water in the tempering process to facilitate wheat milling. Clear flour and bran flour obtained have strong dough structures, and better thermal stability assisted by SAEW with low microbiota. Therefore, SAEW is a novel disinfectant cum tempering agent for grain processing industries (Chen et al., 2020).

In meat and poultry processing, EW has wide applications as a disinfectant. SAEW effectively reduces the microbial load in chicken carcasses in the slaughter line at a concentration of twice the acidic electrolyzed water (AEW). A novel spray cabinet is used to deliver the disinfectant at a pressure of 0.3 MPa and this method is applicable in the large-scale industry (Wang et al., 2018). Similarly, NEW is ten times more effective than NaClO and the major advantage is that does not initiate the formation of oxidized compounds. The presence of Trihalomethanes in chicken meat due to the oxidation of organic matter during NaClO is a serious matter of concern. Application NEW to chicken up to 50 mg/L does not produce trihalomethanes and at this concentration cell lysis is caused by swelling and morphological changes in meat. During storage, pH, and color are the main characteristics of chicken meat and NEW maintains the pH and color of meat more than NaClO (Hernández-Pimentel et al., 2020).

The combination of AEW with an organic acid, levulinic acid to lettuce caused the inactivation of *E. coli* and *Listeria innocua Seeliger* by damaging cell membranes leading to the diminishing of cell diameter to 2.12 and 1.24 μm^2 (Zhao et al., 2019). The formation of *Bacillus cereus* biofilm on the surface of spinach leaves, beet, and lettuce leaves is a serious concern and hurdle disinfection technology involving SAEW with ultrasound and mild heat is an eco-friendly technique that attacks the plasma membrane of the microorganisms and cells embedded in the biofilm gets seriously affected with this combination (Hussain et al., 2019). A similar combinational effect of EW and US was seen in mushrooms and EW and ozone in alfalfa seeds and sprouts (Mohammad et al., 2020; Wu et al., 2018). The application of electrolyzed water is summarized in Table 19.1. The low operational cost of EW disinfection is applied to various food products to inactivate microorganisms because of its easy operation. Indeed, it has some following drawbacks (Deng et al., 2020):

a. *On-site production*: Like, ClO_2 it requires on-site generation
b. *Corrosive*: corrosive hazard occurs because of the involvement of strong acid and Cl_2

19.5 ULTRAVIOLET (UV)

UV light is a non-ionizing radiation with a wavelength ranging from 100 to 400 nm, between X-rays and visible rays in the electromagnetic spectrum. UV light is classified based on its action as UVA (it ranges from 315 to 400 nm and causes human skin changes or tanning), UV-B (it ranges from 280 to 315 nm and causes skin cancer, UV-C (it ranges from 180 to 280 nm and has germicidal action), and UV Vacuum (VUV) (ranges from 100 to 180 nm). Among these, UV-C light is used as a disinfectant in food processing plants because of its germicidal action against bacteria, viruses, and protozoa. Sun is a natural source of UV rays. But UV rays are absorbed by ozone before reaching the earth. Some artificial UV-C light sources are mercury vapor light lamps, black lights, fluorescent lamps, incandescent lamps, and laser sources (USFDA, 2020).

UV-C light incidence produces DNA photoproducts at its peak absorption range of 260 nm. The formed photoproducts (pyrimidine 6-4 pyrimidone (6-4PP) and cyclobutane pyrimidine dimers – CPDs) interfere with DNA transcription and replication. These reactions inhibit surface microbial multiplication. While, the indirect germicidal action of UV-C light results in membrane damage, with the production of reactive oxygen species (ROS), hydroxyl ions (OH^-), hydrogen peroxide, singlet oxygen, and superoxide anions. The following factors affect UV light disinfection (Hinds et al., 2019).

Processing factors – UV dosage, time, temperature

a. Equipment factors – UV light source, number and position of UV lamps, distance between the lamps and product
b. Microbial factors – type, species, initial count, the growth phase of microorganism

In fruit processing, the low penetrating UV-C's effect varies depending on the food matrix (type of juice, pulp, processing conditions like clarification). For example, UV-C light is more effective against orange peel microbial spores than orange juice due to the presence of pulp and suspended particles in the juice. Orange juice required a higher UV-C light dosage and longer treatment time (15 times longer) to achieve the exact inactivation rate (2 log units) achieved in orange peel (Colás-Medà et al., 2021).

In the meat processing industry, mesophilic and facultative anaerobic bacteria in beef carcass were inactivated using UV-C light (253.7 nm) from its initial microbial load ($4.45 \pm 0.17 \times 10^3$ CFU/cm^2). The UV-C light disinfection of beef carcasses should be done under a chill chamber. It was found that the efficiency of UV-C light is directly proportional to the meat half carcass in the chill chamber. The 45 minutes of the treatment enhanced the shelf life of meat up to 12 days in refrigeration (Rodionova et al., 2020).

In dairy processing, decontaminating ricotta cheese using UV-C and near-UV-visible light increased its shelf life by 50%. The control cheese samples became unacceptable within 5 days, whereas the microbial load of decontaminated cheese remained within the acceptable range for more than 6 days. UV-C light inactivated cheese-spoiling microbe, *Pseudomonas fluorescens* below its detection limit by degrading the proteins and making them unavailable for the microbe (Ricciardi et al., 2020).

Hurdle technologies of UV-C light disinfection have a synergistic effect on target microorganisms at reduced treatment intensity and energy input. In dairy processing, the combination of ultra-high-pressure homogenization (UHPH) and UV-C light inactivated *Bacillus subtilis* in whole milk and skim milk. UHPH pre-treatment increased the turbidity and absorption coefficient of whole milk and skim milk for UV-C light to transmit through it. Therefore, the hurdling concept of UV-C light with other technologies can be applied to inactivate bacterial spores (Martinez-Garcia et al., 2019). Similarly, the triple technologies combination (Ozone + UV + IR) effectively reduced the *E. coli* concentration in onion flakes and black peppers. Different sequential combinations were applied to spices to inactivate *E. coli*. Among those, ozone followed by UV-C and IR radiations resulted in a 99.99% of inactivation rate of *E. coli*. Ozone with UV-C light took 17 minutes and ozone with IR light took 30 minutes to achieve a complete reduction of *E. coli*. Interestingly, the complete inactivation of *E. coli* in spices was achieved in 12.5 minutes by ozone, UV-C light, and IR radiation in combination (El Darra et al., 2021). The UV light disinfection application in food industries is summarized in Table 19.1.

Though UV-C light disinfection has advantages like low equipment cost, ease of operation, and minimal effect on food quality, it has the following disadvantages during disinfection (Koutchma, 2016):

a. *Low penetration power*: UV-C light has the shortest wavelength (180–280 nm) and disinfectant surface microorganisms
b. *Shadowing*: Agglomeration and clustering of bacterial cells impede light penetration
c. *The gradual recovery of microbes*: Microbes recover in the foods over a period of time

19.6 PULSED LIGHT (PL)

PL is another non-ionizing, non-thermal technology with a wavelength ranging from 200 to 1100 nm between UV and near-infrared waves in the electromagnetic spectrum. FDA permits PL disinfection of foods with short and intense pulses from a xenon lamp source, with pulse duration not longer than 2 milliseconds and treatment intensity not exceeding 12.0 J/cm^2 (Code of Federal Regulations, 2022). PL is effective against spoilage microorganisms such as *E. coli*, *B. cereus*, *Listeria monocytogenes*, *Salmonella enteritidis*, *Pseudomonas aeruginosa*, and *Saccharomyces cerevisiae* present in fruits, vegetables, beverages, milk, meat, fish and its products (Mahendran et al., 2019). Interestingly, the antimicrobial action of PL is 4 to 6 times higher than continuous UV-C light (Pataro et al., 2015).

The disinfection mechanism of PL produces photophysical, photochemical, and photothermal effects on microorganisms. The photophysical effect of PL kills microorganisms by disrupting their cell wall, modifying the cell membrane and cell shape, and initiating the leakage of the intracellular components from the cell leading to microbial death. PL has an enhanced photophysical effect on microorganisms than UV-C light. The photochemical effect of PL hinders microbial growth with DNA absorbing the light at its threshold wavelength of 253.7 nm. PL of wavelength greater than 295 nm can destroy the virus particles in the food. Lastly, the photothermal effect uses PL of a wavelength in the near-infrared region to create "photothermal stress" on microorganisms by increasing the temperature, leading to the rupture of the cell walls and explosion of intra-cellular contents (Cullen et al., 2012).

The following are factors affecting PL disinfection in food industries (Kramer et al., 2017):

a. *Processing factors*: Light source, dosage, time, spectral range
b. *Equipment factors*: Number of flashlights, applied voltage, the distance between the flash lamp and the target
c. *Microbial factors*: Type of microorganism, the sensitivity of microorganism, population, resistivity against pulses

In fruits and vegetable processing, PL application for about 5 seconds on raspberries was effective in maintaining a low microbial profile and biochemical properties of fruits during storage of 10 days. Yet, PL application for about 30 seconds on raspberries failed to maintain microbial safety and biochemical properties during storage (Xu & Wu, 2016). Despite the incomplete microbial reduction in mung bean sprouts, PL was more effective than EW and ClO_2 disinfection. Also, PL can be used in the integrated washing system of fruits and vegetables because of its ability to disinfect transparent liquids and prevent cross-contamination (Kramer et al., 2017).

In meat processing, minimal PL intensity (0.56 J/cm^2) inactivated *L. innocua* in sliced ham, chicken cold cut, and frankfurter sausages. The applied PL fluence was more than 20 times lesser than the FDA allowable limit (maximum 12 J/cm^2) for disinfection (Kramer et al., 2019). On the contrary, higher PL fluence (9.68 ± 0.15 J/cm^2) and temperature ($10°C$) were required to inactivate *campylobacter jejuni* in poultry. Fluence lower than 5 J/cm^2 and temperature lower than $10°C$ were inefficient to create photochemical and photothermal effects on bacteria. Whereas Enterobacteriaceae strains in poultry meat were inactivated at lower fluence (2.82 ± 0.06 J/cm^2) and temperature ($<10°C$). Moreover, the mesophilic bacteria was not affected by an increase in fluence and temperature (Baptista et al., 2022). PL and UV-C disinfections can be used to reduce the risk of human salmonellosis from the egg.

A combination of water-assisted PL (WPL) with 1% hydrogen peroxide (H_2O_2) eliminated maximum bacterial pathogens in berries than WPL treatment alone. Also, this combination acts as a substitute for chlorine washing of fruits. The synergistic effect of PL and mild heat was effective in reducing *L. innocua* present in water, whey, diluted whey, and skimmed whey. The increase in the number of pulses and total fluence increased the inactivation rate of the bacteria. Also, treatment comprising lower voltage and an increased number of pulses increased bacterial inactivation. (Artíguez & Martínez de Marañón, 2015). PL disinfection application in food industries is summarized in Table 19.1. PL disinfection has the following disadvantages (Baptista et al., 2022):

a. *Phyto chemical effect*: Degradation of phytochemical compounds occurs during storage
b. *Shadowing effect*: non-uniform disinfection results in shadowing
c. *Requires optimization*: PL requires optimized parameters to inactivate different microorganisms on the same food surface

19.7   COLD PLASMA (CP)

Plasma is described as the fourth state of matter with a quasi-neutral gaseous particle system composed of electrons, ions, atoms, molecules, UV photons, radicals, and irradiated heat. The interaction of energy and gaseous medium allows the creation of a plasma state via ionization. Following are the classification of plasma based on temperature and mode of production of plasma (Dharini et al., 2023).

The antimicrobial attributes of plasma disinfection are ROS (O_2, O_3, OH, HO_2, H_2O_2, CO_3), reactive nitrogen species (RNS) (NO, NO_2, ROONO, ONOO$^-$, etc.) and UV photons. ROS and RNS produced during plasma disinfection slow down microbial reproduction by damaging the cell wall and oxidizing the polysaccharides, proteins, lipids, and DNA. Cell apoptosis is attained due to the accumulation of oxidative stress and DNA damage (Baskaran & Radhakrishnan, 2021; Misra & Jo, 2017).

CP disinfection efficiency depends on the below-mentioned factors (Laroque et al., 2022):

a. *External factors*: Temperature, RH
b. *Internal factors (Food parameters)*: Type of product, moisture, pH, the surface-to-volume ratio
c. *Processing factors*: Electrode type, the distance between the electrode, gas composition, voltage, frequency, flow rate, treatment time, plasma size, and geometry
d. *Microbial factors*: Type of microbe, strain, initial count, growth stage

PAW disinfection is influenced by the following factors – ORP and pH. In fruit and vegetable processing, PAW with higher ORP inactivated the maximum population of *S. cerevisiae* in grapes by damaging its outer and inner cell membranes. The oxidation capacity of PAW is related to its inactivation efficiency. The pH of PAW decreased during the inactivation and increased after treatment, indicating water was acidic during the PAW disinfection of fruits. And the temperature does not influence the inactivation of the yeast in the fruit (Guo et al., 2017). Apart from the inactivation of pathogens, the eco-friendly and high sterilization efficient PAW disinfection technology maintains the keeping quality of fruits. For example, the *S. aureus* population in fresh-cut kiwi fruit reached below 2 logs CFU/g after PAW disinfection. Additionally, PAW disinfection improved the keeping quality of kiwi fruit by resisting the accumulation of O_2^- and H_2O_2 free radicals in fruit during storage. And it also activated the antioxidant activity of peroxidase (POD), superoxide dismutase (SOD), and catalase (CAT) enzymes (Zhao et al., 2019).

In grain and pulse processing, dielectric barrier discharge (DBD) plasma was used to disinfect the wheat grains. Factors like treatment time, pulse voltage, and frequency influence the disinfection efficiency of the plasma. For example, higher pulse voltage (8 kV) and longer treatment time (10 minutes) resulted in the inactivation of bacterial spores, *Geobacillus stearothermophilus*. The endospore was inactivated by the direct effect of DBD plasma and the thermal, electrical, and mechanical stress created on grains during disinfection did not inactivate the endospores. And the plasma disinfection did not affect the functional properties of grains (Butscher et al., 2016). The microbial load in mung bean sprouts was reduced by immersing the sprouts in PAW for 30 minutes. Besides, its antioxidant activity, total flavonoids, and total phenolics were neither increased nor decreased (Xiang et al., 2019).

In spice processing, two different atmospheric pressure plasmas such as radio frequency plasma jet and microwave driven remote plasma were used to decontaminate the surface of dry and heat liable black peppercorns by inactivating *B. subtilis* spores, *Bacillus atrophaeus* spores, and *Salmonella enertica*. The spores of *Bacillus atrophaeus* were more sensitive to plasma than *B. subtilis* spores. But the microbial inactivation efficiency. But the microbial inactivation efficiency. The complex structure of peppercorns hinders the plasma incidence. CP is productive non-thermal technology in spices decontamination with the advantages of low energy consumption, no waste production, and cost-effectiveness (Hertwig et al., 2015).

In meat processing, plasma disinfection on meat showed longer exposure to FTDBD plasma on packaged chicken breast meat affected its color values. Color value a* of the chicken breast decreased and L* and b* values decreased after the treatment. An off flavor was developed in the meat, but the meat was at an acceptable score limit. However, the treatment did not oxidize lipids (Lee et al., 2016). The common foodborne microorganisms in pork, beef, and chicken – *E. coli O157:H7*, *L. monocytogenes*, and *S. typhimurium* – are inactivated by installing different plasma jet devices. Table 19.1 summarizes plasma disinfection application in food industries.

Different types of plasma have been used in the disinfection of fruits, vegetables, spices, dairy products, and meat products. With several advantages, CP has the following disadvantages (Dharini et al., 2023):

a. *Oxidation of foods*: Fat-rich foods like meat undergoes lipid oxidation
b. *End products*: End products formed by the reaction of reactive species are not completely understood

19.8   ULTRASOUND

Ultrasound (US) is a novel non-thermal technology with disinfection action against microorganisms. The US is generally carried out at a high-frequency range of 20–100 kHz. It is an eco-friendly, non-toxic operation and is often termed as high-intensity ultrasound (HIUS). Acoustic cavitation is the primary mechanism involved in disinfection process. Acoustic cavitation results in shear waves and shock waves (produced by continuous compression and relaxation of micro bubbles) which causes irreversible damage to bacterial cell walls (Nakonechny & Nisnevitch, 2021). The chemical effect of the US lies in the collapsing of DNA strands and unwinding of them. This mechanism is accomplished by microbubble induced hydroxyl group formation at higher temperatures (Deng et al., 2020). However, the treatment efficiency depends on factors such as treatment time, temperature, food matrix, type, and population of microorganisms.

Exposing fresh strawberries to US waves (33 kHz, 66 W for 30 minutes) resulted in a significant and maximum decrease in the microbial population. But, doubling the treatment time from 30 minutes to 60 minutes has resulted in an increased microbial count, decreased antioxidant activity, and changes in physiochemical properties. Prolonged US exposure may result in dehydration, degradation of color, and poor retention of vitamins due to increased cavitation and OH^- radicals generation (Gani et al., 2016). Along with the above-mentioned factors, US power also plays a major role in enhancing disinfection efficiency. This is because US at power 210 W with NEW for 15 minutes brings 4.3 and 4.4 log reductions of *S. typhimurium* and *E. coli O157:H7* on lettuce. Similarly, exposing tomatoes to the same level of US power for 5 minutes brings a complete reduction in the pathogenic population. Therefore, different power and time combination works for different food product disinfection using US (Afari et al., 2016).

In poultry and meat processing, the effect of ultrasound on disinfection is not observed immediately because the mesophiles, psychrophiles, LAB, and *S. aureus* count in chicken meat increased after HIUS treatment. However, psychrophile count in anaerobic packaged meat decreased, and the US has inactivated pathogenic bacteria, such as *S. aureus* and *Salmonella* spp., in poultry breast meat during storage (Piñon et al., 2020).

Combination technologies along with US in disinfection are more efficient in ensuring the microbiocidal safety of foods than ultrasound disinfection alone. For instance, the combination of US with ClO_2 to alfalfa and mung bean sprouts. This combination is preferred over individual US disinfection of sprouts, as only 1.40 ± 0.40 log CFU/g and 1.89 ± 0.51 log CFU/g reduction of *S. enteritidis*, and 1.06 ± 0.51 log CFU/g and 1.23 ± 0.40 log CFU/g of *E. coli* is achieved. While the combination has resulted in an increased reduction of pathogens (Millan-Sango et al., 2017). Table 19.1 summarizes ultrasound disinfection applications in food industries. The US combined with another non-thermal disinfectant technology, PL reduces bacterial spores and yeast population in commercial and natural apple juice. The US did not inactivate the spores in commercial apple juice but remarkedly reduced the yeast population to 6.4 log cycles. With the reduction in bacterial spores and yeast population, the storage life of apple juices is enhanced (Ferrario & Guerrero, 2016).

With the cavitation mechanism, the US inactivates the microbial population in food industries. The following are drawbacks faced during US disinfection (Deng et al., 2020):

a. *Limited capacity*: Ultrasound disinfection becomes ineffective in large-scale application
b. *Low antibacterial efficiency*

19.9 SUMMARY

To produce safe and nutritious food, it is very important to eliminate the pathogenic microorganism that is responsible for foodborne illness. Herein, disinfectants play a major role in eliminating pathogens from foods by hindering their growth. Novel disinfectant technologies such as chlorine dioxide, ozone, electrolyzed water, UV, PL, cold plasma, and ultrasound are emerging technologies with the advantages of non-toxicity, and no residual forming properties over existing ones. Moreover, these emerging technologies are more effective and do not affect the quality of foods. The hurdle concept of these novel technologies is also a welcoming one because of its promising results against targeted microorganisms in terms of efficiency, and also in overcoming current limitations of disinfectants.

REFERENCES

Afari, G. K., Hung, Y. C., King, C. H., & Hu, A. (2016). Reduction of *Escherichia coli* O157:H7 and *Salmonella typhimurium* DT 104 on fresh produce using an automated washer with near neutral electrolyzed (NEO) water and ultrasound. *Food Control*, *63*, 246–254. https://doi.org/10.1016/j.foodcont.2015.11.038

Alvarez-Ordóñez, A., Coughlan, L. M., Briandet, R., & Cotter, P. D. (2019). Biofilms in food processing environments: Challenges and opportunities. *Annual Review of Food Science and Technology*, *10*, 173–195. https://doi.org/10.1146/annurev-food-032818-121805

Animal and Plant Health Inspection Service. (2020). *Disinfection*. U.S. Department of Agriculture. https://www.aphis.usda.gov/aphis/ourfocus/animalhealth/nvap/NVAP-Reference-Guide/Cleaning-and-Disinfection/Disinfection (accessed on 29.10.2023)

Artíguez, M. L., & Martínez de Marañón, I. (2015). Improved process for decontamination of whey by a continuous flow-through pulsed light system. *Food Control*, *47*, 599–605. https://doi.org/10.1016/j.foodcont.2014.08.006

Aslam, R., Alam, M. S., & Kaur, P. (2022). Comparative study on efficacy of sanitizing potential of aqueous ozone and chlorine on keeping quality and shelf-life of minimally processed onion (*Allium Cepa* L.). *Ozone: Science and Engineering*, *44*(2), 196–207. https://doi.org/10.1080/01919512.2021.1904204

Ayranci, U. G., Ozunlu, O., Gezer, H., & Karana, H. (2020). Effects of ozone treatment on microbiological quality and physicochemical properties of Turkey breast meat. *Ozone: Science and Engineering*, *42*(1), 95–103. https://doi.org/10.1080/01919512.2019.1653168

Baptista, E., Borges, A., Aymerich, T., Alves, S. P., Gama, L. T. da, Fernandes, H., Fernandes, M. J., & Fraqueza, M. J. (2022). Pulsed light application for Campylobacter control on poultry meat and its effect on colour and volatile profile. *Foods*, *11*(18). https://doi.org/10.3390/foods11182848

Baskaran, K., & Radhakrishnan, M. (2021). Surface Pasteurization and Disinfection of Dairy Processing Equipment Using Cold Plasma Techniques. In *Non-Thermal Processing Technologies for the Dairy Industry* (pp. 143–156). CRC Press. https://doi.org/10.1201/9781003138716-11

Boopathy, B., Rajan, A., & Radhakrishnan, M. (2022). Ozone: An alternative fumigant in controlling the stored product insects and pests: A status report. *Ozone: Science & Engineering*, *44*(1), 79–95. https://doi.org/10.1080/01919512.2021.1933899

Brodowska, A. J., Nowak, A., & Śmigielski, K. (2018). Ozone in the food industry: Principles of ozone treatment, mechanisms of action, and applications: An overview. *Critical Reviews in Food Science and Nutrition*, *58*(13), 2176–2201. https://doi.org/10.1080/10408398.2017.1308313

Butscher, D., Zimmermann, D., Schuppler, M., & Rudolf von Rohr, P. (2016). Plasma inactivation of bacterial endospores on wheat grains and polymeric model substrates in a dielectric barrier discharge. *Food Control*, *60*, 636–645. https://doi.org/10.1016/j.foodcont.2015.09.003

Centre for Disease Control and Prevention. (2016). *Guideline for Disinfection and Sterilization in Healthcare Facilities (2008)*. Centre for Disease Control and Prevention. https://www.cdc.gov/infectioncontrol/pdf/guidelines/disinfection-guidelines-H.pdf (accessed on 29.10.2023)

Chen, Y.-X., Guo, X.-N., Xing, J.-J., Sun, X.-H., & Zhu, K.-X. (2020). Effects of wheat tempering with slightly acidic electrolyzed water on the microbial, biological, and chemical characteristics of different flour streams. *LWT*, *118*, 108790. https://doi.org/10.1016/j.lwt.2019.108790

Code of Federal Regulations. (2022). *Pulsed Light for the Treatment of Food*. https://www.accessdata.fda.gov/scripts/cdrh/cfdocs/cfcfr/CFRSearch.cfm?fr=179.41

Colás-Medà, P., Nicolau-Lapeña, I., Viñas, I., Neggazi, I., & Alegre, I. (2021). Bacterial spore inactivation in orange juice and orange peel by ultraviolet-C light. *Foods*, *10*(4). https://doi.org/10.3390/foods10040855

Colgecen, I., & Aday, M. S. (2015). The efficacy of the combined use of chlorine dioxide and passive modified atmosphere packaging on sweet cherry quality. *Postharvest Biology and Technology*, *109*, 10–19. https://doi.org/10.1016/j.postharvbio.2015.05.016

Cullen, P. J., Tiwari, B. K., & Valdramidis, V. P. (2012). Novel Thermal and Non-Thermal Technologies for Fluid Foods. In Cullen, P. J., Brijesh, K. T., and Vasilis P. V.(Eds.), *Novel Thermal and Non-Thermal Technologies for Fluid Foods* (1st ed.). Elsevier. https://doi.org/10.1016/C2009-0-61145-4

Deng, L.-Z., Mujumdar, A. S., Pan, Z., Vidyarthi, S. K., Xu, J., Zielinska, M., & Xiao, H.-W. (2020). Emerging chemical and physical disinfection technologies of fruits and vegetables: A comprehensive review. *Critical Reviews in Food Science and Nutrition*, *60*(15), 2481–2508. https://doi.org/10.1080/10408398.2019.1649633

Dharini, M., Jaspin, S., & Mahendran, R. (2023). Cold plasma reactive species: Generation, properties, and interaction with food biomolecules. *Food Chemistry*, *405*(PA), 134746. https://doi.org/10.1016/j.foodchem.2022.134746

El Darra, N., Xie, F., Kamble, P., Khan, Z., & Watson, I. (2021). Decontamination of *Escherichia coli* on dried onion flakes and black pepper using infra-red, ultraviolet and ozone hurdle technologies. *Heliyon*, *7*(6), e07259. https://doi.org/10.1016/j.heliyon.2021.e07259

Feliziani, E., Lichter, A., Smilanick, J. L., & Ippolito, A. (2016). Disinfecting agents for controlling fruit and vegetable diseases after harvest. *Postharvest Biology and Technology*, *122*(2015), 53–69. https://doi.org/10.1016/j.postharvbio.2016.04.016

Ferrario, M., & Guerrero, S. (2016). Effect of a continuous flow-through pulsed light system combined with ultrasound on microbial survivability, color and sensory shelf life of apple juice. *Innovative Food Science and Emerging Technologies*, *34*, 214–224. https://doi.org/10.1016/j.ifset.2016.02.002

Gani, A., Baba, W. N., Ahmad, M., Shah, U., Khan, A. A., Wani, I. A., Masoodi, F. A., & Gani, A. (2016). Effect of ultrasound treatment on physico-chemical, nutraceutical and microbial quality of strawberry. *LWT*, *66*, 496–502. https://doi.org/10.1016/j.lwt.2015.10.067

Guan, J., Lacombe, A., Tang, J., Bridges, D. F., Sablani, S., Rane, B., & Wu, V. C. H. (2021). Use of mathematic models to describe the microbial inactivation on baby carrots by gaseous chlorine dioxide. *Food Control*, *123*(July 2020), 107832. https://doi.org/10.1016/j.foodcont.2020.107832

Guo, J., Huang, K., Wang, X., Lyu, C., Yang, N., Li, Y., & Wang, J. (2017). Inactivation of yeast on grapes by plasma-activated water and its effects on quality attributes. *Journal of Food Protection*, *80*(2), 225–230. https://doi.org/10.4315/0362-028X.JFP-16-116

Han, G. D., Kwon, H., Kim, B. H., Kum, H. J., Kwon, K., & Kim, W. (2018). Effect of gaseous chlorine dioxide treatment on the quality of rice and wheat grain. *Journal of Stored Products Research*, *76*, 66–70. https://doi.org/10.1016/j.jspr.2018.01.003

Hernández-Pimentel, V. M., Regalado-González, C., Nava-Morales, G. M., Meas-Vong, Y., Castañeda-Serrano, M. P., & García-Almendárez, B. E. (2020). Effect of neutral electrolyzed water as antimicrobial intervention treatment of chicken meat and on trihalomethanes formation. *Journal of Applied Poultry Research*, *29*(3), 622–635. https://doi.org/10.1016/j.japr.2020.04.001

Hertwig, C., Reineke, K., Ehlbeck, J., Knorr, D., & Schlüter, O. (2015). Decontamination of whole black pepper using different cold atmospheric pressure plasma applications. *Food Control*, *55*, 221–229. https://doi.org/10.1016/j.foodcont.2015.03.003

Hinds, L. M., O'Donnell, C. P., Akhter, M., & Tiwari, B. K. (2019). Principles and mechanisms of ultraviolet light emitting diode technology for food industry applications. *Innovative Food Science and Emerging Technologies*, *56*(April), 102153. https://doi.org/10.1016/j.ifset.2019.04.006

Huang, L., Luo, X., Gao, J., & Matthews, K. R. (2019). Influence of water antimicrobials and storage conditions on inactivating MS2 bacteriophage on strawberries. *International Journal of Food Microbiology*, *291*, 67–71. https://doi.org/10.1016/j.ijfoodmicro.2018.11.009

Hussain, M. S., Kwon, M., Park, E. Ji, Seheli, K., Huque, R., & Oh, D. H. (2019). Disinfection of *Bacillus cereus* biofilms on leafy green vegetables with slightly acidic electrolyzed water, ultrasound and mild heat. *LWT*, *116*(September), 108582. https://doi.org/10.1016/j.lwt.2019.108582

Joshi, K., Mahendran, R., Alagusundaram, K., Norton, T., & Tiwari, B. K. (2013). Novel disinfectants for fresh produce. *Trends in Food Science and Technology*, *34*(1), 54–61. https://doi.org/10.1016/j.tifs.2013.08.008

Kocharunchitt, C., Mellefont, L., Bowman, J. P., & Ross, T. (2020). Application of chlorine dioxide and peroxyacetic acid during spray chilling as a potential antimicrobial intervention for beef carcasses. *Food Microbiology*, *87*, 103355. https://doi.org/10.1016/j.fm.2019.103355

Koutchma, T. (2016). Application of infrared and light-based technologies to meat and meat products. *Emerging Technologies in Meat Processing: Production, Processing and Technology*, 131–147. https://doi.org/10.1002/9781118350676.ch5

Kramer, B., Wunderlich, J., & Muranyi, P. (2017). Pulsed light decontamination of endive salad and mung bean sprouts in water. *Food Control*, *73*, 367–371. https://doi.org/10.1016/j.foodcont.2016.08.023

Kramer, B., Wunderlich, J., & Muranyi, P. (2019). Inactivation of *Listeria innocua* on packaged meat products by pulsed light. *Food Packaging and Shelf Life*, *21*, 100353. https://doi.org/10.1016/j.fpsl.2019.100353

Laroque, D. A., Seó, S. T., Valencia, G. A., Laurindo, J. B., & Carciofi, B. A. M. (2022). Cold plasma in food processing: Design, mechanisms, and application. *Journal of Food Engineering*, *312*(April 2021). https://doi.org/10.1016/j.jfoodeng.2021.110748

Lee, H., Yong, H. I., Kim, H. J., Choe, W., Yoo, S. J., Jang, E. J., & Jo, C. (2016). Evaluation of the microbiological safety, quality changes, and genotoxicity of chicken breast treated with flexible thin-layer dielectric barrier discharge plasma. *Food Science and Biotechnology*, *25*(4), 1189–1195. https://doi.org/10.1007/s10068-016-0189-1

Liu, C., Chen, C., Jiang, A., Sun, X., Guan, Q., & Hu, W. (2020). Effects of plasma-activated water on microbial growth and storage quality of fresh-cut apple. *Innovative Food Science and Emerging Technologies*, *59*, 102256. https://doi.org/10.1016/j.ifset.2019.102256

Mahendran, R., Ramanan, K. R., Barba, F. J., Lorenzo, J. M., López-Fernández, O., Munekata, P. E. S., Roohinejad, S., Sant'Ana, A. S., & Tiwari, B. K. (2019). Recent advances in the application of pulsed light processing for improving food safety and increasing shelf life. *Trends in Food Science and Technology*, *88*, 67–79. https://doi.org/10.1016/j.tifs.2019.03.010

Martinez-Garcia, M., Sauceda-Gálvez, J. N., Codina-Torrella, I., Hernández-Herrero, M. M., Gervilla, R., & Roig-Sagués, A. X. (2019). Evaluation of continuous UVC treatments and its combination with UHPH on spores of Bacillus subtilis in whole and skim milk. *Foods*, *8*(11), 10–22. https://doi.org/10.3390/foods8110539

Megahed, A., Aldridge, B., & Lowe, J. (2020). Antimicrobial efficacy of aqueous ozone and ozone–lactic acid blend on Salmonella-contaminated chicken drumsticks using multiple sequential soaking and spraying approaches. *Frontiers in Microbiology*, *11*(December), 1–11. https://doi.org/10.3389/fmicb.2020.593911

Meireles, A., Giaouris, E., & Simões, M. (2016). Alternative disinfection methods to chlorine for use in the fresh-cut industry. *Food Research International*, *82*, 71–85. https://doi.org/10.1016/j.foodres.2016.01.021

Millan-Sango, D., Sammut, E., Van Impe, J. F., & Valdramidis, V. P. (2017). Decontamination of alfalfa and mung bean sprouts by ultrasound and aqueous chlorine dioxide. *LWT*, *78*, 90–96. https://doi.org/10.1016/j.lwt.2016.12.015

Misra, N. N., & Jo, C. (2017). Applications of cold plasma technology for microbiological safety in meat industry. *Trends in Food Science and Technology*, *64*, 74–86. https://doi.org/10.1016/j.tifs.2017.04.005

Mohammad, Z., Kalbasi-Ashtari, A., Riskowski, G., Juneja, V., & Castillo, A. (2020). Inactivation of Salmonella and Shiga toxin-producing *Escherichia coli* (STEC) from the surface of alfalfa seeds and sprouts by combined antimicrobial treatments using ozone and electrolyzed water. *Food Research International*, *136*(June), 109488. https://doi.org/10.1016/j.foodres.2020.109488

Nakonechny, F., & Nisnevitch, M. (2021). Different aspects of using ultrasound to combat microorganisms. *Advanced Functional Materials*, *31*(44), 2011042. https://doi.org/10.1002/adfm.202011042

Nie, M., Wu, C., Xiao, Y., Song, J., Zhang, Z., Li, D., & Liu, C. (2020). Efficacy of aqueous ozone combined with sodium metasilicate on microbial load reduction of fresh-cut cabbage. *International Journal of Food Properties*, *23*(1), 2065–2076. https://doi.org/10.1080/10942912.2020.1842446

Panigrahi, C., Mishra, H. N., & De, S. (2020). Effect of ozonation parameters on nutritional and microbiological quality of sugarcane juice. *Journal of Food Process Engineering*, *43*(11), 1–14. https://doi.org/10.1111/jfpe.13542

Park, S. Y., Jung, S. J., & Ha, S.-Do. (2018). Synergistic effects of combined X-ray and aqueous chlorine dioxide treatments against *Salmonella* Typhimurium biofilm on quail egg shells. *LWT*, *92*(October 2017), 54–60. https://doi.org/10.1016/j.lwt.2018.02.010

Pataro, G., Sinik, M., Capitoli, M. M., Donsì, G., & Ferrari, G. (2015). The influence of post-harvest UV-C and pulsed light treatments on quality and antioxidant properties of tomato fruits during storage. *Innovative Food Science and Emerging Technologies*, *30*, 103–111. https://doi.org/10.1016/j.ifset.2015.06.003

Piñon, M. I., Alarcon-Rojo, A. D., Renteria, A. L., & Carrillo-Lopez, L. M. (2020). Microbiological properties of poultry breast meat treated with high-intensity ultrasound. *Ultrasonics*, *102*. https://doi.org/10.1016/j.ultras.2018.01.001

Rajkovic, A., Tomasevic, I., De Meulenaer, B., & Devlieghere, F. (2017). The effect of pulsed UV light on *Escherichia coli* O157:H7, *Listeria monocytogenes*, *Salmonella typhimurium*, *Staphylococcus aureus* and *Staphylococcal enterotoxin* A on sliced fermented salami and its chemical quality. *Food Control*, *73*, 829–837. https://doi.org/10.1016/j.foodcont.2016.09.029

Ricciardi, E. F., Pedros-Garrido, S., Papoutsis, K., Lyng, J. G., Conte, A., & Del Nobile, M. A. (2020). Novel technologies for preserving ricotta cheese: Effects of ultraviolet and near-ultraviolet–visible light. *Foods*, *9*(5). https://doi.org/10.3390/foods9050580

Rodionova, K. O., Paliy, A. P., Palii, A. P., Yatsenko, I. V, Fotina, T. I., Bogatko, N. M., Mohutova, V. F., Nalyvayko, L. I., Ivleva, O. V, & Odyntsova, T. A. (2020). Effect of ultraviolet irradiation on beef carcass yield. *Ukrainian Journal of Ecology*, *2020*(2), 410–415. https://www.ujecology.com/abstract/effect-of-ultraviolet-irradiation-on-beef-carcass-yield-54251.html

Rutala, W. A., & Weber, D. J. (2019). Disinfection, sterilization, and antisepsis: An overview. *American Journal of Infection Control*, *47*, A3–A9. https://doi.org/10.1016/j.ajic.2019.01.018

Sarooei, S. J., Abbasi, A., Shaghaghian, S., & Berizi, E. (2019). Effect of ozone as a disinfectant on microbial load and chemical quality of raw wheat germ. *Ozone: Science and Engineering*, *41*(6), 562–570. https://doi.org/10.1080/01919512.2019.1642181

Singh, S., Maji, P. K., Lee, Y. S., & Gaikwad, K. K. (2021). Applications of gaseous chlorine dioxide for antimicrobial food packaging: A review, *Environmental Chemistry Letters*, *19*(1), 253–270. https://doi.org/10.1007/s10311-020-01085-8

Song, H., Lee, J. Y., Lee, H. W., & Ha, J. H. (2021). Inactivation of bacteria causing soft rot disease in fresh cut cabbage using slightly acidic electrolyzed water. *Food Control*, *128*(May), 108217. https://doi.org/10.1016/j.foodcont.2021.108217

USFDA. (2020). *Ultraviolet (UV) Radiation*. FDA.

Vaka, M. R., Sone, I., Álvarez, R. G., Walsh, J. L., Prabhu, L., Sivertsvik, M., & Fernández, E. N. (2019). Towards the next-generation disinfectant: Composition, storability and preservation potential of plasma activated water on baby spinach leaves. *Foods*, *8*(12). https://doi.org/10.3390/foods8120692

Van Haute, S., Tryland, I., Escudero, C., Vanneste, M., & Sampers, I. (2017). Chlorine dioxide as water disinfectant during fresh-cut iceberg lettuce washing: Disinfectant demand, disinfection efficiency, and chlorite formation. *LWT*, *75*, 301–304. https://doi.org/10.1016/j.lwt.2016.09.002

Wang, H., Qi, J., Duan, D., Dong, Y., Xu, X., & Zhou, G. (2018). Combination of a novel designed spray cabinet and electrolyzed water to reduce microorganisms on chicken carcasses. *Food Control*, *86*, 200–206. https://doi.org/10.1016/j.foodcont.2017.11.027

Wei, X., Verma, T., Danao, M. G. C., Ponder, M. A., & Subbiah, J. (2021). Gaseous chlorine dioxide technology for improving microbial safety of spices. *Innovative Food Science and Emerging Technologies*, *73*(May), 102783. https://doi.org/10.1016/j.ifset.2021.102783

Wu, S., Nie, Y., Zhao, J., Fan, B., Huang, X., Li, X., Sheng, J., Meng, D., Ding, Y., & Tang, X. (2018). The synergistic effects of low-concentration acidic electrolyzed water and ultrasound on the storage quality of fresh-sliced button mushrooms. *Food and Bioprocess Technology*, *11*(2), 314–323. https://doi.org/10.1007/s11947-017-2012-2

Xiang, Q., Liu, X., Liu, S., Ma, Y., Xu, C., & Bai, Y. (2019). Effect of plasma-activated water on microbial quality and physicochemical characteristics of mung bean sprouts. *Innovative Food Science and Emerging Technologies*, *52*, 49–56. https://doi.org/10.1016/j.ifset.2018.11.012

Xu, W., & Wu, C. (2016). The impact of pulsed light on decontamination, quality, and bacterial attachment of fresh raspberries. *Food Microbiology*, *57*, 135–143. https://doi.org/10.1016/j.fm.2016.02.009

Yeom, W., Kim, H., Beuchat, L. R., & Ryu, J. H. (2021). Inactivation of *Escherichia coli* O157:H7 on radish and cabbage seeds by combined treatments with gaseous chlorine dioxide and heat at high relative humidity. *Food Microbiology*, *99*(April), 103805. https://doi.org/10.1016/j.fm.2021.103805

Zhang, C., Cao, W., Hung, Y. C., & Li, B. (2016). Disinfection effect of slightly acidic electrolyzed water on celery and cilantro. *Food Control*, *69*, 147–152. https://doi.org/10.1016/j.foodcont.2016.04.039

Zhao, Y., Chen, R., Liu, D., Wang, W., Niu, J., Xia, Y., Qi, Z., Zhao, Z., & Song, Y. (2019). Effect of nonthermal plasma-activated water on quality and antioxidant activity of fresh-cut kiwifruit. *IEEE Transactions on Plasma Science*, *47*(11), 4811–4817. https://doi.org/10.1109/TPS.2019.2904298

Zhao, L., Zhao, M. Y., Phey, C. P., & Yang, H. (2019). Efficacy of low concentration acidic electrolysed water and levulinic acid combination on fresh organic lettuce (*Lactuca sativa* Var. Crispa L.) and its antimicrobial mechanism. *Food Control*, *101*, 241–250. https://doi.org/10.1016/j.foodcont.2019.02.039

20 Emergence of Hybrid Technologies for Processing of Foods

Kevin Britto, Chaitanya Chivate, Prakyath Shetty, and Ashish Rawson

20.1 INTRODUCTION

There are several processing techniques that result in a longer shelf life for microbial and enzyme-inactivated products. These methods, which can often be done using high temperatures, have some drawbacks, such as changes to the sensory properties of the food. Thermal processing is particularly popular among those who want foods with long shelf lives due to its ability to kill viruses and bacteria quickly. However, this process can lead to undesirable changes in the flavor, color, texture, and other characteristics of food items (Demirdöven & Baysal, 2009; Hoover, 1997).

Utilizing multiple independent treatments in succession or simultaneously is a necessary step when it comes to food preservation. This is primarily because single treatments are often not enough to guarantee the safety of the food. Combining treatments can also make it possible to use individual treatments more sparingly, improving the quality and taste of the food. Moreover, combining treatments results in a synergistic effect which helps to improve the overall preservation benefit of the various preservation procedures used. This synergy is preferred as it helps to maintain the quality and safety of the food. For example, by combining two methods of preserving food, such as irradiation and high-pressure processing, the combined effect of both methods has a higher impact than either would have if used alone. The combination approach can also help to reduce wastage as it helps to ensure that food remains fresh for longer periods of time. This is especially beneficial for countries which may experience famine or lack access to a wide variety of food products due to geographic isolation or limited resources. Overall, combining different treatments can have numerous benefits when it comes to preserving food while ensuring quality and safety standards are met. Food preservation using combination approaches is therefore an effective way to enhance storage times while helping to reduce wastage and improve nutritional value. Combining technologies can be beneficial in many ways. However, in some cases, it can also have a negative impact. This is especially true when the technology being combined fails to meet certain criteria or if the use of multiple technologies together complicates the process. In short, while combining technology can be beneficial in many ways, there are also some situations where it can have a negative impact on the workflow (Leistner & Gorris, 1995; Raso & Barbosa-Cánovas, 2003).

Emerging Hybrid technologies combine non-thermal and thermal processing techniques which lead to improved food quality, reduction of processing time, and increased microbial and enzyme inactivation (Kumar et al., 2023a, 2023b; Mathews et al., 2023; Modupalli et al., 2022; Naik et al., 2023; Rawson et al., 2011; Sengar et al., 2020; Sunil et al., 2017). This new approach provides a great opportunity to reduce costs and energy consumption while increasing the safety of food products. Additionally, it can provide more efficient use of equipment throughout the process chain, which may result in improved product consistency. The wide range of applications makes it attractive for many different types of food production. Hybrid technology is the way forward in the food industry, as it promises to produce food that is safe, nutritious and of high quality. It also contributes

DOI: 10.1201/9781003359302-20

to a more sustainable form of food production that is both environmentally friendly and economically viable (Zenker et al., 2003).

20.2 DIFFERENT HYBRID TECHNOLOGIES USED IN FOOD PROCESSING APPLICATION

By combining hybrid technology approaches, researchers can create more effective solutions that improve energy efficiency. The use of hybrid technology approaches in many industries is becoming increasingly popular due to their ability to produce high-quality results with greater efficiency. Furthermore, the use of such approaches provides flexibility and scalability across various industries. The combining of different technologies is an effective way to increase energy efficiency while at the same time maximizing cost savings. By utilizing conventional, and emerging technology approaches, businesses can achieve greater success in their energy efficiency initiatives. Overall, it is an effective way to improve energy efficiency while simultaneously reducing costs and environmental impacts (Khadhraoui et al., 2021).

20.2.1 ULTRASOUND-ASSISTED CONVENTIONAL EXTRACTION

This section presents the most typical combinations of ultrasound and other conventional techniques. It is important to note that these methods offer various advantages, including a higher level of accuracy and greater detail than either method alone.

20.2.1.1 Ultrasound-Assisted Soxhlet Extraction: Sono-Soxhlet

Soxhlet extraction is a process which works on a simple iterative percolation of condensed vapors from a boiled substance. Typically, n-hexane is used as the solvent, however, there are some drawbacks associated with this method. These include a lengthy operation period that can take a number of hours, large amounts of solvent being required, and evaporation taking place during the extraction. At the end of the extraction process, there is also an additional step of concentration. Unfortunately, thermolabile analytes cannot be measured using Soxhlet extraction due to their nature. Despite this, it remains a popular choice for many laboratories due to its effectiveness and simplicity. The use of ultrasound in solid-liquid extraction is highly beneficial as it has the ability to reduce extraction times and increase mass transfer. Sono-Soxhlet (Figure 20.1) combines the advantages of both ultrasound and Soxhlet extraction, resulting in a thorough, accurate, and fast sample extraction. This method ensures that more metabolites are successfully migrated from solid matrix to solvent. In addition, the repetitive nature of Soxhlet extraction is maintained while still allowing for efficient and rapid sample processing. Hence, this method provides a reliable and effective approach for efficient sample extraction. Ultrasound also brings its own advantages to the process such as improved reproducibility and increased selectivity, making it an ideal choice for many applications. AlJuhaimi et al. demonstrated the high efficacy of ultrasound as a pre-treatment before Soxhlet extraction for peanut oil, with a 51.50% extraction yield after 10 minutes and a decrease in oleic acid content from 57.10% to 56.69% after 30 minutes (Al Juhaimi et al., 2018). This finding has opened up possibilities for increased research into the potential of combining ultrasound and Soxhlet extraction, not just for peanut oil but also other types of oils, such as sunflower and extra virgin olive oil. This new combined approach has been designed to address the issues posed by traditional enzymatic hydrolysis methods, such as long processing times, higher temperatures, and larger solvent consumption. It has been suggested that this combined approach could reduce the number of Soxhlet cycles, processing time, and temperature required while also providing an efficient way to pre-treat and extract different oils through extrusion. The Sono-Soxhlet method improves mass transfer through ultrasound while maintaining the benefits of Soxhlet extraction, resulting in reduced extraction time. This technique enables precise and efficient specimen collection. Moreover, it has been successfully

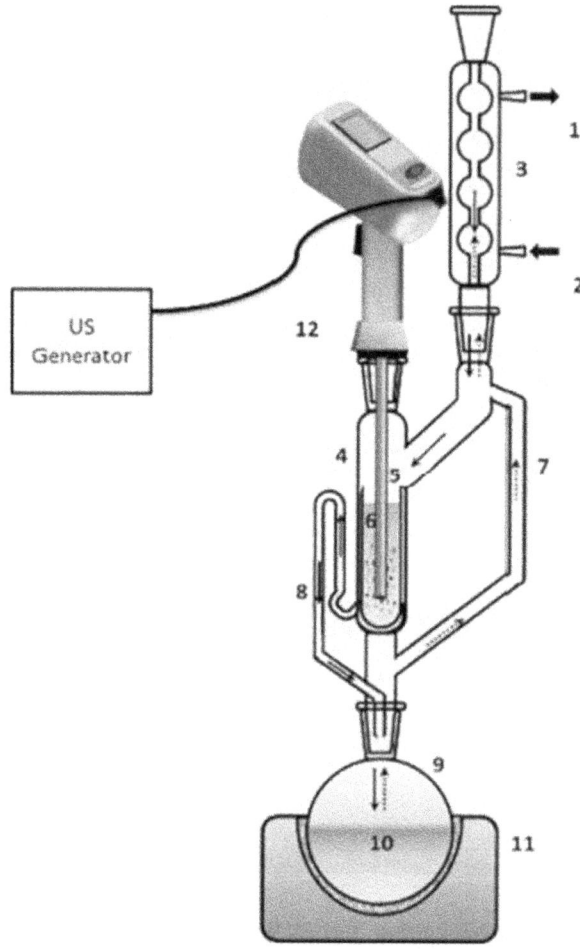

FIGURE 20.1 Ultrasound-assisted Soxhlet extraction. (Sono-Soxhlet reproduced from Khadhraoui et al., 2021.)

employed to extract oil content and fatty acid composition from oily materials such as beans, sausages, cheese, and pastry products (Khadhraoui et al., 2021; Singla & Sit, 2022).

20.2.1.2 Ultrasound-Assisted Clevenger Distillation: Sono-Clevenger

The flavor and fragrance industry has grown exponentially in recent years, resulting in an increase in demand for novel extraction techniques. To meet the growing demand, researchers have come up with the idea of combining Clevenger and alembic distillation with ultrasound technology. This combination of technologies is known as Sono-Clevenger, a device specifically designed to extract essential oils from plant materials. This method holds many advantages over traditional methods, such as increased efficiency, reduced processing times, and improved yields. Sono-Clevenger is capable of operating at both atmospheric and low pressure, making it even more feasible for industrial use. As such, this cutting-edge technology has huge potential applications in the Clevenger extraction industry. With the continual advancements of ultrasound and Clevenger distillation technologies, Sono-Clevenger opens up a world of possibilities for the extraction industry. The demand for natural products has been on the rise over the years, resulting in an increased need for advanced and modern extraction methods. Among them, Sono-Clevenger has emerged as a popular process.

It is the combination of Clevenger refining with ultrasonic cavitation and is applied at ambient or reduced pressure. This method is used to selectively separate essential oils from plant components, thereby allowing for better flavor, aroma, and extraction. Furthermore, it is much more efficient than traditional methods such as Clevenger or alembic refining alone, making it a preferred choice of many researchers. The use of Sono-Clevenger has helped in achieving greater extraction efficiency while maintaining the quality of the essential oils (Chemat et al., 2017; Khadhraoui et al., 2021; Singla & Sit, 2021; Vivek et al., 2016).

20.2.1.3 Ultrasound-Assisted Enzymatic Extraction

The Clevenger apparatus is the industry-standard for efficient and ecofriendly extraction of plant material. Utilizing the ultrasound-assisted enzyme extraction (UAEE) method, this apparatus often produces high yields with minimal energy use. In order to obtain the best yields and reduce degradation, it is vital to select the ideal operating conditions for scalable and economical extractions. By taking advantage of the ability of ultrasound to both activate and denature enzymes, UAEE proves to be an effective extraction technique. Optimizing certain parameters allows for successful outcomes to be achieved in each situation. Therefore, the combination of Clevenger apparatus and UAEE has become a reliable and cost-effective extraction process for many industries. This combination has revolutionized extractions as it offers great yields with minimal energy use; making it an excellent choice for various applications (Görgüç et al., 2019; Khadhraoui et al., 2021; Singla & Sit, 2021). Amiri-Rigi et al. utilized UAEE to extract lycopene from tomato residues in the tomato industry (Amiri-Rigi et al., 2016). In this study, micro-emulsification extraction was applied to tomato residue after UAE and EAE to increase the partition of lycopene. While Tchabo et al. investigated the effects of UAE and EAE for extracting phytochemical components such as flavonoids, phenolics, and anthocyanins (Tchabo et al., 2015). Furthermore, compared to UAE, EAE, or traditional extraction techniques, UAEE extracted samples showed superior extraction and productivity of phyto-bioactive mulberry components. Thus, ultrasound technology provides an effective solution for improving the efficiency of food processing operations. In order to extract juice from banana pulp, enzymatic treatment and ultrasound treatment were combined by Bora et al. which led to an increase in juice production and a reduction in its viscosity. The combination treatment also resulted in juice with higher total soluble solids and improved clarity (Bora et al., 2017).

20.2.2 Extraction Using Green Solvents

This section presents the most promising combinations of ultrasound and green solvents to achieve the desired outcome. Utilizing ultrasound technology in combination with safe and eco-friendly solvents can provide a cost-effective, efficient, and sustainable solution for many applications.

20.2.2.1 Ultrasound-Assisted Hydrotropic Extraction

The hydrotropy effect occurs when organic compounds that are poorly soluble or insoluble in water exhibit increased solubility in aqueous solutions due to the presence of highly water-soluble organic salts, which are commonly referred to as hydrotropes. The ability of an organic solute such as alcohols, aldehydes, esters, hydrocarbons, ketones, and lipids to dissolve more easily is influenced by the concentration of hydrotropes. This phenomenon depends on the type of the solute as well as the hydrotrope. While hydrotropes were initially researched in biochemistry, several chemical investigations have been conducted on their function and usefulness. Hydrophilic and hydrophobic components make up hydrotropes. They slowly self-aggregate but do not form micelles. They are characterized by a concentration threshold known as the minimum hydrotropic concentration (MHC), at which the solubility of hydrophobic substances increases dramatically. Hydrotropes have been known for their ability to dissolve drugs, but they have so much more potential. Their ability to act as fragrance extraction agents has made them a sought-after option in the production of perfumes and cosmetics. As well as this, hydrotropes can be used in distillation, providing an effective way to separate near-boiling liquid mixtures. The selectivity and high solubilization capacity of hydrotropic solubilization has made it an ideal choice for extracting water-insoluble bioactive molecules. As such, hydrotropes are highly sought after in the field of liquid-liquid extraction. With multiple applications, hydrotropes are becoming increasingly valuable in the world of drug and liquid extraction (Khadhraoui et al., 2021; Thakker et al., 2018).

20.2.2.2   Ultrasound-Assisted Ionic Liquid Extraction (UAILE)

Ionic liquids (ILs) constitute a significant group of salts that are formed by the combination of organic cations and organic or inorganic anions, resulting in a collection of non-molecular solvents. In 1914, Paul Walden discovered the first ever IL when ethylamine was neutralized with strong nitric acid to produce ethylammonium nitrate $[EtNH_3]$ $[NO_3]$. These compounds have melting points that are lower than their original constituents at temperatures around 100°C and have many applications. For example, ILs are used as hydrotropes to enhance the solubility of poor-soluble drugs, and also for extraction purposes. They have also been explored as a green alternative to synthesize materials with low environmental impact. In conclusion, ILs are an important class of salts with a wide range of applications that can provide beneficial solutions to many industries. Until 2007, the use of this particular solvent for the extraction of bioactive substances was relatively unknown. However, since that time, there has been a growing interest in employing this solvent (IL) as an extraction medium. This can be seen clearly from the rapidly increasing number of publications on this topic. The advantages of using IL for extraction include its non-toxicity, cost-effectiveness, and its ability to extract a wide range of bioactive components. It is not surprising, therefore, that IL has become a popular choice for many researchers interested in the extraction of bioactive substances. IL has the potential to further advance our understanding of bioactive compounds and their beneficial properties (Benvenutti et al., 2019; Murador et al., 2019).

20.2.2.3   Ultrasound-Assisted DES or NADES Extraction (UADESE or UANADESE)

In 2003, Abbott et al. (2003) made the first discovery of deep eutectic solvents (DESs), which are solid compounds that combine to form a eutectic mixture with a melting point lower than those of the individual components (Figure 20.2). This is primarily because of the presence of hydrogen bond acceptor (HBA) and hydrogen bond donor (HBD) molecules, which together form intermolecular hydrogen bonds (HB). The most common HBA molecule mentioned in scientific literature is choline chloride, which is capable of forming hydrogen bonds with the majority of HBD molecules. As a result, DESs have become essential components in many industries. Choline chloride is a highly effective derivative of vitamin B4, offering an inexpensive and biodegradable solution for animal feeding. It has become immensely popular in the market due to its wide availability and low costs. This has made choline chloride one of the most sought-after substances in the agricultural space (Khadhraoui et al., 2021; Mbous et al., 2017).

Due to their various physicochemical properties, including low volatility, high viscosity, non-inflammability, and thermal and chemical stability, DESs and ILs are extensively employed in the production of environmentally friendly solvents. In addition, compared to ILs, DESs offer the advantage of simpler storage and synthesis processes and the less expensive raw components used. As a result, DESs have become a potential choice for the synthesis of green solvents due to their non-toxic nature.

This attention to natural sources of DESs has opened up research into the development of natural deep eutectic solvents (NADESs). This class of DESs follows the principles of green chemistry proposed by Anastas and Warner, as NADESs are typically composed of multiple substances. These compounds can be formed from such natural sources as amino acids, carbohydrates, and lipids. The end result is a combination that has a melting point lower than any of its individual components. As NADESs offer an alternative to synthetic compounds, research in this field is continuously advancing (Anastas & Warner, 1998; Shishov et al., 2017).

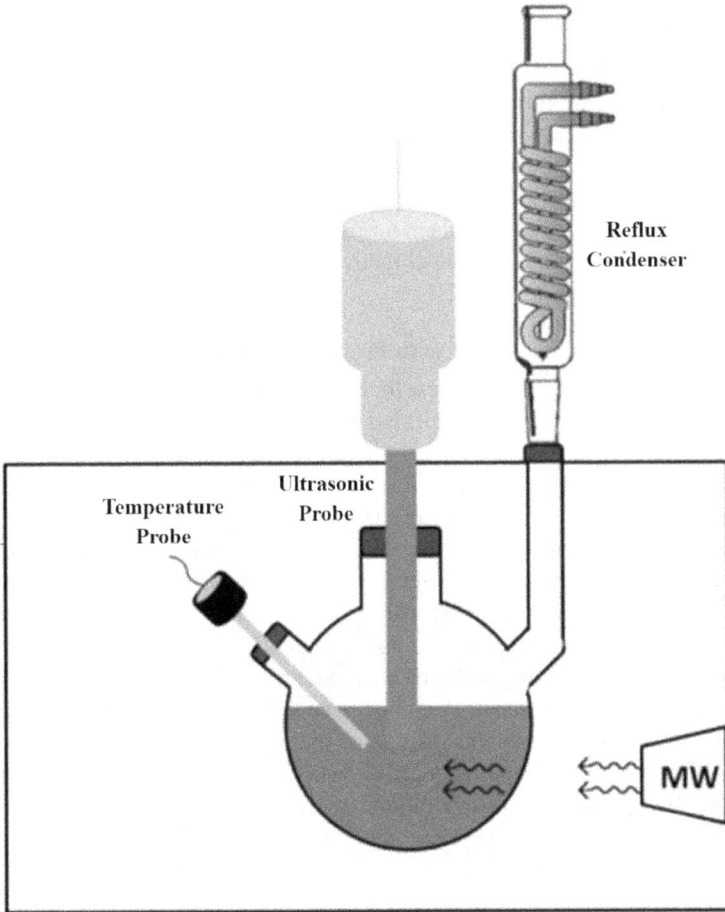

FIGURE 20.2   Ultrasound-assisted DES. (Reproduced from Khadhraoui et al., 2021.)

20.2.3   INNOVATIVE EXTRACTION USING ULTRASOUND AND OTHER NON-THERMAL TECHNIQUES

20.2.3.1   Combination of Ultrasound with Mechanical Pressing Extrusion

Mechanical pressing is becoming a more prevalent way of extracting juice and oil in the food sector. The process involves the use of one or more rotating screws to generate pressure, which pushes materials through a die. The traditional extrusion method only involves thermomechanical phenomena; however, reactive extrusion is becoming more commonplace as it induces chemical reactions within the extruder. This method is allowing for an increased variety of products within the food industry and for companies to explore new possibilities when it comes to using extrusion technology. With its ability to control the physical and chemical properties of food products, reactive extrusion has become an invaluable tool in the food industry. Without the use of solvents, this method of extraction has proven to be a viable alternative due to its superior quality and lack of solvent residues (Chemat et al., 2015; Khadhraoui et al., 2021). Moreover, recent research has focused on combining extrusion with other solvent extraction methods in order to increase extraction yields. Extrusion-solvent extraction has been gaining traction as several studies have shown that ultrasonic (US) application can significantly enhance oil recovery in terms of kinetics and yields. The integration of extrusion with conventional solvent extraction process is thus an intriguing field to explore for optimal extractions (Amirante et al., 2017; Milenkovic et al., 2018).

Zhang et al. and Yang et al. employed ultrasonics as a pre-treatment method to enhance enzymatic hydrolysis and generate fermentable sugars from rice husk without the use of chemicals (Y. Zhang et al., 2020). Yang et al. discovered that ultrasonic pre-treatment resulted in a 38.2% increase in reducing sugars (381.59 mg/g rice husk) compared to extrusion alone (276.11 mg/g rice husk) (Yang et al., 2012). Moreover, researchers have suggested that ultrasonic-assisted extrusion has been employed to enhance nanoparticle dispersion in polymer matrices (A. Ávila-Orta et al., 2019). Ultrasound is a powerful tool to enhance the performance of pre-treatment processes, and, therefore, it has been adopted by many researchers for different applications. The use of ultrasound can be cost-effective and efficient, making it an attractive option for pre-treatment in various industries (Mukherjee et al., 2020; Singla & Sit, 2021).

20.2.3.2   Combination of Ultrasound with Pressurized Liquids (PL)

Pressurized-liquid extraction (PLE) is a process for extracting substances that involves using an extraction solvent under high pressure and temperature but below their critical points. This technique is known by various names, including pressurized hot solvent extraction (PHSE), accelerated solvent extraction (ASE), high pressure high temperature solvent extraction (HPHTSE), high pressure solvent extraction (HPSE), and subcritical solvent extraction (SSE). By applying pressure and temperature, the physicochemical properties of the solvent such as viscosity, surface tension, density, dielectric constant, and diffusivity, are significantly altered in this method (Álvarez-Rivera et al., 2020; Andreu & Picó, 2019).

PLE is a highly efficient method for the extraction of a wide range of compounds from solid and liquid samples. The combination of extraction techniques with ultrasound (US) is being widely explored as it has been demonstrated to increase the accessibility of inner structures that are otherwise difficult to access. This is primarily due to US' capability to disrupt cells and thin membranes, which significantly enhances the solvents' characteristics. Subcritical solvents such as water, ethanol, hydroalcoholic combinations, and others have been used in combination with ultrasound, as it allows for the control of pressure and temperature parameters that enable the extraction of bioactive molecules. Therefore, this combination of extraction technique and ultrasound seems to be of significant interest for numerous applications involving solvent extraction since it can help overcome biological barriers (Khadhraoui et al., 2021; Liu et al., 2020; Sumere et al., 2018). Chen et al. successfully employed sonication as a pre-treatment and subcritical water extraction as separate procedures to extract pectin-enriched material from sugar beet pulp. The resulting extracts exhibited enhanced functional properties such as pasting temperature, gelatinization temperature, enthalpy, and larger extraction yields (H. M. Chen et al., 2015).

The combined application of subcritical solvent extraction and ultrasonic treatment in the extraction of phenolic chemicals from pomegranate peel proved to be successful, as demonstrated by the results. It was found that larger extraction yields can be obtained with a shorter extraction time and at a lower temperature, making this method an efficient one. Liu et al. also reported the successful application of this combination in the extraction of red pepper seed oil, with leftover propane being removed using ultrasonic treatment (Yao et al., 2020). This shows that US is an effective method to achieve complete removal of solvents from extracted oils. Hence, it can be concluded that through subcritical solvent extraction combined with ultrasonic treatment, efficient and complete extraction of phenolic chemicals from pomegranate peel (Sumere et al., 2018) as well as red pepper seed oil can be achieved. Red pepper oil of high quality and strong oxidation stability was produced by the total combination. These instances illustrate the substantial advantages of using US and subcritical solvent extraction, regardless of whether they are combined in-line or in situ, in terms of time, extraction yields, and energy savings (Khadhraoui et al., 2021).

20.2.3.3   Combination of US with Supercritical Fluids (SCF)

Supercritical fluid extraction (SFE) is a method of extraction that involves using a solvent in a supercritical state under high pressure. A fluid is said to be in a supercritical state (Pc) when it is simultaneously compressed above its critical pressure and heated above its critical temperature (Tc). SCFs exhibit unique properties that lie between those of liquids and gases; their diffusivity is intermediate, and their viscosity is similar to that of gases (Chemat et al., 2017; Khadhraoui et al., 2021). As a result, SCFs can dissolve specific chemicals like liquids and penetrate solid matrices like gases. Moreover, SCF density can be easily adjusted by varying temperature or pressure. SCF solvents are effective because they offer a certain level of selectivity and can recover molecules quickly. Among the SCFs, carbon dioxide (CO_2) is the most widely used due to its non-toxic, non-flammable, chemically inert, and easily separable after extraction (E.S. Dassoff, 2019; D. Panda, 2019).

Additionally, SFE employing CO_2 is more suited to the extraction of thermolabile chemicals than other standard methods because it uses moderate temperatures. An increase in the extraction yield of flavorful components from ginger rhizomes was achieved by combining US and CO_2. SEM examinations of the treated ginger rhizomes suggested that US-induced cellular damage may be responsible for this performance enhancement (Balachandran et al., 2006). As a result, the earliest stages of extraction of the targeted chemicals have a twofold diffusivity. With the aid of ultrasound, supercritical CO_2 extraction from agave bagasse also produced a higher extraction yield of antioxidants and saponins. In this instance, the adoption of a multiplate ultrasonic transducer enabled the extraction process to be intensified by US positive contribution while avoiding the compaction issue within the extraction cell (Santos-Zea et al., 2019).

The combination of supercritical CO_2 extraction and ultrasound pre-treatment used by Liu et al. is highly effective in obtaining iberisamara seed oil, resulting in a yield of more than 25% higher than that of supercritical CO_2 extraction alone. In addition, it improves the oil's physio-chemical characteristics and antioxidant activity (K. Li et al., 2020). Gomez-Gomez et al. investigated the stability of lipid emulsions produced by this technique and its impact on eliminating highly resistant microorganisms such as gram-positive bacteria or molds (Gomez-Gomez et al., 2020). The use of high-pressure ultrasound with supercritical CO_2 extraction has become a popular method for extracting oil from various sources due to its high efficiency and ability to maintain the desirable properties of the extracted oil. It is evident that this combined process is a valuable tool for efficiently and effectively extracting oil from a wide range of sources (Singla & Sit, 2021).

20.2.3.4 Combination of US and Instant Controlled Pressure Drop (DIC) Process

Thermomechanical processing is employed for DIC extraction, which typically involves creating a vacuum, injecting steam into the material for a few seconds, and then rapidly reducing the pressure to near vacuum. This pressure drop leads to several outcomes: The biomaterial is immediately cooled, cell walls are swollen or ruptured, resulted in increased porosity and improved accessibility and extraction kinetics, and the moisture within the treated material is quickly vaporized, leading to the expansion of the treated material (Allaf et al., 2014; Chemat et al., 2017; Khadhraoui et al., 2021).

Ultrasound (US) coupling has been demonstrated to be an effective method for enhancing the internal transfer of solute within the pore. Allaf et al. (2013) used a combination of DIC pre-treatment and UAE to enhance the extraction of antioxidants from orange peels, and found that this combination yielded the most antioxidants with the best kinetics (Figure 20.3) (Allaf et al., 2013). Furthermore, the use of DIC pre-treatment in conjunction with UAE yielded acceptable levels of vitamin B1, rebaudioside A, and vitamin B6 from Steviarebaudiana (Bertoni) leaves. Even more impressive was that these three bioactive components had significantly higher yields after swelling drying by DIC when compared to other methods (Mannaï et al., 2019). The potential applications of combining US coupling with DIC pre-treatment and UAE are vast, and further research into such combinations can help unlock new opportunities in extraction technologies (Khadhraoui et al., 2021).

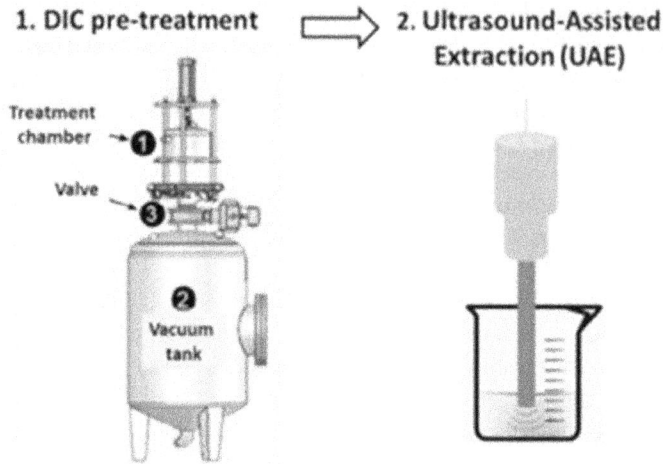

FIGURE 20.3 Combination of US and DIC (instant controlled pressure drop) process. (Reproduced from Khadhraoui et al., 2021.)

20.2.3.5 Combination of Ultrasound(US) with Pulsed Electric Field (PEF)

The PEF approach involves applying repeatedly brief, high-voltage pulses to a substance that is sandwiched between two electrodes. Electropermeabilization, commonly known as "electropora-tion," or the electrical breakdown of cellular membranes, is what propels PEF efficiency in solid-liquid extraction (Khadhraoui et al., 2021). The application of an electric field generates pores within the membrane, enhancing its permeability and resulting in increased extraction kinetics. The combination treatment of PEF and US produced the highest levels of total flavonoids, total phenolics, anthocyanin contents, condensed tannins, and antioxidant activity in DPPH compared to other treatments (Bozinou et al., 2019; Manzoor et al., 2019; Nowacka et al., 2019). Despite its many advantages, PEF also has some drawbacks which limit its effectiveness, including the inability to work with conductive materials and the need for it to be supplemented by other methods in order to obtain desired yields of specific compounds from raw materials. To illustrate this point, a com-bination of PEF and ultrasound (US) treatment was used to successfully extract different bioactive components from almond seeds, such as flavonoids, phenolics, antioxidant activity, condensed tan-nins, anthocyanins, and volatile chemicals. This successful PEF and US combination serves as an example of the potential of using various methods together to enhance extraction yields. Since the combination of PEF and US can provide higher yields than either process alone, it is important to consider such combinations when dealing with complex mixtures (Manzoor et al., 2019).

20.2.3.6 Combination of Ultrasound with UV-C Radiation

UV-C radiation is a non-thermal technique that utilizes short-wavelength, high-energy ultraviolet (UV) rays lasting for a few hundred microseconds. Plant cells respond to biotic and abiotic stressors by creating secondary metabolites, which act as defense chemicals. High-intensity light, particu-larly in the UV range, imposes an abiotic stress on plant cells (Hossain et al., 2015; Kim et al., 2019). Thus, using UV-C treatment before solid-liquid extraction can significantly increase the production of secondary metabolites. Furthermore, UV-C rays can dissolve cell walls physically, enhancing the release of cell contents into the surrounding extraction solvent during solid-liquid extraction (Kim et al., 2019). Combining UV-C with US can be an effective strategy for accelerating extraction and improving efficiency (Khadhraoui et al., 2021).

Combining UV-C with Ultrasound may be a very effective strategy for accelerating extrac-tion and increasing effectiveness. The impact of concurrent US postharvest treatment and UV-C radiation on tomato bioactive components during a 28-day storage period was examined by Esua et al. in 2019. Results showed that this non-thermal treatment boosted levels of lycopene, vitamin C, total phenols, and antioxidant activity in both the hydrophilic and lipophilic families (Esua et al., 2019). An interesting study done by Lima and colleagues revealed the effects of UV-C radiation on the quality of cherry tomatoes. In the study, tomato-irradiated powder was provided to the UAE. It was noted that lycopene content in tomato extracts from irradiated plants was 5.8 times higher than in control samples. This illustrates that UV-C radiation not only increases the amount of extraction but also increases its quality (Lima et al., 2019). In addition, it is quite interesting to note that combining ultrasound (US) and UV-C concurrently or in different steps has not yet been fully utilized despite having enormous potential. It can be assumed that this combination would further amplify the extraction rate and quality of the cherry tomatoes (Khadhraoui et al., 2021).

20.2.3.7   Combination of Ultrasound with CPC and CCC

A crucial step in the extraction of natural compounds from complex material is the effective separation and isolation of these compounds. Different techniques can be utilized to achieve this goal. To achieve this, various chromatographic methods such as liquid chromatography (LC), gas chromatography (GC), and supercritical fluid chromatography (SFC) have been used. Countercurrent chromatography (CCC) is a member of the LC family, which utilizes a liquid mobile phase without any solid support. CCC offers a significant advantage by preventing the sample from permanently adhering to a solid support, but its liquid stationary phase necessitates the design of specific hydrostatic and hydrodynamic columns (Berthod & Faure, 2015; T. Chen et al., 2014; Friesen et al., 2015; Khadhraoui et al., 2021a). The simultaneous extraction and isolation of natural chemicals from white peony roots was achieved by Zhang et al. using a combination of ultrasound-assisted dynamic extraction technique with CPC and CCC (Y. Zhang et al., 2015).

20.2.3.8   Ultrasound Combined with High Pressure Processing

High-pressure processing is created when pressures between roughly 100 and 1000 MPa are applied at room temperature (HPP). However, pressure treatment alone is unable to significantly inactivate spores or reduce the enzymatic activity of some enzymes. It is mostly used to deactivate pathogens coupled with vegetative microorganisms, yeasts, and molds. Temperature changes should be kept to a minimum since ultrasound treatment requires that the HHP pre-treatment be performed at a processing temperature of 5–8°C. In addition, ultrasound therapy combined with high-pressure therapy should result in a decrease in microbial activity. According to reports, the nutritional value and flavor of food aren't significantly impacted by the ultrasound-high pressure treatment (US-HPP). Applying US-HPP results in more inactivation of microorganisms and enzymes and can be used as an additional method of food preservation. In comparison to UHP and ultrasound pre-treatment alone (Pedras et al., 2014; Singla & Sit, 2021). UHP-US pre-treatment on vacuum freeze-dried strawberry slices resulted in the reduction in drying time and energy consumption while significantly improving quality (L. Zhang et al., 2020). US-HPP processing showed significant results in cranberry juice. The treatment had a strong preservation effect on fructo-oligosaccharides (FOS), and there were no significant changes in instrumental color analysis, pH, soluble solids, bioactive components, organic acids, or antioxidant capacity after treatment (Gomes et al., 2017). Lee et al. conducted a test to determine the appropriate parameters for ultrasound-high pressure treatment (US-HPP) in terms of temperature rise and *E. coli* inactivation (Dong-Un, 2009). Combining US-HPP led to increased E. coli inactivation. US-HPP treatment can be used as an additional food preservation method that results in increased inactivation of microorganisms and enzymes.

20.2.3.9   Ultrasound Combined with Liquid/Solid Phase Micro-Extraction

Two commonly used pre-treatment techniques are solid-phase microextraction (SPME) and liquid-phase microextraction (LPME). While LPME is more user-friendly, it often necessitates a greater number of samples and organic solvents, resulting in longer processing times. However, combining ultrasound with LPME/SPME can offer several benefits, including high repeatability, high extraction efficiency, low solvent consumption and time, low cost, simple operation, and minimal environmental impact. This combination is used to identify and preconcentrate a wide range of substances from water, juice, and other liquids, including heavy metals, organic medicines, food dyes, and hazardous colors (Singla & Sit, 2021). The researchers investigated the percentage of Azure B adsorption onto CNTs/Zn: ZnO@Ni2P-NCs at various ultrasound times ranging from 2 to 10 minutes. They found that by using 0.031 g of CNTs/Zn: ZnO@Ni2P-NCs, a sonication time of 8.2 minutes, an initial Az-B concentration of 26.1 mg L^{-1}, and a pH of 6.3, 98.84% Azure B adsorption was achieved (Alipanahpour Dil et al., 2019). Dil et al. investigated the use of ultrasound-assisted dispersive SPME with activated carbon loaded with silver-zinc oxide nanoparticles (Ag-ZnO-NP-AC) for the identification of Rose Bengal (RB) and Methyl Green (MG) colors in water and industrial effluent. The study found that at pH 6.7, with 1.6 mg of Ag-ZnO-NP-AC mass, 3.3 minutes of ultrasonication time, and 160 L of DMF, 99.37% and 99.27% recovery of RB and MG, respectively, was achieved (Alipanahpour Dil et al., 2017). Asfaram et al. also utilized ultrasound-assisted dispersive SPME to identify quercetin in Nasturtium officinale extract and fruit juice, as well as Allura Red (AR) in water samples and fruit juice (Asfaram et al., 2018). Therefore, the use of ultrasound in combination with LPME/SPME is beneficial due to its numerous advantages in terms of efficiency, cost, and environmental impact.

20.3 COMBINATION OF ULTRASOUND WITH MICROWAVE TECHNOLOGY IN FOOD PROCESSING

20.3.1 ULTRASOUND/MICROWAVE-ASSISTED DRYING

Dehydration has emerged as a highly attractive food processing method as it allows for convenient storage, transportation, and low-cost production of fruits and vegetables. Furthermore, dehydrated products have unique characteristics that differentiate them from the original ones, resulting in a more diverse market. The drying process can be broken down into three stages. In the first stage, water transport mechanisms are initiated during the induction phase, drying rate will increase and attain peak in second stage. During the final stage, also known as the falling rate period, The water evaporating from the surrounding surface exceeds the amount of water that moves from within to the surface. Even though convection drying can lead to extended shelf life, it often results in physical, chemical, and physio-chemical changes that reduce the quality of the product and may not meet current consumer standards.

Ultrasound-assisted convective drying (US-assisted) allows for a reduction in drying time and temperature for thermolabile products, making it an advantageous method for preserving bioactive chemicals. The "sponge" effect of US ensures that the product retains its sensory qualities and energy consumption. Furthermore, studies have revealed that there is no substantial negative effect of US on the quality of the final product, or its shelf life. Cavitation and streaming are two aspects of utmost importance to be taken into account when employing ultrasound technology, as they are integral components in maintaining product quality. With the aid of cavitation, streaming, and "sponge" effects, US-assisted convective drying can be accomplished at lower temperatures and for shorter durations, which is crucial for preserving thermolabile bioactive compounds, sensory attributes, and energy consumption. The final product quality and shelf life of the dehydrated product have recently been studied. There have been no reported substantial detrimental effects of US on quality. The majority of studies have demonstrated that foods may be produced that have acceptable microbiological counts, colors, textures after being rehydrated, vitamin contents, and little Maillard reaction products. It has also been used with other physical methods in an effort to lessen the drawbacks of using US power for drying. In order to investigate hybrid drying technologies, US and microwaves have been combined (MW).

Kroehnke et al. studied the drying kinetics of carrots under US and MW intensified convective drying. They found that the drying rates were comparable to MW alone, and the latter even consumed more energy. These outcomes may be due to low substrate permeability and a significant airflow rate. However, in materials with high stiffness and little porosity, such as carrots, US may be less effective (Kroehnke et al., 2018). In a study on strawberries, Szadzinska et al. investigated convective drying with MW (100 W) and US (200 W). They observed a 0.5 kWh increase in total energy consumption but an improvement in heat and mass transfer efficiency (Szadzińska et al., 2020). MW produces a "heating effect," and US creates a "vibration effect," which works together to accelerate the drying rate. Cavitation, streaming, and "sponge" effects, all of which are critical for thermolabile bioactive chemicals, sensory qualities, and energy consumption, make US-assisted convective drying feasible at a lower temperature and for a shorter duration. Moreover, this energy-efficient process can be maintained for an extended period of time due to the choice of optimal parameters for drying. Ultrasound has been proven to reduce the drying time significantly when compared to conventional drying methods, making it a viable option for dehydration needs. The benefits of ultrasound technology are further enhanced by its ability to be used with multiple types of materials, such as fruits and vegetables, thereby providing additional versatility in using this technique. In summary, convective drying with MW and US offers a cost-effective and efficient method for dehydration that can be applied to a wide range of materials (Chemat et al., 2019, 2020; Misra et al., 2018).

20.3.2   ULTRASOUND/MICROWAVE-ASSISTED VACUUM FRYING

Deep-frying is a popular cooking method used to produce a variety of food products both on an industrial and household scale. It is known for its distinct flavor, however, it also presents other unacceptable qualities. In order to reduce the amount of oil used in fried foods and still retain the quality, researchers have come up with various techniques using technology. This not only helps reduce the amount of oil but also gives people healthier options while enjoying the same flavor they get from frying. Therefore, with these technological advancements, it is now possible to fry food without compromising on taste or quality. The use of vacuum frying (VF) technology produces high-quality goods with lower oil content. However, there are some drawbacks, such as the use of a backward heating method, low efficiency, especially at low temperatures, and relatively high oil uptake in products. To address these challenges, the use of ultrasound (US) and combined US and microwave (MW) in VF has shown promising results in improving efficiency, including energy efficiency (by shortening the frying time and increasing effective moisture diffusivity), reducing frying temperature, and lowering oil uptake (Albertos et al., 2016; Su et al., 2016, 2018).

Application of US and MW in vacuum VF (USMWVF) is recommended to improve drying efficiency and moisture evaporation rate. The application of Ultrasound (US) and MW during the VF process has a synergistic effect in increasing the quality criteria. This can be seen through the reduction of the frying time for a variety of products such as apple slices, pumpkin chips, and mushroom chips. By using an ultrasound-microwave vacuum frying (UMWVF) technique, it has been determined that it is superior to the traditional MWVF technique. This is because of its ability to improve sensory and nutritional qualities during storage by reducing oil uptake, speeding up moisture evaporation, and enhancing the drying effectiveness. As a result, UMWVF is preferred for usage in food frying due to its positive effect on quality standards (Devi et al., 2021; X. Zhang et al., 2020).

20.3.3   ULTRASOUND/MICROWAVE-ASSISTED EXTRACTION

USMW-AE technology has demonstrated greater efficiency in extracting high-value chemicals from various matrices when compared to the use of a single technology alone. The mechanical effects of US, which involve damaging cell walls in a sequence of local erosion, shear forces, sonoporation, fragmentation, capillary effect, and detexturation, lead to increased penetration of the solvent into the cellular material. The ultrasound or microwave-assisted extraction method is a highly adaptable process that can be used in situ, in line, constantly, or intermittently. Intermittent extraction is more energy efficient since the US or MW irradiation stops intermittently during the extraction process and this on/off cycle would be repeated. This makes it possible to obtain more consistent and better quality extractions that are not hindered by lack of energy. The combination of these two methods offers unparalleled flexibility allowing users to customize their extraction processes to their specific needs(Chemat et al., 2017; Vinatoru et al., 2017; Wen et al., 2020).

The quick sample preparation method has become more cost-effective with the help of recent process intensification technologies. Using non-metallic horns at moderate strength, along with simultaneous twofold irradiation, can have additional and synergistic effects toward the extraction phenomenon of vegetal matrix. UMAE has been found to extract a greater yield of phenolic compound and antioxidant from brown macroalgae when compared to UAE and MAE used separately (Garcia-Vaquero et al., 2020). X. Wang et al. also observed a 2.454% higher yield of flavonoid components from Eucommia ulmoides leaves with UMAE, compared to ultrasonic and microwave extraction (Wang et al., 2020).

Bagherian et al. and Liew et al. conducted experiments to investigate the effects of ultrasonic-assisted microwave (USMW) as a pre-treatment step for the extraction of pectin from grapefruit and pomelo peel, respectively. Bagherian et al. found that using USMW-AE with HCl and microwave heating at 450 W for 10 minutes resulted in a higher yield (31.88%) compared to MW-AE (27.81%), US-AE (17.92%), and traditional methods (19.26%) (Bagherian et al., 2011). Similarly, Liew et al.'s study found that USMW-AE obtained the maximum yield under the ideal conditions when citric acid was used, followed by MWUS-AE, MW-AE, and US-AE. Both studies concluded that USMW is an effective pre-treatment step for extraction compared to other methods such as MW-AE and US-AE (Liew et al., 2016). Furthermore, this demonstrates that USMW can be a promising tool for improving the efficiency of extraction processes while reducing time and energy consumption.

20.3.4   Ultrasound and Microwave-Assisted Enzymatic Protein Hydrolysis

Since they have beneficial health-promoting properties like antibacterial, antioxidant, anticarcinogenic, hypocholesterolemic, antihypertensive, and immunomodulatory activity, bioactive peptides are a current research area. Bioactive peptides, due to their unique properties, have the potential to improve the quality of food by acting as an emulsifier, preservative, sweetener, color preservative, acidity regulator, taste enhancer, thickening, and foam-forming agent. High-intensity and low-frequency ultrasound has been used for its capability to alter the solubility, gelling, foaming, and emulsifying properties of proteins. Moreover, this process can also be used in numerous unit processes. The use of this technology to affect bioactive peptides has numerous potential applications and could revolutionize the food industry (Chew et al., 2018; Faustino et al., 2019; Görgüç et al., 2020; Hu et al., 2019).

20.4   SUMMARY

Combination of technologies or hybrid technologies presents an interesting area of research with a promising industrial application. Utilizing physical and/or chemical phenomena that are different from those used in individual processes, high process intensification is made possible by the integration of various approaches within a process. By combining the technologies such as ultrasound, HPP, and PEF with various cutting-edge techniques such as microwave, extrusion, SCF, subcritical and pressure liquids, DIC, UV or infra-red (IR) radiations, CCC, or centrifugal partition chromatographs (CPC), it is possible to achieve several benefits, including reduced equipment size, faster response to process control, quicker start-up times, increased production, and elimination of certain process steps. Additionally, combining these techniques can result in optimal performance as well as make them more economical and environmentally friendly. This technique may also enable more effective resource management, which would save costs, increase sustainability, and lessen environmental impact. As a result, the creation of such combinations is strongly promoted because they have a multitude of advantages, including higher product quality and increased process effectiveness. However, further research, especially from a pilot scale and industrial scale, needs to be conducted to make these hybrid technologies feasible for industry use.

REFERENCES

Abbott, A. P., Capper, G., Davies, D. L., Rasheed, R. K., & Tambyrajah, V. (2003). Novel solvent properties of choline chloride/urea mixtures. *Chemical Communications*, *1*, 70–71. https://doi.org/10.1039/b210714g

Al Juhaimi, F., Uslu, N., & Özcan, M. M. (2018). The effect of preultrasonic process on oil content and fatty acid composition of hazelnut, peanut and black cumin seeds. *Journal of Food Processing and Preservation*, *42*(1). https://doi.org/10.1111/jfpp.13335

Albertos, I., Martin-Diana, A. B., Sanz, M. A., Barat, J. M., Diez, A. M., Jaime, I., & Rico, D. (2016). Effect of high pressure processing or freezing technologies as pretreatment in vacuum fried carrot snacks. *Innovative Food Science and Emerging Technologies*, *33*, 115–122. https://doi.org/10.1016/j.ifset.2015.11.004

Alipanahpour Dil, E., Ghaedi, M., & Asfaram, A. (2017). Optimization and modeling of preconcentration and determination of dyes based on ultrasound assisted-dispersive liquid-liquid microextraction coupled with derivative spectrophotometry. *Ultrasonics Sonochemistry*, *34*, 27–36. https://doi.org/10.1016/j.ultsonch.2016.05.013

Alipanahpour Dil, E., Ghaedi, M., Asfaram, A., Mehrabi, F., & Sadeghfar, F. (2019). Efficient adsorption of Azure B onto CNTs/Zn:ZnO@Ni2P-NCs from aqueous solution in the presence of ultrasound wave based on multivariate optimization. *Journal of Industrial and Engineering Chemistry*, *74*, 55–62. https://doi.org/10.1016/j.jiec.2018.12.050

Allaf, T., Besombes, C., Tomao, V., Chemat, F., & Allaf, K. (2014). Coupling DIC and ultrasound in solvent extraction processes. *Food Engineering Series*, 151–161. https://doi.org/10.1007/978-1-4614-8669-5_8

Allaf, T., Tomao, V., Ruiz, K., Bachari, K., ElMaataoui, M., & Chemat, F. (2013). Deodorization by instant controlled pressure drop autovaporization of rosemary leaves prior to solvent extraction of antioxidants. *LWT*, *51*(1), 111–119. https://doi.org/10.1016/j.lwt.2012.11.007

Álvarez-Rivera, G., Bueno, M., Ballesteros-Vivas, D., Mendiola, J. A., & Ibáñez, E. et al. (2020). *Pressurized Liquid Extraction*. Elsevier. https://doi.org/10.1016/B978-0-12-816911-7.00013-X

Amirante, R., Distaso, E., Tamburrano, P., Paduano, A., Pettinicchio, D., & Clodoveo, M. L. (2017). Acoustic cavitation by means ultrasounds in the extra virgin olive oil extraction process. *Energy Procedia*, *126*, 82–90. https://doi.org/10.1016/j.egypro.2017.08.065

Amiri-Rigi, A., Abbasi, S., & Scanlon, M. G. (2016). Enhanced lycopene extraction from tomato industrial waste using microemulsion technique: Optimization of enzymatic and ultrasound pre-treatments. *Innovative Food Science and Emerging Technologies*, *35*, 160–167. https://doi.org/10.1016/j.ifset.2016.05.004

Anastas, P. T., & Warner, J. C. (1998). Green chemistry: Theory and practice. *Chemical Society Reviews*, *39*(1), 301–312. https://doi.org/10.1039/B918763B

Andreu, V., & Picó, Y. (2019). Pressurized liquid extraction of organic contaminants in environmental and food samples. *TrAC – Trends in Analytical Chemistry*, *118*, 709–721. https://doi.org/10.1016/j.trac.2019.06.038

Asfaram, A., Ghaedi, M., Abidi, H., Javadian, H., & Zoladl, M., Sadeghfar, F. (2018). Synthesis of Fe3O4@ CuS@ Ni2P-CNTs magnetic nanocomposite for sonochemical-assisted sorption and pre-concentration of trace Allura Red from aqueous samples prior to HPLC-UV detection: CCD-RSM design. *Ultrasonics Sonochemistry*, *44*, 240–250. https://doi.org/10.1016/j.ultsonch.2018.02.011

Asfaram, A., Ghaedi, M., Javadian, H., & Goudarzi, A. (2018). Cu- and S- @SnO₂ nanoparticles loaded on activated carbon for efficient ultrasound assisted dispersive μSPE-spectrophotometric detection of quercetin in nasturtium officinale extract and fruit juice samples: CCD-RSM design. *Ultrasonics Sonochemistry*, *47*, 1–9. https://doi.org/10.1016/j.ultsonch.2018.04.008

Ávila-Orta, C. A., González-Morones, P., Agüero-Valdez, D., González-Sánchez, A., G. Martínez-Colunga, J., M. Mata-Padilla, J., & J. Cruz-Delgado, V. (2019). Ultrasound-assisted melt extrusion of polymer nanocomposites. *Nanocomposites – Recent Evolutions*. https://doi.org/10.5772/intechopen.80216

Bagherian, H., Ashtiani, F. Z., Fouladitajar, A., & Mohtashamy, M. (2011). Chemical engineering and processing: Process intensification comparisons between conventional, microwave- and ultrasound-assisted methods for extraction of pectin from grapefruit. *Chemical Engineering & Processing: Process Intensification*, *50*(11–12), 1237–1243.

Balachandran, S., Kentish, S. E., Mawson, R., & Ashokkumar, M. (2006). Ultrasonic enhancement of the supercritical extraction from ginger. *Ultrasonics Sonochemistry*, *13*(6), 471–479. https://doi.org/10.1016/j.ultsonch.2005.11.006

Benvenutti, L., Zielinski, A. A. F., & Ferreira, S. R. S. (2019). Which is the best food emerging solvent: IL, DES or NADES? *Trends in Food Science and Technology*, *90*, 133–146. https://doi.org/10.1016/j.tifs.2019.06.003

Berthod, A., & Faure, K. (2015). Separations with a liquid stationary phase: Countercurrent chromatography or centrifugal partition chromatography. *Analytical Separation Science*, 1177–1206. https://doi.org/10.1002/9783527678129.assep046

Bora, S. J., Handique, J., & Sit, N. (2017). Effect of ultrasound and enzymatic pre-treatment on yield and properties of banana juice. *Ultrasonics Sonochemistry*, *37*, 445–451. https://doi.org/10.1016/j.ultsonch.2017.01.039

Bozinou, E., Karageorgou, I., Batra, G., Dourtoglou, V. G., & Lalas, S. I. (2019). Pulsed electric field extraction and antioxidant activity determination of moringa oleifera dry leaves: A comparative study with other extraction techniques. *Beverages*, *5*(1). https://doi.org/10.3390/beverages5010008

Chemat, F., Abert Vian, M., Fabiano-Tixier, A. S., Nutrizio, M., Režek Jambrak, A., Munekata, P. E. S., Lorenzo, J. M., Barba, F. J., Binello, A., & Cravotto, G. (2020). A review of sustainable and intensified techniques for extraction of food and natural products. *Green Chemistry*, *22*(8), 2325–2353. https://doi.org/10.1039/c9gc03878g

Chemat, F., Abert-Vian, M., Fabiano-Tixier, A. S., Strube, J., Uhlenbrock, L., Gunjevic, V., & Cravotto, G. (2019). Green extraction of natural products. Origins, current status, and future challenges. *TrAC – Trends in Analytical Chemistry*, *118*, 248–263. https://doi.org/10.1016/j.trac.2019.05.037

Chemat, F., Fabiano-Tixier, A. S., Vian, M. A., Allaf, T., & Vorobiev, E. (2015). Solvent-free extraction of food and natural products. *TrAC – Trends in Analytical Chemistry*, *71*, 157–168. https://doi.org/10.1016/j.trac.2015.02.021

Chemat, F., Rombaut, N., Sicaire, A. G., Meullemiestre, A., Fabiano-Tixier, A. S., & Abert-Vian, M. (2017). Ultrasound assisted extraction of food and natural products. Mechanisms, techniques, combinations, protocols and applications. A review. *Ultrasonics Sonochemistry*, *34*, 540–560. https://doi.org/10.1016/j.ultsonch.2016.06.035

Chen, H. M., Fu, X., & Luo, Z. G. (2015). Properties and extraction of pectin-enriched materials from sugar beet pulp by ultrasonic-assisted treatment combined with subcritical water. *Food Chemistry*, *168*, 302–310. https://doi.org/10.1016/j.foodchem.2014.07.078

Chen, T., Liu, Y., Zou, D., Chen, C., You, J., Zhou, G., Sun, J., & Li, Y. (2014). Application of an efficient strategy based on liquid-liquid extraction, high-speed counter-current chromatography, and preparative HPLC for the rapid enrichment, separation, and purification of four anthraquinones from *Rheum tanguticum*. *Journal of Separation Science*, *37*(1–2), 165–170. https://doi.org/10.1002/jssc.201300648

Chew, L. Y., Toh, G. T., & Ismail, A. (2018). Application of proteases for the production of bioactive peptides. *Enzymes in Food Biotechnology: Production, Applications, and Future Prospects*, 247–261. https://doi.org/10.1016/B978-0-12-813280-7.00015-3

Dassoff, E. S. & Li, Y. O. (2019). Mechanisms and effects of ultrasound-assisted supercritical CO_2 extraction. *Trends Food Science Technology*, *86*, 492–501. https://doi.org/10.1016/j.tifs.2019.03.001

Demirdöven, A., & Baysal, T. (2009). The use of ultrasound and combined technologies in food preservation. *Food Reviews International*, *25*(1), 1–11. https://doi.org/10.1080/87559120802306157

Devi, S., Zhang, M., & Mujumdar, A. S. (2021). Influence of ultrasound and microwave-assisted vacuum frying on quality parameters of fried product and the stability of frying oil. *Drying Technology*, *39*(5), 655–668. https://doi.org/10.1080/07373937.2019.1702995

Dong-Un, L. (2009). Effects of combination treatments of nisin and high-intensity ultrasound with high pressure on the functional properties of liquid whole egg. *Food Science and Biotechnology*, *18*(6), 1511–1514.

Esua, O. J., Chin, N. L., Yusof, Y. A., & Sukor, R. (2019). Effects of simultaneous UV-C radiation and ultrasonic energy postharvest treatment on bioactive compounds and antioxidant activity of tomatoes during storage. *Food Chemistry*, *270*, 113–122. https://doi.org/10.1016/j.foodchem.2018.07.031

Faustino, M., Veiga, M., Sousa, P., Costa, E. M., Silva, S., & Pintado, M. (2019). Agro-food byproducts as a new source of natural food additives. *Molecules*, *24*(6). https://doi.org/10.3390/molecules24061056

Friesen, J. B., McAlpine, J. B., Chen, S. N., & Pauli, G. F. (2015). Countercurrent separation of natural products: An update. *Journal of Natural Products*, *78*(7), 1765–1796. https://doi.org/10.1021/np501065h

Garcia-Vaquero, M., Ummat, V., Tiwari, B., & Rajauria, G. (2020). Exploring ultrasound, microwave and ultrasound-microwave assisted extraction technologies to increase the extraction of bioactive compounds and antioxidants from brown macroalgae. *Marine Drugs*, *18*(3), 1–15. https://doi.org/10.3390/md18030172

Gomes, W. F., Tiwari, B. K., Rodriguez, Ó, de Brito, E. S., Fernandes, F. A. N., & Rodrigues, S. (2017). Effect of ultrasound followed by high pressure processing on prebiotic cranberry juice. *Food Chemistry*, *218*, 261–268. https://doi.org/10.1016/j.foodchem.2016.08.132

Ghasemkhani, N., Morshedi, A., Poursharif, Z., Aghamohammadi, B., Akbarian, M., & Moayedi, F. (2014). Microbiological effects of high pressure processing on food; *JBES*, V4, N4, April, P133-145 https://innspub.net/microbiological-effects-of-high-pressure-processing-on-food/

Gomez-Gomez, A., Brito-de la Fuente, E., Gallegos, C., Garcia-Perez, J. V., & Benedito, J. (2020). Non-thermal pasteurization of lipid emulsions by combined supercritical carbon dioxide and high-power ultrasound treatment. *Ultrasonics Sonochemistry*, *67*. https://doi.org/10.1016/j.ultsonch.2020.105138

Görgüç, A., Bircan, C., & Yılmaz, F. M. (2019). Sesame bran as an unexploited by-product: Effect of enzyme and ultrasound-assisted extraction on the recovery of protein and antioxidant compounds. *Food Chemistry*, *283*, 637–645. https://doi.org/10.1016/j.foodchem.2019.01.077

Görgüç, A., Gençdağ, E., & Yılmaz, F. M. (2020). Bioactive peptides derived from plant origin by-products: Biological activities and techno-functional utilizations in food developments – A review. *Food Research International*, *136*. https://doi.org/10.1016/j.foodres.2020.109504

Hoover, D. G. (1997). Minimally processed fruits and vegetables: Reducing microbial load by nonthermal physical treatments. *Food Technology*, *51*(6), 66–71.

Hossain, M. B., Aguiló-Aguayo, I., Lyng, J. G., Brunton, N. P., & Rai, D. K. (2015). Effect of pulsed electric field and pulsed light pre-treatment on the extraction of steroidal alkaloids from potato peels. *Innovative Food Science and Emerging Technologies*, *29*, 9–14. https://doi.org/10.1016/j.ifset.2014.10.014

Hu, W., Chen, S., Wu, D., Zheng, J., & Ye, X. (2019). Ultrasonic-assisted citrus pectin modification in the bicarbonate-activated hydrogen peroxide system: Chemical and microstructural analysis. *Ultrasonics Sonochemistry*, *58*. https://doi.org/10.1016/j.ultsonch.2019.04.036

Khadhraoui, B., Ummat, V., Tiwari, B. K., Fabiano-Tixier, A. S., & Chemat, F. (2021). Review of ultrasound combinations with hybrid and innovative techniques for extraction and processing of food and natural products. *Ultrasonics Sonochemistry*, *76*, 105625. https://doi.org/10.1016/j.ultsonch.2021.105625

Kim, S. W., Ko, M. J., & Chung, M. S. (2019). Extraction of the flavonol quercetin from onion waste by combined treatment with intense pulsed light and subcritical water extraction. *Journal of Cleaner Production*, *231*, 1192–1199. https://doi.org/10.1016/j.jclepro.2019.05.280

Kroehnke, J., Szadzińska, J., Stasiak, M., Radziejewska-Kubzdela, E., Biegańska-Marecik, R., & Musielak, G. (2018). Ultrasound- and microwave-assisted convective drying of carrots – process kinetics and product's quality analysis. *Ultrasonics Sonochemistry*, *48*, 249–258. https://doi.org/10.1016/j.ultsonch.2018.05.040

Kumar, S., Rawson, A., Kumar, A., CK, S., Vignesh, S., & Venkatachalapathy, N. (2023a). Lycopene extraction from industrial tomato processing waste using emerging technologies, and its application in enriched beverage development. *International Journal of Food Science & Technology*, *58*(4), 2141–2150.

Kumar, S., Thirunavookarasu, S. N., Sunil, C. K., Vignesh, S., Venkatachalapathy, N., & Rawson, A. (2023b). Mass transfer kinetics and quality evaluation of tomato seed oil extracted using emerging technologies. *Innovative Food Science & Emerging Technologies*, *83*, 103203.

Leistner, L., & Gorris, L. G. M. (1995). Food preservation by hurdle technology. *Trends in Food Science and Technology*, *6*(2), 41–46. https://doi.org/10.1016/S0924-2244(00)88941-4

Li, K., Li, Y., Liu, C. L., Fu, L., Zhao, Y. Y., Zhang, Y. Y., Wang, Y. T., & Bai, Y. H. (2020). Improving interfacial properties, structure and oxidative stability by ultrasound application to sodium caseinate prepared pre-emulsified soybean oil. *LWT*, *131*(136). https://doi.org/10.1016/j.lwt.2020.109755

Liew, S. Q., Ngoh, G. C., Yusoff, R., & Teoh, W. H. (2016). Sequential ultrasound-microwave assisted acid extraction (UMAE) of pectin from pomelo peels. *International Journal of Biological Macromolecules*, *93*, 426–435. https://doi.org/10.1016/j.ijbiomac.2016.08.065

Lima, A. R., Cristofoli, N. L., Veneral, J. G., Fritz, A. R. M., & Vieira, M. C. (2019). Optimization conditions of UV-C radiation combined with ultrasound-assisted extraction of cherry tomato (Lycopersicon esculentum) lycopene extract. *International Journal of Food Studies*, *8*(2). https://doi.org/10.7455/ijfs.v8i2.603

Liu, H. M., Yao, Y. G., Ma, Y. X., & Wang, X. D. (2020). Ultrasound-assisted desolventizing of fragrant oil from red pepper seed by subcritical propane extraction. *Ultrasonics Sonochemistry*, *63*. https://doi.org/10.1016/j.ultsonch.2019.104943

Mannaï, A., Jableoui, C., Hamrouni, L., Allaf, K., & Jamoussi, B. (2019). DIC as a pretreatment prior to ultrasonic extraction for the improvement of rebaudioside A yield and preservation of vitamin B1 and B6. *Journal of Food Measurement and Characterization*, *13*(4), 2764–2772. https://doi.org/10.1007/s11694-019-00197-2

Manzoor, M. F., Zeng, X. A., Rahaman, A., Siddeeg, A., Aadil, R. M., Ahmed, Z., Li, J., & Niu, D. (2019). Combined impact of pulsed electric field and ultrasound on bioactive compounds and FT-IR analysis of almond extract. *Journal of Food Science and Technology*, *56*, 2355–2364. https://doi.org/10.1007/s13197-019-03627-7

Mathews, A., Tangirala, A. S., Kumar, S., Anandharaj, A., & Rawson, A. (2023). Extraction and modification of protein from sesame oil cake by the application of emerging technologies. *Food Chemistry Advances*, 100326.

Mbous, Y. P., Hayyan, M., Hayyan, A., Wong, W. F., Hashim, M. A., & Looi, C. Y. (2017). Applications of deep eutectic solvents in biotechnology and bioengineering – promises and challenges. *Biotechnology Advances*, *35*(2), 105–134. https://doi.org/10.1016/j.biotechadv.2016.11.006

Milenkovic, D., Milosavljevic, M., & Bojic, A. (2018). Ultrasound-assisted extraction of sunflower oil from the cake after sunflower seed pressing. *Journal of Agricultural Sciences, Belgrade*, *63*(2), 195–204. https://doi.org/10.2298/jas1802195m

Misra, N. N., Martynenko, A., Chemat, F., Paniwnyk, L., Barba, F. J., & Jambrak, A. R. (2018). Thermodynamics, transport phenomena, and electrochemistry of external field-assisted nonthermal food technologies. *Critical Reviews in Food Science and Nutrition*, *58*(11), 1832–1863. https://doi.org/10.1080/10408398.2017.1287660

Modupalli, N., Krisshnan, A., Sunil, C. K., Chidanand, D. V., Natarajan, V., Koidis, A., & Rawson, A. (2022). Effect of novel combination processing technologies on extraction and quality of rice bran oil. *Critical Reviews in Food Science and Nutrition*, 1–23. ISSN:1040-839. https://doi.org/10.1080/10408398.2022.2119367

Mukherjee, A., Roy, K., Jana, D. K., & Hossain, S. A. (2020). Qualitative model optimization of almond (*Terminalia catappa*) oil using Soxhlet extraction in type-2 fuzzy environment. *Soft Computing*, *24*(1), 41–51. https://doi.org/10.1007/s00500-019-04158-1

Murador, D. C., Braga, A. R. C., Martins, P. L. G., Mercadante, A. Z., & de Rosso, V. V. (2019). Ionic liquid associated with ultrasonic-assisted extraction: A new approach to obtain carotenoids from orange peel. *Food Research International*, *126*. https://doi.org/10.1016/j.foodres.2019.108653

Naik, M., Natarajan, V., Thangaraju, S., Modupalli, N., & Rawson, A. (2023). Assessment of storage stability and quality characteristics of thermo-sonication assisted blended bitter gourd seed oil and sunflower oil. *Journal of Food Process Engineering*, *46*(6), e14070.

Nowacka, M., Tappi, S., Wiktor, A., Rybak, K., Miszczykowska, A., Czyzewski, J., Drozdzal, K., Witrowa-Rajchert, D., & Tylewicz, U. (2019). The impact of pulsed electric field on the extraction of bioactive compounds from beetroot. *Foods*, *8*(7). https://doi.org/10.3390/foods8070244

Raso, J., & Barbosa-Cánovas, G. V. (2003). Nonthermal preservation of foods using combined processing techniques. *Critical Reviews in Food Science and Nutrition*, *43*(3), 265–285. https://doi.org/10.1080/10408690390826527

Rawson, A., Tiwari, B. K., Patras, A., Brunton, N., Brennan, C., Cullen, P. J., & O'Donnell, C. (2011). Effect of thermosonication on bioactive compounds in watermelon juice. *Food Research International*, *44*(5), 1168–1173.

Santos-Zea, L., Gutiérrez-Uribe, J. A., & Benedito, J. (2019). Effect of ultrasound intensification on the supercritical fluid extraction of phytochemicals from Agave salmiana bagasse. *Journal of Supercritical Fluids*, *144*, 98–107. https://doi.org/10.1016/j.supflu.2018.10.013

Sengar, A. S., Rawson, A., Muthiah, M., & Kalakandan, S. K. (2020). Comparison of different ultrasound assisted extraction techniques for pectin from tomato processing waste. *Ultrasonics Sonochemistry*, *61*, 104812.

Shishov, A., Bulatov, A., Locatelli, M., Carradori, S., & Andruch, V. (2017). Application of deep eutectic solvents in analytical chemistry. A review. *Microchemical Journal, 135*, 33–38. https://doi.org/10.1016/j.microc.2017.07.015

Singla, M., & Sit, N. (2021). Application of ultrasound in combination with other technologies in food processing: A review. *Ultrasonics Sonochemistry, 73*, 105506. https://doi.org/10.1016/j.ultsonch.2021.105506

Singla, M., & Sit, N. (2022). *Raising challenges of ultrasound-assisted processes and sonochemistry in industrial applications based on energy efficiency* (pp. 349–373). https://doi.org/10.1016/B978-0-323-91937-1.00017-7

Su, Y., Zhang, M., Adhikari, B., Mujumdar, A. S., & Zhang, W. (2018). Improving the energy efficiency and the quality of fried products using a novel vacuum frying assisted by combined ultrasound and microwave technology. *Innovative Food Science and Emerging Technologies, 50*, 148–159. https://doi.org/10.1016/j.ifset.2018.10.011

Su, Y., Zhang, M., Zhang, W., Adhikari, B., & Yang, Z. (2016). Application of novel microwave-assisted vacuum frying to reduce the oil uptake and improve the quality of potato chips. *LWT, 73*, 490–497. https://doi.org/10.1016/j.lwt.2016.06.047

Sumere, B. R., de Souza, M. C., dos Santos, M. P., Bezerra, R. M. N., da Cunha, D. T., Martinez, J., & Rostagno, M. A. (2018). Combining pressurized liquids with ultrasound to improve the extraction of phenolic compounds from pomegranate peel (*Punica granatum* L.). *Ultrasonics Sonochemistry, 48*, 151–162. https://doi.org/10.1016/j.ultsonch.2018.05.028

Sunil, C. K., Kamalapreetha, B., Sharathchandra, J., Aravind, K. S., & Rawson, A. (2017). Effect of ultrasound pre-treatment on microwave drying of okra. *Journal of Applied Horticulture, 19*(1), 58–62.

Szadzińska, J., Mierzwa, D., Pawłowski, A., Musielak, G., Pashminehazar, R., & Kharaghani, A. (2020). Ultrasound- and microwave-assisted intermittent drying of red beetroot. *Drying Technology, 38*(1–2), 93–107. https://doi.org/10.1080/07373937.2019.1624565

Tchabo, W., Ma, Y., Engmann, F. N., & Zhang, H. (2015). Ultrasound-assisted enzymatic extraction (UAEE) of phytochemical compounds from mulberry (*Morus nigra*) must and optimization study using response surface methodology. *Industrial Crops and Products, 63*, 214–225. https://doi.org/10.1016/j.indcrop.2014.09.053

Thakker, M. R., Parikh, J. K., & Desai, M. A. (2018). Ultrasound assisted hydrotropic extraction: A greener approach for the isolation of geraniol from the leaves of *Cymbopogon martinii*. *ACS Sustainable Chemistry and Engineering, 6*(3), 3215–3224. https://doi.org/10.1021/acssuschemeng.7b03374

Vinatoru, M., Mason, T. J., & Calinescu, I. (2017). Ultrasonically assisted extraction (UAE) and microwave assisted extraction (MAE) of functional compounds from plant materials. *TrAC – Trends in Analytical Chemistry, 97*, 159–178. https://doi.org/10.1016/j.trac.2017.09.002

Vivek, K., Subbarao, K. V., & Srivastava, B. (2016). Optimization of postharvest ultrasonic treatment of kiwifruit using RSM. *Ultrasonics Sonochemistry, 32*, 328–335. https://doi.org/10.1016/j.ultsonch.2016.03.029

Wang, X., Peng, M. J., Wang, Z. H., Yang, Q. L., & Peng, S. (2020). Ultrasound-microwave assisted extraction of flavonoid compounds from *Eucommia ulmoides* leaves and an evaluation of their antioxidant and antibacterial activities. *Archives of Biological Sciences, 72*(2), 211–221. https://doi.org/10.2298/ABS191216015W

Wen, L., Zhang, Z., Sun, D. W., Sivagnanam, S. P., & Tiwari, B. K. (2020). Combination of emerging technologies for the extraction of bioactive compounds. *Critical Reviews in Food Science and Nutrition, 60*(11), 1826–1841. https://doi.org/10.1080/10408398.2019.1602823

Yang, C. Y., Sheih, I. C., & Fang, T. J. (2012). Fermentation of rice hull by Aspergillus japonicus under ultrasonic pretreatment. *Ultrasonics Sonochemistry, 19*(3), 687–691. https://doi.org/10.1016/j.ultsonch.2011.11.015

Yao, Y., Pan, Y., & Liu, S. (2020). Power ultrasound and its applications: A state-of-the-art review. *Ultrasonics Sonochemistry, 62*, 104722. https://doi.org/10.1016/j.ultsonch.2019.104722

Zenker, M., Heinz, V., & Knorr, D. (2003). Application of ultrasound-assisted thermal processing for preservation and quality retention of liquid foods. *Journal of Food Protection, 66*(9), 1642–1649. https://doi.org/10.4315/0362-028X-66.9.1642

Zhang, Y., Li, T., Shen, Y., Wang, L., Zhang, H., Qian, H., & Qi, X. (2020). Extrusion followed by ultrasound as a chemical-free pretreatment method to enhance enzymatic hydrolysis of rice hull for fermentable sugars production. *Industrial Crops and Products, 149*. https://doi.org/10.1016/j.indcrop.2020.112356

Zhang, L., Liao, L., Qiao, Y., Wang, C., Shi, D., An, K., & Hu, J. (2020). Effects of ultrahigh pressure and ultrasound pretreatments on properties of strawberry chips prepared by vacuum-freeze drying. *Food Chemistry*, *303*. https://doi.org/10.1016/j.foodchem.2019.125386

Zhang, Y., Liu, C., Li, J., Qi, Y., Li, Y., & Li, S. (2015). Development of "ultrasound-assisted dynamic extraction" and its combination with CCC and CPC for simultaneous extraction and isolation of phytochemicals. *Ultrasonics Sonochemistry*, *26*, 111–118. https://doi.org/10.1016/j.ultsonch.2015.02.017

Zhang, X., Zhang, M., & Adhikari, B. (2020). Recent developments in frying technologies applied to fresh foods. *Trends in Food Science and Technology*, *98*, 68–81. https://doi.org/10.1016/j.tifs.2020.02.007

21 Environmental Impact of Novel Non-Thermal Technologies

*Subaitha Z. Afrose, P. Santhoshkumar,
Shubham Nimbkar, and J. A. Moses*

21.1 INTRODUCTION

A primary objective of the food processing sector is to ensure food safety and security. Over the years, as the global population is increasing, food production must also be increased to meet this demand. Furthermore, owing to increased awareness, consumer demand for highly nutritive, fresh, safe foods with prolonged shelf life. Many conventional treatments were practiced to prevent and eliminate microbial proliferation and prolong shelf life. The application of heat is one of the most common conventional treatments which prevent microbial spoilage of foods, thus ensuring stability and safety. Pasteurization, sterilization by retorting, and drying are some of the approaches to name. Though these technologies have been commercially well-implemented, the negative effects such as loss of nutrition quality, texture, color, and consumer acceptance cannot be ignored (Putnik et al., 2018). Furthermore, higher energy requirement which is fulfilled through the use of fossil fuels, and the generation of air pollutants and effluents create a huge environmental impact that needs to be addressed (Kumar et al., 2021).

According to the 2030 agenda for sustainable development of the United Nations, most of the sustainable development goals (SDGs) are directly or indirectly associated with the global food system. Considering the aforementioned concerns, researchers moved toward non-thermal technologies as a promising approach to food processing. Non-thermal technologies have the potential to replace conventional technologies by achieving consumers' needs with high retention of nutrition quality, food safety, and improved shelf life (Režek et al., 2018). In addition to the consciousness of food product safety and security, the impact of processing technologies on the environment, such as carbon footprints, waste generation, energy consumption, and recycling of byproducts, is considered, which also helps to reduce cost (Pardo & Zufía, 2012). Carbon footprints and food loss are critical issues to be addressed to lead a sustainable food supply chain (Shabir et al., 2023). These factors impact economic balance and resource management having a significant impact on both economic and environmental aspects. By considering these social benefits in economic evaluations and decision-making, non-thermal technologies are seen as better than the traditional alternative. These concerns have paved the way for researchers to employ emerging non-thermal technologies such as pulse electric field (PEF), cold plasma, high-pressure processing (HPP), ultrasound, ozone, and pulse light in the food industries as an alternative to conventional thermal technologies considering the sustainability of technology with a perspective of environmental impact, food quality, and safety (Figure 21.1) (Jadhav et al., 2021). The present chapter discusses the role of selective non-thermal food processing methodologies in the establishment of a sustainable food supply chain. Further, the environmental impact of these methodologies studied through life cycle assessment (LCA) is elucidated along with a comparison with conventional thermal processing techniques.

21.2 LIFE CYCLE ASSESSMENT

The entire food supply chain, from raw materials, processing, production, packaging, and transport to consumers, is considered hotspot for achieving sustainability in the food industry. LCA is

DOI: 10.1201/9781003359302-21

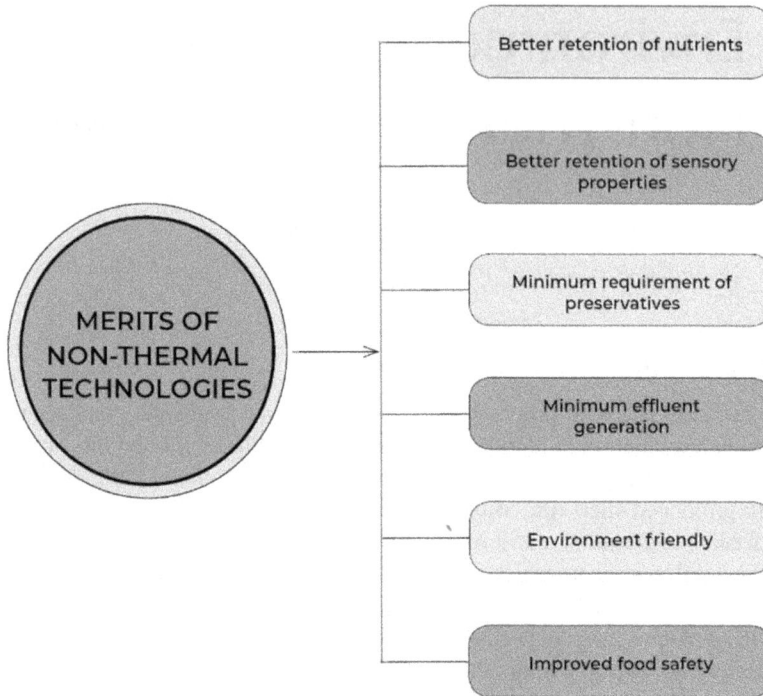

FIGURE 21.1   Advantages offered by non-thermal technologies.

a multifaceted approach used along the food supply chain to evaluate and determine the food product's life cycle and its environmental impact (Kumar et al., 2021). This tool helps to identify the hotspots in the life cycle stages of food products and the mitigation measures to improve the carbon footprint impacts and other environmental burdens. LCA comprises four stages namely goal and scope, inventory analysis, life cycle impact assessment, and interpretation per ISO 14040 (ISO, 2006). This helps to compare and assess the implementation of the technological process of a food product from raw material to final disposal, thus describing the impact on the environment. Since some non-thermal technologies are implemented and processed on a lab scale, which leads to discrepancies in the results compared to the industrial scale, the LCA is a tool that describes environmental performance using non-thermal treatment. It is necessary to consider the adverse environmental effect of these technologies to be effectively exploited in the food industry market to strengthen food security and achieve sustainable goals for society (Priyadarshini et al., 2019).

LCA's food supply chain phases include raw material, processing production, distribution, and consumption. Food loss, carbon footprint, and water resources are all major concerns of related environmental impacts. Apart from LCA, several environmental assessment methodologies such as life cycle costing (LCC), social life cycle assessment (S-LCA), and overall life cycle sustainability assessment (LCSA) can be used, nevertheless, certain peculiar features of LCA make it the most suitable approach for environmental assessment (Notarnicola et al., 2017). However, for economic aspects, LCC is an effective methodology.

ISO 14040 standards regulate the LCA methodology used to study the sustainability of food products in the ecosystem. This LCA methodology helps to identify the hotspots by setting scope and boundaries, mapping and interpreting collected data that helps to reduce or improve environmental burdens. LCA follows a sequential four-step methodology, as depicted in Figure 21.2. The first and foremost step is the identification of the goal or intention of the study as well as the product systems, boundaries, and impact categories. The product systems are classified as cradle-to-gate,

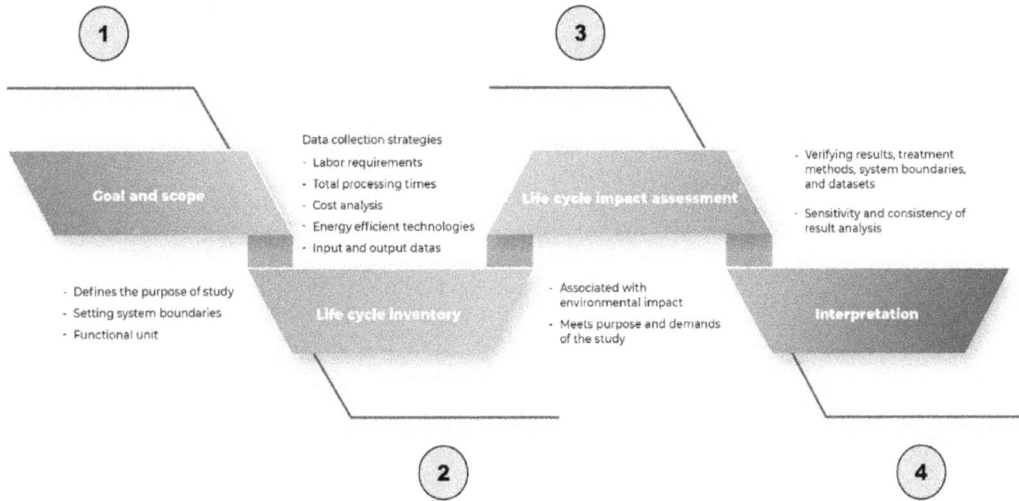

FIGURE 21.2 Four steps of life cycle assessment analysis.

gate-to-gate, and gate-to-disposal based on the system boundaries. In the case of non-thermal technologies where most of the equipment is at the lab or pilot scale, the gate-to-gate approach is the most suitable. Additionally, the functional unit also must be defined. A functional unit is a measurement unit of a chosen food product that is used as a comparison characteristic. As for non-thermal technologies, most of the data are obtained from technologies working at a laboratory scale; it is recommended to approach the gate-to-gate study, which analyzes the production plant's process. However, certain problems are associated with all the steps in performing LCA of non-thermal technologies (Naveed et al., 2022). The life cycle inventory analysis includes the total process's energy and mass balances. The LCA approach includes all resources as inputs and outputs of emissions from the process. The life cycle impact assessment comprises all these inputs and outputs of collecting and transferring into mathematical modeling for analysis. Based on this impact assessment, categories are quantified and help to identify the particular impact category and define its impact level on the environment. Upon methodology and software employed, impact categories such as ozone depletion, global warming, human toxicity, resource depletion, eutrophication, and aquatic acidification influence the environmental impact are assessed as shown in (Figure 21.3) (Raja et al., 2022).

21.3 ENVIRONMENTAL IMPACT OF CONVENTIONAL FOOD PROCESSING METHODS

Food processing or preservation step is responsible for highest environmental impact in entire food supply chain. Conventionally, these processing or preservation steps for effective elimination of microbes and ensuring the safety of foods are heat-mediated. These thermal processing approaches include pasteurization, sterilization, drying, evaporation, etc. These methods commonly use steam as a medium of heat transfer which is generated through burning of fossil fuels having significant environmental impact. A comparative cradle-to-grave LCA was performed comparing the environmental impact of conventional thermal pasteurization (TP) with alternative techniques. TP showed the highest environmental impact in almost all impact categories including global warming potential, photochemical oxidation, water depletion, etc. Higher demand of water and combustion of fossil fuels for steam generation are the reasons behind it (Pardo & Zufía, 2012). Furthermore, during continuous operations such as pasteurization, formation of fouling layers leads to reduced efficiency

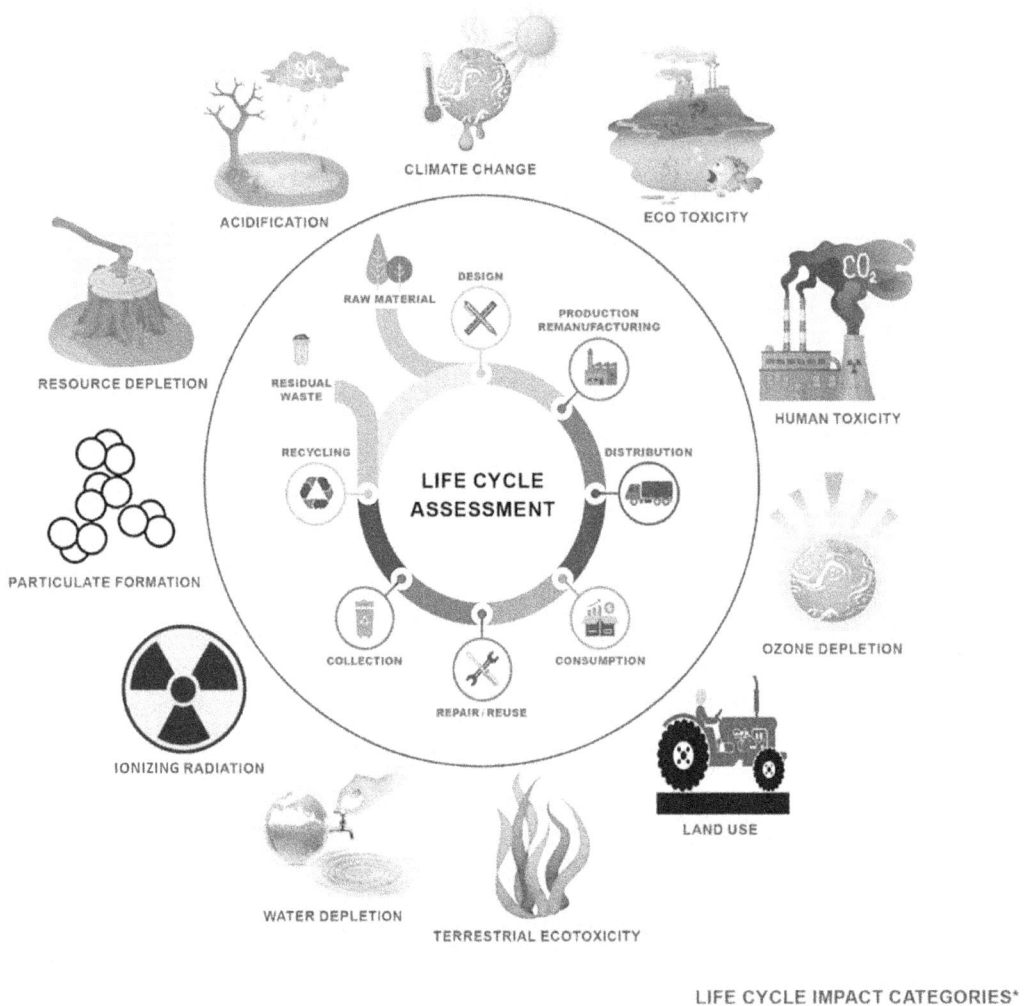

CLIMATE CHANGE

ACIDIFICATION

ECO TOXICITY

RESOURCE DEPLETION

HUMAN TOXICITY

PARTICULATE FORMATION

OZONE DEPLETION

IONIZING RADIATION

LAND USE

WATER DEPLETION

TERRESTRIAL ECOTOXICITY

DESIGN

RAW MATERIAL

PRODUCTION REMANUFACTURING

RESIDUAL WASTE

RECYCLING

DISTRIBUTION

LIFE CYCLE ASSESSMENT

COLLECTION

CONSUMPTION

REPAIR / REUSE

LIFE CYCLE IMPACT CATEGORIES*

FIGURE 21.3   Life cycle assessment and its major impact categories.

and necessitates frequent cleaning-in-place which again adds to increased energy consumption and wastewater generation (Zouaghi et al., 2019). However, opposite trend was observed in another LCA study by Aganovic et al. (2017). A comparison between TP with pulsed electric field (PEF) and HPP revealed that the technologies showed no significant difference in the environmental impact. Surprisingly, conventional TP showed the least energy requirement compared to the other two. Thermal processing is an industrially adopted and optimized process with effective heat recovery (70–95%) which resulted in higher efficiency.

Another LCA was conducted to evaluate the environmental impact of ready to eat canned meat products (Perez-Martinez et al., 2018). The impact assessment revealed that meat production and processing are the two stages having maximum environmental impact. Moreover, processing has more maximum impact in categories such as ozone depletion, human toxicity, photochemical oxidation, freshwater and marine ecotoxicity, metal depletion, fossil depletion, and urban land occupation. It is essential to note that the step of canning followed by sterilization has the greatest contribution (more than 75%) to the impact in aforementioned categories. While the use of tinplate cans has a negative impact in canning step, autoclaving offers a negative impact in sterilization step. Higher consumption of water and fossil fuels is the major reason behind this impact. In a recent study on peach canning, can washing, steam generation by boiler, and peeling are considered to be the hotspots on environmental impact due to heavy resource consumption and wastewater generation (Drosou et al., 2023).

21.4 ENVIRONMENTAL IMPACT OF NON-THERMAL PROCESSING TECHNIQUES

21.4.1 PULSE ELECTRIC FIELD PROCESSING

PEF finds numerous applications in the food processing sector such as preservation, enzyme inactivation, extraction, etc. PEF is sustainable technology as it consumes less energy and water compared to its conventional counterparts. Toward environmental and economic sustainability, crucial areas such as gas emissions, water usage, reduction and valorization of food wastes, ozone depletion, energy consumption, throughput efficiency, and profit outcomes are considered (Thirukumaran et al., 2022).

In the tomato processing industry, pasteurized steam is the most popular technology for peeling tomatoes. This is associated with higher water and energy consumption. Arnal et al. (2018) performed a cradle-to-grave LCA to analyze the environmental impact of the incorporation of PEF as a pre-treatment. The trials were conducted at an industrial scale using ReCiPe software, eight impact categories were considered among 18 factors. It was concluded that the incorporation of PEF technology reduced the steam requirement by 20% compared to conventional pressurized steam processing which is reflected as an improvement in environmental aspects by 17–20%, especially in the advantage of ozone and fossil fuel decline.

In a study, Pappas et al. (2023) examined the extraction of polyphenols from freeze-dried moringa leaves and demonstrated the minimal environmental impact of PEF in terms of electricity consumption. Five techniques, PEF, microwave-assisted extraction, ultrasound extraction, boiling water, and maceration, were compared, and LCA analysis was done using SimaPro eco invent software. Here, electricity usage and moringa leaves consider the main parameter. The cradle-to-gate approach of environmental impact assessment results in PEF and microwave-assisted extracted the best extraction techniques resulting in environmentally friendly, greener technology. This can be more if photovoltaic arcs are renewable sources used. The impact categories were human health, marine ecosystem, freshwater ecosystem, and resources. Using food waste and byproducts from food processing is becoming important to improve economic performance and guarantee environmental sustainability. The compounds recovered from the food wastes could have the potential to be employed in different food and biotechnological applications.

The development of green extraction techniques, microwave, ultrasound, and PEF as a significant approach in recovering byproducts from food wastes as an alternative to traditional treatments such as Soxhlet extraction, liquid-liquid extraction, and mechanical shaking. Among these, PEF, as a novel technique, provides high-efficiency and better-quality products with less processing time, solvent usage, and energy usage (Naveed et al., 2021).

21.4.2 HIGH-PRESSURE PROCESSING

HPP is one of the non-thermal eco-friendly technologies with little negative environmental impact. It has the potential to provide sustainable food without compromising the safety and quality of products ensuring the clean label of foods. Also, HPP ensures the preservation of foods without the addition of chemical preservatives (Nabi et al., 2021). In other aspects such as energy and water consumption, HPP stood superior compared to the conventional techniques. In a comparative study on fish soup pasteurization, HPP and radiofrequency in combination with thermal solar energy were compared with conventional TP. HPP was observed to save up to 75% of water compared to the conventional method (Muñoz et al., 2022). However, it is essential to note that the energy efficiency of HPP strongly depends on the pressure and time combinations used. Optimization of these parameters considering the safety of the product will bring the best results (Aganovic et al., 2017).

Cacace et al. (2020) used LCA and LCC approaches to compare HPP's environmental and economic effects, TP, and modified atmospheric packaging. The study used orange juice and Parma ham slices, which were processed using conventional TP and MAP. According to economic considerations, thermal direct and indirect processes are preferable to HPP since, according to the LCC approach, HPP's initial investment and maintenance costs are higher than TP's. The ReCiPe impact assessment method was used in this study to calculate the eight impact categories such as ozone formation, ozone depletion, global warming, terrestrial ecosystem, human health toxicity, aquatic toxicity, resource depletion, and water and energy consumption. Marine eutrophication, ozone depletion, excessive energy consumption, and mineral resources generated high impacts on the environment, with HPP having a high influence on ionizing radiation. Heinz (2016) examined the LCA study of milk processed by ultra-HPP and conventional thermal treatment. Based on ReCiPe midpoint methods, ultra-HPP exhibits better environmental performance, such as a reduction in the carbon footprint of up to 31% and improved energy use of 34%, which results in low environmental impact because of low electric consumption. Further upscaling of the technology with improved efficiency of reduced carbon footprints was also examined.

Generally, HPP is a more environmentally friendly technique than TP for processing juice since it uses electricity rather than steam to lessen its negative environmental effects. In the case of Parma ham processing, HPP is more efficient than MAP in terms of both economic and environmental aspects. Despite HPP's high initial investment and maintenance cost, LCC methods show a significantly higher value for MAP than HPP. Considering all impact categories, MAP has fewer environmental benefits than HPP, which highly influences environmental impact except for stratospheric ozone and ionizing radiation. The high environmental impact of MAP might be due to the volume of primary packaging used in MAP.

21.4.3 Ozone Technology

Ozonation is gaining importance in the food industry by replacing conventional operations or technology in processing since ozone does not require heat energy and can be used during the process time use; adopting ozone technology cuts down transportation costs, energy use, low operating cost with less environmental impact leaving no residues in the finished products (Senol Ibanoglu, 2023).

Ozonated (OZ) water can reduce the microbial population and increase the shelf life of fresh fruits and vegetables. In this line, Raila et al. (2020) compared and evaluated the effect of ozonated water and tap water in terms of carrot quality and environmental impact. SimaPro software was used in this study to assess environmental factors such as CO_2 emissions, SO_2, and eutrophication-related emissions. OZ water significantly reduced the microbial population with 1.8-fold increase in the shelf life of carrots. Based on the LCA of two technologies, the replacement of the cooling tunnel by ozone spray reduced the environmental impact by reducing emissions of CO_2, SO_2, and aquatic eutrophication. This demonstrated the utilization of ozone to promote cleaner operations, lowering energy and waste production while maintaining quality and food security. And also recommends using ozone technology as one of the steps in the supply chain process (Glowacz et al., 2015).

21.4.4 Ultrasound Processing

Recently, along with other applications, ultrasonication has emerged as a novel disruptive technology for extraction applications. The extraction of essential components requires the high-temperature use of numerous solvents and their residues/byproducts, which imposes a corresponding burden on the environment as well as the use of energy, a time-consuming process and high operating costs, all of which have a significant negative impact on the economy (Ali et al., 2021). Compared to conventional treatments, the environmental performance of greener technology, such as ultrasound-assisted extraction, is assessed based on an LCA study. Implementing any technology in the food industry should be environmentally and economic sustainability.

The conventional Soxhlet and maceration techniques, ultrasound-assisted extraction, employ greener technology which is advantageous over energy consumption and processing time. Solvent usage emits low CO_2 and has a positive impact on the environment. The energy requirement for solvent extraction of fat and oil from oleaginous seeds using maceration, Soxhlet, and ultrasound extraction are 6, 8, and 0.25 kWh, respectively. Then, taking into consideration the burning of fossil fuels to obtain 1 kWh, concerning CO_2 emissions in the environment, Soxhlet extraction stands out high 6400 g CO_2 emissions for 100 g of an extracted solid compound, followed by maceration 3600 g of 100 g of solid material extracted and low in case of ultrasound-assisted extraction (200 g of CO_2/100 g) (Chemat et al., 2016).

To extract the functional food components phenolics from seaweed using ultrasound and conventional technologies. The LCA study of ReCiPe 2016 at the midpoint level was performed for environmental assessment (Priyadarshini et al., 2022). The following impact categories: fine particulate matter formation, human health, ozone depletion, marine toxicity, eutrophication effects, terrestrial acidification and ecotoxicity, usage of water, and resource utilization were considered. Based on impact categories, ultrasound, compared to conventional solvent extraction technologies, low environmental concerns were generated for ozone formation and particulate matter, which is six times and eight times having a low impact, respectively, affecting human health and terrestrial ecosystems. In addition, no global effect and land use impact was observed in both extraction techniques. According to the economic perspective, the cost of operation at the laboratory scale of ultrasound is high owing to the ultrasound yield of 35.1%, comparable to the conventional 11.2% yield, whose operational cost is low. As the operating cost and raw materials are major impact factors for the cost of manufacturing economic concern, ultrasound has a lower environmental impact than conventional solvent extraction, with electricity as a major contributing factor.

21.4.5 SUPERCRITICAL FLUID EXTRACTION (SFE)

The extraction of bioactive compounds from food byproducts is becoming an important sector in the nation. Converting food waste into valuable compounds is required for greater economic and to ensure a sustainable environment. More specifically, supercritical fluid extraction has a major focus which increases efficiency, protects human health, improves resource management, and reduces environmental concerns (Villacís-chiriboga et al., 2021).

Syrpas et al. (2018) reported that compared to conventional hexane extraction of lipophilic products from cyanobacteria, lab-scale supercritical fluid exhibits a high environmental impact of six-fold times higher than conventional hexane process covering impact categories of aquatic toxicity, human health, and ecosystem damage due to emission of CO_2 of high energy demand. The LCA methodology using Gabi software was studied, where 1 g of oil extract is considered as a functional unit in this study. Based on the LCA results, improvements in SFE should be done to increase the efficiency of the SFE process to conventional hexane treatment. Among the impact categories, aquatic toxicity was the main impact which was threefold times higher than hexane treatment by the release of diethylene glycol into the water stream whereas, in the case of human health toxicity, carcinogens had a dominant impact influence the environment by release of chloroethylene 13.63 g equivalent of SFE and 1.46 g hexane into the air.

The environmental sustainability of the supercritical fluid process can be enhanced by improving energy efficiency and using alternative energy sources. Scaling up supercritical fluid can improve, and the overall performance can be anticipated since the data is based on the laboratory level. Hence, it is crucial to use energy efficiently to improve the supercritical fluid technology in an eco-friendly way. Manufacturing and operation costs are high compared to conventional process treatments. The integration of supercritical fluid with other process purification reduces the operational cost, decreases gas emissions, and hence reduces economic aspects (Lee & Soh, 2019). Meoli et al. (2023) used non-thermal plasma emerging technology due to short processing time and zero waste production as pre-treatment in biorefinery biomass treatment which fixes the problems associated with waste generation and fossil fuel consumption.

21.5   SUMMARY

Food production accounts for significant environmental impact and intervention at processing stage is mandatory; this demands an alternative to conventional food processing technologies. Non-thermal technologies have proven their superiority in terms of the safety, quality, and preservation effect compared to conventional technologies. Further, the environmental impact of these technologies has been assessed by several researchers and compared with conventional approaches. This is assessed through LCA-based assessment tools. Studies have suggested that implementation of non-thermal technologies provides environmental benefits through low energy consumption and water usage, valorization of byproducts, process efficiency, and short processing time. However, some studies also reported increased environmental impact and increased cost by non-thermal techniques. As these techniques are at still at laboratory scale, there are certain issues with conducting LCA. Also, effective scale-up and optimization of the process parameters on industrial scale becomes important. It is also essential to note that the major source of energy for non-thermal technologies is electricity which can be sourced from renewable sources further reducing the environmental impact. Overall, scale-up and optimization is the key for unlocking the potential of non-thermal technologies.

REFERENCES

Aganovic, K., Smetana, S., Grauwet, T., Toepfl, S., Mathys, A., Van Loey, A., & Heinz, V. (2017). Pilot scale thermal and alternative pasteurization of tomato and watermelon juice: An energy comparison and life cycle assessment. *Journal of Cleaner Production*, *141*, 514–525. https://doi.org/10.1016/j.jclepro.2016.09.015

Ali, A., Wei, S., Liu, Z., Fan, X., & Sun, Q. (2021). Non-thermal processing technologies for the recovery of bioactive compounds from marine by-products. *LWT*, *147*(April), 111549. https://doi.org/10.1016/j.lwt.2021.111549

Arnal, Á. J., Id, P. R., Pataro, G., Ferrari, G., Ferreira, V. J., Ana, M. L., & Ferreira, A. (2018). Implementation of PEF treatment at real-scale tomatoes processing considering LCA methodology as an innovation strategy in the agri-food sector. https://doi.org/10.3390/su10040979

Cacace, F., Bottani, E., Rizzi, A., & Vignali, G. (2020). Evaluation of the economic and environmental sustainability of high pressure processing of foods. *Innovative Food Science and Emerging Technologies*, *60*(March 2020), 102281. https://doi.org/10.1016/j.ifset.2019.102281

Chemat, F., Rombaut, N., Sicaire, A., Meullemiestre, A., & Abert-vian, M. (2016). Ultrasound assisted extraction of food and natural products. Mechanisms, techniques, combinations, protocols and applications. *Ultrasonics—Sonochemistry*. https://doi.org/10.1016/j.ultsonch.2016.06.035

Drosou, F., Kekes, T., & Boukouvalas, C. (2023). Life cycle assessment of the canned fruits industry: Sustainability through waste valorization and implementation of innovative techniques. *AgriEngineering*, *5*(1), 395–412. https://doi.org/10.3390/agriengineering5010026

Glowacz, M., Colgan, R., & Rees, D. (2015). Postharvest biology and technology influence of continuous exposure to gaseous ozone on the quality of red bell peppers, cucumbers and zucchini. *Postharvest Biology and Technology*, *99*, 1–8. https://doi.org/10.1016/j.postharvbio.2014.06.015

Heinz, V. (2016). Life cycle assessment of emerging technologies: The case of milk ultra-high pressure homogenisation. *Journal of Cleaner Production*. https://doi.org/10.1016/j.jclepro.2016.11.059

ISO. (2006). 14040. *Environmental management—life cycle assessment—principles and framework*, 235–248. ISBN: 0580489930; http://www.cscses.com/uploads/2016328/20160328110518251825.pdf (accessed on 29.10.2023)

Jadhav, H. B., Annapure, U. S., & Deshmukh, R. R. (2021). Non-thermal technologies for food processing. *Frontiers in Nutrition*, *8*(June), 1–14. https://doi.org/10.3389/fnut.2021.657090

Kumar, A., Sriraksha, M. S., & Ravishankar, C. N. (2021). Sustainability of emerging green non-thermal technologies in the food industry with food safety perspective: A review. *LWT*, *151*(November 2021), 112140. https://doi.org/10.1016/j.lwt.2021.112140

Lee, L. Y., & Soh, S. H. (2019). Process Economics and Feasibility of Subcritical & Supercritical Fluid Processing. In *Innovative Food Processing Technologies: A Comprehensive Review*. Elsevier. https://doi.org/10.1016/B978-0-08-100596-5.22931-4

Meoli, C. M., Iervolino, G., & Procentese, A. (2023). Non-Thermal Plasma as a Biomass Pretreatment in Biorefining Processes. *Processes*, *11*(2), 536. https://doi.org/10.3390/pr11020536

Muñoz, I., de Sousa, D. A. B., Guardia, M. D., Rodriguez, C. J., Nunes, M. L., Oliveira, H., Cunha, S. C., Casal, S., Marques, A., & Cabado, A. G. (2022). Comparison of different technologies (conventional thermal processing, radiofrequency heating and high-pressure processing) in combination with thermal solar energy for high quality and sustainable fish soup pasteurization. *Food and Bioprocess Technology*, *15*(4), 795–805.

Nabi, B. G., Mukhtar, K., Arshad, R. N., Radicetti, E., Tedeschi, P., Shahbaz, M. U., Walayat, N., Nawaz, A., Inam-Ur-raheem, M., & Aadil, R. M. (2021). High-pressure processing for sustainable food supply. *Sustainability (Switzerland)*, *13*(24). https://doi.org/10.3390/su132413908

Naveed, R., Abdul-Malek, Z., Roobab, U., Abdullah, M., Naderipour, A., Imran, M., Bekhit, A. E., Liu, Z., & Muhammad, R. (2021). Trends in food science & technology pulsed electric field : A potential alternative towards a sustainable food processing. *Trends in Food Science & Technology*, *111*(January), 43–54. https://doi.org/10.1016/j.tifs.2021.02.041

Naveed, R., Abdul-Malek, Z., Roobab, U., Modassar, M., Nawaz, A., Re, A., Imran, M., Khan, N., Manuel, J., & Muhammad, R. (2022). Nonthermal food processing: A step towards a circular economy to meet the sustainable development goals. *Food Chemistry: X, 16*(June). https://doi.org/10.1016/j.fochx.2022.100516

Notarnicola, B., Sala, S., Anton, A., McLaren, S. J., Saouter, E., & Sonesson, U. (2017). The role of life cycle assessment in supporting sustainable agri-food systems: A review of the challenges. *Journal of Cleaner Production*, *140*, 399–409. https://doi.org/10.1016/j.jclepro.2016.06.071

Pappas, V. M., Samanidis, I., Stavropoulos, G., & Athanasiadis, V. (2023). Analysis of five-extraction technologies' environmental impact on the polyphenols production from Moringa oleifera leaves using the life cycle assessment tool based on ISO 14040. *Sustainability*, *15*(3), 2328. https://doi.org/10.3390/su15032328

Pardo, G., & Zufía, J. (2012). Life cycle assessment of food-preservation technologies. *Journal of Cleaner Production*, *28*, 198–207. https://doi.org/10.1016/j.jclepro.2011.10.016

Perez-Martinez, M. M., Noguerol, R., Casales, B. I., Lois, R., & Soto, B. (2018). Evaluation of environmental impact of two ready-to-eat canned meat products using life cycle assessment. *Journal of Food Engineering*, *237*(November 2018), 118–127. https://doi.org/10.1016/j.jfoodeng.2018.05.031

Priyadarshini, A., Rajauria, G., O'Donnell, C. P., & Tiwari, B. K. (2019). Emerging food processing technologies and factors impacting their industrial adoption. *Critical Reviews in Food Science and Nutrition*, *59*(19), 3082–3101. https://doi.org/10.1080/10408398.2018.1483890

Priyadarshini, A., Tiwari, B. K., & Rajauria, G. (2022). Assessing the Environmental and Economic Sustainability of Functional Food Ingredient Production Process. *Processes*, *10*(3), 445. https://doi.org/10.3390/pr10030445

Putnik, P., Kresoja, Ž., Bosiljkov, T., Jambrak, A. R., Barba, F. J., Lorenzo, J. M., Roohinejad, S., Granato, D., Putnik, P., Kresoja, Ž., Bosiljkov, T., & Jambrak, A. R. (2018). Comparing the effects of thermal and non-thermal technologies on pomegranate juice quality: A review. *Food Chemistry*, *279*, 150–161. https://doi.org/10.1016/j.foodchem.2018.11.131

Raila, A., Zvirdauskien, R., & Naujokien, V. (2020). *The influence of ozone technology on reduction of carrot loss and environmental IMPACT Simona Paulikiene*. 244. https://doi.org/10.1016/j.jclepro.2019.118734

Raja, V., Dutta, S., Murugesan, P., & Anandharamakrishnan, J. A. M. C. (2022). Electricity production using food waste: A review. *Environmental Chemistry Letters*, *21*, 839–864. https://doi.org/10.1007/s10311-022-01555-1

Režek, J. A., Vukušić, T., Donsi, F., Paniwnyk, L., & Djekic, I. (2018). Three pillars of novel nonthermal food technologies: food safety, quality, and environment. *Journal of Food Quality*, *2018*, 1–18. https://doi.org/10.1155/2018/8619707

Senol Ibanoglu (2023). *Applications of ozonation in the food industry*. https://doi.org/10.1016/B978-0-12-818717-3.00003-2

Shabir, I., Kumar, K., Hussain, A., Kumar, V., Fayaz, U., Srivastava, S., & Nisha, R. (2023). Carbon footprints evaluation for sustainable food processing system development: A comprehensive review. *Future Foods*, *7*(June 2023), 100215. https://doi.org/10.1016/j.fufo.2023.100215

Syrpas, M., Bukauskait, J., & Ba, L. (2018). Recovery of lipophilic products from wild cyanobacteria (Aphanizomenon flos-aquae) isolated from the Curonian Lagoon by means of supercritical carbon dioxide extraction. *Algal Research, 35*(November 2018), 10–21. https://doi.org/10.1016/j.algal.2018.08.006

Thirukumaran, R., Kumar, V., Priya, A., Krishnamoorthy, S., Ramakrishnan, P., Moses, J. A., & Anandharamakrishnan, C. (2022). Resource recovery from fish waste: Prospects and the usage of intensified extraction technologies. *Chemosphere, 299*(July 2022), 134361. https://doi.org/10.1016/j.chemosphere.2022.134361

Villacís-chiriboga, J., Vera, E., Camp, J. Van, & Ruales, J. (2021, February). *Valorization of byproducts from tropical fruits: A review, Part 2: Applications, economic, and environmental aspects of biorefinery via supercritical fluid extraction.* https://doi.org/10.1111/1541-4337.12744

Zouaghi, S., Frémiot, J., André, C., Grunlan, M. A., Gruescu, C., Delaplace, G., Duquesne, S., & Jimenez, M. (2019). Investigating the effect of an antifouling surface modification on the environmental impact of a pasteurization process: An LCA study. *ACS Sustainable Chemistry and Engineering, 7*(10), 9133–9142. https://doi.org/10.1021/acssuschemeng.8b05835

22 Status of International Regulations for Non-Thermal Processing of Foods

M. Dharini and R. Mahendran

22.1 INTRODUCTION

In the current and the end of last (20th) century, consumers have become more cautious about the food that they eat, their fresh like properties, and the energy and nutrients provided by the same. Their preference has been shifted to more natural like processed products that can provide more nutrients. As commercially available thermal methods of processing are reported to reduce the quality by degrading nutrients, many novel non-thermal food processing methods are introduced for processing foods to ensure safety and quality. Though seeking for newer technologies for improving foods has been followed for more than 100 years, the last three decades have seen the increase in the invention of more novel technologies (Barbosa-Cánovas & Bermu, 2022). However, before incorporating these novel processing methods in commercial production of foods, it is mandatory to assess the safety and the changes it imparts on the food material. For betterment and commercialization, it is also necessary for these novel technologies to meet the certain regulations (Prakash, 2013). There are regulations already available for thermal food processing methods, and it is necessary for non-thermal technologies to comply with the safety regulations issued by governmental bodies for specific foods. Besides complying with the available guidelines, additionally, there is a lack of information on safe processing conditions and regulatory guidance for processing foods using novel non-thermal technologies (Hernández-Hernández et al., 2019). The reason for this could be the use of electricity, pressure, or sound waves for non-thermal processing instead of heat as in thermal method (Pathak & Leong, 2021). The following are the general regulations that are available globally for processing foods.

22.2 REGULATORY GUIDELINES FOR NOVEL PROCESSING OF FOODS

There are various regulations for processing foods; owing to globalization and international trade, several national and international regulations have been laid and must be complied for commercialization nationally and internationally (Marone, 2016). These regulatory bodies work in collaboration or independently on a common goal which is to formulate a processing strategy, manage methodologies, quality, and risk assessment involved in processing. The major organization that are involved in forming regulation internationally are Food and Agricultural Organization (FAO), World Health Organization (WHO), and World Trade Organization (WTO); yet, these organizations don't address on issues such as processing (Trienekens & Zuurbier, 2008). Though the WTO doesn't involve in forming regulatory guidelines, it aids in controlling some of the safety measures (Meulen et al., 2022). The FAO and WHO together formed a Codex Alimentarius Council, which is an organization for making food-related policies on global market (Trienekens & Zuurbier, 2008). When it comes to new product or process development, it must pass through series of regulations before getting commercialized. In the year 2004, Institute of Food Technologists (IFT) along with European Federation of Food Science and Technology launched a Global Harmonization Initiative (GHI) to generalize the food regulations and legislations globally (Barbosa-Cánovas & Bermu, 2022).

DOI: 10.1201/9781003359302-22

Additionally, the environmental impact of any food techniques, products, and processing can be assessed using the life cycle assessment (LCA). Recently, the assessment by LCA revealed that the Ozone technology and Pulsed electric field had less impact on the environment (Scally et al., 2023).

In the United States, the management of food safety and risk assessments were carried out by local, state, and central government and also the private sectors. However, the regulations were laid are specific which originated majorly from the Food and Drug Administration (FDA) or the US Food Safety Inspection services, Department of Agriculture. The regulatory guidance provided by FDA is majorly on the protection of public health by maintaining the safety of foods (produced domestically and imported), drugs, medical equipment, and other biological products. Yet, there are few other federal agencies that can lay guidelines based on the commodities processed (Smith, 2007). The FDA doesn't directly address the regulations for processing rather it deals with the use of substances, methods, or the components used in the process (Sepu, 2002). Further, they are also responsible for motivating the innovation in the stream of medicine and food science and technology. In the case of new product development, it must be approved by the FDA through safety evaluation and risk assessment. Whereas, when it comes to novel processing methods, the organization must be transparent with the FDA regarding the procedure followed from beginning to end. If any changes have been made in a process such as inclusion of a commodity or new process, then a new regulatory submission must be produced to the FDA (as it might affect the end product). Through Hazard Analysis Critical Control Point (HACCP) and Good Manufacturing Practices (GMP), FDA ensured the safety of foods in the industries during all the stages of processing (Barbosa-Cánovas & Bermu, 2022). Regarding the safety of food ingredients and food packaging, the Center for Food Safety and Applied Nutrition (CFSAN) in FDA is responsible (Pedro, 2022).

In the case of European Union (EU), the major regulatory body that was involved in the production, processing, and trade of food and feed commodities is the Food Law Regulation 178/2002 (Pedro, 2022). This general food law provides regulations to ensure food and feed safety by establishing certain standards, principles, and responsibilities. Apart from the Food Law Regulation, the European Food Safety Authorities (EFSA based on Italy) provides several scientific advices on common and new risks related to foods (Barbosa-Cánovas & Bermu, 2022). EFSA is composed of 10 scientific panels that deal with different roles and responsibilities. The panel of importance to this chapter is the panel on Nutrition, Novel Foods and Food Allergens. The Novel Food Regulation Act (EC 2015/2283) embodies rules for novel food products and emerging technologies that are used for processing foods (Lyng et al., 2020). The regulation EC 2015/2283 comes into form in the year 2018; according to which the term "novel food" refers to any food that has not been consumed within EU before May 1997 (Lyng et al., 2021). The Novel Food Regulations Act by EU was drafted mainly to apply regulations to genetically modified food materials. However, the use of novel methods by non-thermal food processing led to the consideration in this regulation.

Such that the novel food may belong to categories such as food produced from algae, fungi, or any other microorganism or food produced by tissue culture method or food processed from a new method which was not used before 1997 (that can have an impact on the nutrition and quality of food), or new food produced from marine source (Pedro, 2022). Thus, if any company or organization has to sell their novel food product, they must apply for authorization of their novel food under Regulation EU 2015/2283 (Lyng et al., 2021). Applications of novel foods are accepted only if it fall under the category of novel food. The received application is passed to the EFSA for safety assessment; based on the assessment, if the crew decides on a positive note, then the product can be added on the EU's list of novel foods.

In Canada, the major organization that enacts the regulatory guidelines is Canadian Food Inspection Agency (CFIA); it published two regulatory modernization initiatives such as Safe Food for Canadian Regulations (SFCR) and Act (SFCA) for the reinforcement of food safety system in Canada; the aim of these regulations was to improve food system which can be impacted by the parameters, volume, and complexity of processing.

Food regulatory guidelines in Australia and New Zealand have undergone maximum reform in 30 years. Australia, New Zealand, and the surrounding territories combinedly implemented the food regulations and standards with an aim to protect public health and food safety; it embodiments were implemented as Food Standards Code, Food Standards Australia New Zealand (FSANZ) (Pathak & Leong, 2021; Szabo et al., 2008). The "codes" framed by these regulatory bodies (Australia, New Zealand, and territories) must be complied by the foods sold (both produced locally or imported). The major objective of these codes is to ensure safety of foods by dictating how food must be made and defined. Any changes in these codes must be proved by the FSANZ organization and can be published in the Food Standards Gazette. Implementation of novel processing methods is encouraged by the FSANZ, however, it must be amended with the codes. Standard 1.5.1 deals with the sale of novel foods and food ingredients. This code embodies few processing methods, new inclusion of novel processing methods must get prior approval from the FSANZ and it can't be sold before including in the Food Standards Gazette (Szabo et al., 2008).

Though there are regulatory bodies in Africa, the regulations laid are still in the armature stage which are not written scientifically (Pedro, 2022). However, most of these regulations mainly involves reducing the risk of foodborne illnesses. Only a few countries in Africa have regulatory guidelines and frameworks regarding the development of innovative novel foods. The following are the regulations that are specific for each non-thermal processing technology.

22.3 REGULATIONS FOR APPLYING SPECIFIC NON-THERMAL TECHNOLOGIES

22.3.1 High Pressure Processing (HPP)

HPP is a most widely used novel emerging technology that has been commercialized for processing of foods such as seafoods, fruits, fresh cuts, and juices. This high pressure induced processing, in general, is considered globally as a novel technique for processing foods. It is mandatory to get prior approval for commercialization such that the processed food doesn't give off any toxic compounds and doesn't degrade the nutrients in food.

In the United States, there are no sanitary regulations available for applying HPP techniques on foods. The US FDA has approved the commercialized usage of HPP equipment as a non-thermal pasteurization technology (Huang et al., 2017). HPP, in combination with thermal processing, is approved by FDA in the United States as an alternative technique for thermal sterilization and has been known as pressure-assisted thermal sterilization (PATS) (Barbosa-Cánovas & Bermu, 2022). Further, FDA approval requires the pressure-time combination and its corresponding log reduction; preferably, 5 to 6 log reduction is necessary immediately after processing and remains the same during the storage. However, in 2003, the Food Safety Inspection Service of the United States Department of Agriculture (USDA) approved the use of HPP for the reduction of *Listeria monocytogenes* in pre-packed ready to eat meat products (Serment-Moreno et al., 2021).

However, in EU, the EU Regulation (258/97/EC) which relates to the novel foods and ingredients laid down regulations for high pressure processed foods. Nevertheless, HPP processed foods have been considered in EU as ones equal to the fresh food products; HPP didn't have any significant changes in the food product. Thus, in Europe, HPP foods are not considered as novel foods and are approved and authorization is not required for commercialization (Poji et al., 2018). This statement by EU Regulation was similarly reflected by French Agency for Food, Environmental and Occupational Health and Safety (ANSES) organization in the year 2010; it states that any food material processed by HPP up to a pressure of 600 MPa and exposure time of 5 minutes at or below ambient temperature might be considered not to alter the components in food (Serment-Moreno et al., 2021). Akin to which Canada also approved HPP for commercialization in the year 2016 by Health Canada (Alexandre & Saraiva, 2018). Further, the use of pressure vessel must also comply with the regulation of the pressure equipment directive for the use on food materials.

22.3.2 Irradiation

Irradiation is a novel processing method that uses ionizing radiation for various benefits such as microbial reduction, sprouting inhibition, pest control, and increase in shelf life of foods. Radiation sources such as gamma radiation, high energy electrons (equal to and below 10 MeV), and X-rays (below 5 MeV) are approved for food application by the US FDA.

In irradiation processing of foods, major parameters such as dosage, exposure time, source, type of food, and effect on food are studied by regulatory bodies for their commercialization (Bashir et al., 2021). International regulatory bodies such as FDA, WHO, and International Atomic Energy Agency (IAEA) have approved irradiation as an effective technology for food preservation (Hernández-Hernández et al., 2019). These regulatory bodies have recommended the use of less than 10 kGy dose of ionizing radiation for the safe use on food materials (Niemira & Gao, 2012). US FDA regards irradiation process as a food additive; has formed a set of regulations specific for irradiation of foods and food packaging substances under the Food Additives Amendments and Code of Regulation 21 CFR Part 179 (Swer et al., 2021). Further, The Code of Regulation 9 CRF 424.22 (c) by the USDA and FSIS governs the use of irradiation for refrigerated or frozen meat and poultry products for reducing the microbial load and ensuring safety. International Organization for Standardization has also laid down legislations for food irradiation; ISO/ASTM 51204 and 51431 are the regulations for dosimetry in food processing industry, whereas the regulation ISO 14470 deals with the process and validation of irradiation in food industry (Swer et al., 2021).

Similarly in EU, certain irradiated food products (10 kGy) are approved by the member states (Watkins, 2012). EU regulatory body EFSA has also published that irradiation of foods as a reliable and feasible alternative technology for preserving food materials (Fan & Regional, 2023). The common food commodities that are approved by all the EU countries are dried spices, condiments, and seasoning; yet few other EU countries like Belgium, Italy, France, and the UK have framed their own national regulatory guidelines for the sale of irradiated fresh meat, poultry, seafoods, fruits and vegetables, and grains.

Further, when it comes to packing the irradiated food commodities, US regulatory authorities have made it clear that the consumer must be notified about the irradiation process. Thus, it is mandatory to label the Radura symbol on the packaging of commercially available products processed by ionization (Rahman, 2017). In EU it is not necessary to label the retail package with the Radura symbol, however, the manufacturer must indicate that the food has been irradiated in the package (Fan & Regional, 2023). For treating pre-packed foods by irradiation, it must comply with the regulations provided by 21 CFR 179.45; it reports that the packaging material must have both physical and chemical stability and less interaction with food material (Kumar & Han, 2012).

22.3.3 Ultrasound

Ultrasound is a high-frequency sound wave that is not audible to human hearing. Power ultrasound is of primary importance in food sonoprocessing as this frequency produces cavitations (bubbles) (Rastogi, 2011) which creates immense chemical and physical changes in food and, in turn, has boundless applications in food industry. Ultrasound-assisted processes involve physical (mechanical and sonoluminescence) and chemical effects (sonochemistry) in food. The major use of ultrasound in food processing is the preservation and microbial safety. It has also been widely used in changing the properties of food material thereby producing a modified food product. These changes in food composition are mainly due to the effect of ultrasound on macro and micro molecules (biomolecules) in the food material.

The HACCP for processing foods using ultrasound was designed as a first step for regulatory approval (Chemat & Hoarau, 2004). Though there are regulatory guidelines for the equipment to be used for diagnostic ultrasound, there are no regulations for the processing of foods. There are numerous studies that have been conducted on ultrasound processing of foods, yet some of the processing conditions are still not clear. It must be addressed, and the harmonized regulation must be produced for the commercialization of this technology (Bermudez-Aguirre & Niemira, 2022). However, it has been reported that ultrasound processing of liquid foods yields a US FDA recommended 5 log reduction (Tiwari & Mason, 2012). According to the Australian Regulatory Law, any process that complies with The Code doesn't need any changes and can be approved for use on foods (Szabo et al., 2008). The defoaming application of ultrasound doesn't have any impact on the food material; thus, it doesn't require special permission within The Code. An Australian company, Cavitus, uses ultrasound equipment for the reduction of foams in beer and juice production.

22.3.4 Pulsed Electric Field

In pulsed electric field processing of foods, electrodes (electrically charged electrodes) are placed directly inside the liquid or semi-solid food. High electric field intensities in the form of pulses (15 kV/cm) are applied to liquid foods through the electrodes dipped inside the liquid food. Any novel food processing method might have both merits and demerits; the regulatory acceptance deals with whether the risk assessed is acceptable or not. In the case of PEF processing of food, the risk concerns arise from the safety issue based on the reaction that takes place during processing and the contact of food with the equipment or hardware part.

In European countries, the EC regulation has approved PEF for commercialization based on the fact that it didn't cause any substantial modification in foods compared to fresh foods. Thus, it was considered as substantial equivalence to fresh produce and are not regarded as novel foods; also, it is assumed safe for consumption. The Novel Food Regulation was not basically intended for novel processing; however, with respect to verbiage Article 1 (f), processing can be added in it (Keener et al., 2001).

In the case of the United States, the regulatory body FDA issued a no objection certificate for PEF processing of foods to a company called PurePulse (Rahman, 2017). No objection certificate doesn't mean that the technology is approved rather FDA has no reason to reject the processing methods under a specific condition of processing. Further in the year 2007, a juice processing company, Genesis, based in Oregon has acquired the FDA approval for PEF processing of fruit juices (Smith, 2007)

However, in PEF, the major food safety issue that needs to be addressed is the possible leaching out of metal from the electrodes to the food material during processing. According to the EC Commission regulation No 1881/2006 and Codex STAN 193-1995, the amount of heavy metal in any food material must be under a limit for approval. Codex STAN 247-2005 also states that heavy metals in fruit juices must comply with STAN 193-1995 for approval (Pataro & Ferrari, 2020). Further, EC directive No 98/83/EG also provided the maximum limits for many heavy metals in the drinking water which can be complied for PEF-treated liquid foods.

22.3.5 Ultraviolet Light Technology

Ultraviolet light as a mono or polychromatic has been used for treating foods. Electromagnetic rays of short wavelengths are used for processing; the major mechanism that takes place in UV radiation is photothermal and photochemical changes (Onyeaka & Nwabor, 2022). Further, when treated on light foods, it can cause photolysis of water which will aid in generation of reactive species within the liquid. UV technology has been approved as Novel food by many countries including EU, Canada, Australia, and China around the world for the treatment of foods (Koutchma, 2022). In Canada, the apple cider juice (fresh or fermented) treated with UV rays, CiderSure 3500 UV light was assessed to have no impact on human health and no toxicities were found in processed juice (Britain & Zealand, 2014). Thus, Canada approved it for the consumption of humans. EU has also approved the usage of UV light (253.7 nm) for processing of post-pasteurized milk in the year 2016. The assessment of UV-treated milk revealed that it increased the availability of vitamin D_3

in milk after processing (Sunita & Sharma, 2022). Thus, EU EFSA approved UV rays as specified by the applicant for commercialization. Additionally, EU EFSA has also approved UV-treated mushrooms, bread, and baker's yeast for commercialization. Regarding the approval of bread, it was reported that the pre-baked rolls and breads treated with UV rays were found to increase the vitamin D_3; thus, EFSA approved usage of UV rays (240–315 nm) for the dosage of 10–50 mJ/cm^{-2} for an exposure time of 5 seconds. In India, the governmental organization FSSAI has also approved UV treatment of milk using a specific equipment "SurePure UV system" in the year 2013 (Koutchma, 2021). Milk processing UV rays has also been approved in Israel based on the Israelian Regulation Standard 82 to the petitioner AseptoRay company for the processing of milk between the wavelength 30 and 200 nm (Koutchma, 2021).

In the United States, the FDA considered UV processing as radiation and added it to the Food Additives section (CFR 12 Part 179). Similar to EU, US FDA in the year 2000 approved the use of UV light at 253.7 nm (low pressure mercury lamps) for the reduction of microbial load in fruit juices (Koutchma, 2022). The application for approval was initiated by the US-based California Day-Fresh Foods under the Federal Food, Drug and Cosmetic Act (Section 201); they appealed it as a radiation process for the treatment of foods. However, the approval doesn't restrict the exposure time, as it solely depends on the machine used, type of liquid processed and quantity. FDA approved the usage of UV processed baker's yeast, as UV treatment aided in increasing the availability of vitamin D_2.

22.3.6 Ozone Technology

Ozone is a triatomic oxygen that has a potential to inactivate insects of all stages; thus, it has been used as a potential disinfestation technique in storage of grains. Further, the major advantage of using ozone technology is that it won't leave any residue after processing unlike the commercial phosphine fumigation technique. Ozone technology can also be used for activation water; this activated ozonized water can be used as surface disinfectant for fresh fruit, vegetables, and meats. Oxygen or air is used for processing gas for the generation of ozone molecules; this ozone has reduced half-life, thus it gets disintegrated into oxygen molecules.

US FDA has recognized ozone as GRAS for treatment of foods (Keener et al., 2021); additionally, it has been stated that there are no restrictions on the dosage and labeling requirements for ozone processed foods (Keener et al., 2021). Though the FDA has approved it as GRAS, the process for getting this approval was a long journey; it took approximately 20 years. At first, FDA approved ozone for treating drinking water in the year 1982, yet it was not approved for foods. Then, in 1997, FDA allowed for GRAS self-affirmation for ozone processing of foods (Keener et al., 2021). The GRAS self-affirmation is an approval in which the company or organization who wants approval for the use of ozone technology must take full responsibility for the risks that occur in processed foods. GRAS self-affirmation is the shortest technique for getting approval for any novel foods; the process is not dependent on the FDA, and the legal procedure of review or approval is also not based on the FDA. Apart from ozone's application on foods, the ozone generator must also comply with the US Environmental Protection Agency (EPA) (Rasco, 2017).

22.3.7 Cold Plasma

Cold plasma is the ionized gas that is used for processing foods for reducing microbial load, property modification, quality maintenance, and antinutritional and allergen reduction. The reactive species present in the plasma gas is the reason for the desirable change that plasma impacts on the food material. Despite all its beneficial nature of plasma on food processing with numerous applications, it has not yet been approved by any regulatory bodies around the globe (Scally et al., 2023). This could be due to the presence of some major challenges in using plasma on food materials despite various applications of plasma in food processing (Keener et al., 2021).

The USFDA has not yet recognized cold plasma as a commercial technology for treating food materials (Sonawane et al., 2020). Being a novel technology, regulatory guidelines are necessary for its application, however, to date to the best of our knowledge, there has not been any guideline or legislation from a government body framed for the use of plasma in food materials. Providing safety guidelines for both process and product parameters is essential for the futuristic commercialization of this technology in food processing (Gavahian et al., 2019). For the approval of cold plasma for meat and poultry products in industries, it is required to get approval from the US FDA-FSIS. In the case of grains, the general approval must be acquired from the USDA-Federal Grain Inspection Service (FGIS); however, the approval for the process efficiency and environmental impact are provided by US EPA and the packaging and labeling requirements and the generation of residual by-products are provided by the FDA (Rasco, 2017). The assessment of novel technology by EFSA is done by a method called PROmoting METHods for Evidence Use in Scientific assessments (PROMETHUS) (Bourke et al., 2018).

The reactive species in plasma oxidizes food components such as proteins, carbohydrates, and lipids to form toxic substances such as aldehydes (Cullen et al., 2018). Apart from the toxicity generation, the presence of nitrogen reactive species such as nitrates and nitrites in plasma-treated liquid samples could pose threat to food. Their concentration could increase in plasma-treated drinking water and fruit juices above the WHO standard recommendation (50 mg/L nitrate and 3 mg/L nitrite) (Muhammad et al., 2018). In the past few years, scientists have started working on the toxicity studies of food, owing to the disapproval from US FDA for the commercialization of this technology for food processing (Keener et al., 2021). However, the studies done so far are not sufficient to understand the concentration of plasma, and exposure time that can be used for treatment without producing toxicity in the sample has not yet been reported. The US FDA regards ozone processing of foods as a direct additive to food, gives a mirage of hope for the approval of plasma in the future. Nevertheless, unlike ozone technology where only one reactive species of O_3 in the gas is applied, the complex nature of plasma (which contains numerous reactive species) makes it hard for moving toward the commercialization of this technology. Therefore, the working scientists from Food Safety and Technology of the German Research Foundation (DFG) and the Senate Commission on Food Safety started collecting information and works on plasma processing of foods. Based on their recommendations for plasma treatment, since then, many studies have emerged on biochemical, microbial, and toxicological aspects (Cullen et al., 2018). Therefore, these kinds of regulatory guidelines from authorized bodies can be useful in the future for improving the standards of research in this area of food processing; this, in turn, can lead to the ray of light for the commercialization of the technology.

22.4   FUTURE ASPECTS

Though there are numerous regulations around the world in the food processing sector, it is necessary for all the regulatory bodies to work together to develop a harmonization in regulation. Global harmonization of regulation could aid in developing common regulation which might increase the possibility of commercialization of most of the non-thermal processing technologies (Barbosa-Cánovas & Bermu, 2022). Apart from the United States and EU, major other countries in Africa still don't have proper regulations for processing technologies; global harmonization could help in improving the quality and safety of foods by incorporating non-thermal processing technologies even in under developed and developing countries.

22.5   SUMMARY

The current technologies which we are commercially using in this century are highly sophisticated compared to the ones used in the last century. The regulations and safety guidelines for these commercial technologies were developed based on the decades of ceaseless works conducted to tackle major safety issues. Food safety management and regulations have been transformed from basic microbial safety to high-end science-based risk assessment and management. Additionally, as seen in the UV light processing, the approval of these processes for obtaining commercialization is a complex process. Though these non-thermal processing technologies have numerous beneficial effects on food materials and are found to be more fruitful compared to thermal processing methods, commercialization of these technologies is still in the armature stage for most of the food products. The major reason could be the possibility of toxicity generation in processed foods either from the equipment used or from the result of interaction with food material. Standardizing the design and components used for processing equipment, formulation of products and ingredients, and studying the interaction of novel processes with different food materials could help toward optimizing the process for various food materials. This might pave the way for obtaining regulatory approval for the commercialization of novel technologies in the food processing sector.

REFERENCES

Alexandre, E. M. C., & Saraiva, J. A. (2018). Environmental Footprint of Emerging Technologies, Regulatory and Legislative Issues. *Innovative Technologies for Food Preservation Inactivation of Spoilage and Pathogenic Microorganisms*, 255–276. https://doi.org/10.1016/B978-0-12-811031-7.00008-X

Barbosa- Cánovas, G. V, & Bermu, D. (2022). *Novel Food Processing Technologies and Regulatory Hurdles*. In *Ensuring Global Food Safety* (pp. 221–228). Elsevier. 221–228. https://doi.org/10.1016/B978-0-12-816011-4.00006-9

Bashir, K., Jan, K., Kamble, D. B., Maurya, V. K., Jan, S., & Swer, T. L. (2021). 2.08 History, Status and Regulatory Aspects of Gamma Irradiation for Food Processing. In *Innovative Food Processing Technologies: A Comprehensive Review* (Vol. 2). Elsevier. https://doi.org/10.1016/B978-0-08-100596-5.23051-5

Bermudez-Aguirre, D., & Niemira, B. A. (2022). Pasteurization of foods with ultrasound: The present and the future. *Applied Sciences (Switzerland)*, *12*(20). https://doi.org/10.3390/app122010416

Bourke, P., Ziuzina, D., Boehm, D., Cullen, P. J., & Keener, K. (2018). The potential of cold plasma for safe and sustainable food production. *Trends in Biotechnology*, *36*(6), 615–626. https://doi.org/10.1016/j.tibtech.2017.11.001

Britain, G., & Zealand, N. (2014). Regulatory Aspects of UV Technology Commercialization. In *Preservation and Shelf Life Extension, UV Applications for Food*. https://doi.org/10.1016/B978-0-12-416621-9.00009-1

Chemat, F., & Hoarau, N. (2004). Hazard analysis and critical control point (HACCP) for an ultrasound food processing operation. *Ultrasonics Sonochemistry*, *11*(3–4), 257–260. https://doi.org/10.1016/j.ultsonch.2004.01.016

Cullen, P. J., Lalor, J., Scally, L., Boehm, D., Milosavljević, V., Bourke, P., & Keener, K. (2018). Translation of plasma technology from the lab to the food industry. *Plasma Processes and Polymers*, *15*(2), 1–11. https://doi.org/10.1002/ppap.201700085

Fan, X., & Regional, E. (2023). *Food Irradiation*. 1–13. https://doi.org/10.1016/B978-0-12-822521-9.00079-4

Onyeaka, H. N., & Nwabor, O. F. (2022). Green technology in food processing and preservation. In *Food Preservation and Safety of Natural Products* (pp. 87–118). Elsevier. https://doi.org/10.1016/B978-0-323-85700-0.00011-3

Gavahian, M., Chu, Y. H., & Jo, C. (2019). Prospective applications of cold plasma for processing poultry products: Benefits, effects on quality attributes, and limitations. *Comprehensive Reviews in Food Science and Food Safety*, *18*(4), 1292–1309. https://doi.org/10.1111/1541-4337.12460

Hernández-Hernández, H. M., Moreno-Vilet, L., & Villanueva-Rodríguez, S. J. (2019). Current status of emerging food processing technologies in Latin America: Novel non-thermal processing. *Innovative Food Science and Emerging Technologies*, *58*, 102233. https://doi.org/10.1016/j.ifset.2019.102233

Huang, H. W., Wu, S. J., Lu, J. K., Shyu, Y. T., & Wang, C. Y. (2017). Current status and future trends of high-pressure processing in food industry. *Food Control*, *72*(12), 1–8. https://doi.org/10.1016/j.foodcont.2016.07.019

Keener, K. M., & Illera, A. E. (2021). Perspectives with Respect to Regulations, Consumer Acceptance and Sustainability. In *Innovative Food Processing Technologies: A Comprehensive Review* (Vol. 1). Elsevier. https://doi.org/10.1016/b978-0-08-100596-5.23009-6

Keener, L. (2001). Validating the Safety of Foods Treated by Pulsed Electric Fields. In *Food Preservation by Pulsed Electric Fields*. Woodhead Publishing Limited. https://doi.org/10.1533/9781845693831.2.178

Koutchma, T. (2022). Validation of Light-Based Processes. In *Validation of Food Preservation Processes Based on Novel Technologies*. INC. https://doi.org/10.1016/B978-0-12-815888-3.00005-1

Koutchma, T. (2021). 2.20 Global Regulations on Ultraviolet and Pulsed Light Technology for Food Related Applications. In *Innovative Food Processing Technologies: A Comprehensive Review* (Vol. 2). Elsevier. https://doi.org/10.1016/B978-0-08-100596-5.22945-4

Kumar, P., & Han, J. H. (2012). Packaging Materials for Non-Thermal Processing of Food and Beverages. In *Emerging Food Packaging Technologies: Principles and Practice* (pp. 323–334). https://doi.org/10.1533/9780857095664.3.323

Lyng, J. G., Lalor, F., & Pedros-Garrido, S. (2020). Legislative Considerations Relating to the Uptake of Emerging Food Processing Technologies within the European Union. *Innovative Food Processing Technologies: A Comprehensive Review, 1*, 318–328. https://doi.org/10.1016/b978-0-12-815781-7.00021-4

Marone, P. A. (2016). Food Safety and Regulatory Concerns. In *Insects as Sustainable Food Ingredients* (Vol. 2012). Elsevier Inc. https://doi.org/10.1016/B978-0-12-802856-8.00007-7

Meulen, B. Van Der, Card, M. M., Din, A., Fortin, N. D., Mahmudova, A., Bilgin, F. K., Lederman, J., Poto, M. P., Maister, B., & Gökçe H. (2022). *Food regulation around the world*. https://doi.org/10.1016/B978-0-12-816011-4.00001-X

Muhammad, A. I., Liao, X., Cullen, P. J., Liu, D., Xiang, Q., Wang, J., Chen, S., Ye, X., & Ding, T. (2018). Effects of nonthermal plasma technology on functional food components. *Comprehensive Reviews in Food Science and Food Safety, 17*(5), 1379–1394. https://doi.org/10.1111/1541-4337.12379

Niemira, B. A., & Gao, M. (2012). Irradiation of Fluid Foods. In *Novel Thermal and Non-Thermal Technologies for Fluid Foods*. Elsevier Inc. https://doi.org/10.1016/B978-0-12-381470-8.00007-4

Pataro, G., & Ferrari, G. (2020). Chapter 13. Limitations of Pulsed Electric Field Utilization in Food Industry. In *Pulsed Electric Fields to Obtain Healthier and Sustainable Food for Tomorrow*. INC. https://doi.org/10.1016/B978-0-12-816402-0.00013-6

Pathak, R., & Leong, T. (2021). Consumer Acceptability of Ultrasonically Processed Foods. In *Innovative Food Processing Technologies: A Comprehensive Review* (Vol. 1). Elsevier. https://doi.org/10.1016/B978-0-12-815781-7.00024-X

Pedro, S. (2022). *Regulations on the Use of Emerging Technologies and Bio-Marine Food Products*. https://doi.org/10.1016/B978-0-12-820096-4.00013-4

Poji, M., Mi, A., & Tiwari, B. (2018). Eco-innovative technologies for extraction of proteins for human consumption from renewable protein sources of plant origin. *Trends in Food Science & Technology 75*(March), 93–104. https://doi.org/10.1016/j.tifs.2018.03.010

Prakash, A. (2013). Non-Thermal Processing Technologies to Improve the Safety of Nuts. In *Improving the Safety and Quality of Nuts*. Woodhead Publishing Limited. https://doi.org/10.1533/9780857097484.1.35

Rahman, S. (2017). Minimal Processing Methods. In *Food Processing Technology* (4th ed.). Woodhead Publishing Series in Food Science, Technology and Nutrition, 431–512. https://doi.org/10.1016/B978-0-08-100522-4.00007-9

Rasco, B. (2017). Laws and Regulations for Novel Food Processing Technologies. In *Ultrasound: Advances in Food Processing and Preservation*. Elsevier Inc. https://doi.org/10.1016/B978-0-12-804581-7.00020-8

Rastogi, N. K. (2011). Opportunities and challenges in application of ultrasound in food processing. *Critical Reviews in Food Science and Nutrition, 51*(8), 705–722. https://doi.org/10.1080/10408391003770583

Scally, L., Ojha, S., Durek, J., & Cullen, P. J. (2023). *Principles of Non-Thermal Plasma Processing and Its Equipment*. https://doi.org/10.1016/B978-0-12-818717-3.00011-1

Sepu, D. R. (2002). *Food Processing by Pulsed Electric Fields: Treatment Delivery, Inactivation Level, and Regulatory Aspects. 388*, 375–388. https://doi.org/10.1006/fstl.2001.0880

Serment-Moreno, V., Queiro, R. P., & Gonza, M. (2021). *1.03 Food and Beverage Commercial Applications of High Pressure Processing* (Vol. 1). https://doi.org/10.1016/B978-0-12-815781-7.00009-3

Smith, M. (2007). Regulatory acceptance. In *Food Preservation by Pulsed Electric Fields* (pp. 352–357). Woodhead Publishing Limited. https://doi.org/10.1533/9781845693831.3.352

Sonawane, S. K., T, M., & Patil, S. (2020). Non-thermal plasma: An advanced technology for food industry. *Food Science and Technology International, 26*(8), 727–740. https://doi.org/10.1177/1082013220929474

Sunita, T., & Sharma, A. P. M. (2022). Chapter 7 – Light-Based Processing Technologies for Food. In *Current Developments in Biotechnology and Bioengineering* (1st ed.). Elsevier Ltd. https://doi.org/10.1016/B978-0-323-91158-0.00004-1

Swer, T. L., Mukhim, C., Bashir, K., Rani, S., & Kamble, D. B. (2021). Irradiation and Legal Aspects. In *Innovative Food Processing Technologies: A Comprehensive Review* (Vol. 2). Elsevier. https://doi.org/10.1016/B978-0-08-100596-5.22994-6

Szabo, E. A., Porter, W. R., & Sahlin, C. L. (2008). Outcome based regulations and innovative food processes: An Australian perspective. *Innovative Food Science & Emerging Technologies*, *9*(2), 249–254. https://doi.org/10.1016/j.ifset.2007.12.001

Tiwari, B. K., & Mason, T. J. (2012). Ultrasound Processing of Fluid Foods. In *Novel Thermal and Non-Thermal Technologies for Fluid Foods*. Elsevier Inc. https://doi.org/10.1016/B978-0-12-381470-8.00006-2

Trienekens, J., & Zuurbier, P. (2008). Quality and safety standards in the food industry, developments and challenges. *International Journal of Production Economics*, *113*, 107–122. https://doi.org/10.1016/j.ijpe.2007.02.050

Watkins, D. (2012). Regulatory and Legislative Issues for Non-Thermal Technologies. An EU Perspective. In *Novel Thermal and Non-Thermal Technologies for Fluid Foods*. Elsevier Inc. https://doi.org/10.1016/B978-0-12-381470-8.00015-3

23 Sustainable Processing Using Non-Thermal Techniques

Malini Buvaneswaran and V. R. Sinija

23.1 INTRODUCTION

Sustainability is defined as "to endure", but this term does not fully capture the sense when it is used in different sectors to address the impact of human activity on society at the economical and environmental levels. A food system is a key element and illustrates nicely the world's environmental and other sustainability challenges. In this food is a key factor, the way of addressing the effect of nutritional science on environmental impact when formulating population dietary can be the best choice to express the sustainable indicator. According to the United Nations' Millennium Ecosystem Assessment, 15 of 24 world's ecosystems are used unsustainably in that food is the major source of degradation (Lang & Barling, 2013). The food system consists of several elements including environment, people, inputs, processes, infrastructure, institutions, and activities related to food production, processing, distribution, consumption of foods, and outcome of these activities (Caron et al., 2018). So, sustainable food production helps to obtain sustainable development goals (SDGs) in line with technological innovations.

In the food system, food security and food sustainability are interlinked with each other or we can say food sustainability is the fifth dimension of food security (Lang & Barling, 2013). Sustainability represents the extension of a time frame of stability or precondition for long-term food security. In the World Food Conference of 1974, food security was defined as "the availability at all times of adequate world food supplies of basic foodstuffs to sustain a steady expansion of food consumption and to offset fluctuations in production and prices" (United Nations, 1974). It also consists of other important parts such as sustainable agriculture, sustainable economy, sustainable food production, and sustainable diets. So the notion of sustainable development or sustainable processing was first introduced by the Brundtland Commission in an international discussion in 1983, defined as "development that meets the need of the present without compromising the ability of future generations to meet their own needs which is the assessment of the robustness of the food system" (Berry et al., 2015). Tiwari et al. (2013) explain the concepts of sustainable development in food processing, they are:

i. The raw materials involved in food processing should be produced on an ongoing basis without affecting the environment, social or economic form.
ii. The production process should not be reliant, long-term, and finite energy usage.
iii. The produced products should not have an adverse effect on human health.

To feed the growing population food production will be increased to 70% in 2050 (Kashyap & Agarwal, 2019). The increasing demand for ready-to-eat, processed foods, and ultra-processed foods, and most of the food is processed in large, centralized factories; which indirectly generates a more carbon footprint in environments in terms of energy or water usage, waste generation, and usage of non-renewable energy sources, which needs to be identified and sustainable processing technologies should be incorporated in food processing system (Baker et al., 2020; Kumar & Chakabarti, 2019). This chapter explains the impact of food processing on carbon footprint and how a non-thermal technology helps to improve sustainable food processing.

DOI: 10.1201/9781003359302-23

23.2 SUSTAINABLE PROCESS

23.2.1 NEED FOR SUSTAINABLE PROCESS

It is necessary to think about environmental concerns, social responsibility, and economic viability while shaping future food processing techniques. The change in economic conditions such as climbing ingredient costs and production costs including energy requirements and environmental conditions highly influence food processing. Consumption of renewable resources at a rate higher than the regeneration rate creates insufficient or no available resources for future generations. Therefore, it is necessary to find new processing techniques or alter the existing techniques to meet the needs of the present society without compromising future sources that have to be insisted on in the food industry. To obtain sustainability, the indicators supported are carbon footprint, energy audit, and nutritional indices, hence, this mono-dimensional method is not enough to fulfill the complex inherent in sustainability. The key drivers for sustainable processing in the food system are food security, population health, social justice, global change, resource depletion, environmental impact, and eco-labeling. The core element of food security is to match the food supply to the nutrition demand in the population, where sustainability plays a major role to ensure the food supply can last for centuries. The controlling factor in the food industry is the customer demand for foods to fulfill their nutritional demand, where the population plays the role. The uncontrollable growth in the global population rapidly increases the food demand. To maintain this demand and supply we need to look forward to a sustainable approach in the coming days. Global change and resource depletion are factors related to environmental resources such as air, water, and land. On the other hand, food processing techniques are the sources of the environmental impact and its impact is still not clearly evaluated. The direct impacts including the discharge of waste in the air, water, and soil, and indirect impacts, such as transport energy emissions, are still in the early stage of the study. The whole food system (from farm to table) contributes 62–75% of climate change (Jianyi et al., 2015). When we speak about the food system two major inseparable sectors are agriculture (food production) and food processing. Most of the studies are only concentrated on the impact of food production on a sustainable approach (Anser et al., 2021; Jianyi et al., 2015), in the modern world, most agricultural and animal products are undergone processing techniques that consume abundant energy and water (Karwacka et al., 2020). So, analyzing the effect of food processing on sustainability is also an unavoidable thing (Karwacka et al., 2020). For example, for producing 1 kg of rye bread the estimated carbon footprint is 731 g CO_2 equivalents, where agricultural production is the primary hotspot and processing and distribution stages are secondary hotspots (Jensen & Arlbjorn, 2014). During the processing of ready-to-eat products, the industrial and retail portions are responsible for 25–38% (nearly 1.63–2.35 kg CO_2 eq.) of climate change impacts in Finland (Virtanen et al., 2011).

Some analyzing techniques, such as life cycle assessment (LCA), are used to study carbon footprint. In the Paris agreement on climate195 countries have agreed to the 2030 Agenda for Sustainable Development, the goals related to the food system are "End hunger, achieve food security and improved nutrition, and promote sustainable agriculture" (Goal 2), "Take urgent action to combat climate changes and its impact" (Goal 13) (Caron et al., 2018). The concern about environmental issues, rural impoverishment, vulnerability, and human rights creates attention towards processing technology, dependency on imported foods, the impact of food on health, and risk evolving between food production and distribution chain. So, the transformation of a food system is required, and it depends on well-adopted processes, enlightened policies, and local-to-global integration.

23.2.2 THE ECONOMICS INVOLVED: SHOULD WE INVEST OR NOT?

The food industry is one of the promising industries and essential to the global economy to the health and wealth of its citizens, where nearly 30% of global energy was utilized in different forms of food supply system. Nearly 17% of fossil fuels are used by developed countries for the production, processing, and packaging of food products. Meat and dairy products contributed to 24% of consumers' environmental, but only 6% represents the consumer's spending. The variation in quantity and quality of food causes an environmental impact. According to the European Commission's assessment food accounts for one-third of greenhouse gas emissions (Lang & Barling, 2013). In the food supply chain, food production, preservation, distribution, storage, and dumping of food waste in landfill greatly contributed to greenhouse gas emissions. The rising population and nutrition demand create unforeseen consequences, which have complex interactions with environmental footprints. These effects of food processing imply the different dimensions of sustainability. So Tiwari et al. (2013) claimed that the usage of indicators of sustainability that is easy to calculate and appeal to the government and the public is not enough to address the many dimensions of sustainability in food processing. Recently, the concern about the lack of food security due to environmental degradation including climate change, economic growth, and population growth turned the focus toward food processing. Many countries have taken action to implement processing techniques that help in sustainability and corporate social responsibilities. Combining the ecological footprint of different food processing techniques and their economic analysis helps to choose the better processing technique from a sustainability point of view. Collins & Fairchild (2007) analyzed the footprints of food and drink in Cardiff, results show that by increasing the proportion of organic food and drink consumption by 87.9%, the reduction in carbon footprint can be reduced up to 33.9–22.9% which increases the average mean expenditure per resident up to 31.2%.

23.3 CARBON FOOTPRINT

In the food system, the carbon footprint is a valuable indicator to evaluate the sustainable food processing system. Carbon footprint refers to the emission of greenhouse gases (GHG) during the different operations involved in food processing. Carbon footprint includes environmental footprints like water and energy use, emission of greenhouse gases during food processing, usage of renewable energy, and waste disposal (Jianyi et al., 2015). The term "cradle to grave" analysis represents the quantification of the carbon footprint from the start of the food production process to the final consumption point and waste generation. It will help to analyze the effect of food processing on the environment. To analyze the carbon footprint analytical models such as LCA, game theory, and simulation models are used (Zhu et al., 2018).

23.3.1 THE CARBON FOOTPRINT OF FOOD PROCESSING OPERATIONS IN TERMS OF ENERGY

Among the industrial sector, the food industry is one of the major electricity consumers (approximately 12%), but electricity is not appropriately used in many food processing techniques. In the United States, 6% of total energy is used by food industries, which is why food industries are considered the third largest energy consumers (Toepfl et al., 2006). Energy requirements in the food sector increased from 14.4% in 2002 to 15.7% in 2007 in the national energy budget, in that controlling the temperature of processed food under storage conditions occupies a major portion (Zhu et al., 2018). According to FAO (Food and Agricultural Organization), in 2050, food production will be increased by 70%, but energy production only increased by one-third. This implies a huge gap between energy requirement and food production, so the usage of renewable energy sources like solar, wind, and geothermal energy needs to be appreciated. Also in developing countries, policies like industrial energy efficiency recommend reducing energy inputs. Because in developing countries, inefficient energy usage is high, for example, in African countries energy requirements for agri-food chain are 55%, but in the United States it is around 15.7% only. According to the LCA calculated carbon foodprint for frozen spinach (from farm to storage) is 0.43 kg CO_2 eq./kg, but the actual value was 0.49 kg CO_2 eq./kg, in that 25 to 65% of energy difference was identified in processing (Sanjuan et al., 2014). This energy requirement hugely varied based on the type of energy source used, the size of the industry, the type of processing technique involved, and the type of technologies used. For example, in the United Kingdom, 13.68 MJ/kg of fuel or 1.48 MJ/kg of electricity are required for French fries; 0.43 MJ/kg of electricity or .53 MJ/kg of fuel are required for jam production (Clairand et al., 2020). The common energy sources of thermal, electrical, and mechanical energy release greenhouse gases during energy production or usage. Based on the place and application of energy usage, the emission of greenhouse gases during energy usage is divided into two groups. (i) Direct use: energy used in the on-farm processing operations, and (ii) Indirect use: energy used in the food industry, storage, and transportation. The utilization of fuel oil in cassava starch plants produced 539 kg CO_2 eq/t of starch when it was replaced by electricity (212 MJ/t starch), the greenhouse gas emission was reduced to 105 kg CO_2 eq/t (Tran et al., 2015). In the United States, to prepare concentrate and diced tomato, the estimated environmental footprint was 0.83 and 0.16 kg CO_2/kg, respectively (Karwacka et al., 2020). The main energy-consuming technologies in food industries are drying (0.5–0.75 MJ/kg), storage (1–3 MJ/kg), food and beverage processing (50–100 MJ/kg), cooking (5–7 MJ/kg), evaporation (2.5–2.7 MJ/kg), corn milling (18.3×10^{16} J), cane sugar (13.5×10^{16} J) and beet sugar (6.8×10^{16} J) production, and soybean oil milling (6×10^{16} J) (Clairand et al., 2020; Toepfl et al., 2006).

23.3.2   THE CARBON FOOTPRINT OF FOOD PROCESSING OPERATIONS IN TERMS OF WATER RESOURCES

Water scarcity, usage, and pollution is the major environmental concern that requires an environmental solution, food sector is also one of the high water consumable sector in various processing techniques. Water consumption in food processing can be divided based on the usage: (i) process function (used in the processing line) and (ii) non-process function (used in washing or as an energy source) (Asgharnejad et al., 2021). The analysis of water footprint (volume of freshwater used) is useful in processing insofar as it helps in decision-making regarding plant management and full fill environment policy (Ruini et al., 2013). For example, in a cassava starch production plant, if the plant has a capacity of 100–200 t starch/day means the estimated water usage was 1 m³/t starch, in case the plant has a capacity of 1–2 t starch/day then the water usage was 21–63 m³/t starch, which was 60% higher than larger plants. So, the analysis of the carbon footprint of water resources helps to select/decide the process criteria and techniques used for the particular food production (Tran et al., 2015). The major consumers of water are food (dairy, beverage industry) and drink industry, in England and Wales, 157 Mm³ of water is used by the food industry. Most of the food industries consume blue water (surface water or groundwater) more than green water, which also causes more grey (impingement of pollutants into freshwater) water footprints (Hoekstra, 2017). The main factor involved in carbon footprint is the emission of greenhouse gases from the wastewater produced by the processing plant. In the Netherlands, 2400 liters and 200 liters of water are used to produce 150 g of hamburger and 200 mL of milk, respectively. Irrigation and intensive livestock production are the major causes of water pollution (Lang & Barling, 2013). In food processing, to determine the carbon footprint many factors should be considered, the primary factor is energy usage. Around 30% of global energy demand is consumed by food industries, which is the major source of greenhouse gas emissions (Karwacka et al., 2020). For producing 1 kg of Barilla pasta water footprint lies between 1.34 and 2.85 l of water, in this cultivation phase contributes 90% of the water footprint (Ruini et al., 2013). To produce 586 kWh of energy by pump 1 million tons of water was used in the United Kingdom, it produces 289 kg CO_2. Recently, business players like Coca-Cola, SABMiller, Nestle, Unilever, Heineken, AB InBev, and PepsiCo engaged in sustainable water resource approaches by analyzing water footprint (Antonelli & Ruini, 2015).

23.3.3   THE CARBON FOOTPRINT OF FOOD PROCESSING OPERATIONS IN TERMS OF OTHER NON-RENEWABLE RESOURCES

Hence, 81% of the total energy utilization of the world is derived from non-renewable sources by consumption of natural gas (22.1%), coal (271%), and oil (31.9%); so they are considered the backbone of the power system. Only 1.7% of world energy was derived from renewable sources such as solar, wind, and geothermal. The consumption of non-renewable in food processing have a positive impact on the carbon footprint, the overconsumption of non–renewable resource cause environmental degradation, but limited usage of renewable improves the environmental quality. In BRIC-T (Brazil, Russia, India, China, South Africa, and Turkey), Russia consume higher non-renewable source (5940), and India comes at the lowest level of 350, even with a 1% augmentation in non-renewable source creating 0.551% ecological footprint (Usman & Makhdum, 2021). The long-run utilization of non-renewable sources positively affects the ecological footprint because it emits more carbon into the environment. So the requirement for sustainable energy utilization techniques is required in the food processing technological progression. A non-renewable energy requirement in food processing creates environmental damage and pollution.

23.3.4 Waste Generation and Its Effect on the Carbon Footprint of the Process

In the food system, food loss and food waste are interrelated, food loss at the retail and consumption stages are considered food waste, and food waste occurred during the different processing stages in the food industry. The waste generated in agro-food industries is disposed of or recovered by anaerobic digestion, product extraction, bioenergy production, incorporation, or dumped in land, sea, or sewer. Which ultimately causes a carbon footprint (Morales-polo et al., 2018). United Nations Environment Program reported that 10–12% of cereals and oil seeds in India and 25% of cereals in African countries are lost during post-harvest processing. For perishable products like fruits, vegetables, and roots this post-harvest losses were reported up to 50%. Nearly 28.57 million tonnes of solid food waste were generated worldwide in 2020, which cause 8% of anthropogenic greenhouse gas emissions and third largest emitter of greenhouse gases. The waste generation in the food industry has an interrelation between land, carbon, and water footprint. In the food sector, 39% of processed products are wasted globally (Buvaneshwaran et al., 2022). To analyze the impact of food waste on the environment either top-down approaches or bottom-up approaches are used (Tonini et al., 2018). In the United States, nearly 30% of processed foods are thrown away each year, which represents 4% of US energy use in other terms quarter of all water use. To generate the food waste nearly 23–24% of land, fertilizers, fresh water, and energy were used. India reported food losses in 2013 were 58 Mt, consisting of 115 billion m^3 of water (105 billion m^3 (direct water) + 9 billion m^3 (indirect water)), which cause a 9.5 Mha land footprint and 64 Mt CO_2 eq. carbon footprint (Kashyap & Agarwal, 2019). Food products that have short shelf life have a higher risk for spoilage and need special care, 31% of greenhouse gas emission was contributed by food in European Union (Zhu et al., 2018). From food waste reported bluewater footprint is 250 billion m^3, which was 3.6 times higher than the footprint created by total USA consumption (Hoekstra, 2017). Compare to agricultural products, animal-based food waste creates more carbon footprints, for example, while comparing beef and bread production beef produces relatively less waste (3.8%) than bread (10%), but the emission of the carbon footprint was much higher in beef waste (29 kg CO_2/kg) than bread waste (0.7 kg CO_2/kg). Likewise, waste from bananas, lettuce, and chicken has a global warming potential of 1.4 kg CO_2/kg, 0.3 kg CO_2/kg, and 3.1 kg CO_2/kg, respectively (Eriksson et al., 2015). Based on the type of waste disposal carried out the global warming potential can vary, in the case of chicken waste disposed of in a landfill has a potential of 3.1 kg CO_2/kg, if it is composted or anaerobically digested the potential was reduced to 0.1 kg CO_2/kg. The alternative waste management system can reduce the annual emission of up to 135301 Mt CO_2 which is equal to the greenhouse gas emitted by 28,500 cars (Karwacka et al., 2020). So, the analysis of LCA helps to choose better waste disposal techniques. In Australia, total foot waste represents 6% of national greenhouse gas emissions (57,507 Gg CO_2 eq.) and 9% of total water use (Tonini et al., 2018; Zhang et al., 2019).

TABLE 23.1

Effect of Food Processing on Carbon Dioxide Emission

Food Processing	Carbon Dioxide Emission	Food Processing	Carbon Dioxide Emission
Coffee roasting	27% of total emission	Dried plums (traditional method)	5.96 kg CO_2/kg of product
Coffee milling	2% of total emission	Butter	12.1 kg CO_2/kg of product
Coffee from capsule expresso machine	46% of total emission	Plant products	3.3 kg CO_2/kg of product
Coffee bean packaging	36% of total emission	Pasture milk	15 kg CO_2/kg of product
Tea brewing	40–70% of total emission	Cheese	16.9 kg CO_2/kg of product
Steamed plum	5.50 kg CO_2/kg of product	Unsweetened almond milk	0.71 kg CO_2/kg of product
Sliced bread	0.113 kg CO_2/kg of product	Baby cereal porridge	0.53 kg CO_2/kg of product

Source: Karwacka et al. (2020).

23.3.5 ROLE OF CARBON FOOTPRINT IN DESIGNING A SUSTAINABLE NON-THERMAL FOOD PROCESSING

The effect of conventional processing techniques on carbon dioxide emission was mentioned in Table 23.1. Due to increased awareness and visibility of the environment, economic and social issues in the food system require sustainable food processing techniques. Consumer preference towards sustainable food aspects such as energy, emission, and social responsibility creates new trends in sustainably efficient processing technologies. The main concern of these technologies is to improve the sustainable benefits in economic, environmental, and social dimensions (Zhu et al., 2018). Most of the thermal processing affects nutritional security by changing their nutritional value (Thangaraju et al., 2023). In the food industry, energy conversion is a vital SDG, so replacing conventional techniques with novel non-thermal and novel heating techniques is required to reduce the environmental impact. Most of the conventional techniques in food processing create a linear (one-way) economy, but the implementation of non-thermal techniques helps to promote a circular economy pattern. Where regeneration processes like resource production-consumption-regeneration can be implemented, this will help to eliminate waste, reduce emissions, and lower energy consumption (Naveed et al., 2022). Generally, non-thermal techniques denote the processing of food without direct heat, it will be helpful to satisfy (Goal 9: Industry, Innovation, and Infrastructure; Goal 13: Lower Environmental Footprint) the SDGs approved by United Nations in 2015. The non-thermal techniques can obtain sustainable processing by lowering processing costs, energy utilization, waste generation, and enhancing resource utilization.

23.4 ROLE OF NON-THERMAL TECHNOLOGIES TO ATTAIN SUSTAINABILITY

Consumer demand, existing regulations to attain the sustainable environment goals, economic need for low waste or by-product generations, and the need for resource efficient food systems lead to the necessity of novel food processing techniques in the food industry. Due to the utilization of less energy and water during processing and less energy impact during storage non-thermal technologies grab special attention in sustainable processing. Also, this technology helps to reduce the solid waste generation and valorization of biomass resources. Many researchers found that technological innovations help to reduce the carbon footprint, enhance energy efficiency, and initiate green financing options which help to improve sustainable economic growth (Anser et al., 2021).

23.4.1 HIGH-PRESSURE PROCESSING

High-pressure processing (HPP) is used as the alternative for heat treatment in packaged treatment, due to the effect of high pressure (100–600 MPa) and temperature combination the microbial safety and quality of the products are assured. Even HPP exhibits a significant impact on global warming and acidification due to electricity consumption, which is 20% lower than thermal pasteurization techniques (Pardo & Zufía, 2012). In terms of water footprint compared to other preservation techniques HPP has a lower contribution. HPP can be used as a preservation technique in most commodities such as fruit and vegetable, dairy, egg processing, seafood, infant food, minimally processed products, jam, jelly, meat, and meat products. In food products, quality and safety are the two key factors; HPP is a green technology that can satisfy the key factors in food processing. Also, HPP assures clean labeling and reduce the incorporation of additives and chemical preservative which is satisfying consumer demand. It maintains the color and texture of food materials to contribute to environmental sustainability by recovering the valuable compounds in the food products or waste that contribute to economic sustainability. To ensure food security, HPP improved the food yield (Nabi et al., 2021). Overall, HPP contributes to food security, nutrition security, environmental production, food waste reduction, food safety, and low energy and water consumption. In milk processing, Valsasina et al. (2017) reported that high-pressure techniques reduce the carbon footprint by up to 88% with better efficiency. Compare to other non-thermal techniques, HPP has the potential industrial application in vegetable processing (33%), juice and beverage processing (12%), meat processing (30%), and seafood processing (15%). In 2011, 60 companies installed 125 HPP units and produced 250 different HPP processed products (Picart-palmade et al., 2019). Cacace et al. (2020) compare the sustainability approach between HPP and thermal pasteurization of orange juice, the results show HPP exhibits a lower effect on global warming, ozone depletion, freshwater eutrophication, and other environmental effects than the direct and retort thermal pasteurization; but the effect on ionizing radiation is higher in HPP technology. The cost of processing was higher in HPP processing, for example, the cost of orange juice pasteurization by HPP is seven times higher than conventional thermal processing and three times higher than PEF treatment (Picart-palmade et al., 2019). It shows that not all the less expensive techniques are environmental friendly.

23.4.2 PULSED ELECTRIC FIELD (PEF)

In a PEF, the biological cell is exposed to high-intensity electric pulses that induce electro permeabilization and lead to cell death. In apples and potatoes to attain total permeabilization 1–5 kJ/kg electrical energy was required in a PEF, in contrast for mechanical 20–40 kJ/kg, for enzymatic 60–100 kJ/kg, for thermal processing >100 kJ/kg of energy are needed for degradation of plant tissue. Compare to conventional pasteurization PEFs require only 20 kJ/kg of energy. The studies carried out on carrot juice and grape juice denoted that PEF treatment significantly reduces the energy requirement in fruit juice pressing. The best example to state PEF is a sustainable process, in fruit or vegetable juice production the conventional enzymatic size reduction process provides the undesirable side effect of enzymes and alters the pectin structures, but the electro permeabilization effect obtained by PEF can overcome both demerits and also the energy efficiency of the process (Toepfl et al., 2006). PEF also exhibits economic and ecological benefits, for example, in sugar processing, PEF-treated pulp exhibits better water binding capacity; it can be used in animal feeds, and in most applications, the by-products can be used for valuable compound extraction. In PEF, high electric field strength (20–40 kV/cm) is required to obtain effective inactivation, in industrial scale processing of 5 ton/h the required energy will be 100–1000 kJ/kg, hence 2 to 3 million US$ investment is required for the treatment. Because of the structural damage in solid foods during PEF, this process is mostly used for liquid food processing like fruit and vegetable juices, milk, carbonated beverages, beer, etc. In food processing, PEF has several applications, the energy requirement for microbial inactivation is 15–40 kV/cm to 1000 kJ/kg, sludge disintegration is 10–20 kV/cm to 50–200 kJ/cm, an induction stress response is 0.5–1.5 kV/cm to 0.5–5 kJ/kg, electro permeabilization of biological cells is 0.7–3 kV/cm to 1–20 kJ/kg. To obtain enzyme inactivation, 1000–40,000 kJ/L of specific energy is required, but for microbial inactivation only 50–1000 kJ/L is sufficient. While comparing HPP and PEF, the cost of processing HPP was much higher (between 0.77 and 3.15 US$) than PEF processing, it is because of the higher power requirement to rise the pressure (Picart-palmade et al., 2019).

23.4.3 MEMBRANE PROCESSING

Membrane processing is considered a better alternative technique to obtain sustainable processing. The membrane process provides a solution to major sustainable problems like water scarcity, energy crisis, and non-renewable resource usage, and promotes resource conservation and environment-friendly technology. Membrane technology has an impact on 17 goals (e.g., Goals 2, 6, 7, 8, 12, 13 and 14) proposed by SDGs by United Nations (Drioli et al., 2021). Membrane processing is operated under continuous mode which will eliminate the cleaning procedures, and the lower operating temperature of <80°C reduces the higher energy requirement. Membrane processing has potential advantages like low energy requirement than other filtration techniques, no need for additional chemicals, less space requirement, easily automated and electrifiable (Song et al., 2020). The setup of membrane processing is modular so scaling up to the industrial level is much easy compare to other non-thermal techniques. It has applications in unit operations like concentration/clarification, purification, fractionation, and extraction. In the agri-food sector, membrane processing plays a major role in waste management. Membrane separation is considered the best suitable technique to recover, separate, and initial fractionation of organic compounds from waste (Papaioannou et al., 2022). Membrane processing is a well-known technique in water treatment, it has a productive energy requirement; for example, in seawater desalination pressure driven membrane processing required the same cost as other water treatment, but reverse osmosis has lower operating cost due to the excessive osmotic pressure. So membrane processing is considered sustainable technology in water treatment and also reduces the concern about freshwater depletion (Yusuf et al., 2020). Membrane processing has a potential application in beverage industries, they can replace the conventional filtration technique in juice clarification or filtration lines. The process driven force used in membrane processing is the water vapor pressure gradient so it required less thermal energy during the processing. In winemaking, membrane technology can be used in separation/purification, clarification, stabilization, concentration, and de-alcoholization (Rayess & Mietton-Peuchot, 2016). Due to the higher energy consumption by other techniques, membrane technology was adopted to maintain the alcohol content in grape must (Rayess & Mietton-Peuchot, 2016). After the long use of membrane in reverse osmosis technique, it can be reused in either ultra or microfiltration process. The involvement of nanotechnology in membrane process further enhanced its sustainability, quantum dots are used to synthesis of sustainable membranes for the long run. Also, the membrane processing can be assist with low-grade heat like industrial waste heat, solar energy, etc.

23.4.4 PULSED LIGHT PROCESSING

In pulsed light processing, food materials are introduced to high-intensity light for surface decontamination and other purposes of work. This technique was approved by the US Food and Drug Administration (FDA) in 1996 for food production and processing. In this technology, microbial inactivation is achieved by irreversible damage in DNA and the photothermal effect on microbes. It considers being sustainable processing because the process does not generate any carcinogenic, toxic by-products, does not require any chemicals for the process, not need thermal energy. When we compare the energy efficiency of the food processing technology for microbial inactivation, pulsed light processing has more energy efficient than others. The

D value required to inactivate microbial populations like enteric bacteria, yeast, and enteric virus lies between 2–8, 2.3–8, and 5–30 mJ/cm^2, respectively (Koutchma, 2009). Pulsed light is considered a clean and sustainable source of energy and the processing system exhibits minimal impact on the environment. The photo light technique is one of the technologies which produce low water footprints (Hoekstra, 2017). Still, research work needs to focus on finding energy consumption and carbon footprint emission of different pulsed light-treated products, very minimum data about the impact of pulsed light on the environment was available. Plastic packaged food materials are commercially interested in pulsed light treatment, but now the negative impact of plastics on carbon footprint requires replacing plastic with compostable or recyclable material. Pulsed light also has a potential application in water treatment, but validation and optimization of the techniques need to be done (Rowan, 2019).

Apart from this, non-thermal techniques such as cold plasma, ozone, and ultrasound techniques are emerging in food processing which will help in sustainable food processing. Atmospheric cold storage is suited for existing packaging materials, non-solvent requirement makes it environment friendly, low cost, and less time consumption (Chakka et al., 2021). Ozone treatment requires no solvents or chemical compounds for processing, it is considered sustainable processing because it has low toxicity, minimal residual effect, and no waste generation. Ultrasound is used for waste valorization and extraction of valuable compounds without affecting the environment (Buvaneshwaran et al., 2022). By the application of different non-thermal processing techniques in food processing, a sustainable processing system can be achieved.

23.5 LIFECYCLE ASSESSMENT OF NON-THERMALLY PROCESSED FOODS

The LCA method is widely used to analyze the environmental footprint of food products by taking their entire life cycle "from field to plate", which is used to elaborate efficiency of the selected technology in the industry (Karwacka et al., 2020). The most widely used environmental footprint includes carbon, water, nitrogen, and energy footprint. Song et al. (2020) conducted a study to analyze consumer preference for the consumption of non-thermally processed fruits and vegetables in six European countries, study concluded that ensuring food safety and enhanced food quality enhances the purchase rate. The lack of knowledge and the higher unit cost are considered a hurdle for bringing non-thermally processed products to the commercial market. The LCA of non-thermally processed foods is listed in Table 23.2. The LCA assessment of thermal, PEF, and HPP of tomato and watermelon juice preservation was reported by Aganovic et al. (2017). The highest energy consumption of 20 kWh/L was observed in HPP processing followed by a PEF and thermal process. Compare to watermelon tomato juice processing cause more environmental footprint, the PET bottles used for the packaging cause 85% of the environmental footprint. Cradle-to-grave LCA analysis was done for HPP and modified atmospheric packaging (MAP) techniques by Pardo and Zufía (2012), results show both emerging techniques have lower energy demand, water requirement, and CO_2 emission. But when we focus only on the processing stage emerging techniques require higher energy demand than conventional techniques like autoclave and microwave processing. The adaptation of a PEF in potato processing can reduce 85% and 95% of energy and water consumption, in tomato peeling it can improve the environmental indicators to 17-20%, also this technique can be used in the natural colorant extraction process (Gómez et al., 2019). While comparing the production cost of orange juice, the HPP treatment cost is 7 times higher than thermal processing and the PEF cost is 2.5 times higher than thermal processing. In the same processing, the estimated CO_2 was 90,000 kg, 700,000 kg, and 773,000 kg by thermal pasteurization, PEF, and HPP, respectively. The electro membrane approach in cranberry juice processing shows a 20.6% more eco-friendly nature than salt precipitation techniques without reuse, recovery of a chemical solution by reusing it can be further increased (Cacace et al., 2020). The membrane desalination of seawater can reduce 25–71% of the impact on the environment.

TABLE 23.2

Life Cycle Assessment of Non-Thermally Processed Food

Non-Thermal Technique	Type of Process	Key Findings	References
Ultra high-pressure processing	Milk pasteurization	• Lower energy consumption • Less environmental impact and water consumption • 90% energy recovery • 44% reduction in climate change	(Valsasina et al., 2017)
	Pasteurization of tomato and watermelon juice	Higher energy requirement: 0.20 kW/L	(Aganovic et al., 2017)
Pulsed electric field (PEF)	Pasteurization of tomato and watermelon juice	Energy requirement – 0.12 kWh/L	(Wakeland et al., 2012)
	Tomato peeling	Compare to thermo physical peeling the environmental indicators increased to 17–20%	(Arnal et al., 2018)
Membrane processing	Cranberry juice	Electrodialysis with a bipolar membrane is 20.6% more eco-friendly than the salt precipitation technique	(Faucher et al., 2019)
Reverse osmosis	Desalination process	Less than 1% of CO_2 emission Reusable membranes are environment friendly	(Lawler et al., 2015)

23.6 PACKAGING REQUIREMENT OF NON-THERMAL PROCESSES

In non-thermal processing, most packaged foods are undergone treatment condition, so the treatments have a significant impact on the packaging material which affect or improve the product quality. So, there is a need to study the packaging requirement for non-thermally processed food. Apart from that most of the currently available packaging materials cause more carbon footprint, so it is necessary to develop packaging materials that are environmentally free and can be used for the non-thermal process.

23.6.1 CURRENT PACKAGING TECHNIQUES AVAILABLE

Aseptic food packaging is considered the better choice for non-thermally processed food, a wider range of packaging materials can be used for this and they do not need to withstand the high temperature. The non-thermally processed foods have a long shelf life so they have to be stored under refrigerated or ambient conditions for a long period. Based on the type of material packaging material can be glass (types I and II), metal (aluminum foil, can, tinplate, tin-free steel), paper and paperboard (kraft paper, greaseproof paper, fiberboards, solid boards), synthetic materials (polyolefin, polyester, polyvinyl chloride, polyamide, ethylene vinyl alcohol), biodegradable (PHA, PHB, PLA, PCL, PBS, and PHH) and edible coating (chitin, chitosan, agar, starch, corn zein, wheat gluten) are used for food packaging system (Verma et al., 2021). For effective packaging modified paper, coated glass and metal containers, or metal foils, plastic, and synthetic materials have been in current use.

A variety of plastic-based packaging materials was used based on their surface property, barrier property, composition, and stiffness. Other than this, biopolymer packaging materials and edible packaging are used in different products. Till now plastic and paper laminated materials widely used in non-thermal processing. Glass bottles, thermoformed plastic containers (lid material: nylon/Al/LDPE; base material: HIPS/PVDC/LDPE), PET, LDPE, HDPE bottles, PP tubes are used for PEF treated fruit juice, chocolate milk, and fortified fruit beverages; PE bag, plastic bag with EVOH, nylon/PE pouches, PE/PET bottles, modified acetonitrile methyl acrylate copolymers, plastic films, laminated PE-PA foil are has been used in HPP treated products like sausages, ham, kimchi, salsa, liver, turkey meat, and orange juice; steel tray, plastic films, PE bags, LDPE pack, vacuum pack, PVDC film had been used for irradiated foods (Min & Zhang, 2005). PVC-based packaging material is commonly used for pulsed light treatment due to its heat resistance. Active packaging technique is mostly preferred to store non-thermally processed foods to preserve fresh-like quality. In some cases, like irradiation modified, atmospheric packaging techniques are required.

23.6.2 Carbon Footprint of Packaging

In most food materials, primary, secondary, and tertiary packaging materials are used for various applications. In food processing next to the effect of processing on the environmental footprint (50–80%), the application of packaging techniques exhibits a larger carbon footprint (20–40%) on the environment (Pardo & Zufía, 2012). In the category of global warming, primary packaging (64%) contributes a huge percentage followed by processing technique (19%) and secondary/tertiary packaging (15.74%) (Cacace et al., 2020). The production and disposal of packaging material after usage are considered the primary hotspot, which was getting increased by each year. In the United States, from the food system, the packaging is responsible for 10% of greenhouse gas emissions. Cartons and aluminum-based packaging materials are the major contributors to environmental impact, and aseptic cartons and plastic are the lower contributors. In carbonated beverages, aluminum packages contribute 80% of carbon emissions (Kumar & Chakabarti, 2019). In tomato and watermelon juice pasteurization plants maximum of 85% of environmental impact was associated with PET bottle production (Aganovic et al., 2017). Plastic-based packaging materials (99%) are of fossil fuel origin and cause white pollution, landfill depletion, and high energy use (Mahalik & Nambiar, 2010).

Paiano et al. (2020) analyzed the carbon footprint of water and beverage packaging material, the results revealed that PET 0.5, PET 1.5, glass, and aluminum have carbon footprints of 0.13, 0.18, 0.05, and 0.18 kg CO_2 eq/person/per day, respectively. In the canned tuna industry LCA packaging material cause 20–40% of the total production carbon footprint. The carbon footprint of canned tuna can be reduced to 10% and 22% when the retort cups are replaced by metal cans and retort pouches, respectively (Ingrao et al., 2015). Likewise, polypropylene cups with aluminum covers lead to 10% of soygurt (soya yoghurt) carbon footprint, packaging material in bread and ham cause 1–3%, plastic bags in frozen food cause 1%, metal cans in canned carrot cause 68%, and plastic bag in frozen carrot cause 2% of total carbon dioxide emission (Poovarodom et al., 2012). Compare to retort glass or cans plastic containers cause small environmental effects. LCA in Nestle also showed the same result due to the packaging processing steps. In the beer brewing industry, when beer bottles are replaced by PET bottles instead of glass the carbon footprint varied from 127–192 to 103–169 kg CO_2 h/L. The major hotspot in the beer industry was primary packaging material, even though the existing packaging term changed to recycled glass or PET bottles only 56–60 or 80 kg CO_2 h/L reduction in carbon footprint was obtained (Cimini & Moresi, 2018). In some cases, it was dependent on packaging material, size of the packaging, processing step, energy use, recycling capacity, and all. Some packaging materials have a very low carbon footprint, for example, poly (lactic acid) provides a zero carbon footprint (Ncube et al., 2020).

23.6.3 Development of Sustainable Packaging for Non-Thermal Processing

The development of sustainable packaging represents the development of packaging using recyclable materials or which can reduce water use, landfill, potential to reuse made from renewable energy, no air pollution, no greenhouse gas emission, etc. Sustainable packaging system focused to produce smart, active, and intelligent packaging which can be biologically degraded or biobased polymers. Combining eco-friendly packaging material and non-thermal processing technology improves sustainable processing. But the biobased polymers can't be used in non-thermal processing technology because, based on the processing principle of non-thermal techniques, they cause adverse effects on packaging materials like oxidative degradation, chain distribution, depolymerization, etc. (Table 23.3). Commonly available biobased polymers are poly (lactic acid) (PLA), poly (butylene adipate-co-terephthalata) (PBAT), thermoplastic starch (TPS), poly

TABLE 23.3

Effect and Application of Non-Thermal Processing Techniques on Packaging Materials

Packaging Material	Non-Thermal Processing				
	High-Pressure Processing	Pulsed Electric Field	Membrane Processing	Pulsed Light Processing	Irradiation
Poly (lactic acid) (PLA)	Impacts morphology, change in structure/crystallinity, change in gas and oxygen permeability	Improvement in surface proper	No change	No changes	Improved degradation
Poly (butylene adipate-co-terephthalate) (PBAT)	No change	Not applicable due to less conductivity	No change	Photodegradation	Not appropriate changes at >300 kGy
Thermoplastic starch (TPS)	No change	Not applicable due to less conductivity	No change	No change	Increased moisture resistivity, improve the tensile strength
Poly (3-hydroxybutyrate-co-3-hydroxyvalerate) (PHBV),	No change	Not applicable due to less conductivity	No change	No change	Oxidative degradation
Cellulose acetate (CA)	Change in surface property	Not applicable due to less conductivity	No change	No change	No change
Polyhydroxyalkanoate (PHA)	No change	Not applicable due to less conductivity	No change	Higher surface properties	No change
Edible coating (EC)	Increased water resistance and tensile strength in starch blend film and gelatin-based film	Effective in fruits and vegetables	No change	Vaseline coating on whole shell egg	Enhance cross-linking in protein-based EC, Improved mechanical property, thermal stability, and water vapor permeability
Nano packaging	The addition of nanoparticles in biopolymers increases the oxygen and water vapor transfer rate	Non-toxic electrically conductive nanofillers (grapheme oxide + multiwall carbon)	No change	No change	No change

(3-hydroxybutyrate-co-3-hydroxyvalerate) (PHBV), cellulose acetate (CA), and polyhydroxyal-kanoates (PHA). Other than biobased polymers, edible coating, nanopackaging materials, and active packaging are used in non-thermal processing. So, the development of packaging material for non-thermally processed foods should be based on the characteristics of the food, process parameters, and type of packaging material to be used (Gabric et al., 2022).

In PEF treatment, the packaging material should contain equal or more electrical conductivity than the packaged product, but most of the biobased films and edible coating contain very less electrical conductivity (0.1–2 S/m), so they cannot be used for the PEF treatment. Nanobased films like nanofillers which consist of graphite oxide or multi-walled carbon nanotubes with natural biopolymers improve the electrical conductivity of the films. For example, PBAT+confining nanofillers have an electrical conductivity of 338 S/m, but the addition of nanosized fillers inversely affects the transmitted light intensity and reduces the optical transparency of the packaging material. Electrically conductive bio nanocomposites and nanobased packaging materials have the future applications in PEF treatment. But still, research work needs to develop a suitable packaging material for PEF treatment.

Also, non-invasive methods like ultrasound affect the packaging materials by accelerating hydrolysis or transesterification reactions. Most of the non-thermal techniques affect the barrier properties of the packaging materials, in that the radiation technique mainly changes the functional properties of the material and the formation of free radicals during radiation accelerates the cross-linking and chain scission reactions, so the packaging material should not transmit the radiolysis products. Irradiation techniques mainly used active packaging to accelerate to release of active agents. From the studies, an edible coating is considered the suitable packaging technique for pulsed light treatment. Modified atmospheric packages provide better results in irradiated foods. During HPP disruption in flexible packaging material was identified, which alters the package structure and protection properties, especially in aluminum foil lamination (Morris et al., 2007). So, the development of packaging material based on the working conditions and principle of non-thermal techniques enhanced the usability of the packaging material in line processing, shelf life, product safety, and quality.

23.7 SUMMARY

Sustainability becomes the internal part of food processing, so the technologies used in the processing line should satisfy zero discard/loss/waste. Sustainable food processing techniques significantly reduce the impact on the environmental footprint and reduce energy demand, waste generation, and chemical usage. The increasing demand for processed foods creates a huge negative impact on the environmental footprint in terms of energy usage, water usage, non-renewable source usage, and waste generation. Several non-thermal techniques can provide a solution to this problem. Proper LCA of processed foods describe a hypothetical future situation of the products on the environment, so analyzing the LCA of non-thermally processed foods provide a clear view of the environmental impact of non-thermal processed foods in different kind of products. Which will help to choose the more sustainable processing approach based on the type of product, processing, and requirement criteria. Worldwide, 250 HPP and 40 PEF units are adopted in the food industry, in 65 high-pressure units are operated for juice processing. So, the incorporation of non-thermal techniques provides a path to reach SDGs. Apart from processing the primary packages used for the processed food cause 40–85% of environment impact, developing the sustainable packaging materials and implement them in usage further helps for the sustainable processing.

REFERENCES

Aganovic, K., Smetana, S., Grauwet, T., Toepfl, S., Mathys, A., Van Loey, A., & Volker. (2017). Pilot scale thermal and alternative pasteurization of tomato and watermelon juice: An energy comparison and life cycle assessment. *Journal of Cleaner Production*, *141*, 514–525. https://doi.org/10.1016/j.jclepro.2016.09.015

Anser, M. K., Khan, M. A., Abdelmohsen, A., Aldakhil, A. M., Voo, X. H., & Zaman, K. (2021). Relationship of environment with technological innovation, carbon pricing, renewable energy, and global food production. *Economics of Innovation and New Technology*, *30*(8), 807–842. https://doi.org/10.1080/10438599.2020.1787000

Antonelli, M., & Ruini, L. F. (2015). Business engagement with sustainable water resource management through water footprint accounting: The case of the Barilla company. *Sustainability*, *7*, 6742–6758. https://doi.org/10.3390/su7066742

Arnal, Á. J., Royo, P., Pataro, G., Ferrari, G., Ferreira, V. J., Ana, M. L.-S., & Ferreira, A. G. (2018). Implementation of PEF treatment at real-scale tomatoes processing considering LCA methodology as an innovation strategy in the agri-food sector. *Sustainability*, *10*(4), 979. https://doi.org/10.3390/su10040979

Asgharnejad, H., Nazloo, E. Khorshidi, Larijani, M. Adani, Hajinajaf, N., & Rashidi, H. (2021). Comprehensive review of water management and wastewater treatment in food processing industries in the framework of water-food-environment nexus. *Comprehensive Reviews in Food Science and Food Safety*, *20*(5), 4779–4815. https://doi.org/10.1111/1541-4337.12782

Baker, P., Sievert, K., Hadjikakou, M., Russell, C., Huse, O., Backholer, K., Bell, C., Scrinis, G., Worsley, A., Friel, S., & Lawrence, M. (2020). Ultra-processed foods and the nutrition transition: Global, regional and national trends, food systems transformations and political economy drivers. *Obesity Reviews*, *21*(12), e13126. https://doi.org/10.1111/obr.13126

Berry, E. M., Dernini, S., Burlingame, B., Meybeck, A., & Conforti, P. (2015). Food security and sustainability: Can one exist without the other? *Public Health Nutrition*, *18*(13), 2293–2302. https://doi.org/10.1017/S136898001500021X

Buvaneshwaran, M., Radhakrishnan, M., & Natarajan, V. (2022). Influence of ultrasound-assisted extraction techniques on the valorization of agro based industrial organic waste: A review. *Journal of Food Process Engineering*, *2021*, e14012. https://doi.org/10.1111/jfpe.14012

Cacace, F., Bottani, E., Rizzi, A., & Vignali, G. (2020). Evaluation of the economic and environmental sustainability of high pressure processing of foods. *Innovative Food Science and Emerging Technologies*, *60*(January), 102281. https://doi.org/10.1016/j.ifset.2019.102281

Caron, P., Loma-Osorio, G. F. De, Nabarro, D., Hainzelin, E., Guillou, M., Andersen, I., Arnold, T., Astralaga, M., Beukeboom, M., Bickersteth, S., Bwalya, M., & Caron, P. (2018). Food systems for sustainable development: Proposals for a profound four-part transformation the four parts. *Agronomy for Sustainable Development*, *38*(4), 1–12. https://doi.org/10.1007/s13593-018-0519-1

Chakka, A. K., M.S., S., & Ravishankar, C. N. (2021). Sustainability of emerging green non-thermal technologies in the food industry with food safety perspective: A review. *LWT – Food Science and Technology*, *151*(November 2021), 112140. https://doi.org/10.1016/j.lwt.2021.112140

Cimini, A., & Moresi, M. (2018). Mitigation measures to minimize the cradle-to-grave beer carbon footprint as related to the brewery size and primary packaging materials. *Journal of Food Engineering*, *236*, 1–8. https://doi.org/10.1016/j.jfoodeng.2018.05.001

Clairand, J., Briceño-León, M., Escrivá-Escrivá, G., & Pantaleo, A. M. (2020). Review of energy efficiency technologies in the food industry: Trends, barriers, and opportunities. *IEEE Access*, *8*, 48015–48029. https://doi.org/10.1109/ACCESS.2020.2979077

Collins, A., & Fairchild, R. (2007). Sustainable food consumption at a sub-national Level: An ecological footprint, nutritional and economic analysis. *Journal of Environmental Policy & Planning*, *9*(1), 5–30. https://doi.org/10.1080/15239080701254875

Drioli, E., Macedonio, F., & Tocci, E. (2021). Membrane science and membrane engineering for a sustainable industrial development. *Separation and Purification Technology*, *275*(May), 119196. https://doi.org/10.1016/j.seppur.2021.119196

Eriksson, M., Strid, I., & Hansson, P. (2015). Carbon footprint of food waste management options in the waste hierarchy e a Swedish case study. *Journal of Cleaner Production*, *93*(April 15), 115–125. https://doi.org/10.1016/j.jclepro.2015.01.026

Faucher, M., Henaux, L., Chaudron, C., Mikhaylin, S., Margni, M., Chaudron, C., Mikhaylin, S., Margni, M., & Bazinet, L. (2019). Electromembrane approach to substantially improve the ecoefficiency of deacidified cranberry juice production: Physicochemical properties, life cycle assessment and ecoefficiency score. *Journal of Food Engineering*, *273*, 109802. https://doi.org/10.1016/j.jfoodeng.2019.109802

Gabric, D., Kurek, M., Scetar, M., Brncic, M., & Galic, K. (2022). Effect of non-thermal food processing techniques on selected packaging materials. *Polymers*, *14*(23), 5069.

Gómez, B., Munekata, P. E. S., Gavahian, M., Barba, F. J., Martí-Quijal, F. J., Bolumar, T., Cezar, P., Campagnol, B., Tomasevic, I., & Lorenzo, J. M. (2019). Application of pulsed electric fields in meat and fish processing industries: An overview. *Food Research International, 123*(March), 95–105. https://doi.org/10.1016/j.foodres.2019.04.047

Hoekstra, A. Y. (2017). *Water footprint assessment: Evolvement of a new research field.* https://doi.org/10.1007/s11269-017-1618-5

Ingrao, C., Tricase, C., Cholewa-Wójcik, A., Kawecka, A., Rana, R., & Siracusa, V. (2015). Polylactic acid trays for fresh-food packaging: A carbon footprint assessment. *Science of the Total Environment, 537,* 385–398. https://doi.org/10.1016/j.scitotenv.2015.08.023

Jensen, J. K., & Arlbjorn, J. S. (2014). Product carbon footprint of rye bread. *Journal of Cleaner Production, 82*(November 1), 45–57. https://doi.org/10.1016/j.jclepro.2014.06.061

Jianyi, L., Yuanchao, H., Shenghui, C., Jiefeng, K., & Lilai, X. (2015). Carbon footprints of food production in China (1979–2009). *Journal of Cleaner Production, 90*(March 1), 97–103. https://doi.org/10.1016/j.jclepro.2014.11.072

Karwacka, M., Ciurzy, A., Janowicz, M., & Janowicz, M. (2020). Sustainable development in the agri-food sector in terms of the carbon footprint: A review. *Sustainability, 12*(16), 6463.

Kashyap, D., & Agarwal, T. (2019). Food loss in India: Water footprint, land footprint and GHG. *Environment, Development and Sustainability, 22,* 2905–2918. https://doi.org/10.1007/s10668-019-00325-4

Koutchma, T. (2009). Advances in ultraviolet light technology for non-thermal processing of liquid foods. *Food and Bioprocess Technology, 2*(2), 138–155. https://doi.org/10.1007/s11947-008-0178-3

Kumar, S. N., & Chakabarti, B. (2019). *Energy and Carbon Footprint of Food Industry.* Springer, Singapore. https://doi.org/10.1007/978-981-13-2956-2

Lang, T., & Barling, D. (2013). Nutrition and sustainability: An emerging food policy discourse. *Proceedings of the Nutrition Society, December 2012,* 1–12. https://doi.org/10.1017/S002966511200290X

Lawler, W., Alvarez-Gaitan, J., Leslie, G., & Le-Clech, P. (2015). Comparative life cycle assessment of end-of-life options for reverse osmosis membrane. *Desalination, 357,* 45–54. https://doi.org/10.1016/j.desal.2014.10.013

Mahalik, N. P., & Nambiar, A. N. (2010). Trends in food packaging and manufacturing systems and technology. *Trends in Food Science & Technology, 21*(3), 117–128. https://doi.org/10.1016/j.tifs.2009.12.006

Min, S., & Zhang, Q. H. (2005). Packaging for Non-Thermal Food Processing. In *Innovation in Food Packaging* (pp. 482–500). Elsevier. https://doi.org/10.1016/B978-012311632-1/50059-0

Morales-Polo, C., Castro, M. M. C., & Soria, B. Y. M. (2018). Reviewing the anaerobic digestion of food waste: From waste generation and anaerobic process to its perspectives. *Applied Sciences, 8*(10), 1804. https://doi.org/10.3390/app8101804

Morris, B. C., Brody, A. L., & Wicker, L. (2007). Non-thermal food processing/preservation technologies: A review with packaging implications. *Packaging Technology and Science, 20*(4), 275–286.

Nabi, B. G., Mukhtar, K., Arshad, R. N., Radicetti, E., Tedeschi, P., Shahbaz, M. U., Walayat, N., Nawaz, A., Inam-Ur-Raheem, M., & Aadil, R. M. (2021). High-pressure processing for sustainable food supply. *Sustainability, 13*(24), 13908. https://doi.org/10.3390/su132413908

United Nations. (1974). Report of the World Food Conference. Rome, 5–16 November, 1974, 1–64

Naveed, R., Abdul-Malek, Z., Roobab, U., Ranjha, M. M. A. N., Jambrak, A. R., Qureshi, M. I., Khan, N., Lorenzo, J. M., & Aadil, R. Muhammad. (2022). Nonthermal food processing: A step towards a circular economy to meet the sustainable development goals. *Food Chemistry: X, 16*(October), 100516. https://doi.org/10.1016/j.fochx.2022.100516

Ncube, L. K., Ude, A. U., Ogunmuyiwa, E. N., Zulkifli, R., & Beas, I. N. (2020). Environmental impact of food packaging materials: A review of contemporary development from conventional plastics to polylactic acid based materials. *Materials, 13*(21), 4994.

Paiano, A., Crovella, T., & Lagioia, G. (2020). Managing sustainable practices in cruise tourism : the assessment of carbon footprint and waste of water and beverage packaging. *Tourism Management, 77*(January 2019), 104016. https://doi.org/10.1016/j.tourman.2019.104016

Papaioannou, E. H., Mazzei, R., Bazzarelli, F., Piacentini, E., Giannakopoulos, V., Roberts, M. R., & Giorno, L. (2022). Agri-food industry waste as resource of chemicals: The role of membrane technology in their sustainable recycling. *Sustainability, 14*(3), 1483.

Pardo, G., & Zufía, J. (2012). Life cycle assessment of food-preservation technologies. *Journal of Cleaner Production, 28,* 198–207. https://doi.org/10.1016/j.jclepro.2011.10.016

Picart-Palmade, L., Cunault, C., Chevalier-Lucia, D., Belleville, M.-P., & Marchesseau, S. (2019). Potentialities and limits of some non-thermal technologies to improve sustainability of food processing. *Frontiers in Nutrition, 5,* 130. https://doi.org/10.3389/fnut.2018.00130

Poovarodom, N., Ponnak, C., & Manatphrom, N. (2012). Comparative carbon footprint of packaging systems for tuna products. *Packaging Technology and Science, 25*(5), 249–257. https://doi.org/10.1002/pts.975

Rayess, Y. El, & Mietton-Peuchot, M. (2016). Membrane technologies in wine industry: An overview of applications and recent developments. *Critical Reviews in Food Science and Nutrition, 56*(12), 2005–2020. https://doi.org/10.1080/10408398.2013.809566

Rowan, N. J. (2019). Pulsed light as an emerging technology to cause disruption for food and adjacent industries – Quo vadis ? *Trends in Food Science & Technology, 88*(March), 316–332. https://doi.org/10.1016/j.tifs.2019.03.027

Ruini, L., Marino, M., Pignatelli, S., Laio, F., & Ridolfi, L. (2013). Water footprint of a large-sized food company: The case of Barilla pasta production. *Water Resources and Industry, 1*(2), 7–24. https://doi.org/10.1016/j.wri.2013.04.002

Sanjuan, N., Stoessel, F., & Hellweg, S. (2014). Closing data gaps for LCA of food products: Estimating the energy demand of food processing. *Environmental Science and Technology, 48*(2), 1132–1140. https://doi.org/10.1021/es4033716

Song, X., Pendenza, P., Navarro, M. D., Garcia, E. V., Monaco, R. Di, & Giacalone, D. (2020). European consumers' perceptions and attitudes towards non-thermally processed fruit and vegetable products. *Foods, 9*(12), 1732.

Thangaraju, S., Shankar, M., Buvaneshwaran, M., & Natarajan, V. (2023). Effect of Processing on the Functional Potential of Bioactive Components. In *Bioactive Components* (pp. 183–207). Springer, Singapore. https://doi.org/10.1007/978-981-19-2366-1_12

Tiwari, B. K., Norton, T., & Holden, N. M. (eds.). (2013). *Sustainable food processing.* John Wiley & Sons. ISBN: 978-1-118-63437-0. https://www.wiley.com/en-in/Sustainable+Food+Processing-p-9781118634370

Toepfl, S., Mathys, A., Heinz, V., & Knorr, D. (2006). Review: Potential of high hydrostatic pressure and pulsed electric fields for energy efficient and environmentally friendly food processing. *Food Reviews International, 22*(4), 405–423. https://doi.org/10.1080/87559120600865164

Tonini, D., Albizzati, P. F., & Astrup, T. F. (2018). Environmental impacts of food waste: Learnings and challenges from a case study on UK. *Waste Management, 76*, 744–766. https://doi.org/10.1016/j.wasman.2018.03.032

Tran, T., Da, G., Moreno-santander, M. A., Vélez-Hernández, G. A., Giraldo-Toro, A., Piyachomkwan, K., Sriroth, K., & Dufour, D. (2015). A comparison of energy use, water use and carbon footprint of cassava starch production in Thailand, Vietnam and Colombia. *Resources, Conservation & Recycling, 100*, 31–40. https://doi.org/10.1016/j.resconrec.2015.04.007

Usman, M., & Makhdum, M. S. A. (2021). What abates ecological footprint in BRICS-T region? Exploring the influence of renewable energy, non-renewable energy, agriculture, forest area and financial development. *Renewable Energy, 179*, 12–28. https://doi.org/10.1016/j.renene.2021.07.014

Valsasina, L., Pizzol, M., Smetana, S., Georget, E., Mathys, A., & Heinz, V. (2017). Life cycle assessment of emerging technologies: The case of milk ultra-high pressure homogenisation. *Journal of Cleaner Production, 142*, 2209–2217. https://doi.org/10.1016/j.jclepro.2016.11.059

Verma, M. K., Shakya, S., Kumar, P., Madhavi, J., Murugaiyan, J., & Rao, M. V. R. (2021). Trends in packaging material for food products: Historical background, current scenario, and future prospects. *Journal of Food Science and Technology, 58*(11), 4069–4082. https://doi.org/10.1007/s13197-021-04964-2

Virtanen, Y., Kurppa, S., Saarinen, M., Katajajuuri, J., Usva, K., Mäenpää, I., Makela, J., Gronroos, J., & Nissinen, A. (2011). Carbon footprint of food approaches from national input-output statistics and a LCA of a food portion. *Journal of Cleaner Production, 19*(16), 1849–1856. https://doi.org/10.1016/j.jclepro.2011.07.001

Wakeland, W., Cholette, S., & Venkat, K. (2012). *Food transportation issues and reducing carbon footprint.* 211–236. https://doi.org/10.1007/978-1-4614-1587-9

Yusuf, A., Sodiq, A., Giwa, A., Eke, J., Pikuda, O., De Luca, G., Di Salvo, J. L., & Chakraborty, S. (2020). A review of emerging trends in membrane science and technology for sustainable water treatment. *Journal of Cleaner Production, 266*, 121867. https://doi.org/10.1016/j.jclepro.2020.121867

Zhang, Y., Ravi, S. K., & Tan, S. C. (2019). Food-derived carbonaceous materials for solar desalination and thermo-electric power generation. *Nano Energy, 65*(June), 104006. https://doi.org/10.1016/j.nanoen.2019.104006

Zhu, Z., Chu, F., Dolgui, A., Chu, C., Zhou, W., & Piramuthu, S. (2018). Recent advances and opportunities in sustainable food supply chain: A model-oriented review. *International Journal of Production Research, 56*(17), 5700–5722. https://doi.org/10.1080/00207543.2018.1425014

Index

Page numbers in *italics* and **bold** refer to figures and tables, respectively.

For Product Safety Concerns and Information please contact our EU
representative GPSR@taylorandfrancis.com
Taylor & Francis Verlag GmbH, Kaufingerstraße 24, 80331 München, Germany